Workboats for the World

P&O
K
SD SEAL
K
SD SHARK

WORKBOATS FOR THE WORLD

THE ROBERT ALLAN STORY

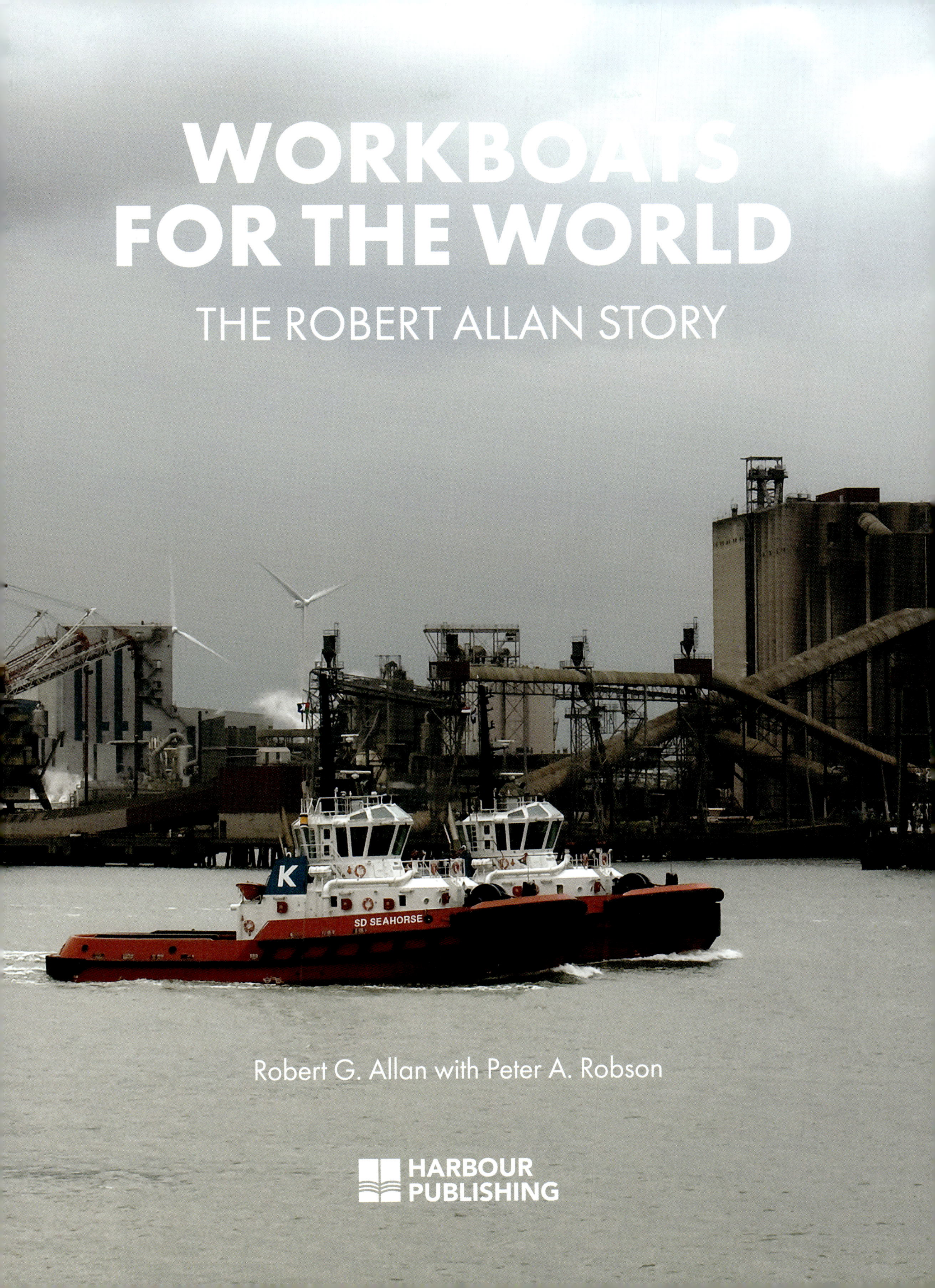

Robert G. Allan with Peter A. Robson

HARBOUR PUBLISHING

Previous pages: A quartet of Kotug's ASD tugs in the Port of Rotterdam.

1 2 3 4 5 — 26 25 24 23 22

Harbour Publishing Co. Ltd.
P.O. Box 219, Madeira Park, BC, V0N 2H0
www.harbourpublishing.com

Indexed by Martin Gavin
Dust jacket design by Dwayne Dobson
Text design by Roger Handling / Terra Firma Digital Arts
Printed and bound in South Korea

Harbour Publishing acknowledges the support of the Canada Council for the Arts, the Government of Canada, and the Province of British Columbia through the BC Arts Council.

Library and Archives Canada Cataloguing in Publication

Title: Workboats for the world : the Robert Allan story / Robert G. Allan with Peter A. Robson.
Names: Allan, Robert G., author. | Robson, Peter A. (Peter Andrew), 1956- author.
Description: Includes index.
Identifiers: Canadiana 20220248826 | ISBN 9781550179873 (hardcover)
Subjects: LCSH: Robert Allan Ltd.—History. | LCSH: Work boats,— British Columbia—Vancouver—Design and construction—History. | LCSH: Naval architecture—British Columbia—Vancouver—History. | LCSH: Naval architects—British Columbia—Vancouver—Biography.
Classification: LCC VM301.R63 A45 2022 | DDC 623.8/109711,-dc23

DEDICATION

I grew up surrounded by the business of my father and grandfather, and as time passed it became very apparent to me the scope and importance of the ship design legacy contained within our company archives. Attempting, in this book, to document that legacy has been a labour of love for me, in the hope that the new owners of Robert Allan Ltd will understand and appreciate the technical foundation upon which the current business is based, and that those who value BC's maritime history will recognize and more easily identify the breadth of the contributions of Robert Allan/Robert Allan Ltd over more than a century of ship design in BC. It was also apparent that none of this success could have been accomplished without the support of the three amazing women to whom this work is fittingly dedicated.

Caroline Malcolm Allan

My grandmother left the comfort of her genteel Chester upbringing to follow her husband to the relative wilderness of the North Shore of Vancouver in 1919. I have often wondered how difficult that must have been for her, without any friends or family for support. She was ever gracious and kind, a true, proper Victorian lady, and bemusedly tolerant of a husband who preferred to dominate every conversation!

Caroline Allan

Annie (Nancy) Grant Allan

My mother's family immigrated to Canada from Elgin, Scotland, in 1930, when she was a young teenager. Her Scots roots ran very deep, and she was a very accomplished trained singer, especially of the songs of her native home. She had a succesful musical career in Vancouver, particularly throughout the 1950s, participating in many choral and operatic productions. These events caused me to be "baby-sat" frequently in the basement office of my father, undoubtedly influencing my long-term career choice! In addition to her own professional pursuits, my mother spent many years supporting the family business in the role of typist/secretary and as the treasurer/bookkeeper—roles for which she had no training but performed most ably. She also played a large role in establishing a "family" environment within the business that I believe continues to this day. Her involvement in the formative years of the company was invaluable.

Enneke Allan

I am sure that when Enneke and I were first married (50 years ago), she had no idea she had hitched herself to a workaholic who was so committed to this family business. That realization was made abundantly obvious when I unexpectedly found myself in charge of the business in 1980/81, at the time we were busy raising three young sons. Enneke gave up a promising teaching career to commit to ensuring we had a healthy and happy home while I was trying to ensure the company survived throughout the 1980s. She has been a pillar of strength to me for all these years and has been "mostly" accepting of my long hours and working weekends. We have, however, also enjoyed the benefits of the extensive worldwide travel that came with the growth in the business of global tug designs.

Thank you, Enn; this journey could not have been made without you!

—R.G. Allan, 2022

Nancy Allan

Enneke Allan

Contents

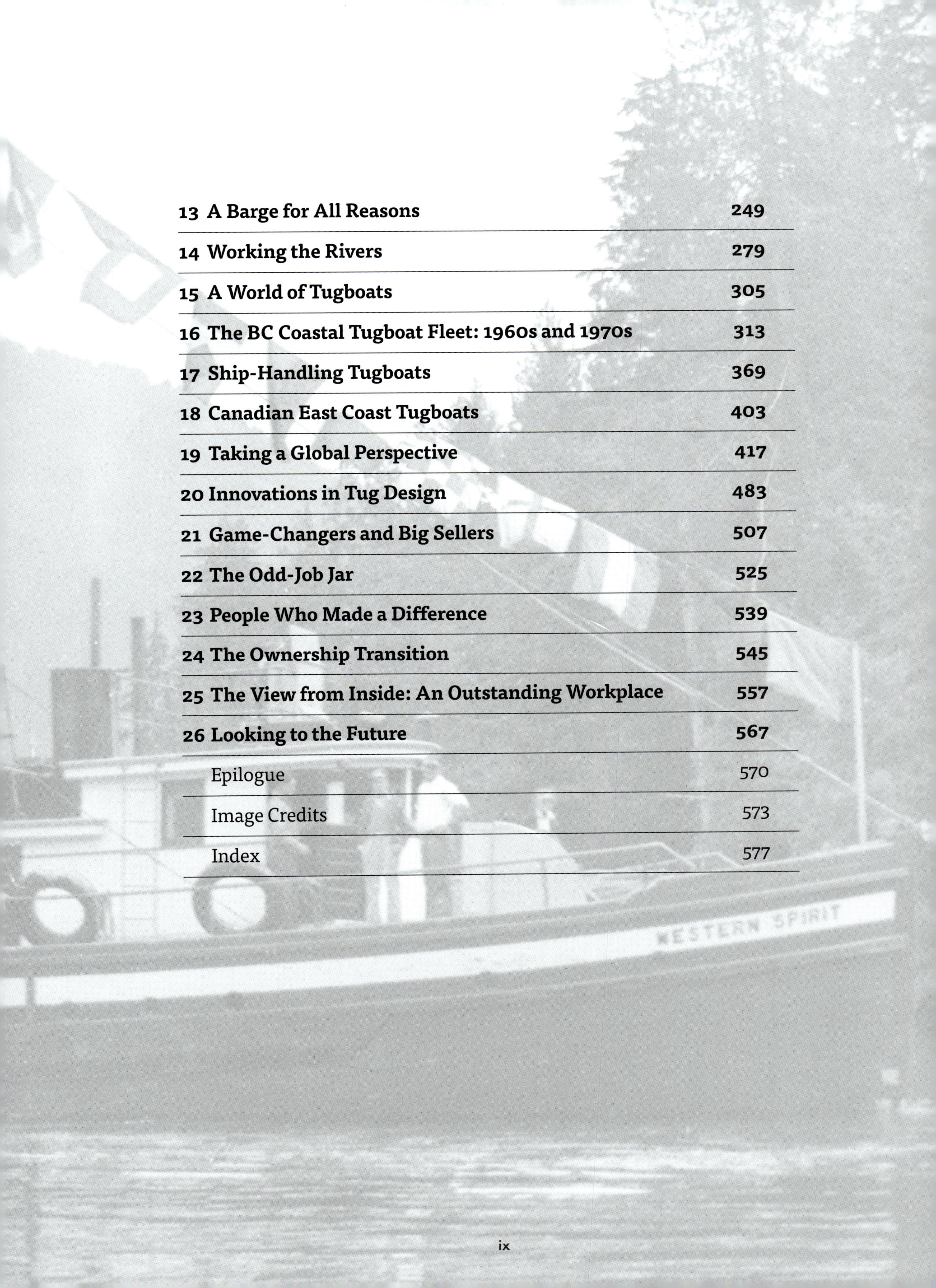
WESTERN SPIRIT

Foreword

By Peter G. Noble

My first encounter with a naval architect by the name of Robert Allan was in the library of the Department of Naval Architecture at the University of Glasgow in the mid-1960s. As part of my studies, I was working on the design of fishing vessels and discovered a wonderful book, *Fishing Vessels of the World*, published by the Food and Agriculture Organization of the United Nations. This tome contained a wealth of information on fishing craft from around the globe, but one paper in particular caught my eye. It was titled "Design and Construction of British Columbia Fishing Vessels," authored by Robert F. Allan of Vancouver.

My connection with Robert Allan, naval architect, may have ended there, but the stars aligned when a notice appeared on the Department of Naval Architecture message board seeking to hire a new graduate naval architect for a position in Vancouver, BC, with Robert Allan Ltd.

An interview with Mr. and Mrs. Allan, who were visiting Scotland, led me to be offered the advertised position, and I started my professional career in Vancouver in the summer of 1967.

Peter G. Noble, Noble Associates Inc., Houston, Texas

The local BC economy was booming with forest products and mining projects, which brought requirements for many tugboat and barge designs, as well as designs for a number of large fishing boats and other vessels, such as trailer ferries. Although I only worked at Robert Allan Ltd for a relatively short period, I have had the honour of knowing three generations of Robert Allans: Robert Allan, Robert F. Allan (Bob) and Robert G. Allan (Rob). Bob Allan was running the business, while his father, the founder of the company and a 1907 Glasgow University graduate, made regular visits to the office to see what projects were underway and to talk with the designers and draftsmen; Rob Allan was studying at the University of British Columbia prior to going to Glasgow University to study naval architecture.

Mr. Robert Allan was in his 80s and was kind enough to invite me to dinner at his house on a few occasions, and I well remember his questioning about things in the "old country," especially the demise of the shipbuilding industry. I also recall he still had a strong Scottish accent and knew more about Burns and his poetry than I ever did!

My professional career has taken me away from BC to other parts of Canada and beyond, but I have kept myself informed of the company's progress, which is so well detailed in this book. I have long held the belief that the Canadian West Coast is a centre of excellence for small commercial craft design and construction, with a number of design companies active in that market. However, while Robert Allan Ltd has maintained its strong position in the British Columbia maritime industry over many decades, providing a plethora of successful designs for tugs, barges, fireboats, ferries, patrol craft, pilot boats, fishing vessels and icebreakers, among other vessel types, what makes the company stand out from other naval architectural design firms is the way the company has "gone global" in the 21st century.

Early adoption of computer tools and recognition that, with modern communications, geographic proximity to the clients and customers was no longer a primary requirement allowed Robert Allan Ltd to expand its client base far beyond the West Coast of Canada, and its long history of continuous development of exceptional tugs of all types has placed the company in a world leadership position in that sector. Hardly a day goes by but that I read on LinkedIn of yet another Robert Allan tug project coming to fruition in some far-flung corner of the world, and I still read the *Western Mariner* magazine, which regularly features their tugs and other craft in their "home" market.

I commend Rob Allan in taking on the mammoth task of recording the excellent near-century history of Robert Allan Ltd and capturing the myriad of designs from the 1930s to the present day. It is clearly a labour of love, with early Robert Allan designs such as the *Meander* and *Fifer* sharing space with modern ship-assist and escort tugs.

I am sure that I, along with many others, will enjoy reading this book from cover to cover and that we will find ourselves drawn back to study again the many fine designs within its pages.

I congratulate Rob on the authorship of this book, and also the whole Robert Allan Ltd team, which I am sure will continue the legacy of being a driving force in the tug and working craft design world in the years ahead.

Preface

On Being a Naval Architect

This book is about a company, about the family that built it, and about the many people whose efforts have helped to make it a Canadian success story in the global maritime world. It is also, of necessity, about the thousands of working vessels around the world which have been designed by this company over the better part of a century.

Robert Allan Ltd is a consulting naval architecture business, designing commercial working vessels to suit the specific needs of our clients, whether governments, private companies, shipyards or private individuals. It serves a diverse set of vessel owners and is often very cyclical in its workload. This is not a business for the faint of heart, but it is one that represents one of the most professionally satisfying careers imaginable.

A consulting naval architect gets to be the consummate integrator of almost every engineering discipline, combining the knowledge of a hydrodynamicist, structural, mechanical and electrical engineers, and wrapping all of that together with the talents of an architect and interior designer. We take the needs of our clients and then develop ideas, sketch concepts, prove the viability of those concepts, and finally develop a design of sufficient detail to be built by a (hopefully!) competent shipyard. With good luck we will see that "paperwork" converted into steel and perform its assigned duties for a happy owner, and ideally look good while doing it. That process is incredibly gratifying when everything goes according to plan, which is thankfully most often the case!

I grew up in a household of naval architects, my father and grandfather practising that craft before me. I spent many hours as a small child in the basement office at my grandfather's house surrounded by drawings, slide rules and French curves,

marine magazines and photographs of the key projects of the day. While I doubt this was deliberate indoctrination, it ultimately had the desired effect. In grade school I can never recall saying that I wanted to be anything other than a naval architect, a profession which the teachers (expecting doctor, lawyer, fireman or policeman as the answer) invariably could not comprehend.

In my teens I had a hobby designing houses, or at least tinkering with that concept, applying my emerging drafting skills. This was fun and was something I enjoyed, especially since if you needed more space, you simply made the floor plate bigger, and it didn't have to float! As I considered this work I recall clearly saying to my father when I was 16 or 17 that "Maybe I'll just become a 'regular' architect instead of going into the ship design business." My father was a very wise man, and he had never told me that I should follow in the family tradition, but he said, "That is certainly an option, and you could do that and become one of the tens of thousands of architects out there who struggle to make a living, or you could stick with naval architecture and get to one day be a 'big fish in a little pond.'" I followed that advice, and I have often felt how prophetic those words were.

After more than a few challenges, Robert Allan Ltd has grown to be one of the most respected and certainly one of the most enduring names in the commercial small ship industry around the world. That success belongs to everyone who has been employed by the company, many of whom have spent the majority of their professional careers doing what we love and enjoying every day doing it.

The tools we use to design ships have changed dramatically over the years. Gone are the big drafting tables, splines and lead weights, French curves and planimeters. The computer has taken over not just our drafting tasks but also has brought to hand some incredibly powerful analytical tools that enable computations undreamed of a generation ago.

As witnessed by the following pages, the output of the company has been prodigious, particularly in the 21st century. That reflects a devotion to doing good and careful work throughout, ensuring that we take care to make certain that the boats built to our designs are as safe as possible for the men and women who work on them.

The sea is a demanding taskmaster and must never be underestimated. It is our professional responsibility to do our very best every day in the tradition established by the creators of this company.

—R.G. Allan

Acknowledgements

Much of the impetus for this book came from my colleague Rollie Webb, an experienced shipbuilder and a definite authority on BC's maritime history. Rollie compiled and gifted me with a comprehensive set of extracts from the archival records of *Harbour and Shipping* and *Western Fisheries* magazines, marking and copying every page, from the late 1920s to well into the 1960s, in which there was a reference to "Robert Allan." This was a treasure trove of information, much of which I was not familiar with, and it unlocked pathways to much of the early history of this company. I am deeply indebted to Rollie for this significant and most generous effort.

Although technical writing is not new to me, I wasn't sufficiently confident in my abilities to put this story together in a manner that would hold the interest of any but the most die-hard of "boat geeks." I was introduced to Peter Robson as someone who could guide me through that process and whose own marine writer background would help to keep the project on course. It has been a distinct pleasure to work with Peter throughout this process, and I thank him for his steady hand in keeping this challenging project on track!

I am most grateful to Mike Fitzpatrick and all the management team at Robert Allan Ltd for their support during this project, not only in terms of the significant financial support necessary to enable the entire project, but for indulging my wishes to ensure that this bit of history be well documented for the benefit of all the employees, present and future.

Ernst Schneider is the in-house graphics artist/designer at Robert Allan Ltd, where he has worked his craft for many years, helping to promote and illustrate our design work, and generating our popular *RAindrops* newsletters. Ernst has been instrumental in carefully scanning and grooming the numerous drawings and illustrations in this book, frequently working from old, often fragile and not always pristine original documents.

John MacFarlane has done more to document the nautical history of British Columbia than anyone can possibly imagine. His *Nauticapedia.ca* contains a wealth of material and was a source of much information to me in my quest to identify many of our early designs. The Facebook page *British Columbia Nautical History*, also coordinated by John, was another wonderful source of both information and contacts with people who have connections with many of our designs. Sincere thanks to John for his unwavering commitment to the cause of BC's maritime history, and the enormous resource that is his life's work.

As will be obvious to the reader, this is a book which will hopefully entertain in both words and drawings but mostly in pictures. While it has been hard to refrain from including a picture of every boat ever built to a Robert Allan design, it is hoped that those chosen are at least representative of the many vessel types covered. In the course of this research, we have often reached out to the many skilled people whose passion is photographing the working craft of the BC coast and sought permission to use their work. Of special note in this group are Rone McLellan and Kirstin Hodge, wife and daughter, respectively, of the late Murray McLellan, publisher of *Harbour and Shipping* magazine. Murray's vast collection of well-documented marine photographs was donated to the Vancouver Maritime Museum (VMM), and Rone and Kirstin graciously gave us unfettered access to those. We are also grateful to Ashlynn Prasad, the librarian and archivist at VMM, who assisted us with having many of these images scanned for use.

Several other archives helped with research and provided images. They include the City of Richmond, the University of British Columbia, the City of Vancouver and the Northwest Territories archives.

Special mention among the photographers is due to Alan Haig-Brown, whose wonderful collection of photos of classic BC fishing vessels from his book *Fishing for a Living* contained many Allan designs, and to Don Macmillan of Cassiar Packing, whose family archives contained many photos of trollers and gillnetters with which I was totally unfamiliar, helping to identify a whole sector of our design work which was very poorly documented. David Rahn, publisher of *Western Mariner* magazine, was also very helpful in tracking down images that appeared in his magazine over the years. We'd also like to acknowledge Michelle Imbeau (Canadian Coast Guard/Fisheries and Oceans Canada), Kate Lessels and Kris Neely (Seaspan), and Don Brown for their help with multiple photos. A full list of photographers and credits is provided following the Epilogue.

Introduction

It is assumed that if you're holding this book in your hands there is a very good chance you know something about boats! The writer has attempted to avoid making this an overly technical book about the science and art of ship design, in the interests of the general appeal of this story about an important BC maritime company, but maritime technology has many unusual terms and odd conventions, and the measurement systems have changed through the course of this story, and so some of these require explanation.

Dimensions

The individual vessel dimensions cited in this book are the overall design dimensions as recorded on our drawings at the time of the design. Accordingly, for reasons of historical accuracy, the dimensions are given in either feet and inches or in metres, depending upon the era of the design or the geography of the owner, but with rare exception not in both units. In the interests of readability, we have also resisted adding converted dimensions in parentheses to the original data.

It is also important to note that these dimensions will very likely differ slightly from those recorded in the Ship Registry, and consequently in sources such as *Nauticapedia*. The length recorded in the registry is "Registered Length," which is defined in the Canada Shipping Act as "96% of the total length on a waterline at 85% of the least moulded depth measured from the top of the keel, or the length from the fore side of the stem to the axis of the rudder stock on that waterline, if that is greater...," so it can be significantly less than the actual length overall.

Steel vessels are designed to the "moulded" lines, namely to the *inside* of the plating, hence measured dimensions, especially beam, are always nominally two plate thicknesses wider than the design dimensions, rarely a large difference. Wooden vessels however are designed to the *outside* of the planking, and so measured and design dimensions should be the same. However, wooden boat building has many variables, and it is not uncommon for surveyors to have mistakenly measured to the outside of a guard rail or similar, so the recorded beam could easily be up to six inches in variance to that shown on the design.

Vessel Names

It is of course quite common when vessels change owners that they also experience a name change. Keeping track of those changes is a challenge and no attempt has been made in this book to do so. In most cases the boats described are referred to by their original owners and original names only. Readers are recommended to consult *Nauticapedia* for very good records of the name changes for many of the vessels described here.

Design Records

For the majority of the boats described in this book, the data given has come directly from our design drawings. Fortunately, the company management has seen fit to maintain and protect the extremely complete set of our historic drawings, especially those from the pre-CAD era, and that record is quite complete, at least from about 1945. It must be noted, however, that it's not uncommon that changes, such as a different engine choice or even dimensional changes, are made to a boat either by the builder or owner during construction, and not always with the knowledge of the naval architect, so it is possible in some circumstances that the "as-built" data may differ from that given here.

In several instances the original drawings (from the pre-CAD era) have been lost or re-used and modified, and it has not always been possible to tie a vessel name to a specific project number in our files. In those instances, we are most grateful for the amazingly complete information set that exists in the archival records of *Harbour and Shipping* magazine, which did a marvellous job of recording the waterfront activities of Vancouver and environs for many decades, providing direct links to follow in our own files.

1

A Brief Family History

The story of Robert Allan Ltd could easily be just about the thousands of vessels we have designed, but those designs couldn't have happened without the creative people behind the pens and pencils and ultimately the computers, nor certainly could they have been developed without our faithful and valued clients, many of whom have entrusted their new design requirements to us for decades. This is fundamentally the story of one family, but it is also about the many talented, hard-working people who have worked in the company over the past many years, and who have contributed greatly to the company success.

In considering this publication, rather than a simple chronological history, it seemed more logical to describe the prodigious output of the organization by boat types or industry sectors and to trace the evolution of those individual design types through the lens of the company responsible for defining them. But it is hard to put those vessels in context without the back story of how the company was formed and how it evolved over nearly a century from a simple one-man consultancy into a world-leading designer of high-performance workboats.

Opposite: "Let it be known that I was not indoctrinated to become a naval architect."

The story of the company is inextricably tied to the general history of the Allan family and the immigration of all my grandparents to Canada, the ultimate *raison d'etre* for the company's existence. So, with the reader's indulgence...

My grandfather Robert Allan was born in Kilbirnie, Ayrshire, Scotland, in 1884, the first child and only son of James and Mary (née Campbell) Allan. His mother, Mary, died when Robert was about nine years old, leaving James to provide and care for a family of five. James was a well-regarded builder in the Kilbirnie area, and so they seem to have fared reasonably well as a middle-class family (albeit with a typically dour Scots countenance!).

Robert Allan (left), his father James Allan and his four sisters (l–r): Kate, Margaret, Mary and Agnes.

They grew up in the Ayrshire countryside of Scottish poet Robbie Burns, a factor which influenced my grandfather's life profoundly. Burns' works became one of his lifelong passions, and he was considered a Burns scholar, able to recite virtually all of Burns' works, which he loved to do, at length and often! This mastery, coupled with his broad Scots accent, was however mostly lost on my unimpressed youthful ears.

At the age of 14, young Robert was apprenticed to the drawing office of the famous Clydeside shipyard of Fairfield Shipbuilding & Engineering. This was the normal process of entering the industry at the time; however, his academic bent resulted in him later earning an Andrew Carnegie bursary to attend the University of Glasgow to study naval architecture. His 1907 graduating class numbered over 100 students, many of whom reportedly had difficulty finding employment in shipbuilding, as academic training in this field was quite new and was looked upon with scepticism by many employers. The traditional apprenticeship career path was still considered the "right way" to learn the trade. Fortunately, my grandfather had both credentials and appears to have had no difficulty gaining employment. He often remarked (with some glee) that those of his classmates unable to get work as naval architects went back to university to study medicine.

After graduation he returned to work at Fairfields until 1912, but later migrated to other well-known British shipyards, including Cammell Laird in Birkenhead

Robert Allan upon graduating as a naval architect from the University of Glasgow, 1907.

(1913–1915), where he worked on the Town class light cruiser HMS *Birkenhead* and the last King George V class dreadnought battleship built for the Royal Navy, HMS *Audacious*. He then moved to Palmers (Yarrow on Tyne) in Newcastle (1915–1919), where he worked on a variety of tankers and motor torpedo boats.

Whilst in the employ of Cammell Laird, Robert was introduced to Caroline Malcolm Brittain, youngest daughter of a rather well-to-do Cheshire family (fortunately with some strong Scottish roots). Caroline's father, William Henry Brittain, had been the general superintendent of Cammell Laird, but he had died before Robert's employment there, and so in spite of that historic industry connection, it is unclear how the couple were introduced, but clearly they did meet, courted and were married in January 1915.

Their only child, my father, Robert Frederick Allan, was born in the northeast of England in September 1916. This accident of an English birth locale was a source of some (albeit often facetious) embarrassment to him, as his Scots roots ran very deep.

In 1919, seeking employment away from the war-ravaged UK, Robert and Caroline immigrated to Canada aboard the Allan Line's *Corsican*, and arrived in Vancouver late that year. He was briefly employed at Coughlin's Shipyard in Vancouver, but it is uncertain whether he had that offer of employment when he left the UK. In

1920 he was hired by Wallace Shipyards in North Vancouver as chief draftsman, with the main commission of designing the *Princess Louise* for the Canadian Pacific Steamship Company.

More Than a Change of Scenery

I have often wondered how my grandmother Caroline viewed this move to the rather rough and undeveloped west coast of BC of the 1920s. She had a very genteel upbringing in a large and stately home, "Crabwall Hall," near Chester, but as a typical Victorian lady she would have abided by the decision of her husband.

In 1926 they bought a large house high on the hill in West Vancouver (in the vicinity of 14th and Mathers), of which my father had many fond memories as a small boy. I am quite certain that this time spent on the still somewhat undeveloped North Shore was rather lonely and isolating for my grandmother, but she remained polite and gracious all her life.

Crabwall Hall, ancestral home of the Brittains.

Below: The Allan family home in the vicinity of 14th and Mathers, in West Vancouver, ca. 1926.

In spite of his demanding professional career, Robert Allan had a significant set of interests outside the drawing office. He was very active in the local St. Andrew's and Caledonian Society, the largest of apparently 14 Scottish societies that existed in Vancouver before 1950, and he served as its president for a number of years in the late '30s. He often lectured on topics such as the Scots language, Scots history and of course Burns' poetry.

–R.G. Allan

Robert Allan on the streets of Vancouver ca. 1940, taken by well-known street photographer Foncie Pulice.

My father, Robert F. (Bob) Allan, who had effectively spent his entire life in Vancouver, worked as a project superintendent at Burrard Dry Dock Ltd during WWII. He joined his father as a partner in the consulting business immediately after the war, and they continued to serve the local market with designs for a wide range of working coastal vessels.

There are of course two sides to the equation of every family, and my mother's kin were no less Scottish and equally as marine-oriented as the Allans. My mother, Annie (Nancy) Grant Masson, was born in Elgin, near Inverness, in the north of Scotland, in August 1916. Her grandfather, Thomas Masson, was a North Sea fisherman from the village of Findhorn (much later to become a New Age mecca), and her father, John Masson, was a seagoing steam engineer, also from Findhorn. In 1930 their family immigrated to Vancouver, from where John spent much of his career as an engineer officer aboard the *Empress of Asia* on the trans-Pacific route to the Far East, until that ship was bombed out from under him and his shipmates in the Straits of Malacca during the Japanese invasion of Singapore in 1942. He was "missing in action" for several months after that incident, eventually landing in Durban, South Africa, before finally returning to Vancouver. He resumed his seagoing career in 1944 as chief engineer on a variety of ships, eventually retiring in 1951.

The family story is that my parents were introduced at a 1941 event of the aforementioned Scottish Society in Vancouver, over which my grandfather was presiding and at which my mother, a professionally trained singer, was presenting a collection of Scots songs, which was her great strength. Apparently, her singing impressed my grandfather more than my father at the time, but he soon realized her many charms! My mother had a fairly distinguished career as a professional classical singer in Vancouver throughout the '40s and '50s and appeared in several operatic and local theatre productions. Her theatrical bent was in marked contrast to my father's quieter and totally rational demeanour; however, they had a long and happy marriage and seemed to each be the perfect foil for the other.

My parents were married in Vancouver in 1943 at St. Andrew's-Wesley Church on Burrard Street, where my mother was the (paid!) soloist. In 1945 they purchased a home on West 27th Avenue near Blenheim Street and raised my sister Betty (1945) and me (1947) there, in the only home I knew until I left for university overseas in 1968. The die was rather firmly cast that I would grow up in a strongly tartan-tinted home tinged with saltwater.

From 1949 onward we spent a large portion of our summers at a small, rustic rented cabin at Sunset Beach in Howe Sound. Life there was great for a small boy: swimming, playing on the beach or the floats or in the woods, and just mucking about in small boats. My father, ever busy in the basement office, would commute to Sunset Beach from Eagle Harbour at weekends in the small wooden "putt-putt" we owned, these being the days before the railroad and Sea-to-Sky highway were built. In 1959 my parents purchased property on Bowyer Island, and we continue to enjoy extended summers there to this day, where the only way to get around is in a small boat. Boat-handling became second nature to me, and the prospect of being a naval architect seems to have been imprinted in my DNA, as it was certainly never forced onto me. Nor was it discouraged!

Young Rob Allan learning some boat-handling skills at Sunset Beach in Howe Sound.

From an early age I remember going with my father to visit many of the numerous small boat builders that existed in the Vancouver area in those days, particularly in Coal Harbour. Looking back, I'm sure this was just subtle indoctrination, but I have fond memories of yards like Menchions, Star Shipyard and John Manly Ltd, and it was impressive to see (and smell!) the lovely wooden creations of those days.

I recall the busy years of the 1960s when I was in high school and the company was growing significantly, with commissions for increasingly larger and more complex vessels. My father, who was now running the office, my grandfather having effectively retired, worked very long hours, almost every evening and weekend, and many clients considered his ear to be available for their ideas at any time of night or day. My father never complained (my mother was another story); he just considered it part of his job, which earned him a great deal of respect in the industry.

I attended UBC for a couple of years in order to gain acceptance to a foreign school of naval architecture, there being none available in Canada at the time. In 1968, after weighing options, I set off to pursue studies in the family tradition. My choices were either in the USA, where the University of Michigan was the leading school of the day, or the UK. The US was embroiled in race riots at the time and so offered little appeal, and the choice of my grandfather's alma mater, the University of Glasgow, became an easy one. This was, however, a grim period for British shipbuilding after the glory days of the pre- and post-war shipbuilding boom. The once-dominant British yards had faded to almost no building activity at all and would not boom again until the development of North Sea oil exploration. Accordingly, I was one of a class of six students, which made for a very high lecturer-to-student ratio!

Another (and admittedly the leading) reason for my choice of Glasgow was more romantic. In the fall of 1966 I had met a very lovely young woman who was a member of a girls' choir that my mother directed in Vancouver. I had been enlisted, begrudgingly, with some of my friends, to play some male roles in an annual musical production that the "Girls Ensemble" put on at Dunbar Heights United Church. Shortly after that I began dating Enneke Spoke, until later in 1967 when her family decided to repatriate to their native Holland. I was told this decision had nothing to do with my dating their daughter, but I was never quite sure! In any case Enneke, who had spent almost her entire life in Canada, had to decide where to pursue her post-secondary studies, and London was her obvious choice. At least London and Glasgow were on the same land mass, so we were able to continue our courtship throughout the years we both studied in the UK, albeit at the lengthy intervals which student budgets permitted.

After graduation in 1971 Enneke and I were married in the Netherlands, with a diverse medley of guests from around the globe. We worked in Basingstoke, England, for two years: Enneke as a primary school teacher and I for the well-known firm of Burness, Corlett & Partners, the world's leading consultants for tug design of the day. It was a fine place to get some basic training in design skills.

In January 1973, however, we headed back home to settle in Vancouver, where I joined my father in Robert Allan Ltd at one of the busiest times the company had ever experienced. And the rest, as they say, is (corporate) history!

2

Evolution of the Business

Starting in Canada

After immigrating to Canada in 1919, Robert Allan worked briefly at J. Coughlan & Sons Shipyard in Vancouver, and then took a position early in 1920 at Wallace Shipyard in North Vancouver as chief draftsman. (Wallace's changed its name to Burrard Dry Dock Co. Ltd in 1921.) He was soon promoted to general superintendent. His initial responsibility was the design of the 331-foot "pocket liner" SS *Princess Louise* for CP Steamships, the first of their fleet not to have been built in Scotland. According to Francis Mansbridge in *Launching History: The Saga of Burrard Dry Dock*, "Robert Allan...outdid himself with the elegant design." This stately vessel served on the BC to Alaska run until 1962, after which she was sold to American interests in 1966 and converted to a floating restaurant in Long Beach, California.

Opposite: The only photo (ca. 1949) in existence with three generations of Robert Allans, (left to right) Caroline Allan, Rob (R.G.) Allan, Bob (R.F.) Allan, Robert Allan and Betty Allan.

Duet, Then Solo

In October 1927 Robert Allan left Burrard Dry Dock and with George Stackhouse entered into a partnership named Allan and Stackhouse. This venture was noted in the *West Van. News* on October 20 of that year.

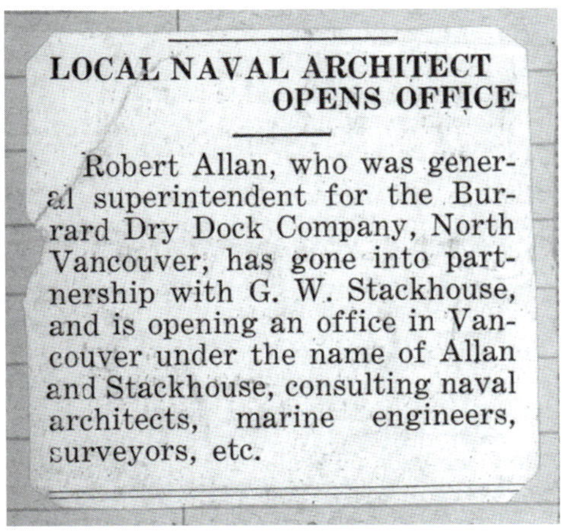

LOCAL NAVAL ARCHITECT OPENS OFFICE

Robert Allan, who was general superintendent for the Burrard Dry Dock Company, North Vancouver, has gone into partnership with G. W. Stackhouse, and is opening an office in Vancouver under the name of Allan and Stackhouse, consulting naval architects, marine engineers, surveyors, etc.

Left: A 1927 newspaper clipping announcing the partnership of Allan and Stackhouse.

There are very few drawings in the archives identified under this banner, and the partnership was relatively short-lived. In March 1929 the first known advertisement for Robert Allan as an independent consulting naval architect appeared in *Harbour and Shipping* magazine, though the design records indicate he started work independently late in 1928.

He operated out of downtown offices at 325 Howe Street through 1930, then relocated to offices at 626 West Pender Street until sometime in 1933, when presumably, in the depths of the Great Depression, he simply could no longer afford to maintain private offices.

One Man in the Basement

Hanging out his shingle as "Robert Allan, B.Sc." (an engineering degree credential, unusual in his day, and of which he was very proud), he operated a struggling business throughout the '30s from a series of home offices, progressively on Elm Street, Dunbar Street and West 1st Avenue. The scant work in these early years consisted of a series of coastal patrol vessels and a few personal yachts. The lovely 67-foot *Meander*, commissioned by Mr. George Kidd in 1933, was the project that really saved his consulting career, providing much-needed income in those tough times. This lovely vessel is described in more detail in Chapter 3.

The impact of the hard times of the '30s are obvious in the words penned in a letter to his friend Frank Gildersleeve, the Vancouver architect who had been tasked with the interior décor of the *Princess Louise* and who had subsequently relocated to the UK. Robert had the habit of drafting every letter in full before writing the final version, and he made use of the back side of many pages in his design notebooks for this purpose. Thankfully, a few of his musings are preserved in this fashion. In 1934, he wrote:

> ...Many thanks for [the *Times* newspaper] which vastly interested us all making me wish I were back once again on the old Clyde. But like many other exiles in Vancouver I am afraid I shall have to remain where I am for lack of the wherewithal to pay my fare.
>
> Boatbuilding here is almost an extinct business [with] only a small amount of work keeping the yards open. Since the completion of Mr. Kidd's yacht Meander there has been nothing much larger than row boats. At the moment all I have to look forward to is [a] small retainer fee from Ian Clark, a lawyer who is acting on behalf of the Government in a suit with Cochlans [sic] over a question of payment for the ships they built in 1919.
>
> Bobby is attending university having obtained a small scholarship which has enabled him to do so. He has made a beautiful model of the Meander; really the best I have ever seen; everything in exact replica even to the furniture inside the cabin. It is all lit up with electricity. Our hope is that Mr. Kidd will purchase it at a decent price...

The 105-foot steel luxury motor yacht *Fifer*, built in 1939 for Capt. William Crawford, was a significant commission that closed out those lean pre-war years on a bit of a high. Throughout the war years of 1939–1945 the elder Allan was kept busy with the design of large fishing vessels for the BC coast, and a few ferries, including the *North Vancouver Ferry #5*, whose main function was moving workers across Burrard Inlet to the busy Burrard Dry Dock shipyard. Ultimately replaced by the bridges crossing the inlet, this vessel ended its days as the Seven Seas Restaurant at the foot of Lonsdale Avenue in North Vancouver.

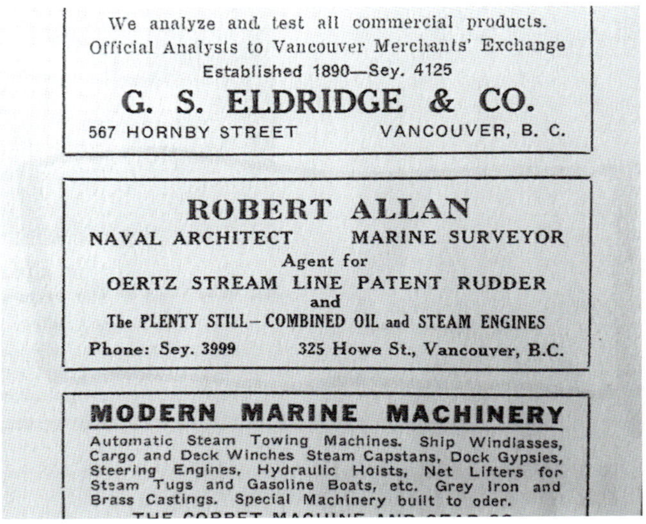

We analyze and test all commercial products.
Official Analysis to Vancouver Merchants' Exchange
Established 1890—Sey. 4125
G. S. ELDRIDGE & CO.
567 HORNBY STREET VANCOUVER, B. C.

ROBERT ALLAN
NAVAL ARCHITECT MARINE SURVEYOR
Agent for
OERTZ STREAM LINE PATENT RUDDER
and
The PLENTY STILL—COMBINED OIL and STEAM ENGINES
Phone: Sey. 3999 325 Howe St., Vancouver, B.C.

MODERN MARINE MACHINERY
Automatic Steam Towing Machines. Ship Windlasses, Cargo and Deck Winches Steam Capstans, Dock Gypsies, Steering Engines, Hydraulic Hoists, Net Lifters for Steam Tugs and Gasoline Boats, etc. Grey Iron and Brass Castings. Special Machinery built to oder.

Above: As general superintendent of Wallace Shipyards, Robert Allan's initial responsibility was designing the S.S. *Princess Louise*.

Left: A 1929 ad in *Harbour and Shipping* for Robert Allan as an independent naval architect.

Above: *North Vancouver Ferry #5* arriving at North Vancouver, ca. 1941.

Right: The Allan residence at 2026 West 63rd Avenue was home to the business from 1944 until 1966.

In 1944 (after finally a few years of reasonable earnings), he settled at 2026 West 63rd Avenue, which became home to the business for most of the next 20 years.

This spacious, comfortable home, which sat on a large but oddly triangular lot, had an undeveloped basement, about half of which was turned into quite pleasant offices, initially with space for about four persons, though it would be several years before that space was challenged and ultimately exceeded.

The Elder Robert Allan

When I met Robert Allan Sr., he had to have been in his mid-80s and the office had been moved to 72nd and Granville. This was when his son Bob Allan had taken over the business. Before then, Bob and his father had operated from their house on 63rd in Marpole. It had a large basement office and fairly large workshop off to the side. I guess the elder Robert Allan had been banished from the basement at the time. Bill Cleaver was one of Robert Allan's earliest draftsmen. Bill had worked in the basement for Bob Allan, but he would tell me that whenever Bob had to go out to a meeting, his father would come down and have a look at what was going on and talk, and when Bob would come back, he had to scurry out. Bill said it was quite a distraction, as he'd occasionally go off quoting Robbie Burns. He was very, very civilized, and I knew his reputation dealing with the Department of Transport and people like that. All the company's plans had to be approved by the Department of Transport, which had a large technical staff of engineers who would look at them and stamp them. Bill told me it was well known that if anyone at Transport had put notations on his drawings, or asked for corrections, he would storm around and berate them, saying, "I'm the naval architect, not you!" He was feisty.

—Tom Ward

Two Men in the Basement

The second Robert Allan was well tutored in the science of naval architecture by his father. Robert F. (Bob) Allan had attended UBC as an engineering student from 1934–1936 until a serious illness caused him to drop out before completing his degree, something he always regretted. He was, however, even in his teens, an accomplished mechanic and ship-modeller. He obtained work at Burrard Dry Dock, which lasted throughout the war, where he finished as project manager for the conversion of the passenger/cargo ship HMCS *Prince Robert* into an armed merchant cruiser. His role in that project was commemorated by an engraved silver cigarette box from "Commanding Officer; Officers & Ship's Company; HMCS Prince Robert" on the occasion of his wedding to Nancy Masson, on June 1, 1943; an odd choice of gift given that he was never a smoker, but presumably a sign of the times.

In 1945 Burrard Dry Dock Ltd had a contract to construct a series of colliers for France as part of the post-war recovery efforts. Robert Allan obtained the contract from Burrard to prepare the construction working drawings for these vessels. That project led to the formal teaming of father and son, working together in the home basement office, and a major new phase in the development of this business.

Robert and Bob Allan in their home office, from a 1955 *Vancouver Province* article.

Work in the post-war years 1945–1950 was patchy, but there was a reasonable number of fishboats, several coastal and inland ferries, a few small tugs and barges, and a lot of work converting wartime marine assets such as minesweepers to coastal freighters or fishing vessels. In 1946 the design for the shallow-draft tug *Sandy Jane* for Yellowknife Transportation Ltd marked the first serious foray into the design of very shallow-draft towboats, a market which became extremely important to the firm in the 1970s (Chapter 14).

The early 1950s continued to provide a modest living for the father-and-son duo. Fishboats large and small continued to be a mainstay of the business, as well as several patrol vessels and the occasional wooden tug. Also at this time, designs were developed for landmark vessels for the Columbia Coast Mission: *Columbia* and *John Antle*, followed in 1959 by the smaller *Alan Greene* (Chapter 6).

There was sufficient work in the mid-1950s that they could afford to hire a draftsman to assist with the workload. The first hire for the Allans, in about 1957, was Derek Cove, who later struck out on his own and formed a series of partnerships, first with Peter Hatfield and later with Eric Dixon.

The late '50s marked the beginning of designs for purpose-built steel barges, particularly for railcar transport, but also the very first purpose-built log-carrying barges. The archives also show the clear emergence of steel rather than wood as the material of choice for all vessel types, though there was still a steady diet of designs for smaller wooden fishboats. The first large steel seiners on the coast, *Bering Sea*, *Attu* and *Caamano Sound*, were designed in 1958 and 1959. Notably, all of these handsome boats are still actively fishing on the BC coast.

Sandy Jane, the first shallow-draft tug designed by Robert Allan.

Attu was one of the first Allan-designed steel seiners on the coast.

As the decade closed, the design of the 125-foot ocean-going tug *Ocean Master* marked a major first for the company, as this was not only a significant vessel in its day but was the first Robert Allan design ever to be built overseas, constructed in the Netherlands at N.V. Scheepswerf Gebr. van der Werf, a shipyard still active and now a leading European builder of superyachts.

It was also evident that as the elder Robert Allan eased into retirement, Bob Allan was emerging firmly as the head of the business.

Mr. Allan was quite impressive. Nobody in the office called him Bob, you called him Mr. Allan. I don't ever remember anybody in the office calling him Bob. You wouldn't dare. He'd done some impressive work and was quite a strong personality, but very technically savvy. He really knew his way in and around boats. I learned a lot from him.

— Hans F. Muhlert

Bob Allan was a gentleman, but certainly very firm and a hard taskmaster. He was very knowledgeable, and if you were working on something and had a problem, he was quite happy to come over and sit next to you and help figure it out. He would work long, long hours, and so did I and the others. Bob paid us very, very well. It was basically a family business. Nancy Allan, Bob's wife, would come in and pay us and sort our benefits and things like that. Nancy was a very good singer. She was taken under the granddad's wing because he was president of the Scottish Society, and she would sing Scottish songs.

— Tom Ward

Too Many Men in the Basement: The First "Real" Office

In the 1960s the consulting business increased significantly on its growing reputation. The introduction of a substantial federal shipbuilding subsidy (the Ships Construction Assistance Regulations, authorized in 1961 and succeeded in 1966 by the Ship Construction Subsidy Regulations) spurred the construction of an entire new generation of steel tugs and barges to replace the earlier generation of steam-powered wooden tugs and wooden scows. This movement began in 1959 but really escalated in 1964 through 1969. There were about 1,220 vessels built under this program before it ended in the mid-1970s.

Bob Allan was fully in charge of the business now and was earning a reputation for designing well-performing tugs of all sizes. The company was formally incorporated as Robert Allan Ltd on September 21, 1962, heralding an era of much greater business formality, not to mention growth. By 1966 there were about eight engineers and naval architects working in the basement, and Bob Allan finally realised that bringing clients in through the back garden gate and down the basement steps didn't create much of a professional impression. That prompted the move in May 1966 into the first real corporate offices, at 1496 West 72nd Avenue, at the south end of Granville Street, an occasion marked with Friday the 13th festivities!

Invitation to the official opening of the first "real" business offices for Robert Allan Ltd.

ROBERT ALLAN LTD.

REQUEST THE PLEASURE OF YOUR COMPANY

AT AN INFORMAL

LAUNCHING PARTY

ON THE OCCASION OF THE

OFFICIAL OPENING OF THEIR NEW OFFICES

on

FRIDAY, MAY 13th, 1966

from 5 to 8 p.m.

at

1496 West 72 Avenue, Vancouver 14, B.C.

R.S.V.P. to 266-6285

The '60s were busy years for Robert Allan Ltd, with many tugs, both large and small, designed for virtually all major local owners, the advent of the self-loading and self-dumping log barges, and many specialized self-unloading barges for the construction industry. Of particular note was the introduction of long-distance ocean-going tug-barge systems for MacMillan Bloedel, operating between BC and California, something quite unique in the world at the time. The significant designs of this era, often referred to as the golden age of BC tugboats, were well described in a paper Bob Allan presented to the 1st International Tug Conference in London in 1969. These superb conferences, which endured as a worldwide biennial event until 2018, would prove to be of significant benefit to Robert Allan Ltd in meeting tug owners and operators from around the globe (see Chapter 19).

The year 1969 was a banner one for the company, with several large log-barge designs, some large tugs, several smaller tugs, ferries and, of particular note, the first few shallow-draft tug and barge designs for the Mackenzie River system—project types that would define the business for much of the following decade.

The Office on 72nd Avenue

In 1972 I registered with Canada Manpower and answered an ad for an office person for Robert Allan Ltd. The pay was $500 a month, which was very good money then. The office was located on 72nd Avenue in Marpole. It was then a "one-girl" office, though that terminology is outdated. The woman who I was to replace was still there. When I started, there were eight people working in the office on the second floor of the building. Besides myself, there were Mr. Allan (Bob Allan), Hans Muhlert, Jack Macfie, Daan Hengeveld, Cyril Lihou, John Dale, Hans Meier and Mrs. Allan (Nancy), who was secretary-treasurer and worked mostly from home. Mr. Allan, Daan Hengeveld and Hans Muhlert were the naval architects, Jack Macfie monitored construction at the shipyards, and the rest were draftsmen, another outdated term. If I recall correctly, Mr. Allan sketched things out and then gave his sketches to the draftsmen. They'd turn them into proper drawings using drafting tools, pencils, erasers and brushes to brush away the eraser crumbs.

—Carol Nilson

A Northern Outlook

The early 1970s were dominated by significant volumes of work for companies like Northern Transportation Co. Ltd (NTCL) and Northern Construction Co. Ltd (NCCL) as the promise of the Mackenzie Valley Pipeline project and the prospect of offshore oil exploration and drilling development in the Beaufort Sea created significant demand for specialized marine equipment of many types. This wave of design and shipbuilding challenges generated many unique and innovative shallow-draft vessel designs, as required for the challenging remote-operating environment. It also created a bountiful workload for local BC shipyards, a level of activity which sadly would not be seen again.

Rob Allan and his father, Bob, ca. 1976, with framed photo of the elder Robert Allan.

The Third Generation Arrives

The heavy workload of the early '70s led Bob Allan to call his son Robert G. (Rob) Allan home from his post-university work in the UK at the beginning of 1973. At this time Robert Allan Ltd was operating with a staff of about 12 people.

There was more than ample work for the design team throughout the '70s: extensive work in northern Canada and a decent amount of coastal tug-barge work. This included the design in 1974 of the four *Canmar Supplier* ice-breaking offshore supply vessels for Dome Petroleum, the first of their type and the first vessels ever built under the then new Canadian Arctic Shipping Pollution Prevention Regulations (CASPPR).

Overseas Venture

In spite of the busy activity during this period, by the mid-1970s Bob Allan could see that the amount of local design work was unlikely to continue at a level sufficient to sustain the size of operation that Robert Allan Ltd had become. In addition to a diminishing volume of local industry work, there was also more competition in the field than ever before, so the shrinking pie was having to support many more naval architecture firms, some of which had been formed by former employees of the company.

In about 1975 Bob Allan had occasion to visit the Philippines and Singapore for specific consulting tasks. He returned totally enamoured by the level of marine activity centred in Singapore and saw that area as an outlet for the talents of the office. A subsidiary company, Robert Allan Far East Ltd, was soon established in Singapore. After several initial attempts to provide design services to that area from Vancouver, it became apparent that the distances, time difference and lack of regular direct contact with potential clients was going to be a barrier to real growth, in spite of the efforts of a very well-connected local agent, Ian Carne. It also became apparent that vessel owners in this area were not accustomed to dealing directly with design firms; they did their boat shopping through local shipyards and through brokers.

To counter this, Robert Allan Ltd entered into a joint venture with Carne's brokerage company and with a local finance company to provide a turnkey "finance-design-build" approach to owners, hiring local shipyards to be the production facilities. After initial resistance from the yards, who saw this as competition rather than potential collaboration, the venture, known as SeaConstruct Pty Ltd, had a reasonable degree of success. Hans Muhlert, a highly skilled naval architect who had joined Robert Allan Ltd in 1970, moved to Singapore in 1978 to lead the design work performed there, with backup support from the Vancouver office. This led to an interesting array of projects, including patrol boats, crew boats, a hydrographic survey launch and shallow-draft supply vessels.

Business in Singapore

We had done a few jobs for Ian Carne, our agent in Singapore. We had designed some ferries, oilfield crew boats and some shallow-draft supply boats for his clients, and it came to the point where we were getting pretty busy with projects over there and that Carne needed some technical help. To make it easier, Mr. Allan formed a joint venture with Ian Carne and an investment company to design and build vessels in Singapore. Someone from our company would have to go over there to set up an office. I thought that sounded interesting, so I spoke up immediately and said, "I'll go!" This was in early 1979.

Most of the vessels we designed were built and used in Singapore, except for some crew boats and shallow-draft supply boats for oil companies working throughout the region. My first job was designing an 80-foot aluminium crew boat for Borneo. It had to be able to get over a shallow sandbar, so it could only have a draft of 2 feet or so, and it had to have waterjet propulsion. That was a challenge. We also did a fleet of 19 patrol boats, about 36 feet long, for the Singapore police. They were pretty fast high-tech boats that could do 32 knots. Most of our work over there was designing and overseeing construction. For us, it was a very competent place to get things built and at a fairly low cost.

I was fascinated by the place because everything was so different—the climate, the language, the environment—and I found that refreshing. It was a very good experience. I was there for three years. I was happy. It was very interesting.

—Hans F. Muhlert

Changing of the Guard

As business opportunities overseas were developing and leading to increased recognition of the firm, fate, as it often does, threw a wrench in the works. Bob Allan's wife, Nancy, who had worked in the company as typist and bookkeeper for many years, was diagnosed with cancer in 1977. Bob cut back on his significant workload to devote much of his time to caring for her through an extended and difficult period, culminating in her death in 1980 on the eve of her 64th birthday. In the midst of this, Bob also was diagnosed with cancer, and succumbed to that illness little more than a year later, in October 1981, just a month after his 65th birthday.

Not the Original Succession Plan

This combined loss of my parents in a short interval, coupled with the loss of leadership in the company, was profound, and at the age of 34 I was left "holding the bag," not quite the succession story I had envisaged. This change came directly on top of one of the largest projects the company had ever undertaken, two 80-metre 15,000 bhp ice-breaking, anchor-handling supply vessels for Gulf Canada Resources Ltd, one built in Vancouver and one in Japan. Shortly after the delivery of these two significant vessels, the heady days of northern work in Canada came to a crashing halt with a worldwide recession in the oil industry. The Singapore operation was wound up, and the Vancouver office went into survival mode.

—R.G. Allan

The Challenging 1980s

The first half of this decade was a serious challenge to the survival of the firm. Work was scarce, and several clients were clearly not certain that the young new president of Robert Allan Ltd was up to the task. However, buoyed by a number of projects from the federal government (primarily intended to support a flagging shipbuilding industry) and with tremendous employee support, the company struggled through. At one point survival was guaranteed for a few months through a federal shared-time wage-support program. The limited work available is perhaps best illustrated by the small number of vessels (an average of three or four per year) built to Robert Allan Ltd designs in that decade.

In contrast to the general gloom of that period, however, in about 1985, the business committed to invest in what was then the emerging technology of computer-aided design and drafting (CADD). This was a large investment at a challenging time, but it provided the opportunity to market our design services as a forerunner in the use of this technology and obtain work that we could not have obtained otherwise. Having CADD capability distinguished Robert Allan Ltd from the competition and created a market advantage. The system comprised a refrigerator-sized computer (located in its own air-conditioned room) and four 9-inch Hewlett-Packard workstations. Finding software suitable for ship design work in a small office was a challenge, but some options were rapidly developing. Before long, the CADD system was updated and working smoothly, and it has continued to evolve and expand in quite dramatic fashion ever since.

Of significance in the late 1980s was a contract for *Eleu*, our first azimuthing stern-drive (ASD or Z-drive) tug built for a client outside of Canada. Built by Martinac Shipbuilders in Tacoma for Hawaiian Tug & Barge, *Eleu* was the harbinger of many similar tugs to follow. She was a derivative of the highly successful 2,400 hp series of Z-drive tugs developed for C.H. Cates & Sons Ltd in the Port of Vancouver. Developed in Europe in the early '70s, these steerable propulsion devices were revolutionizing tug design and operations, and the early work with Cates gave Robert Allan Ltd a significant advantage in developing modern tug designs (Chapter 17). The US was almost a decade slower than Canada in adopting this technology.

Eleu was the first Allan-designed Z-drive tug built for a client outside of Canada.

Looking Better in the '90s

Work in the early '90s remained spotty, but a lot of small consulting tasks and several studies kept the business afloat, perhaps just without a lot of freeboard. One of the most important of those studies was "The Canadian Towing Industry: Tugs, Barges and Their Relationships," prepared for the Canadian Coast Guard Ship Safety Branch in 1991. The work done then is still relevant today, as the industry still struggles to define safety standards, especially for small tugs. That significant report, as with so many comparable works, regrettably appears to sit unopened on government shelves.

In the background was the aftermath of the 1989 *Exxon Valdez* grounding and the resulting demand (and regulations) for tugboats capable of providing significant steering and braking forces to escort tankers in confined waters. Robert Allan Ltd undertook some extensive independent research and model testing into the sort of hull form needed to generate these forces. It was apparent that what had served even as a large and powerful conventional tug in the preceding decade or more was not the right machine for this new and very demanding service. Fortuitously, we devised a solution with an innovative sponsoned hull form, which combined with a large under-keel skeg could generate high indirect forces at the 8–10 knot speeds required for tanker escort while offering a very high standard of stability. This solution was unique, inexpensive and highly effective. Our international tug design "star" began to skyrocket.

At the same time contacts were established, quite unexpectedly, with companies in Turkey looking for designs for a new generation of harbour tugs. Those connections have become a mainstay of the business to this day, and the shipyards who started building Robert Allan Ltd designs in the 1990s have become world leaders in the field of tug-building, with unsurpassed quality of work.

Robert Allan Ltd's business grew rapidly in the second half of that decade, adding staff almost monthly until the firm had a total staff over 50 as we surged forward into the 21st century.

Going Global in the 21st Century

Right Place, Right Time

I recall clearly looking at the rapid expansion of our tug design business in the late '90s and wondering how that rate of growth could possibly be sustained for much longer. However, the growth in world shipping and the increasing size of large ships created even more demand for efficient new tugboats. Robert Allan Ltd was definitely in the right place at the right time to provide the modern designs needed.

— R.G. Allan

The new century was a very busy time for our expanding office. Tug projects were coming in from around the globe and our designs were garnering praise in all quarters. Shipyards in Turkey, China and Brazil were churning out Robert Allan-designed tugboats by the dozens. Large high-performance escort tugs were built for major oil and gas terminals around the world, but particularly in Europe and Scandinavia. The upscaling of shipping worldwide, and particularly the growth in size of large container ships, generated demand for ever larger and more powerful tugs. Business was very good!

In addition to the booming tug design business, the tragic events of 9/11 in 2001 led to a spate of demand for new and more effective emergency response fireboats in the US, with significant homeland security funding suddenly available. Robert Allan Ltd was awarded design contracts for new fireboats in many major US cities (Chapter 10). In parallel, the emergence of large new ports in China resulted in similar commissions there. These were exciting and demanding projects with often very challenging design and performance constraints.

This graph illustrates the growth of the business in terms of relative earnings compared to a starting point of 1980. The rapid expansion of the business in the early years of the 21st century is obvious, with a small downturn in the challenging years of 2008–2010.

Robert Allan Ltd
Normalized Revenue 1978 to 2020

Note: *1980 equals 1*

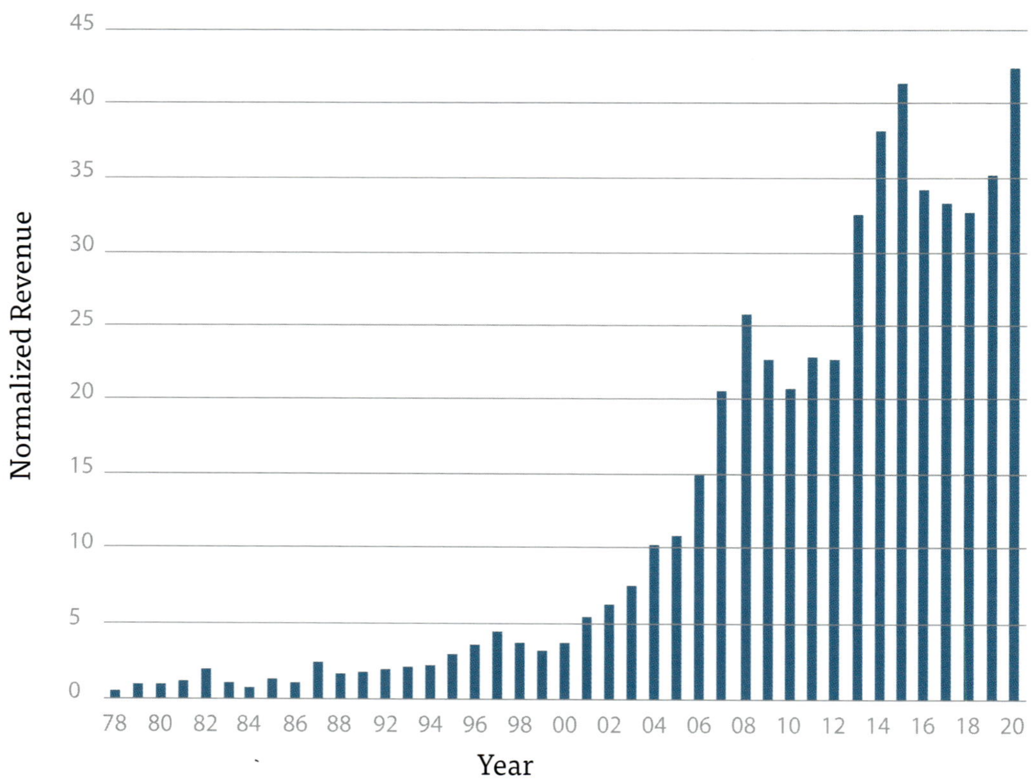

The business grew dramatically between 1978 and 2020.

An Employee-Owned Company for the Future

With the age of, if not the intent for, retirement looming in the mid-2000s, I was faced with the reality that there were no direct successors within my immediate family to carry on this family-owned naval architecture business. At the time, several international businesses were busy buying up small marine consulting firms in BC that had no obvious succession plans. A couple of these buyers were making serious overtures to Robert Allan Ltd, and it was obvious that a decision about the future of the company would have to be taken soon.

I met collectively and individually with a group of the more senior personnel, firstly to just set out the situation regarding long-term succession, and to address the options of either a sale to a large foreign organization, or to reorganize as an employee-owned company. It became immediately apparent that the outright sale option was not favoured and that, to a person, these employees, many of whom had been with Robert Allan Ltd for 20 years or more, liked the idea of going "on their own." The decision was made, and the wheels were set in motion in 2006 to reorganize the company for 100-percent employee ownership, a process concluded in 2008.

At the time of writing there are now a total of 27 employee owners in the company. The transition process has been entirely successful and has gone very smoothly. Robert Allan Ltd has grown and strengthened significantly under the new corporate structure and a new set of business policies. The new structure has removed any concerns I ever had about the long-term stability of the company and its dependence on the presence of, or the personality of, just one individual.

With the centenary of Robert Allan Ltd in the not-too-distant future, I often wonder if my father and grandfather would be pleased with what has become of what they created, but I am also in no doubt about their answer.

—R.G. Allan

FIFER

3

Getting Through the '30s

The record of early work done by Robert Allan as a lone practitioner from 1928 onward is not complete in terms of archived drawings and certainly not for any associated calculations, but considering the time elapsed, it is perhaps remarkable that anything has survived at all. There are a reasonable number of original drawings still extant in the company files, but not always in complete sets, and it is obvious that some drawings migrated over time from job to job, sometimes with fairly extensive modification, so the originals were often lost. However, the best source of the work record for the decade of the 1930s was the archives of *Harbour and Shipping* magazine, which faithfully described everything of interest on the Vancouver waterfront in those years. A reasonably complete picture of the design and shipbuilding activities of that most difficult of decades is therefore possible.

Opposite: MY *Fifer*, the most elegant vessel ever designed by Robert Allan, and a portend of better times ahead after the difficult 1930s.

Job #1 – The *Jeaneva M*

The very first recorded solo commission for Robert Allan, in late 1928, was a 44-foot private yacht, christened as *Jeaneva M*, for Mr. Philip White. This "able and roomy" vessel, built by Lind & Co., was described in the December 1928 issue of *Harbour and Shipping* as follows: "She...is of the bridge deck type with a stateroom forward and the main cabin and galley aft, while an engine of about 60 hp will be installed beneath the bridge deck. She will be planked with fir and cedar, with oak deckhouse and mahogany trim." A later article states that the boat was equipped with a 130 hp Sterling gasoline engine, but the registry records show an 85 hp engine. The *Jeaneva M* was also featured in Peter Vassilopoulos' 1980 book, *Antiques Afloat*, under her new name of *Fusilier*. This commission is interesting, as the design package is considerably more extensive than most done in those days. It comprises the typical arrangement, lines and basic structural drawings, but in addition there are several outfitting detail drawings, such as the section through the forward saloon area. One might guess that in a spirit of entrepreneurship as a new consultant Robert Allan was trying hard to make a favourable impression.

At about the same time, a smaller but similar style of vessel, christened *Corsair*, was designed for Mr. W.K. Sully. Measuring 36 feet in length, and also built by Lind & Co., she was powered by a 30 hp Kelvin gas engine. The owner reportedly completed much of the finishing of this boat himself. This design was used again in the mid-1950s, with subtle changes, for the cruiser *Dorisdee II*. Although the drawing for *Jeaneva M* does not indicate so, both these vessels were reportedly equipped with the patented Oertz rudder, for which Robert Allan was the agent, as noted in his advertisements of the day. The foil-shaped device was claimed to offer less resistance than more conventional plate rudders, with beneficial effects on overall propulsion efficiency, such that power reductions of

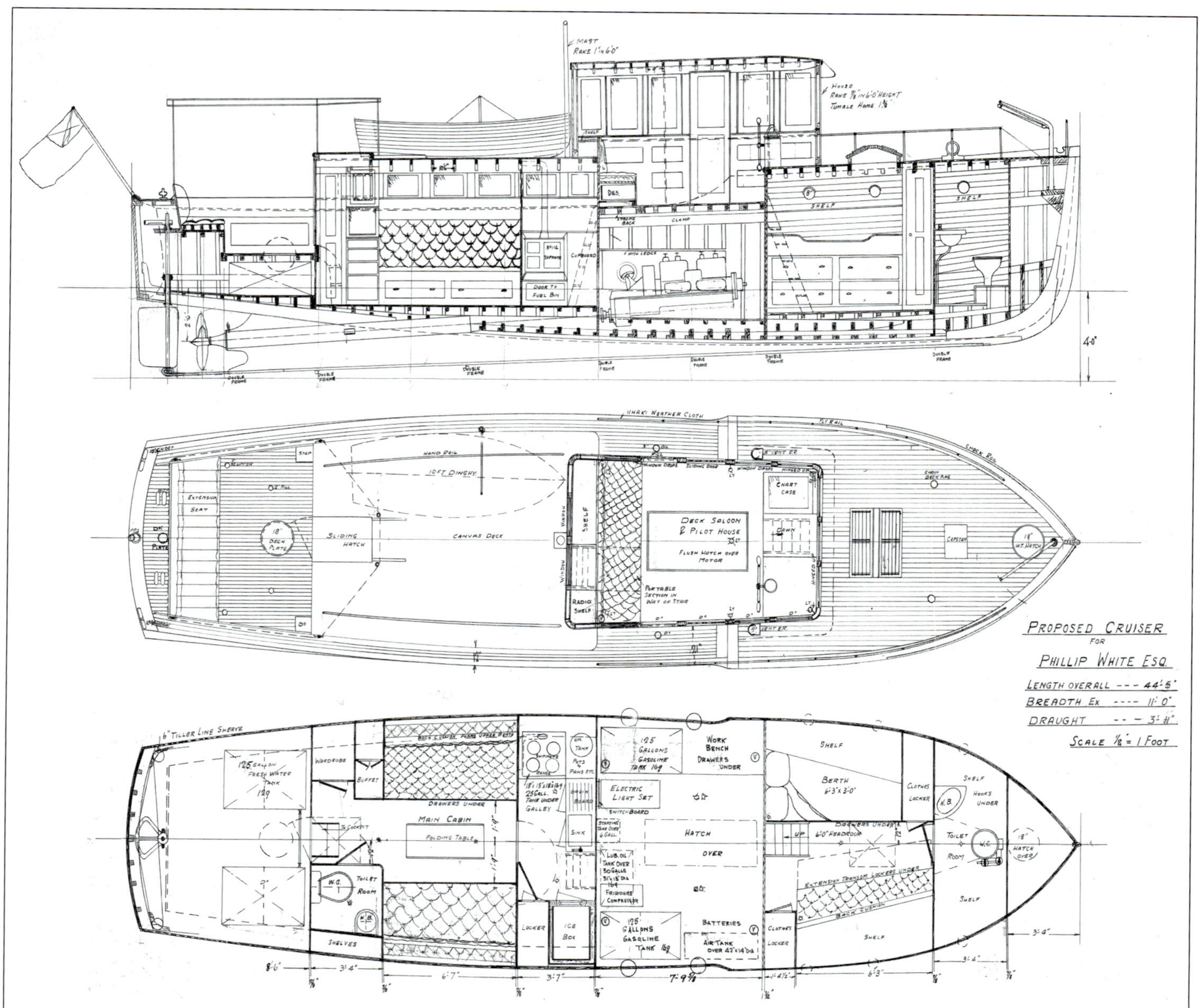

General Arrangement for *Jeaneva M*, the first recorded design to Robert Allan's credit as an independent designer.

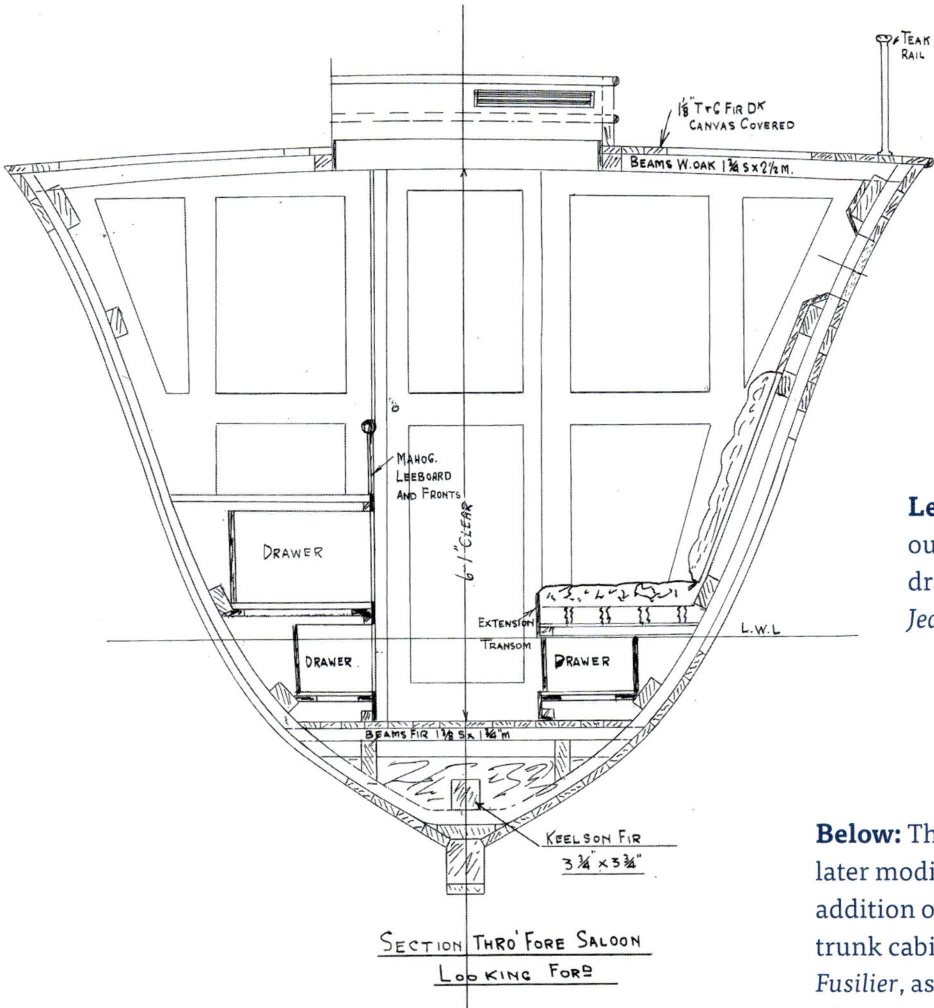

Left: Detailed outfitting section drawing for *Jeaneva M.*

Below: The *Jeaneva M* was later modified with the addition of the forward trunk cabin and renamed *Fusilier*, as shown in this photo from 1976.

up to 15 percent were possible for the same speed. It is not obvious that those claims were substantiated in smaller craft such as these yachts. The Oertz patent shows a two-part rudder with a fixed foil-shaped leading edge, and a "tail" that rotates within the fixed portion, creating a quite efficient wing section. There are no details of the Oertz rudder in any of the older drawing sets, but shortly after, drawings for a similar style of rudder known as the "Hydrogap" appear, and this was used on a number of vessels, including some of the large seiners described in Chapter 4.

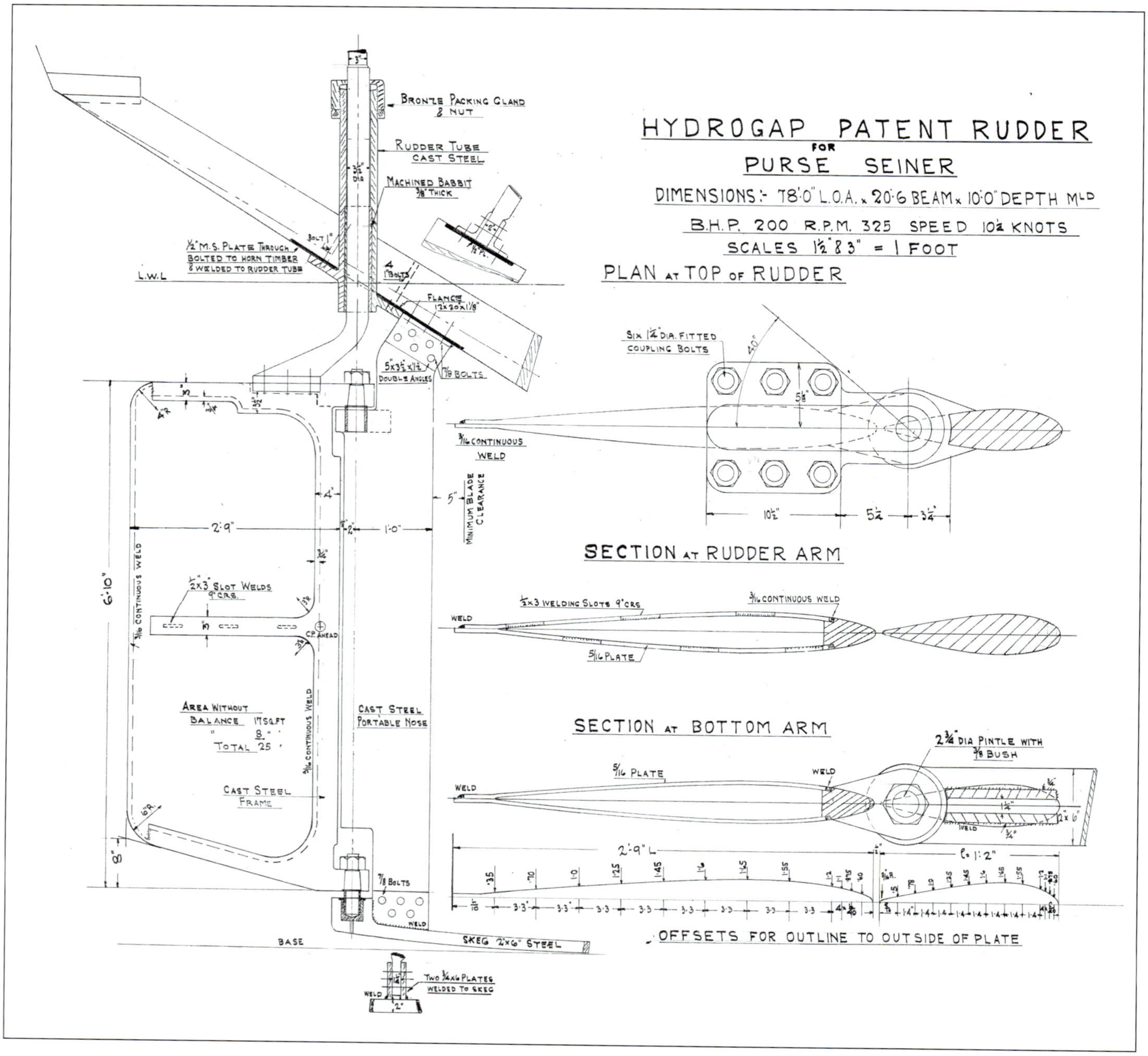

Detail drawing of a "Hydrogap" rudder.

Early Work

The years 1930 and 1931 seemed to have held some promise, bearing in mind that this was in the shadow of the 1929 stock-market crash. There was a sizeable group of coastal patrol vessels built to Allan designs for the federal Department of Fisheries, as well as similar boats for the Pacific Game Department. The most notable of the era was the 1931 design for the quarantine launch *Salucan v*. All these patrol vessels are described in more detail in Chapter 5. The number of vessels built, which includes several to identical designs, belies the fact that only three or four such designs were done over that two-year period, so it is obvious that his time on the drawing board was not heavily challenged.

Another notable vessel of the time, probably designed in 1929, was the 75-by-18-foot seiner *Michael J. Tayler*, built at Vancouver Shipyards for the Tayler Shipping Company. This is the first recorded fishing vessel design by Robert Allan in the archives and certainly was the forerunner of many dozens of big seiner designs to follow a decade later (Chapter 4). Regrettably only the Lines Plan for this particular vessel exists today.

The Motor Yacht *Meander*

The classic yacht *Meander* has always been considered in our office as the project that put Robert Allan firmly on the map as an independent naval architect of note. It wasn't quite "Job #1," but it was certainly his first significant and high-profile commission, and undoubtedly was also the single project that enabled him to survive the difficult Depression years.

In 1933 he was contracted by Mr. George Kidd, a prominent local businessman (obviously less affected by the Depression), to design a luxury yacht. Kidd was president of

Bert Menchions' shipyard crew took a serious pay cut in order to get the contract for *Meander*, a rare bit of work during the Depression.

the Vancouver Power Co., and the chairman of the Vancouver Board of Trade 1934–35. This was a major commission at the time and lifted the designer out of both an economic and an emotional depression, and consequently paved the way for many future years of company success. Kidd was a fellow Scot, which undoubtedly had some influence in the selection of Robert Allan to do this work. He reportedly wanted a ship that would carry him anywhere up and down the Pacific Coast and ultimately to his ancestral home in Glasgow, Scotland. The vessel was built in 1934 at W.R. Menchions shipyard in Coal Harbour, located just east of where the Westin-Bayshore hotel now stands.

Although there is no record of her cost, among old company papers are notes of estimates of very similar vessels, and an educated guess would be about $24,000. Regrettably, it is not known how much Robert Allan was paid for this design, but probably about $400! The original design drawings for *Meander* still reside in the company archives. These are classic design drawings in India ink on pale-blue waxed linen, and the Inboard Profile and Plans drawing in particular is a mass of information detailing the construction, machinery and outfitting of the boat. An enlarged reproduction of that drawing adorns the entry to the company office today, together with the original bronze name-plate for the vessel, which came into company hands only in 2020.

It is a testimony to the skills of the designer and the shipwrights of the day alike that this venerable vessel, still gracing Vancouver's waterfront, was built with just four detailed drawings: a Lines Plan, General Arrangement, Inboard Profile and Deck Plans, and a Midship Section.

Outboard Profile and Main Deck Plan for *Meander*.

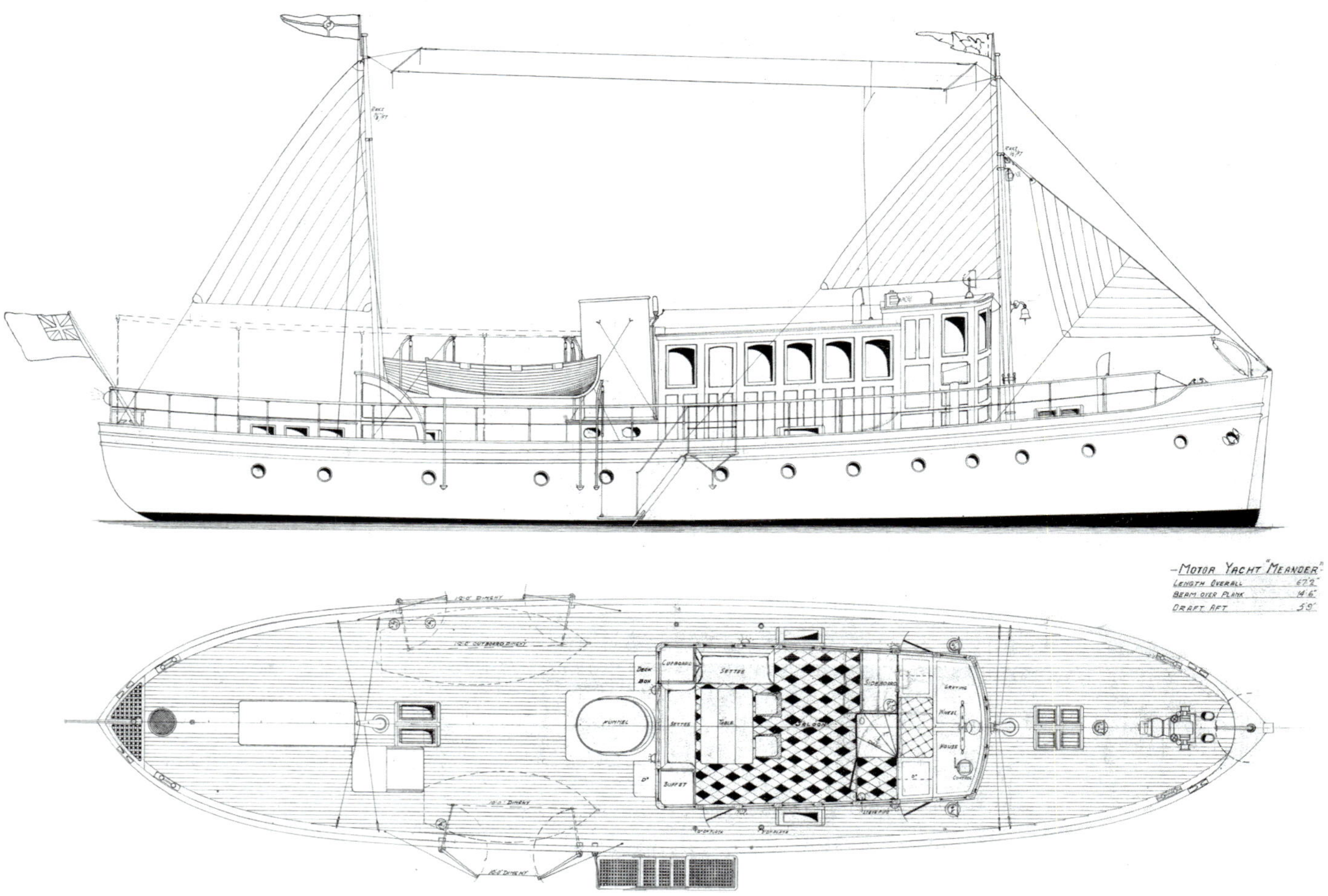

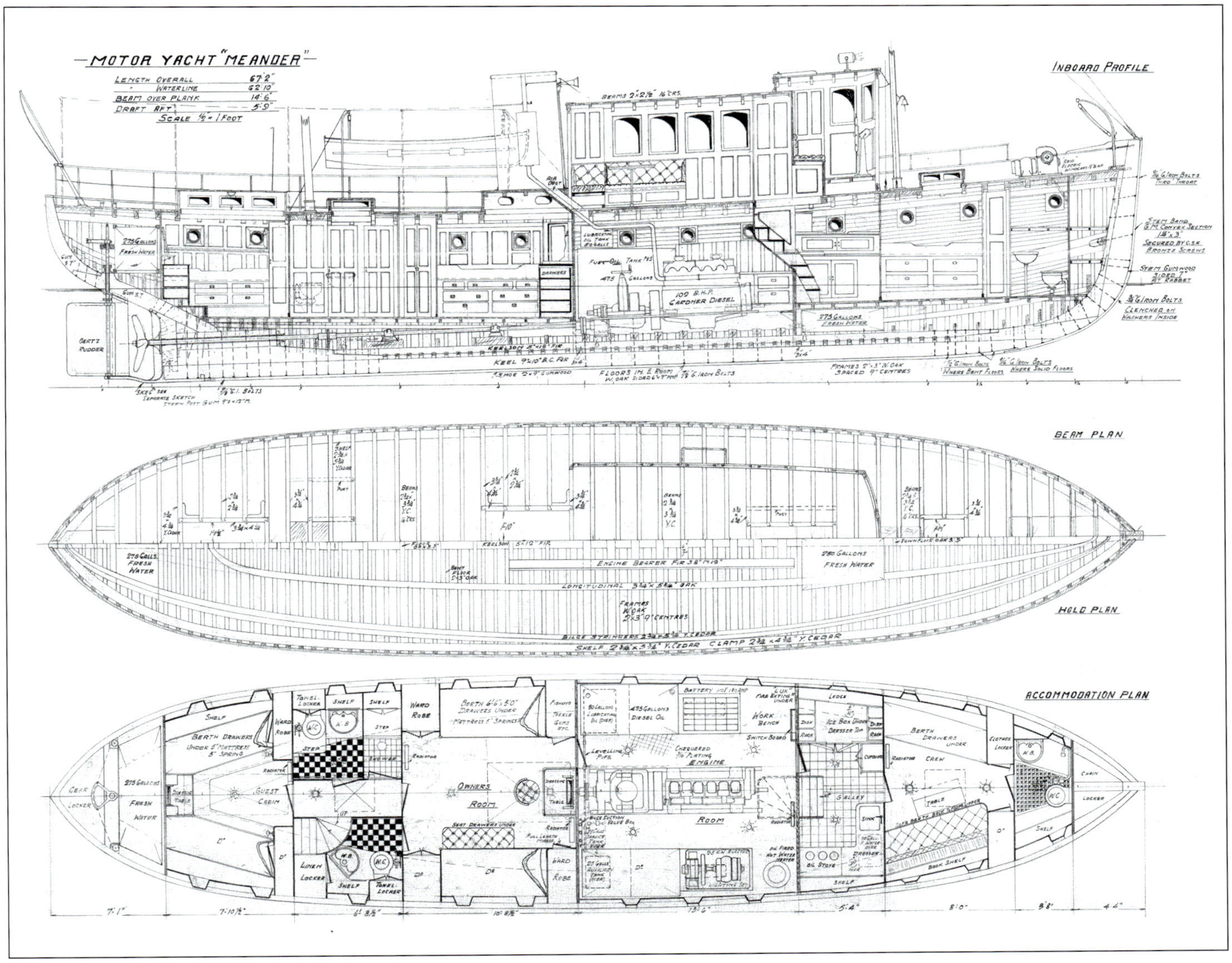

Inboard Profile, Beam Plan and Accommodation Plan drawing for *Meander*.

The commissioning of the yacht on April 24, 1934, was described widely in the press. The date is significant, as the design drawings are dated October 1933, indicating barely six months from start to finish of the vessel, a remarkable feat even by today's standards. It was also noted by Bert Menchions in his memoirs that the shipyard crew made a serious concession in their mid-Depression wages, from $5 to $4 per day, in order to get this rare bit of valued work.

This classic, canoe-sterned yacht was stoutly built from the finest materials available: yellow cedar hull planking; bent 2- by 3-inch white oak frames at 9-inch centres; teak deck planking; and yellow cedar ceiling (inner planking).

Meander is powered by a 6-cylinder Gardner 6L3 diesel engine rated at 102 bhp at 800 rpm, a machine which has been in continuous service ever since. Like many large yachts of the period, *Meander* was "impressed" by the Royal Canadian Navy in 1939 for coastal patrol duty, serving as HMCS *Meander* in the Fishermen's Reserve during WWII. Reportedly, the vessel was returned to the Kidd family at the end of the war, but it soon changed hands.

In 1945 *Meander* was sold to Pacific International Airways (PIA), an apparent subsidiary of Pan Am Airlines based in Alaska, but the boat's logbooks show her owned by and operated for Mr. Edward Lowe of San Francisco, who likely had the PIA

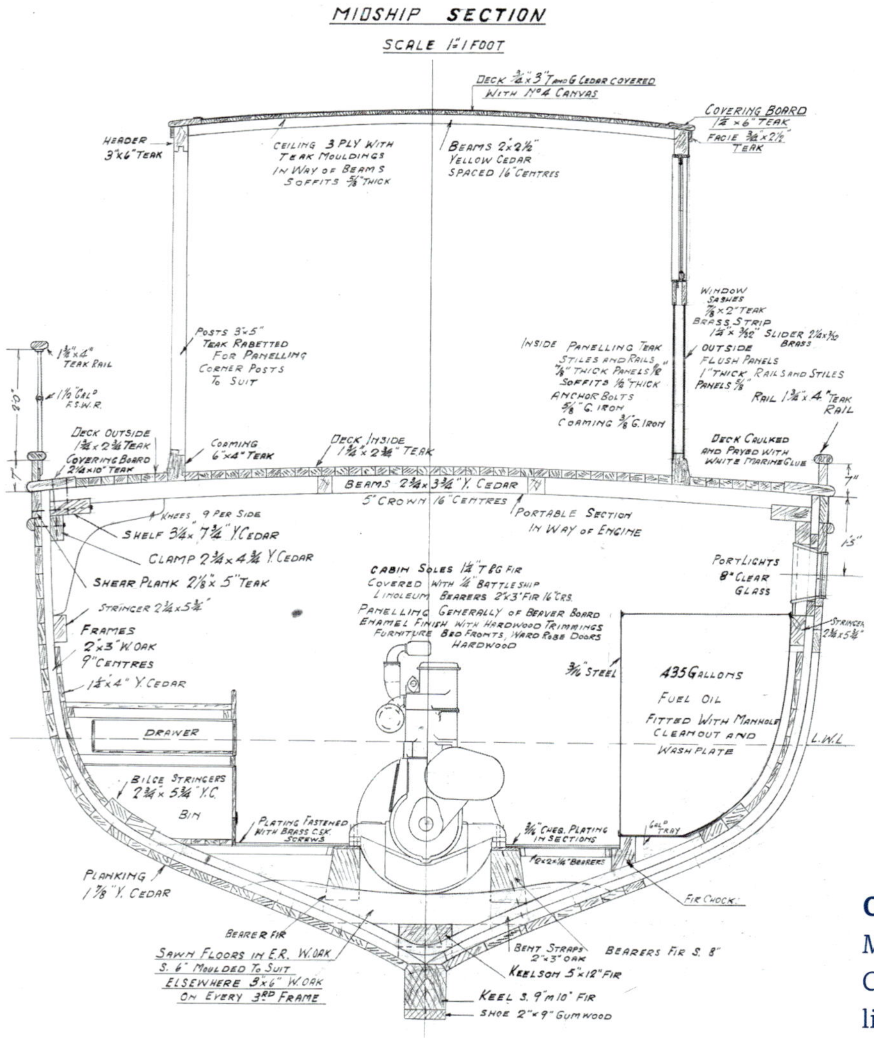

Midship Section drawing for *Meander*.

Opposite: HMCS *Meander* in Royal Canadian Navy livery, ca. 1940.

connection. Her logbooks from this period show the yacht, with the Lowes and friends aboard, cruising the coast extensively from Alaska to San Francisco, but primarily in the Seattle–Vancouver–Gulf Islands areas. It appears that the vessel remained in Canadian registry despite her American owner. Lowe was responsible for adding rolling chocks to the hull (crudely sketched into the back of the 1946 logbook) and the extended fixed canopy over the aft deck.

In 1948 the boat was bought by Sydney & Walter Wilson and used as a charter yacht under the name of Meander Charters Ltd. In 1950 *Meander* was bought by the United Church of Canada, and from 1950 (or '51) to 1963 worked as the United Church missionary vessel *Melvin Swartout II*, working up and down the west coast of Vancouver Island, based initially out of Port Hardy, then in 1953 moved to Bamfield. The logbooks from this period show a busy life programmed around a multitude of Sunday services.

From 1963 to 1972 *Meander* was owned by Frederick Alexander Menzies of West Vancouver, who operated her as a charter yacht. On Mr. Menzies' death, the yacht was bequeathed to his son Robert Menzies, who appears to have sold it almost immediately. In 1972 *Meander* was sold to Art and Norma Clements. The boat was purchased in 1975 by Hedley Rendell and Dennis Feroce. Dennis had fallen in love with the boat, and he was helped with the purchase by Rendell through his company. *Meander* has remained with Dennis ever since, and with him and his wife, Jan Iliffe, since their marriage in 1995. Jan has beautifully restored the interior.

Having bought *Meander* as a young man in 1975, Dennis needed to make some revenue from her, and that consisted largely of chartering as frequently as possible. But these were the '70s, and Greenpeace, a relatively new presence on the coast, with their philosophy of environmental protection, appealed strongly to Dennis. In 1977 he offered the boat to them as the *Greenpeace IX*, and she soon lost her beautiful white hull to the rainbow colours of that organization. In May of that year, the Kitimat Pipeline Company, a consortium of fifteen oil companies, was hosting the annual conference of the North Central Municipal Association of Mayors. The mayors, with their respective entourages, about 250 people in total, were taken aboard the CP Steamship's MV *Princess Patricia* to transit up Douglas Channel to demonstrate how safe this route might be for tanker traffic. Greenpeace was strongly opposed to the concept, and with the support of many coastal First Nations bands and fishermen, organized a maritime protest. *Greenpeace IX* led a flotilla that included the United Church Mission boat *Thomas Crosby 5*, and a large number of coastal fishboats that intercepted and disrupted the voyage of the politicians. The colourful story is well told in a number of sources, including *Greenpeace: The Inside Story*, by Rex Weyler.

Meander at home in False Creek, 2012.

Recovering the *Meander* Model

As an anecdote to the story of the yacht itself, there exists a rather fine model of *Meander* with its own unique history. As described in the introductory section in the senior Allan's letter to Frank Gildersleeve, this model was built by my father, Bob Allan, early in 1934, when he would have been just 18 years old and attending UBC.

As anticipated, the model was sold to the vessel's owner, George Kidd, for an undisclosed amount, but apparently sufficient to enable Dad to complete another year at university! After a fair bit of detective work, I was able to uncover the model in the hands of Mr. Peter Kidd, the great-nephew of the original owner. The model was a bit the worse for wear after nearly 80 years, with a notable gap in her starboard bulwarks and a few other missing small bits, but basically intact and worth restoring. After some negotiation, this model was repatriated to Robert Allan

Before (top) and after: Model of the yacht *Meander* built by 18-year-old Bob Allan in 1934, restored in 2016.

Ltd in 2014 and was beautifully restored by Lucian Ploias, master model-maker at the Vancouver Maritime Museum.

Meander is, at the time of writing, approaching 90 years of age and showing her age a bit, but still as graceful as ever. She has as colourful a history as any boat in British Columbia. Vessels such as this are truly deserving of a formal heritage recognition and associated restoration and preservation.

—R.G. Allan

The restored *Meander* model was welcomed "home" with a gathering of all involved: Peter Kidd and several members of his family, Rob and Enneke Allan, model restorer Lucian Ploias and family, boat owners Dennis and Jan, and Ken Harford, president of Robert Allan Ltd.

The Mid-1930s

There is a paucity of designs recorded in the period between 1932 and 1936, *Meander* being a notable exception. The few designs that were done in this period, however, include some noteworthy projects which helped to set the stage for the future of the company.

Weaver Lake

This simple profile drawing, done on what at first glance appears to be a small scrap of linen drawing paper, about 8 inches by 14 inches, turned out, after considerable research, to be for the very first tugboat ever designed by Robert Allan. The *Weaver Lake*, described as a "shallow-draft towboat" with dimensions of 41 feet 8 inches with an 11-foot beam, was designed in 1934 for Capt. E.C. Merchant and worked primarily on the Fraser River and tributaries. The plethora of tug designs bearing the Robert Allan name that have followed in her modest wake could certainly not have been foreseen in those struggling days. More details of this pioneering tug design follow in Chapter 16.

Hyak

One of only three tugboats designed by Robert Allan in this decade, the *Hyak*, built in 1937 by A.C. Benson, is most notable as the first project completed by Robert Allan for Oswald "Sparky" New, the principal of Coastal Towing Co. Ltd, and ultimately the head of a series of historic BC marine industry companies. Sparky New and later his son Bill were probably the first of many clients whose faithful support

Profile drawing of the first Robert Allan tug design, *Weaver Lake*, 1934.

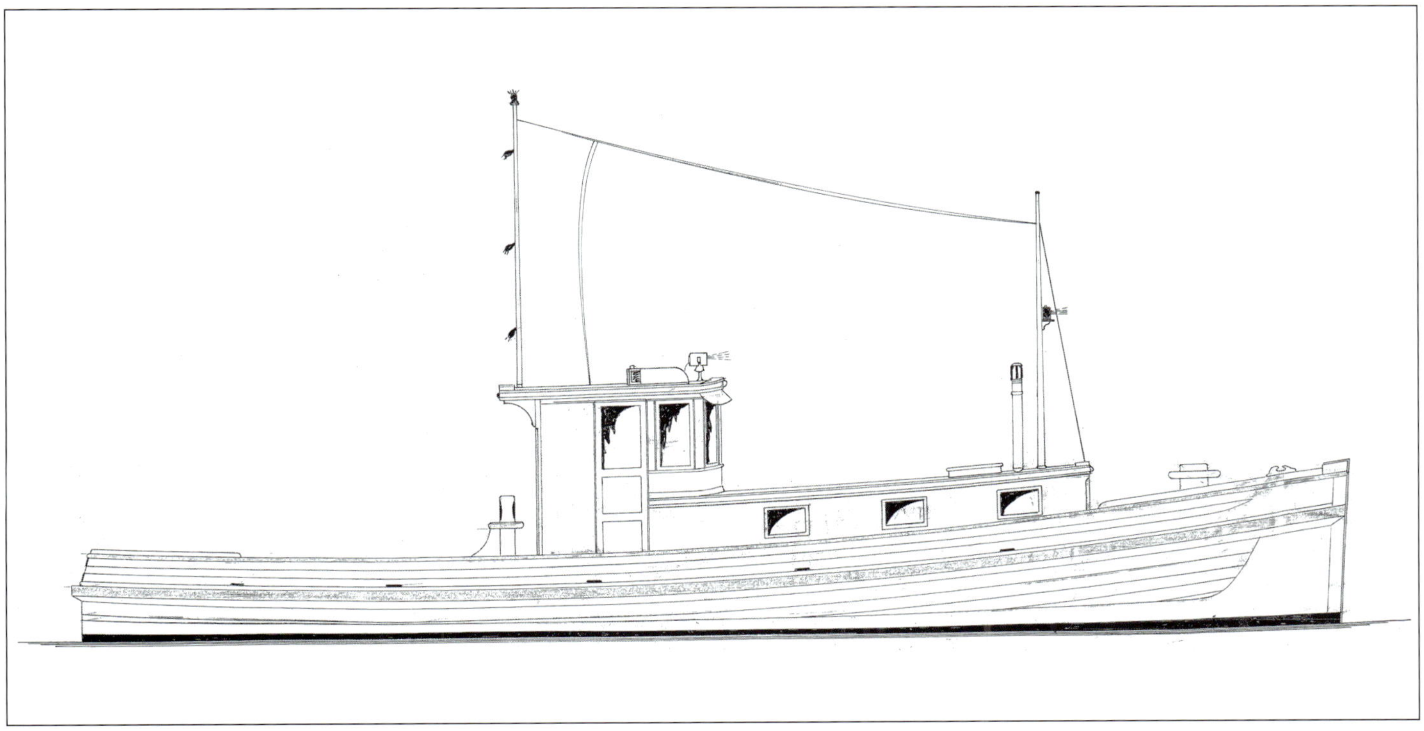

Hyak working with log booms, 1938.

over decades ensured the survival of the fledgling Allan design business. At 52 feet long with a 15-foot beam and 120 hp, the *Hyak* was a pretty large tug for her day. She has had a long and remarkable career under a wide range of owners, including the US Army during WWII, and reportedly is still working the BC coast in 2020.

Black Hawk

One of the most unusual designs in the company archives is that for the 34-foot "Express Hydroplane" *Black Hawk*.

Finding *Black Hawk*

We had very little information about this boat beyond a few small drawings in our archives. Fortunately, its story was well-documented by Don Morison, an amateur maritime historian who contacted me about 25 years ago for information to support his investigations into the history of *Black Hawk*. In researching this book, I tried unsuccessfully to contact Don but eventually heard from his daughter Karen Hutton, who advised that Don was now in his 90s and suffering from dementia, but she dug through his old files and shared with me all the papers and work done in tracing *Black Hawk*. Subsequently, we also determined that Don's article on *Black Hawk* had been published in *Western Mariner* magazine in February 1994. Among the research material was a *Daily Province* clipping dated July 27, 1935, with a photo of the boat under the caption "She Will Do Forty-Five!" and mentioning that a top speed of 50 mph was expected.

— R.G. Allan

Black Hawk's interesting story is best told in the following extracts from Don Morison's monograph:

> To construct the *Black Hawk*, Lloyd purchased two World War I Liberty aeroplane engines that had been used in a rum runner, and commissioned naval architect Robert Allan to design a fast passenger launch using one of the engines. The design that Allan came up with was similar to the British motor gunboats of WWI, but with a step amidships. The construction was double diagonal planking, with yellow cedar on the inside and red cedar on the outside. This gave a round bilge, planing hull with an added step. This resulted in an extremely strong light hull with three layers of sheathing forward and two aft. In all details, Allan stressed strength and lightness in the 33' 4" × 7' 6" hull. The Liberty was coupled to a 22" × 36" propeller through a 16' 6" shaft which was supported at the stern by an outboard rudder mount, a common arrangement for Gold Cup Race Boats of the time. The extremely large pitch of the prop attests to the impressive torque of the Liberty. At 1,800 rpm this gives a theoretical speed of 50 mph. Actually, Lloyd was able to attain this speed during later tune-up runs. The high efficiency hull performance from such an over square propeller says much for Allan's hull design...Lloyd cruised his 12-passenger water taxi at 27 mph, burning 14 gallons per hour. The charge of 90 cents per mile hardly covered the operating costs but was the going rate at the time. Many of his trips were to and from the numerous logging camps around the lower coast. At cruising speed with 10 to 12 loggers onboard, the *Black Hawk* was an impressive sight. The roar of her Liberty combined with the large rooster tail from the prop left a very memorable impression with many coast watchers.
>
> In 1940 Lloyd sold the *Black Hawk* to Reece Riley of Port Alberni. An article in the *Alberni Times* describes the excitement brought to the canal by Reece's new boat, which allowed the mayor and party to picnic in Ucluelet and return to Port Alberni all in a day. The boat was sold to BC Packers in 1950, and it eventually found its way north to Namu. By the early 1960s the Liberty had been replaced with a Hall-Scott and the step had been removed from the hull. Sold to a logger in the Queen Charlottes, the *Black Hawk* was later leased to the Department of Fisheries as a patrol boat.
>
> About 1968, the *Black Hawk* was destroyed by a howling southeaster as she sat on a tidal grid in Queen Charlotte City. Her owner at the time was totally unaware that the wreckage he viewed the next morning was once the king of the British Columbia coast water taxis.

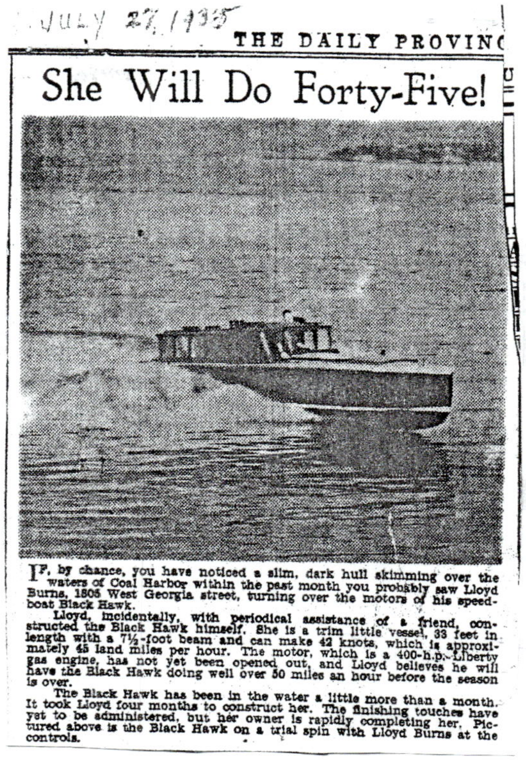

JULY 27 1935

THE DAILY PROVINC

She Will Do Forty-Five!

IF, by chance, you have noticed a slim, dark hull skimming over the waters of Coal Harbor within the past month you probably saw Lloyd Burns, 1805 West Georgia street, turning over the motors of his speedboat Black Hawk.

Lloyd, incidentally, with periodical assistance of a friend, constructed the Black Hawk himself. She is a trim little vessel, 33 feet in length with a 7½-foot beam and can make 42 knots, which is approximately 45 land miles per hour. The motor, which is a 400-h.p. Liberty gas engine, has not yet been opened out, and Lloyd believes he will have the Black Hawk doing well over 50 miles an hour before the season is over.

The Black Hawk has been in the water a little more than a month. It took Lloyd four months to construct her. The finishing touches have yet to be administered, but her owner is rapidly completing her. Pictured above is the Black Hawk on a trial spin with Lloyd Burns at the controls.

Black Hawk caught the attention of *The Daily Province* in 1953.

Black Hawk's construction is quite interesting and must have required considerable skill to complete, with effectively a very narrow double hull in way of the "step," as illustrated in the Structural Sections drawing. This is another vessel for which young Bob Allan built a fine model, however the fate of this handiwork is regrettably unknown.

Top: With a top speed of 42 knots, the *Black Hawk* was called "the king of the BC coast water taxis."

Left: Nineteen-year-old Bob Allan with his model of *Black Hawk*.

Opposite top: Inboard Profile and Deck Plan for *Black Hawk*.

Opposite bottom: Detail Structural Sections of *Black Hawk*, showing the complex construction details in way of the stepped hull.

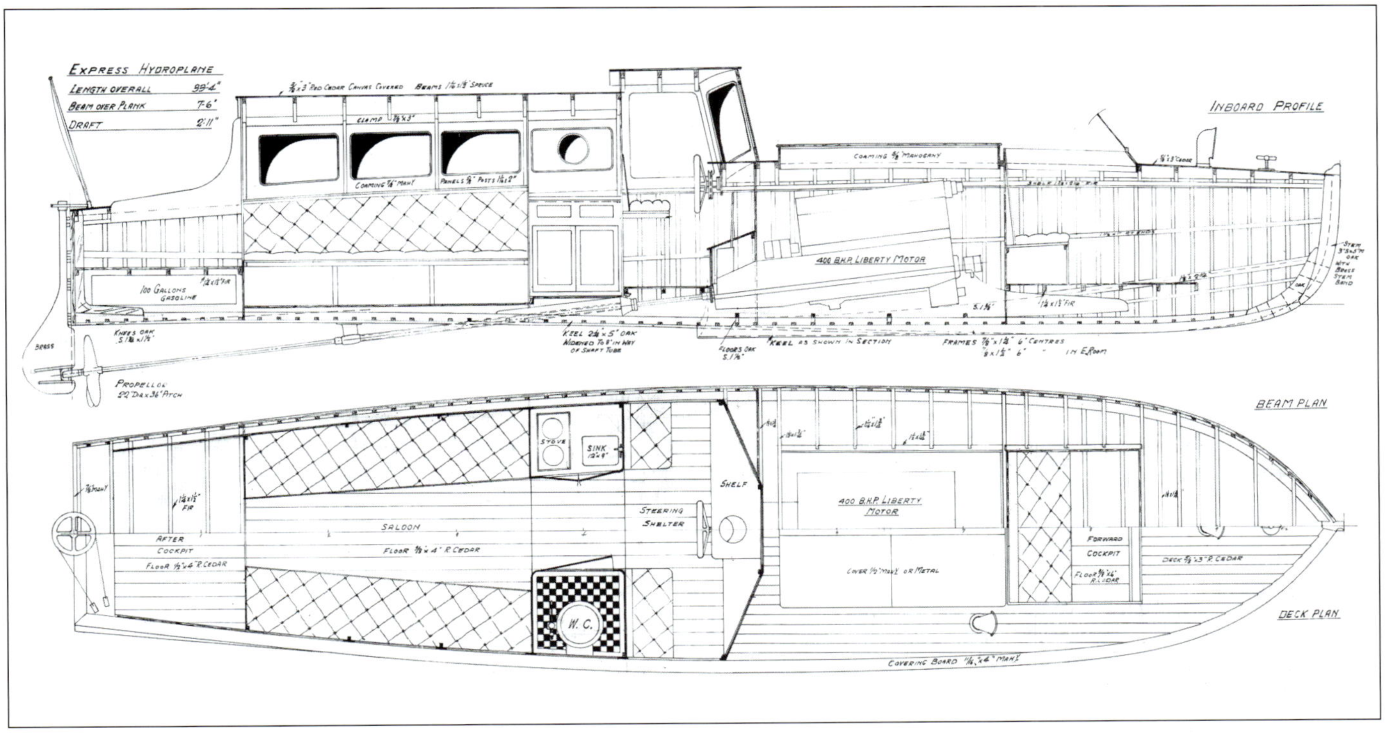
Express Hydroplane
Length overall
Beam over plank
Draft
Inboard Profile
400 B.H.P. Liberty Motor
100 Gallons Gasoline
Propeller
Beam Plan
Deck Plan
Saloon
After Cockpit
Steering Shelter
Shelf
Stove
Sink
W.C.
Forward Cockpit
Covering Board

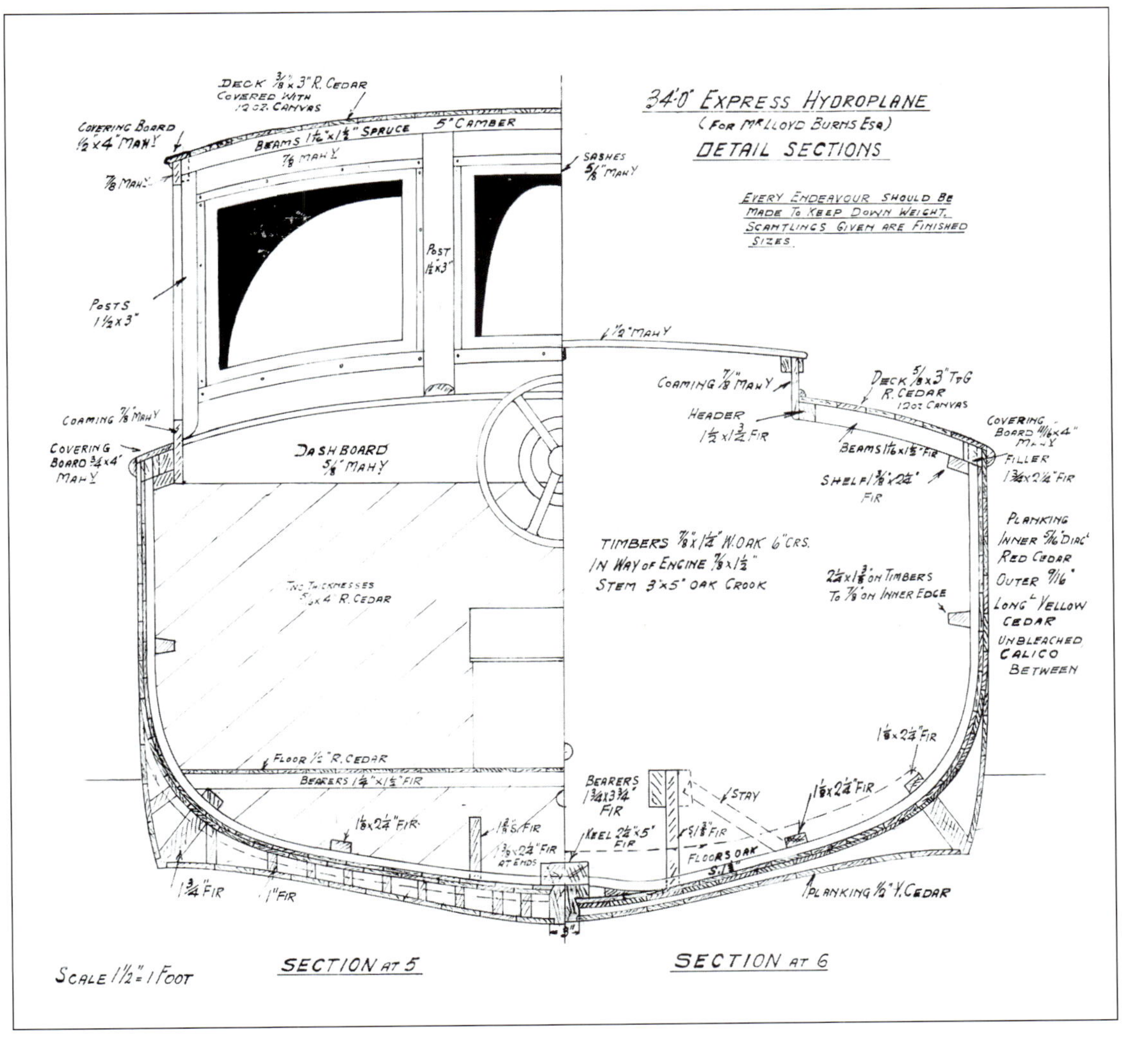
34'0" Express Hydroplane
(For Mr Lloyd Burns Esq)
Detail Sections
Every Endeavour Should Be Made To Keep Down Weight. Scantlings Given are Finished Sizes.
Deck 7/8 x 3" R. Cedar Covered With 12 oz. Canvas
Covering Board 1 1/2 x 4" Mah'y
Beams 1 1/2 x 1 1/2" Spruce
5" Camber
Sashes 5/8" Mah'y
Post 1 1/2 x 3"
Posts 1 1/2 x 3"
Coaming 7/8 Mah'y
Dash Board 5/8" Mah'y
Header 1 1/2 x 1 3/4 Fir
Shelf 1 7/8 x 2 1/4 Fir
Timbers 7/8 x 1 1/2" W. Oak 6" Crs.
In Way of Engine 7/8 x 1 1/2"
Stem 3" x 5" Oak Crook
Planking Inner 5/16 Diag. Red Cedar Outer 9/16 Long'l Yellow Cedar Unbleached Calico Between
Floor 1/2" R. Cedar
Bearers 1 3/4 x 1 3/4 Fir
Stay
Keel 2 1/4 x 5" Fir
Floors Oak
Planking 1/2" Y. Cedar
Scale 1 1/2" = 1 Foot
Section at 5
Section at 6

The Mystery *Fifer*

The Confusion of Name Changes

In the early stages of work on this book, I received a request from Vince and Carrie Stamper, who identified themselves as "...the current custodians of *Fifer*, the original *Deerleap* built in 1928. She, like the 1929 *Deerleap*, were both designed by Robert Allan and built by Hoffar-Beeching." This was not the first time I had received requests about this "other" *Fifer*. The 1939 steel-hulled version of *Fifer* (more about her later) was so ingrained in my psyche as my grandfather's masterpiece that the first time the inference to a previous *Fifer* came up, I found it barely credible. However, there is certainly proof of my grandfather's involvement in her evolution, and so we can at least claim a detailed association with another fine BC yacht.

—R.G. Allan

The 1928 *Fifer*, built as *Deerleap*, then *Bardick*, *Quenca*, then *Fifer* briefly, *Grampian I*, *Kitten F*, *Barbara O'H*, *Bardick* (again) and finally *Fifer* again.

The complex evolution of this particular vessel is well documented by Peter Vassilopoulos in his 1980 book *Antiques Afloat* under the boat name of *Bardick*. An abbreviated extract follows:

> *Bardick* was built (in 1928) by Hoffar-Beeching Shipyards of Vancouver...She was first called *Deerleap*...A year after she was first registered she went back to the shipyard, and was taken over by a new owner...She was given the name *Quenca* on July 26, 1929...In 1936 *Quenca* passed into the hands of the Canada Trust Company and one John Litton Mara as joint executors of the estate of Sir Frank Barnard...She was transferred to Captain William Marr Crawford of the Empire Stevedoring Company...In January 1937 Captain Crawford changed her name to *Fifer* (his original), but one month later, it seems, he decided to build a new boat, and wanting to retain the name *Fifer*, he changed it again—now to *Grampian I*... *Bardick*, then still *Grampian I*, changed hands again...[and was renamed] *Kitten F* in November of 1941...when she was sold to Hollis O'Hanlon in 1967 she was renamed *Barbara O'H*... [The registry records show her later history: in 1972 changed back to *Bardick*; in 1991 changed once again to *Fifer*.]

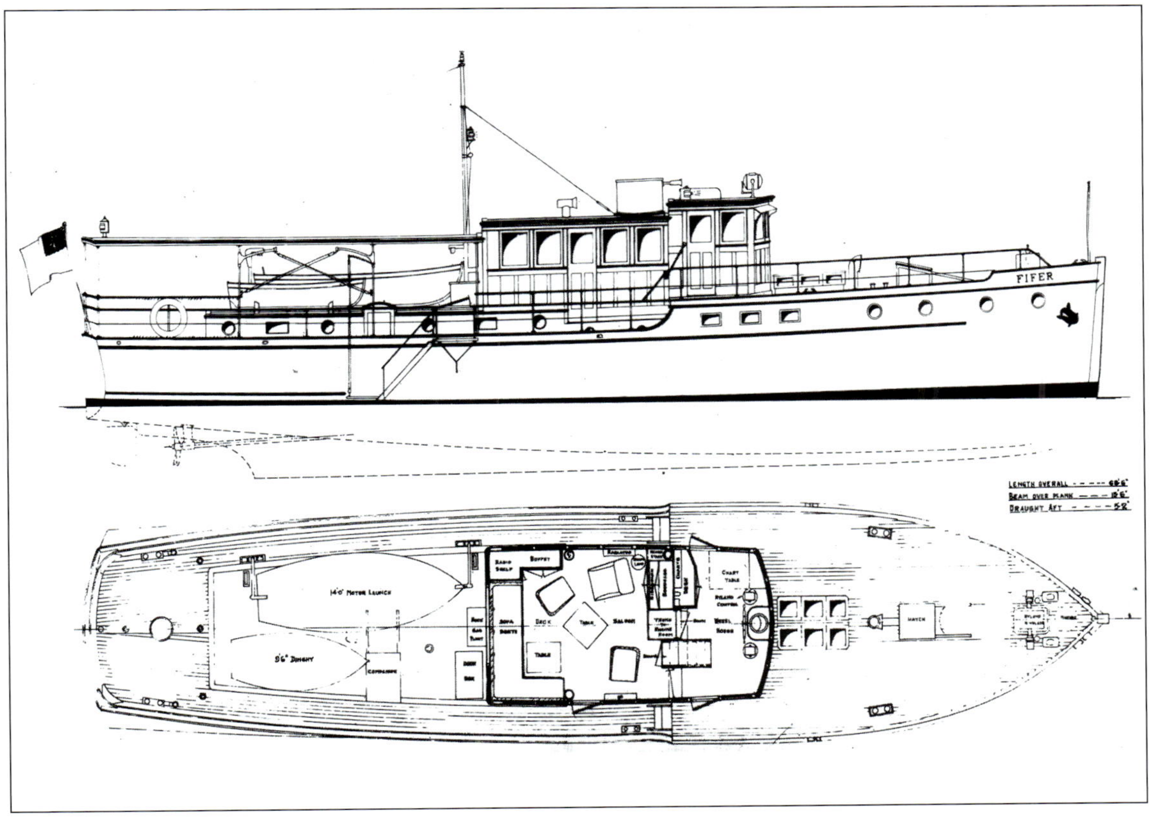

Proof of the pudding? The images of this and other drawings obtained from the present owners clearly show Robert Allan as the designer of record when Crawford took ownership, but unfortunately the original design can't with certainty be traced to him.

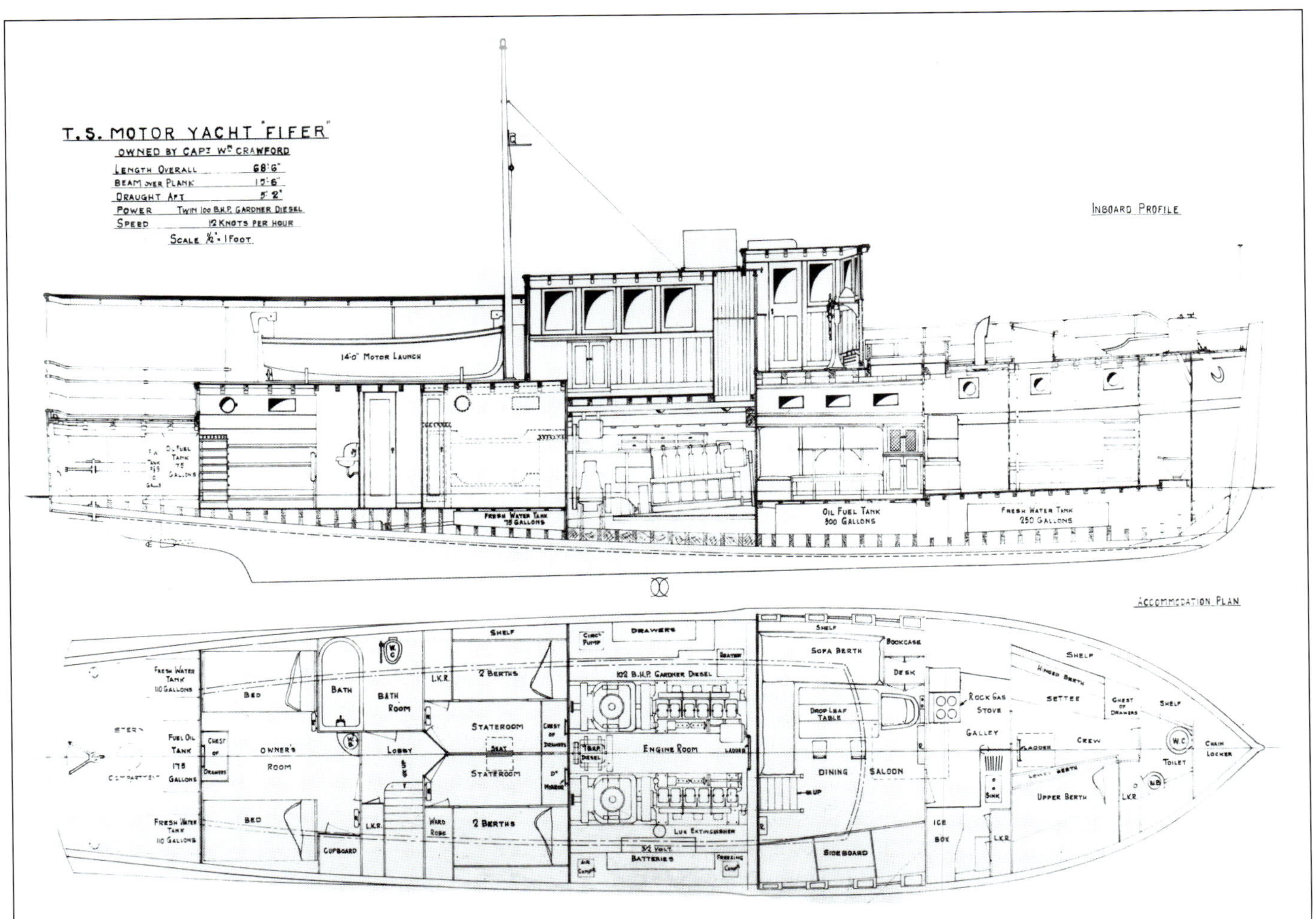

The Motor Yacht *Fifer*

Without question the most elegant vessel ever to emerge from any Robert Allan drawing board was the lovely motor yacht *Fifer*. This steel-hulled ocean-going yacht was commissioned in 1938 by Capt. William Crawford, president of Empire Stevedoring Company in Vancouver, with the intent of cruising back to his native Scotland. According to Francis Mansbridge in *Launching History: The Saga of Burrard Dry Dock*, "On February 1, 1939, Crawford signed a $300,000 agreement with Burrard [Drydock Co. Ltd], and the *Fifer* was launched on November 11 of that year to the skirl of bagpipes."

The design package for *Fifer* was the most comprehensive ever done by Robert Allan up to that time, and included more than 20 drawings defining hull form, complete arrangements, extensive aspects of the steel construction, all critical machinery items, and extensive definition of the elaborate outfitting. The Arrangement drawings shown here give just a hint of the detail as defined.

The fate of this lovely vessel ended up quite differently from that intended by its owner. The war had commenced just a few weeks before the ship was launched, and the Canadian Navy immediately took her over for coastal patrol duties in BC. By the end of the war, Capt. Crawford had died and the yacht passed into the hands of several other wealthy owners, including Clarence Wallace, when he was Lieutenant-Governor of BC, and the same Edward Lowe who had previously owned *Meander*. In 1986 she was sold to the celebrity class-action lawyer Melvin Belli, who tastelessly renamed the boat *Adequate Award*. Thankfully, subsequent owners restored her proper name and have invested in maintaining this classic beauty in true Bristol fashion, albeit in American waters.

Fifer, majestic in BC waters, probably in the 1970s, with her clinker-built lifeboats replaced by Hourston runabouts.

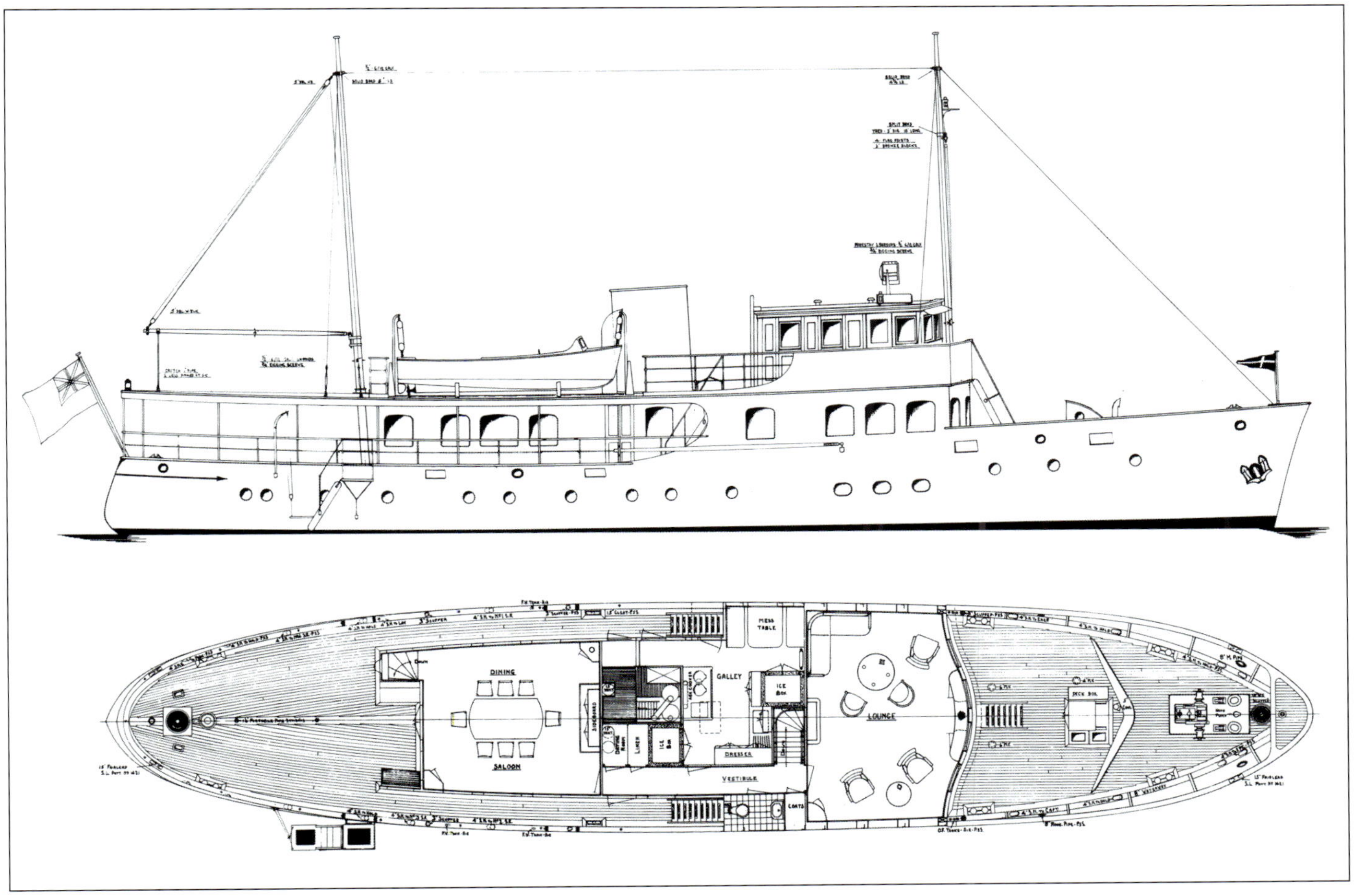

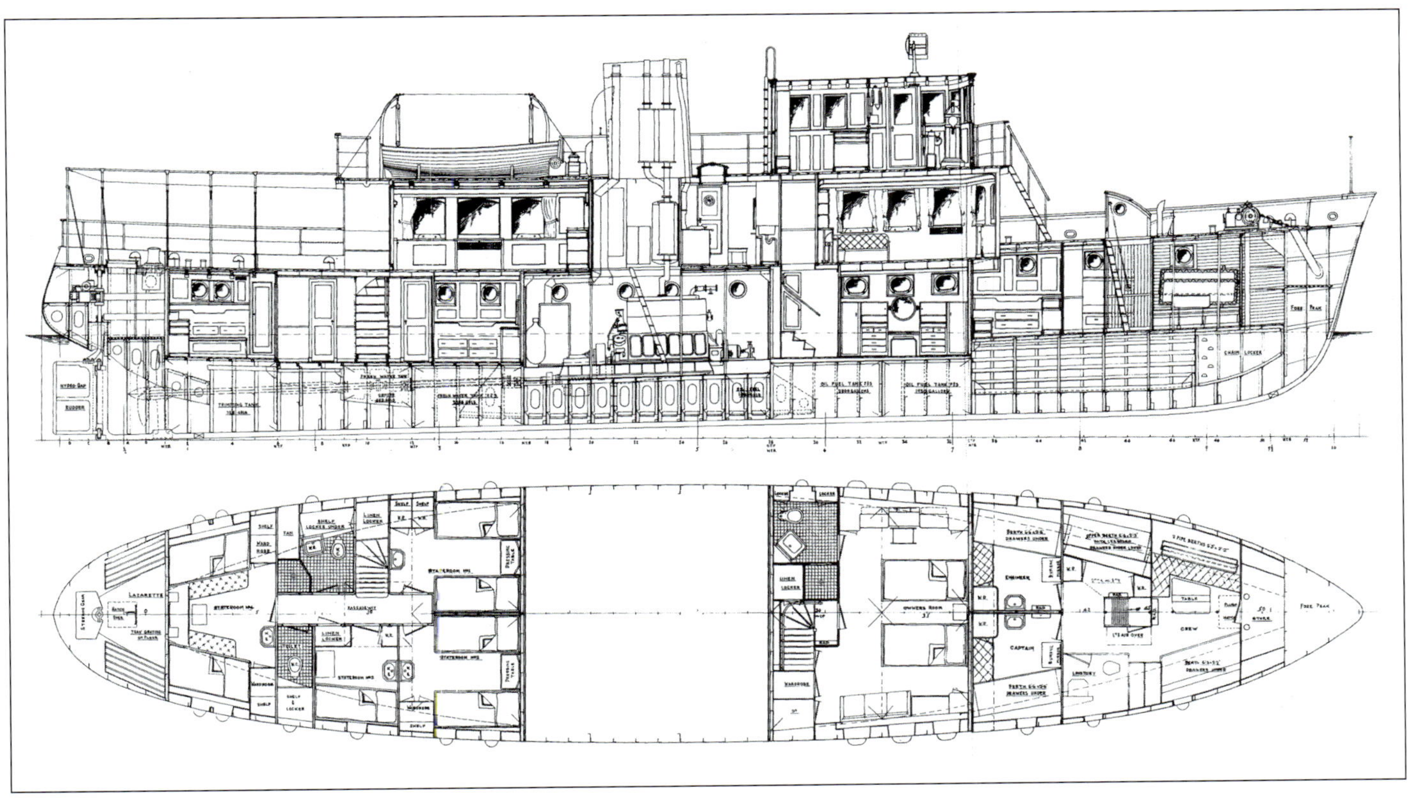

The design package for *Fifer* was the most comprehensive done by Robert Allan up to that time.

A Family Connection

The photo shown here hung prominently in the basement office of my grandfather's home during my boyhood years. It shows *Fifer* in her wartime insignia. In a complete coincidence, my father's cousin Margaret Brown married Frank Walsh, who served much of his wartime years as a naval officer on board *Fifer* guarding the Gulf Islands of BC, undoubtedly one of the better gigs of the war.

—R.G. Allan

Above: HMCS *Fifer*

Left: The crew of HMCS *Fifer* during WWII included my father's "cousin-in-law" Frank Walsh (second from right, front row).

Yorkhome

This toughest of decades for an independent businessman ended on a promise of slightly better times ahead with a few more contracts in 1939 and 1940, which included the *North Vancouver Ferry #5* (described later), some fishboats, a couple of 37-foot fishing industry "scout boats," and the 40-foot "Express Cruiser" *Yorkhome*, built at Vancouver Shipyards for Mr. Roy Hyndman.

The design drawings show a pair of Gray engines of unspecified power. The initial ship registry lists a pair of Ford engines at 170 hp each, but according to the press reports of the day, she was initially built with a trio of Mercury V-8 engines rated at 132 hp each, so her true initial power is somewhat of a mystery, but between 350 and 400 hp. What is certain, however, is that quite soon thereafter she was repowered with a pair of 316 hp V-12 Scripps gasoline engines, with which she made a recorded speed of 38 knots. It can be appreciated that a 50-percent boost in power would have had the desired effect on speed.

A *Harbour and Shipping* article noting her delivery in September 1940 describes her as follows: "*Yorkhome*...is no mere racing shell but a fully equipped cruiser with accommodation for sleeping seven, a galley with Rock gas range and sink, a dinette which is convertible to a double bed at night, also toilet room, hot water heating system and other conveniences including radio telephone...Constructed with a double-planked teak hull and mahogany housework, Mr. Hyndman has tested her out very thoroughly in heavy weather and found her a remarkably good sea boat."

The 40-foot *Yorkhome* had a style reminiscent of motor torpedo boats of the day and a top speed of 38 knots.

Better Times Ahead...

Largely on the strength of the *Fifer* project, the 1930s ended with a hint of better times to come, but they had been very lean years for my grandparents, and indeed for my father, shaping their work ethic for years to come.

NANCEDA
NANCEDA
Esso

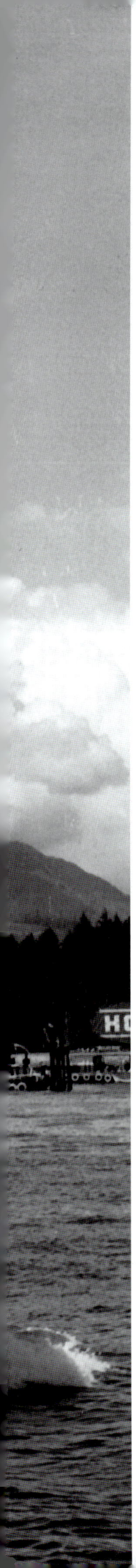

4

Fishing Vessels

In November 1956 Harry Furniss, the boating editor (remember those?) of the *Vancouver Province* newspaper, wrote a very nice feature-length article about the Robert Allan father-son team, though it was primarily a biographical sketch of the elder Allan. In "They Have Designs...on Ships," Furniss wrote: "He [the elder Allan] is credited with having developed the modern fishing boat design in the revolution brought about by increased competition among fishing companies and the need for fast craft that could bring a catch speedily to port, rather than leisurely unload at coastal canneries for processing and further costly transportation."

Opposite: The *Nanceda*, see page 57.

Although the company reputation has largely been built on the tug and barge world, fishboat designs, as indicated above, can be credited with really putting the elder Robert Allan on the local map as a much sought-after naval architect. Trained in large cargo vessel and warship design, his "smaller craft" design credentials since embarking on an independent career had been well established by his work on major yachts such as *Meander* and *Fifer*, but the wartime reconstitution of the BC seiner fleet in 1942–1944 created a niche market in which he seems to have had little competition, and his practice clearly flourished.

Later, in the '50s and early '60s, his son Bob Allan put his own stamp on this market with numerous successful designs for trollers and gillnetters, particularly for local builders who were constructing relatively standard boats for a mass market. As the company moved increasingly into the busy emerging tug and barge market, the fishboat design work slowly dried up, with only a few fishboat designs coming in the 1970s and later.

Robert Allan in his home office in 1956, not quite ready for retirement at age 70.

Early Designs

In the earliest years of his career as an independent designer, only two fishboats appear in the Robert Allan records, and only one of those can be identified. The 75-by-18-foot seiner *Michael J. Tayler* was built in 1929 by Vancouver Shipyards for Tayler Shipping Co. Ltd and thus has the distinction of being the first fishing vessel on record built to a Robert Allan design. Renamed as *Western Spirit* in 1932 under the Nelson Brothers Fisheries banner, she was quite a large vessel for the day, foreshadowing a flotilla of similar but somewhat larger table seiners to follow in the 1940s.

As noted previously, the 1930s were not a prosperous time for boat designers or builders, and there are no fishboats in the company records between 1930 and 1939. *Harbour and Shipping* magazine, however, in March 1939, credits the design of the 80-foot seiners *Midnight Sun* and *Bligh Island* to Robert Allan. By the appearance of these handsome boats, those credits seem justified, but no design record exists, so it is likely that those design drawings migrated onto later projects without a recorded copy being identified with the 1939 builds.

In 1940 the 37-foot 6-inch fisheries "scout boats" *Surf Point* and *Cliff Point* were designed for Canadian Fishing Co. and built by A.C. Benson. The following drawing, rather scant on details and depicting a relatively fast patrol-type vessel, is believed to be the General Arrangement for those boats.

Left: *Michael J. Tayler*, renamed *Western Spirit*, the first fishing vessel designed by Robert Allan.

Below: General Arrangement for an "Express Cruiser," believed to be used for the scout boats *Surf Point* and *Cliff Point*.

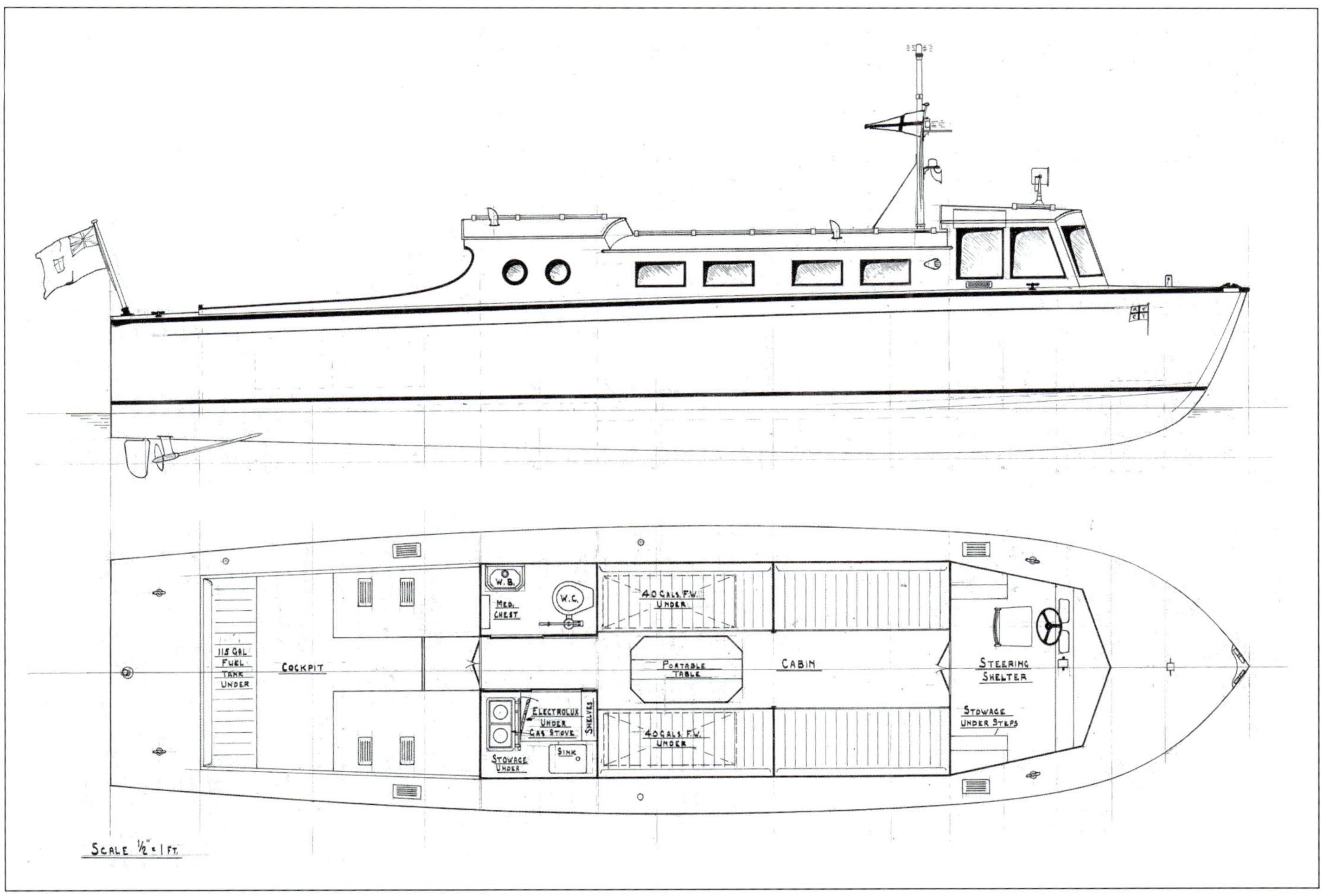

The Subsidy Seiners

In contrast to the previous decade, between 1941 and 1945 Robert Allan's design records show nothing *but* fishboats, with the single exception of an unidentified 32-foot tugboat. This represents an interesting period in BC boatbuilding history, and the reason for this boom in fishboat construction during those busy war years is well told in a series of articles published in *Harbour and Shipping* magazine. In April 1942 the magazine reported:

> On account of the Dominion Government having taken over so many of the larger fishing craft for naval service, including most of the larger Japanese-owned seiners and packers, which the fishing companies had figured on using, and orders having been placed by Britain for all of the B.C. canned salmon made this season, besides Britain's heavy wartime requirements of canned herring, the Dominion Government is being asked by the fishing interests to contribute the difference between regular building prices and the cost of wartime construction of these fishing vessels...To qualify for the subsidy, the boat must be built to government specifications and be 72 to 78 feet long and from 90–110 tons.

By early 1943, this construction program was well underway, and according to *Harbour and Shipping* at that time 11 seiners of 72 to 78 feet in length overall qualified for the subsidy of $165 per gross ton. The subsidy was intended to offset the premium costs for vessel construction during wartime due to shortage of materials, particularly machinery and, presumably, well-qualified labour.

Harbour and Shipping reported in October 1943 that

> Government estimates on the cost of these 72- to 78-foot vessels were reported to have ranged from $48,000 to $53,000, but with steadily increasing costs the actual outlay went somewhat higher, the 75-foot seiners *Southisle* and *Eastisle* being reported to have cost about $55,000 apiece, while the 78-foot seiner *Pacific Sunset* of extra heavy construction is said to have cost about $65,000. The latter vessel was measured at 100 GRT, so the subsidy would have been $16,500 or 25% of the total cost, a rather decent contribution, and more than covering the difference between the highest of the Government's estimates and the final shipyard price.

The majority of the subsidized boats were 78 feet by 21 feet and were fitted with single diesel engines of about 200 hp, though some had as much as 300 hp. The 11 subsidized seiners were recorded as *Western Girl, Cape Russell, Southisle, Eastisle, Pacific Sunset, Mermaid 2, Tatchu, Cape Mark, Western Mariner, Western Commander* and *Nordic Queen*. The majority of these were built in Vancouver by either Menchions or Star Shipyard. All were Robert Allan designs, and there are drawings on file for all of them, however, not always in complete drawing sets. These were all quite similar designs, set up as table seiners, reflecting the technology of the day; however, all of them were later converted to drum seiners. Some of these "subsidy seiners" are shown in the accompanying photos.

Evidently the subsidy process and presumably the boats were so successful that in its March 1943 issue *Harbour and Shipping* reported that "this season the construction of about seventeen new seiners is reported being planned under the government subsidy arrangement, mainly the larger class of vessels of about 78 feet...although about 8 smaller craft are included..." It is not evident from the magazine archives however that those 17 boats were ever realized, at least not under a subsidy program.

Top left: *Western Commander;* **top right:** *Western Girl*, ca. 1950s; **bottom:** *Southisle*, ca. 1950s.

Cape Russell

Post-War Designs

After WWII, the demand for new fishboats diminished but certainly did not disappear. This may have been at least partly attributable to the return of all the appropriated vessels, the "Fisherman's Navy," to the industry as well as a diminished demand for fish with the end of the war. Design work continued at a reasonable pace, however, with a notable trend to rather smaller seiners, probably reflecting the need to keep costs lower. In addition, trollers and gillnetters started to emerge in the design repertoire, and this sector became the preserve of the younger Bob Allan.

The following designs were notable in the period between 1945 and 1950:

1945	*Taplow*	57' × 16' seiner
1947	*Marinet*	60' 6" × 17' 3" seiner
1947	*Saturnina*	42' × 12' halibut troller
1949	*Chief Themlano*	32' 8" × 9' shallow-draft gillnetter
1949	*Western Breeze*	64' × 17' 6" seiner
1949	*Freeland*	66' × 17' combination seiner
1950	*Northview*	68' 7" × 18' 6" seiner
1950	*Louisa C*	64' × 17' 3" seiner

Without question, the star design of this period was the *Nanceda*, built in 1950 by W.R. Menchions. She was built for Louie Percich, the owner of *Marinet*, and Nelson Brothers Fisheries, and at 78 feet by 21 feet beam and with 400 hp, she was a big, handsome boat and very akin to the earlier subsidy seiners, undoubtedly incorporating some lessons learned from that series of successful designs.

Marinet (as a drum seiner) with a full load. She was built for Capt. Louie Percich/ Nelson Brothers Fisheries.

Saturnina, a pretty 42-foot halibut troller built in 1947 by Withey & Sear Boatbuilding. She is recorded as still afloat in 2020.

Below: *Taplow*, built in 1945 by Sterling Shipyards, looking very handsome in her white livery.

The *Nanceda* Drawings

There are, within all the archives in our files, a few drawing sets that stand out not just as impressive designs, but as remarkable works of the draughtsman's art. The *Nanceda* drawings are one of these sets and are perhaps my very favourite. The detailing of wooden construction and the copious structural and mechanical details shown on the General Arrangement drawing can keep one engaged for hours (assuming that one is so inclined!).

—R.G. Allan

Left: Seiner *Nanceda* in Coal Harbour, ready for fishing, ca. 1950.

Below: The intricate General Arrangement drawing for *Nanceda*.

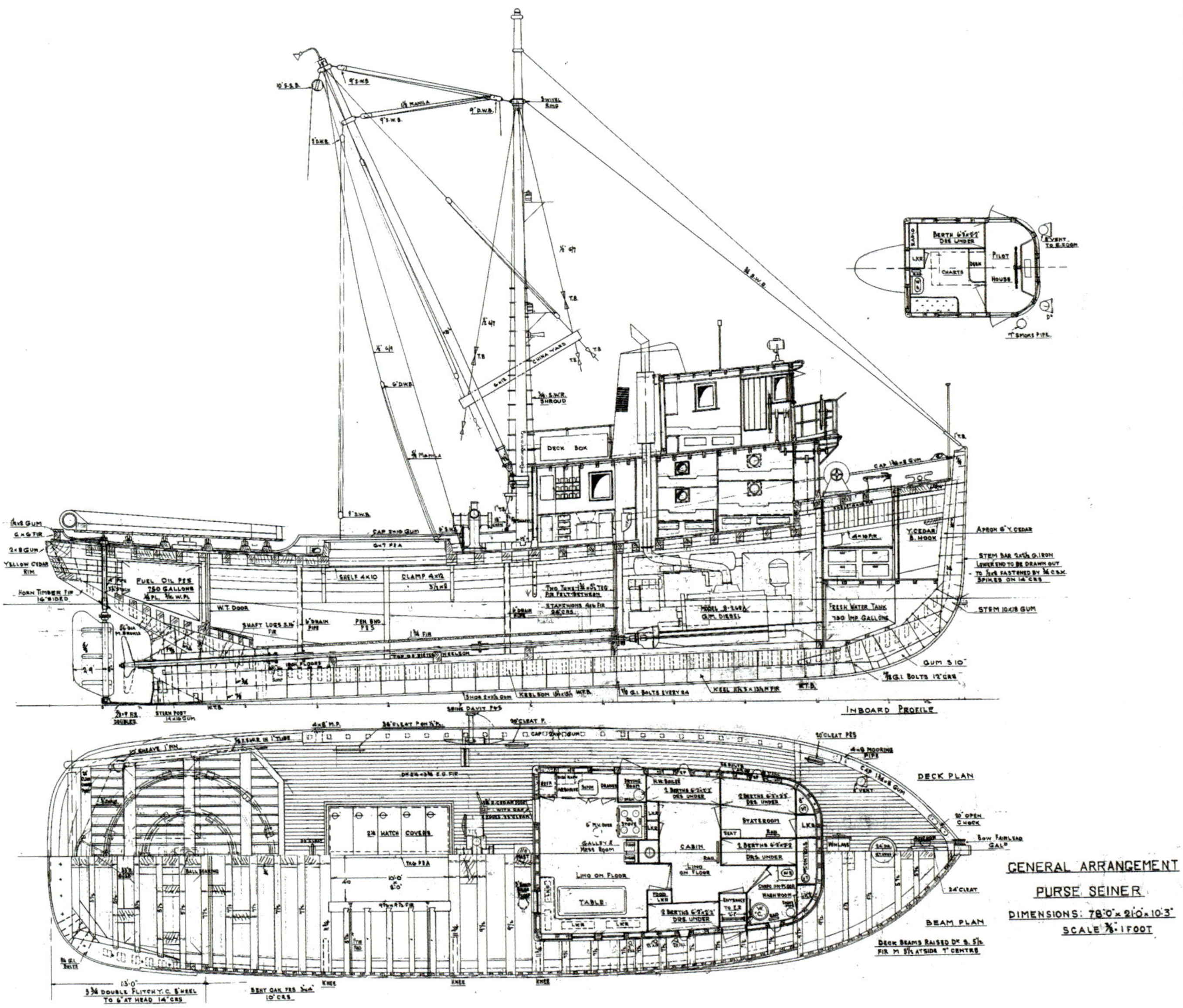

One interesting development credited to Robert Allan in 1951 was the "first published plan of a boat designed as drum seiner," as reported in *Western Fisheries* magazine that year. The *Pacific Mist,* built by Withey's Shipyard on Gabriola Island, was a rather diminutive seiner at 42 feet 6 inches, but clearly designed to address a niche market for smaller boats under 15 Gross Tons (a ludicrous benchmark perpetuated by Transport Canada to this day, allowing vessels under this volumetric measure to avoid inspection). The drum seine was clearly a faster and more efficient way of handling large nets and eliminated the need for high loads on the boom. In this particular design, the drum was set in a large well on deck which, though obviously intended to keep the VCG as low as possible, might have led to problems with net fouling within the well, not to mention water accumulation if the well-drains became plugged. This feature doesn't show up on any later designs, and subsequent seiners showed the drum mounted on the main deck.

Large fishing vessel construction waned somewhat in the early 1950s but with a few notable exceptions. *Kaare* (later *Istra*) was built at Harbour Boatyards Ltd in 1951 for Capt. Otto Novik to a Robert Allan design. The 72-foot-by-19-foot 6-inch seiner was powered by a 300 hp Crossley diesel, then in 1954 she was repowered with a 270 hp Caterpillar diesel, which occasioned a cover photo in the May 1954 edition of *Western Fisheries* magazine. At 71 years of age this vessel appears to still be in excellent condition, as indicated in this 2020 photo, and is still a very fine-looking working vessel. She is also one of the few boats of this era that has not had her original wooden house replaced by some discordant aluminium construct.

Regrettably, only a handful of these sturdy and capable wooden seiners remain today. Quite a few of them, designed to carry salmon on ice, were pressed into service carrying herring in later years, and/or fitted with refrigerated seawater systems. These denser cargoes put the boats much deeper in the water, where formerly dry upper seams allowed water ingress, contributing to their subsequent sinking. *Istra* appears to be the best cared for of those remaining, at least of those visible around Vancouver's lower mainland. The trend to steel seiners began later in the 1950s. The remnants of the wooden fleet are fast disappearing and should be recognized for their important contribution to BC industry, making *Istra* another ideal candidate for heritage designation and preservation.

The last wooden seiner design in our drawing files was done in 1961, for the 68-foot *Carina No. 1*, reflecting a classic West Coast style that didn't alter much over about 20 years. Fortunately, these lovely wooden boats are often the focus of the work of many West Coast artists, such as my friend and relative Doug Harrison, and they are well remembered through images such as those below.

Right: *Istra*, formerly *Kaare*, in Steveston in 2020; **below:** Compact drum seiner *Pacific Mist* at a later stage of her long career, with her drum mounted on deck; **bottom:** Outboard Profile of the 42-foot *Pacific Mist*, first published plans for a drum seiner.

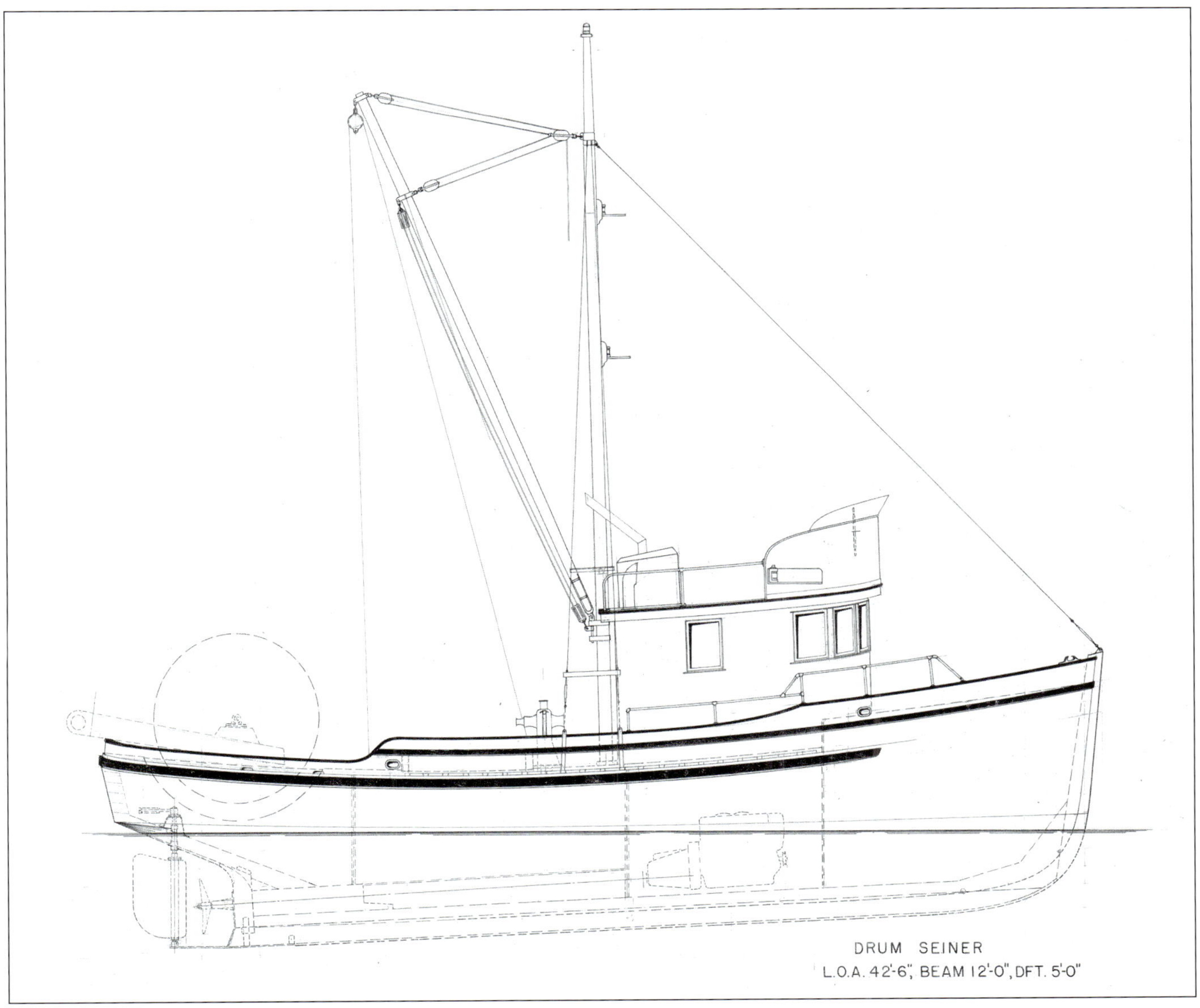

The *New Queen* (named to honour the coronation of Queen Elizabeth II) is another boat of the post-war era, when top-quality wood and associated boatbuilding skills were still available locally.

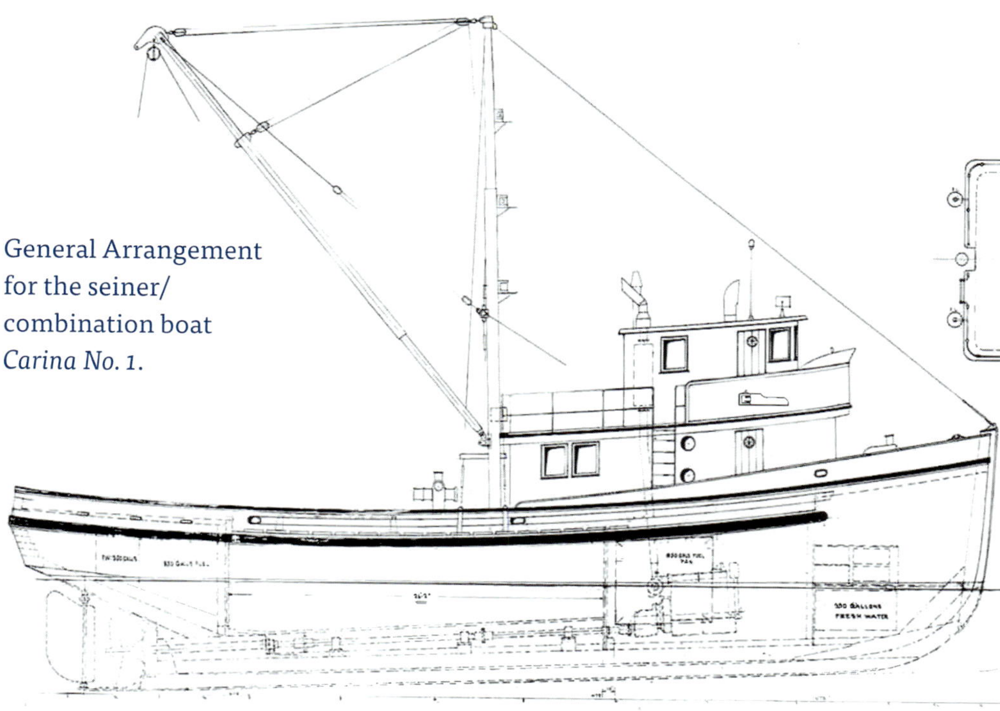

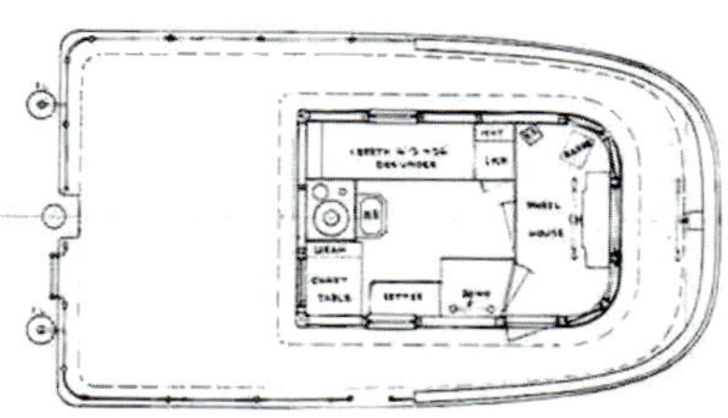

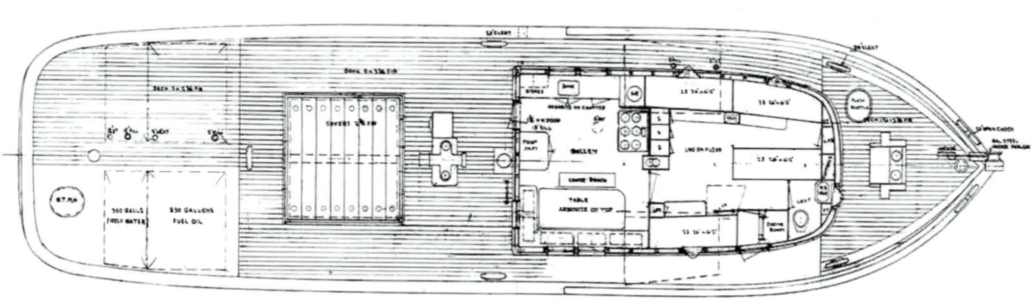

General Arrangement for the seiner/ combination boat *Carina No. 1*.

Doug Harrison's painting of an elegant trio of wooden West Coast seiners hangs in the Robert Allan Ltd offices. *Nanceda* and *Western Warrior* were designed by Robert Allan.

Trollers and Gillnetters

In the mid-1950s the trend to smaller fishboat designs developed. Bob Allan led this design effort, having clearly established close working relationships with several small boat builders in the Lower Mainland. While this may have been good for business at the time, it is impossible these many years later to identify all the boats that were built to any single design, or indeed to any builder-initiated variants thereto. Although many local boatbuilders brought valued business to the design office, the primary boatyards that became significant clients in this time, and with whom Bob Allan continued to work until the mid-1960s, were Sather Boatworks Ltd, Albion Boatworks and Thames Boatworks. Other builder names that appear throughout the record and some of their Allan-designed deliveries include George Kishi (*Roving Beauty*), George Collins (*Silver Lure*), Harbour Boatyards (*Tanza*, et al.), Bob Stolz (at least *Debbie Mac, Darlene Jean, Elsie K No. 1, Emma K* and *Lady Gladys*), Yamanaka Boat Works (*Teraguchi*), Atagi Boatworks, Nakade Boatworks and Deltaga Boatworks.

"Paying for Every Boat"

The subject of "paying for every boat built to a design" can be the bane of a naval architect's existence. I recall clearly my father telling me in the early days of our working together that he had always maintained this dictum to be an important truth to the survival of a consulting business, but I am not sure he had a great deal of success in tracking the number of fishboats built by the various boatyards who were his clients in the 1950s and '60s and, more critically, getting compensated for all of them. It is not uncommon for builders to make a small change and thus declare the variant to be *their* design. It is certain that many fishboats on the West Coast were built to Allan designs or their very close derivatives but without a direct link back to the designer. We have had to challenge a few Chinese shipyards in recent years on this same issue but, by and large, at least since the 1990s, Robert Allan Ltd has been extremely successful in collecting design license fees for every boat built to our designs.

—R.G. Allan

Some internet "trolling" turned up dozens of previously unknown smaller fishboats built to Allan designs. We are especially indebted to Don Macmillan, whose family owned Cassiar Packing Co. Ltd, for a wealth of information and for many photographs related to the early troller and gillnetter fleets on this coast.

There were a wide range of these boats designed and built, from specialized shallow-draft gillnetters to large and sea-kindly West Coast trollers. The images shown here illustrate some of these classic designs.

These classic drawings are favourites with model-makers, who often share photographs of their handiwork with us, such as the following image of a model of *Georgia Girl*.

Opposite top left: Thirty-six-foot troller plan (1951); **opposite top right:** A 43-foot 1-inch canoe-stern troller plan (1962); **opposite bottom:** Inboard Profile and Deck Plan for a 38-foot troller (1950).

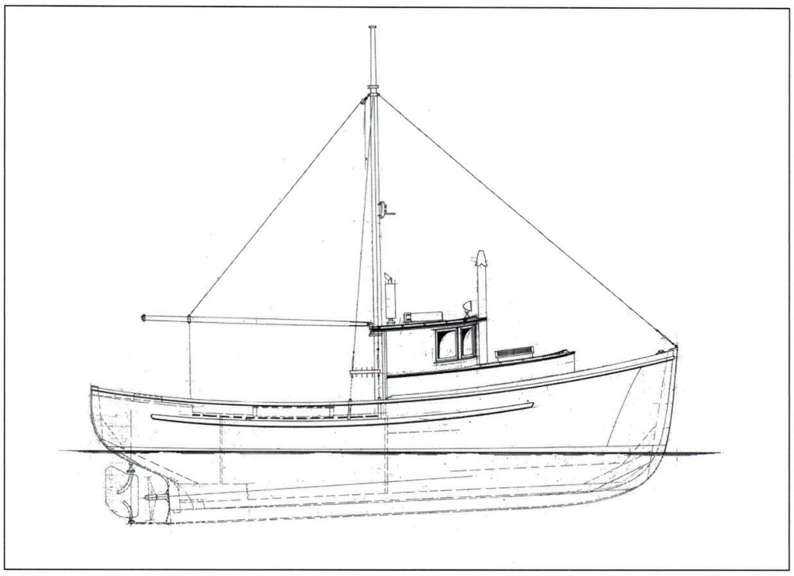

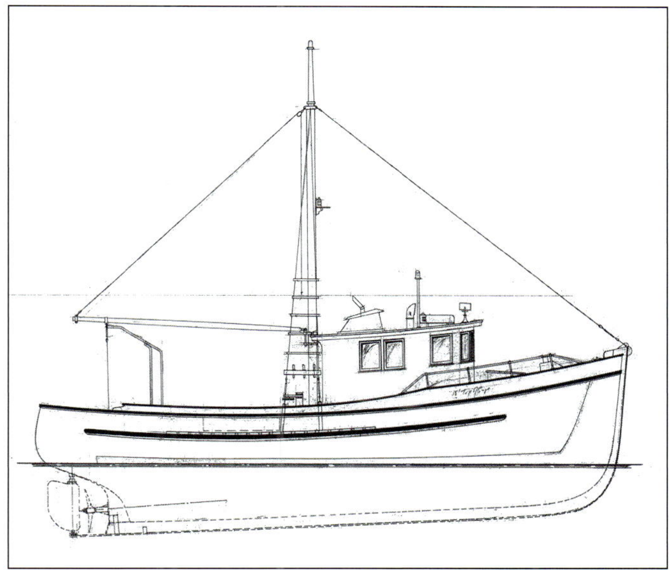

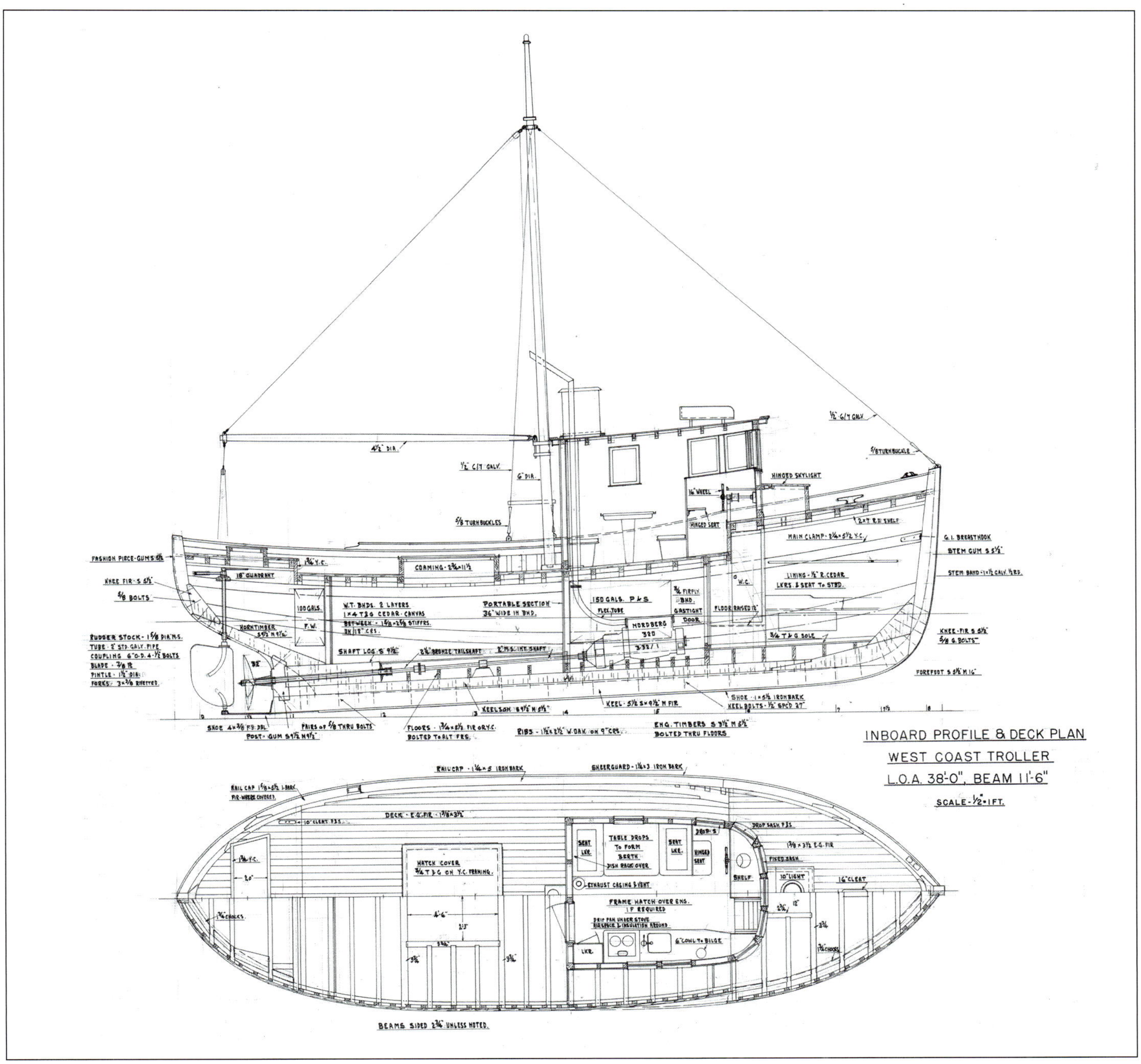
INBOARD PROFILE & DECK PLAN
WEST COAST TROLLER
L.O.A. 38'-0", BEAM 11'-6"
SCALE-1/2"=1FT.

Model of 38-foot troller *Georgia Girl*, built by George Bird.

Below: A 33-foot gillnetter plan (1962), perhaps the *Debbie Mac*.

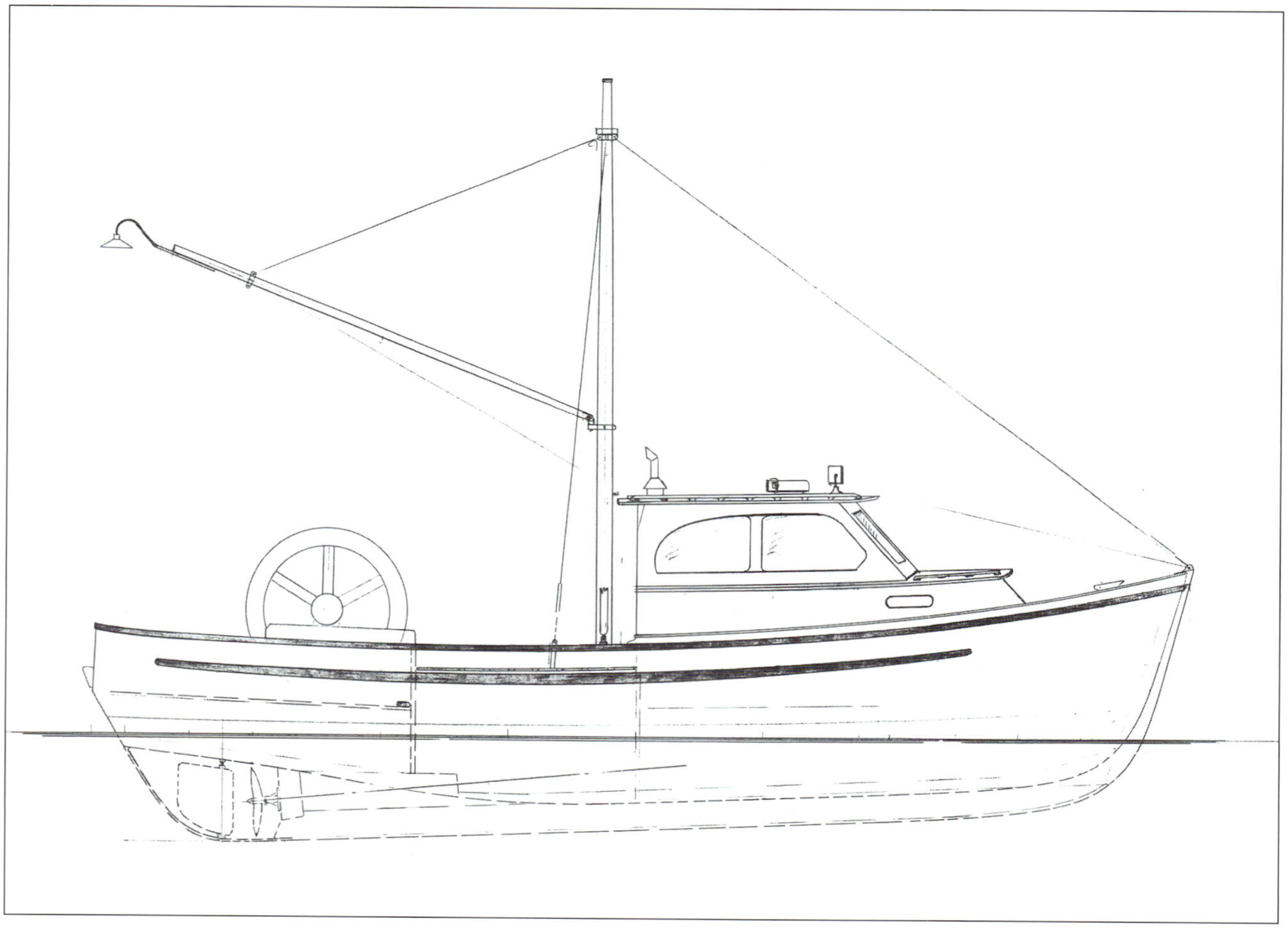

The following photos illustrate but a few of the dozens of these smaller fishboats built to Allan designs, mostly throughout the 1960s.

Top left: *Cassiar 64* at sea.

Top right: *Cassiar 36*, one of an extensive series of pretty wooden gillnetters built by Albion Boatworks for Cassiar Packing.

Above: Traditionally styled gillnetter *Caution Lady* ready for launching at Albion Boatworks, 1963, with owner Gordon Crafton.

Right: *Debbie Mac* with a more "modern" look to the deckhouse.

There are some notables among this fleet of often charming wooden fishing boats, including the lovely 50-foot troller *Karen O*, designed for Axel Ostrom and built by W.R. Menchions & Co. Ltd in 1963.

The *Karen O* looking pristine at nearly 60 years of age in 2020.

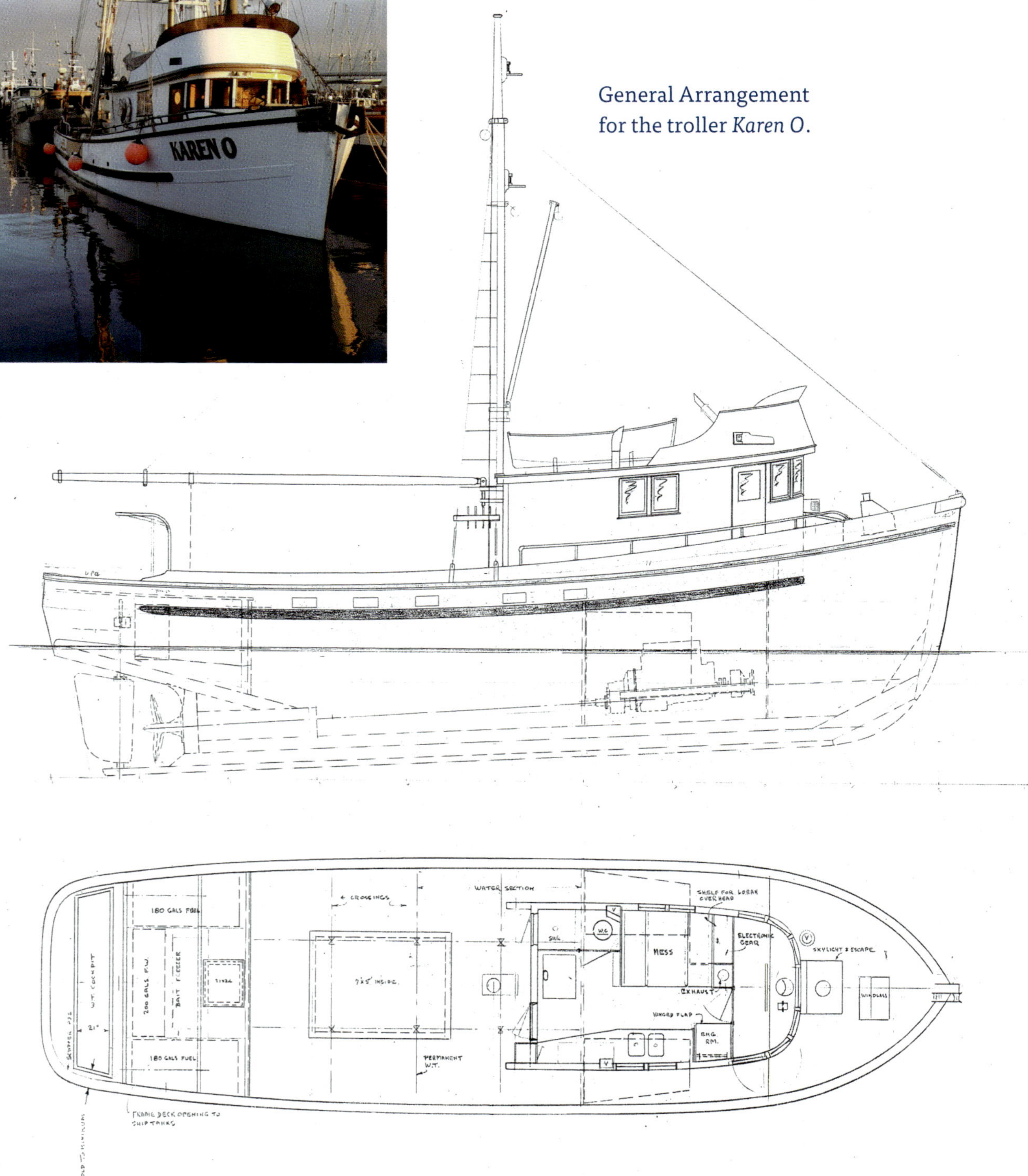

General Arrangement for the troller *Karen O*.

Axel Ostrom and the *Karen O*

While researching the scope of the fishing fleet designed by my forebears, I was directed to the *Westcoast Fisherman* Facebook page created by Alan Haig-Brown, a recognized authority and the author of several very fine books about the West Coast fishing fleet. The participants in this group, including Alan himself, were invaluable in turning up information about boats, owners and builders alike, and provided me with many more photographs than there is room for in this book. One person who contacted me was Al Hoskins, who spoke about "Uncle Axel" (Axel Ostrom), who was still alive and living in care on Vancouver Island at the time. I shared a set of prints from the original drawings for the *Karen O* with Al to present to his uncle, which he did, but sadly Axel passed away soon after, in November 2021. Al told me that Axel was very pleased to have been reacquainted with those drawings in his final weeks.

—R.G. Allan

Thames Trollers

Among the boatbuilders on the BC coast, few can match the reputation for quality workmanship of the late craftsman Cyril Thames. In the 1960s Bob Allan designed several trollers from 39 to 50 feet for Thames Boatworks, in Bowser, on Vancouver Island, and in 1980 Rob Allan designed a smaller 30-footer for him. The 44-footers in particular gather great reviews from the coastal fishers, who usually give credit for the "great lines" of these boats to Cyril, failing to acknowledge that he started by working to a fine design. It is difficult to trace exactly how many boats were built by Thames to these various designs, but at least the following well-regarded boats can be traced to the Allan drawing boards:

Marlaine C.	troller – ~40' × 12' 1"
Cymry	troller – 43' 8" × 12' (later nicely converted to a yacht)
Thetis Prince	troller – 47' 3" Loa × 13' Beam
Thetis Queen	troller – 47' 3" × 13'
Pacific Star	troller – 47' 3" × 13'
Ocean Steele	troller – 47' 3" × 13'
Windward II	troller – 50' × 14' 6"
Eagle Rock	troller – 30' 10" × 10' 6"
Gordon Point	troller – 30' 10" × 10' 6"

The 47-footers, which were quite popular, were known as "44-footers" (their registered length), built to comply with various not always sensible small fishing vessel licensing regulations.

Thames-built *Thetis Queen.*

Below: The 1965 General Arrangement for a Thames 47-foot troller from which numerous versions were built.

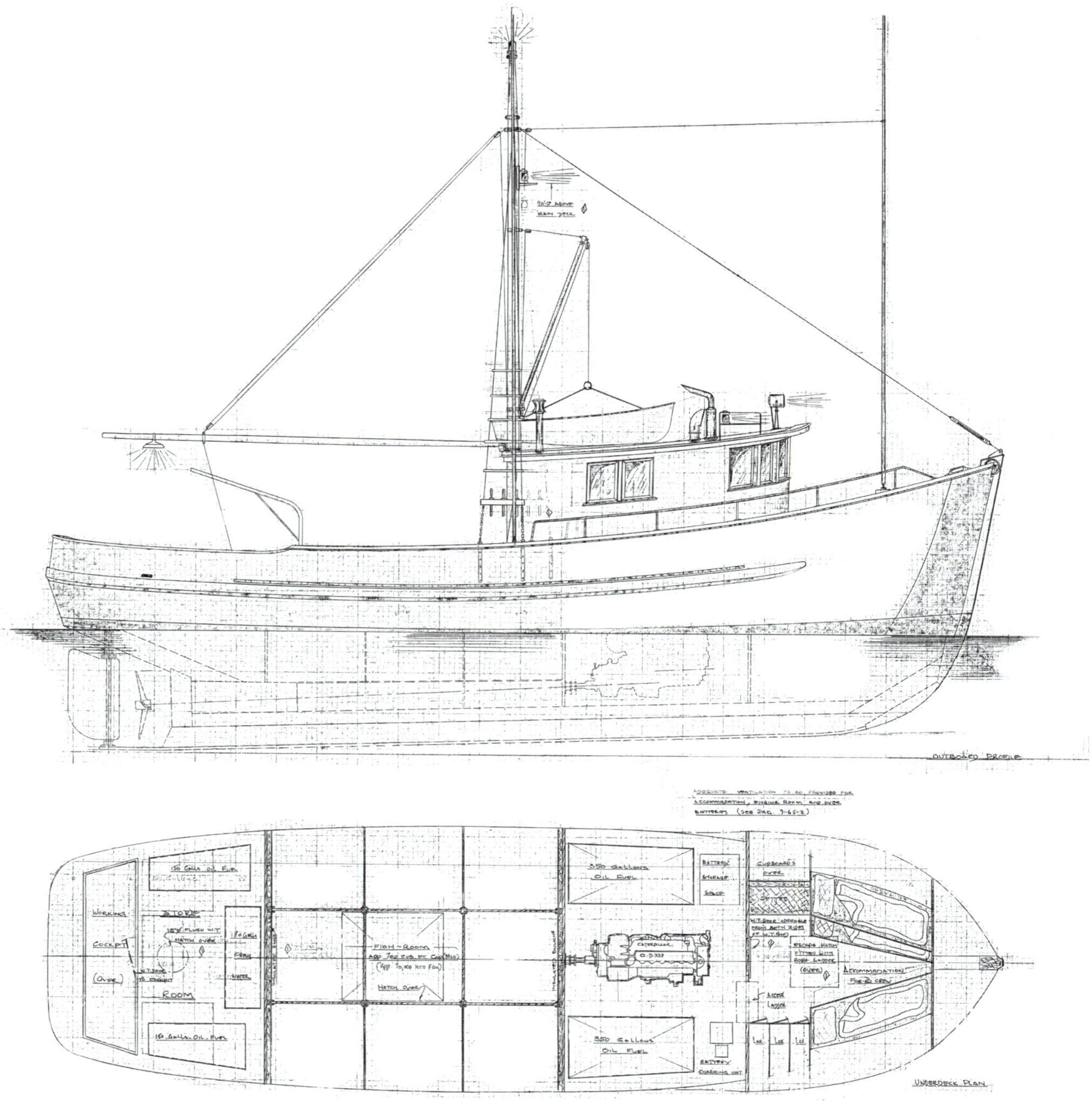

The design and construction of typical wooden BC fishboats are well described in a seminal paper written by Robert F. (Bob) Allan for the Food and Agriculture Organization (FAO) of the United Nations, presented to the World Fishing Boat Congress in 1953. "Design and Construction of British Columbia Fishing Vessels" is perhaps the definitive work describing the typical construction and design details of the day as practised in BC. It is a most useful reference, as wooden commercial vessel construction was definitely already on the wane as the paper was written and within a few years had largely disappeared.

Top left: *Windward II*, a lovely 50-foot West Coast troller designed by Bob Allan and built by Cyril Thames.

Top right: Another Thames-built 47-foot troller, *Pacific Star*.

Right: *Cymry*, a West Coast 44-foot troller as a tasteful yacht conversion.

Shifting to Steel

Steel vessels of all types began to appear in the Allan design files in significant numbers in the mid-1960s, but several builders persisted with wooden fishboats for some years after that, until steel and aluminium took over in the larger boats and fibreglass began to dominate in the smaller fleets. Finding suitable quantities of properly seasoned timber was becoming a challenge, and the costs of building in steel were considerably lower than in wood.

The first recorded steel fishboat from a Robert Allan design was in 1958, so it was a "Bob" Allan design. The 76-foot seiner *Bering Sea* was built by John Manly Ltd for Olaf Rorstad and Lars Iversen in 1959. The boat received a lot of attention in the local press, as it signalled a new era in the fishing fleet. The boat was also notable for having what was at the time considered excessive deckhead height to accommodate the height of her lofty Norwegian owners, evident in the photo below.

Launching ceremony for the *Bering Sea*: John Manly, second from left, then Bob Allan, and Olaf Rorstad and Lars Iversen (in hats) to his left.

Below: *Bering Sea*, the first steel seiner from the Robert Allan drawing boards, built by John Manly Ltd in 1959, in Vancouver harbour.

In 1959–60 the seiners *Attu*, 76 feet by 21 feet, for Capt. Edgar Arnet, and *Caamano Sound*, 77 feet 6 inches by 21 feet 6 inches, for Union Trading Co., were built at Allied Shipbuilders Ltd and Star Shipyard (Mercer's) Ltd, respectively. These three big boats are still actively fishing in BC waters and, though outwardly quite similar, they have quite different hull forms. *Bering Sea* and *Caamano Sound* have similar double-chine forms effectively, with fully developable shapes, whereas the *Attu* has a round bilge form much like earlier wooden boats.

The distinctive steel seiner *Attu*.

General Arrangement for the seiner *Attu*, for Capt. Edgar Arnet.

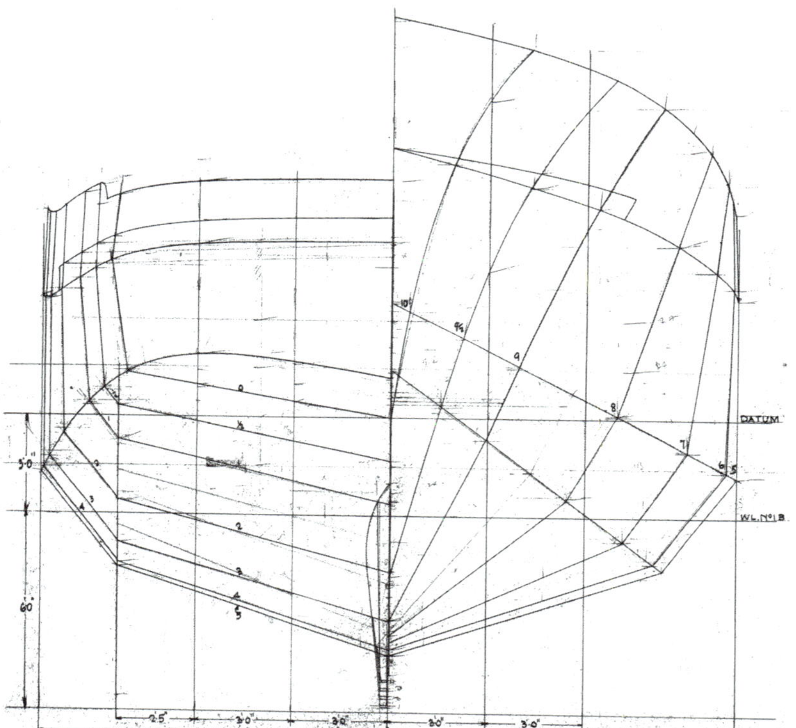

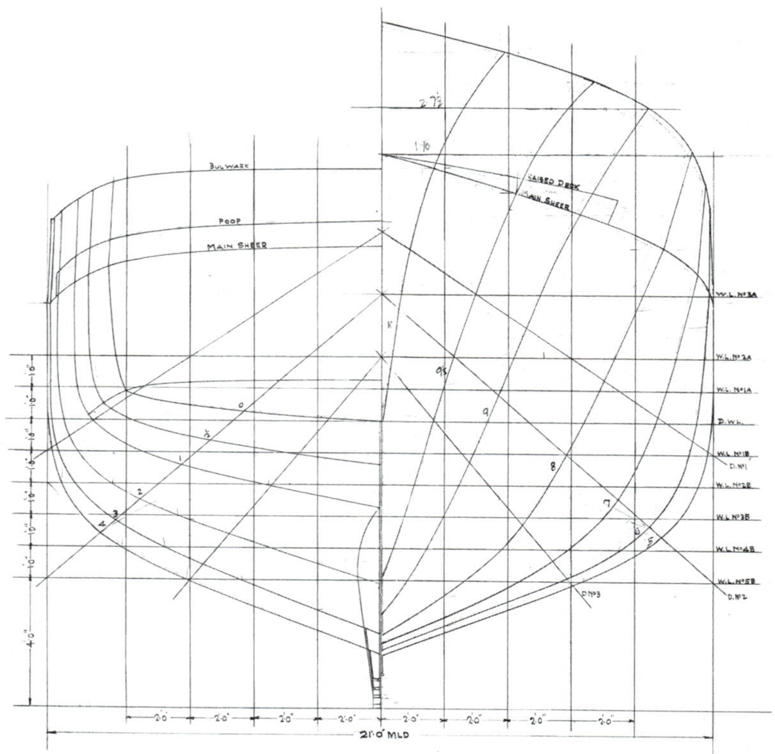

Top: The launching of *Caamano Sound*, 1960; **above left:** double-chine hull form of *Caamano Sound* (*Bering Sea* was very similar); **above right:** Round bilge hull form of *Attu*.

The trend to steel construction soon migrated into smaller boats, beginning with Gerry del Secco's 52-foot seiner *Tiber* and the 60-foot crab boat *Amber III*, both built by John Manly Ltd in 1960. Throughout the balance of the 1960s, fishboats really took a back seat to the booming tugboat business at Robert Allan Ltd, but one steady client for steel boat designs was Berg Shipyard in Blaine, Washington. These commissions included several 58-foot Alaska seiners and some bigger crab boats, like the 76-foot *Alaskan*. The 58-footers were known as "limit seiners," as that was the largest vessel licensed to fish in Alaska at the time, presumably with the intent to conserve fish stocks. Unfortunately, we have been unable to trace the names of boats built by Berg.

A smattering of local steel fishboat designs were done by Robert Allan Ltd throughout the '70s, mostly for boats below 60 feet in length, such as *Viking Leader*, *R.M. Nelson*, *Dawn Treader #1*, *Star Pacific* and *Island Rogue*.

The 52-foot steel seiner *Tiber*, built in 1960 by John Manly Ltd.

Fibreglass Boats for the Alaskan Fishery

In the early 1980s, a new client emerged that brought some interesting challenges to the fishboat design market. Beck Boats Inc. in Marysville, Washington, approached Robert Allan Ltd to design a new generation of fast gillnetters for the Alaska market, to be built in fibreglass. An initial successful series of 32-foot gillnetters led to further designs for some "bow-pickers" and several versions of fast "beach seiners" in the 48-to-50-foot length range. These boats set their nets close to the beach, presumably using the shoreline to aid in entrapping the fish, and thus had to be quite shallow draft. They also had to be fast in order to get quickly to the areas where limited and short-term openings were declared.

The last of these projects was for a 48-foot twin-screw beach seiner with an awe-inspiring 850 hp. To be certain of meeting the 20-plus knot speed requirement of these shallow semi-tunnel stern designs, we convinced Ed Beck to conduct some preliminary model-testing of the hull at what was then the Ocean Engineering Centre, part of the BC Research facilities at UBC. These tests proved the concept, and Beck Boats built quite a few of these very pretty boats, such as the *Kilokak*.

Kilokak, a fast twin-screw Alaska beach seiner built by Beck Boats Inc. to a Rob Allan design.

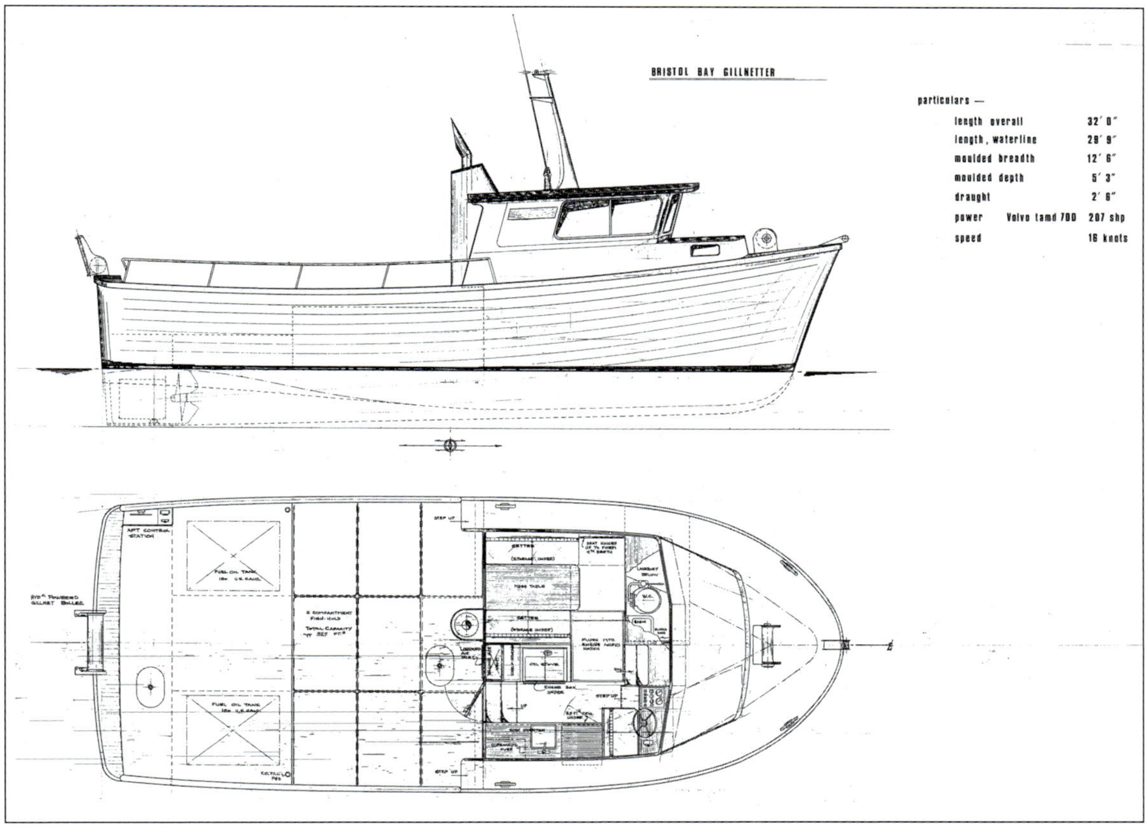

A 32-foot fast gillnetter for Beck Boats Inc.

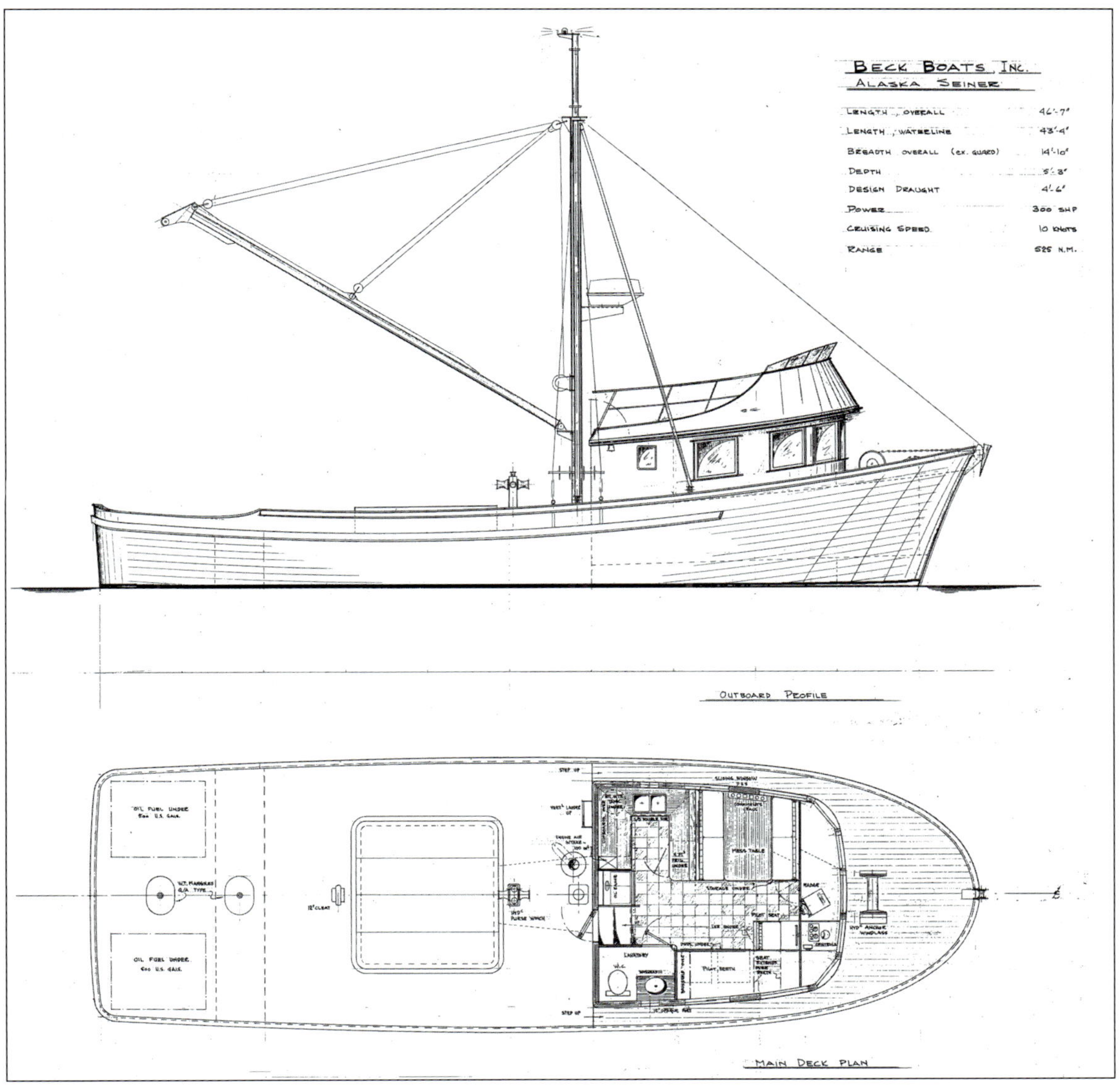

Outboard Profile of first-generation Beck 48-foot beach seiner.

Ed Beck

These projects from Beck Boats were welcome design work in the doldrums of the mid-1980s, and Ed Beck was a really interesting client. He was a rough-around-the-edges character, himself an experienced fisherman and a boatbuilder, and he knew exactly what he wanted to see in a new design. I learned a lot about fibreglass boatbuilding from Ed, and he finished his ruggedly built boats to yacht-like quality. I recall vividly him reviewing the lines plan I was drawing for the first of these fast seiners, and he wasn't satisfied until the flare in the bow was really quite extreme; far more than I would have considered necessary. However, the end result, as illustrated, was both lovely to see and kept these boats very dry at sea!

—R.G. Allan

By the 1990s it was quite clear that the market for Robert Allan Ltd's design services was going in a different direction than that of local fishboats, and none have been recorded subsequently. Given the generally reduced state of the local fishery since then, that was obviously a fortuitous eventuality.

Fishing Vessel Safety

Although fishboats have not been a large part of our business for many years, in 2002 we were called upon by the Workers' Compensation Board of BC (WCB) to assist in their assessment of the tragic sinking at the mouth of the Fraser River

Diagram showing cumulative impacts of various factors on the available righting energy in the fishboat *Cap Rouge*, with the ultimate loss of virtually all stability.

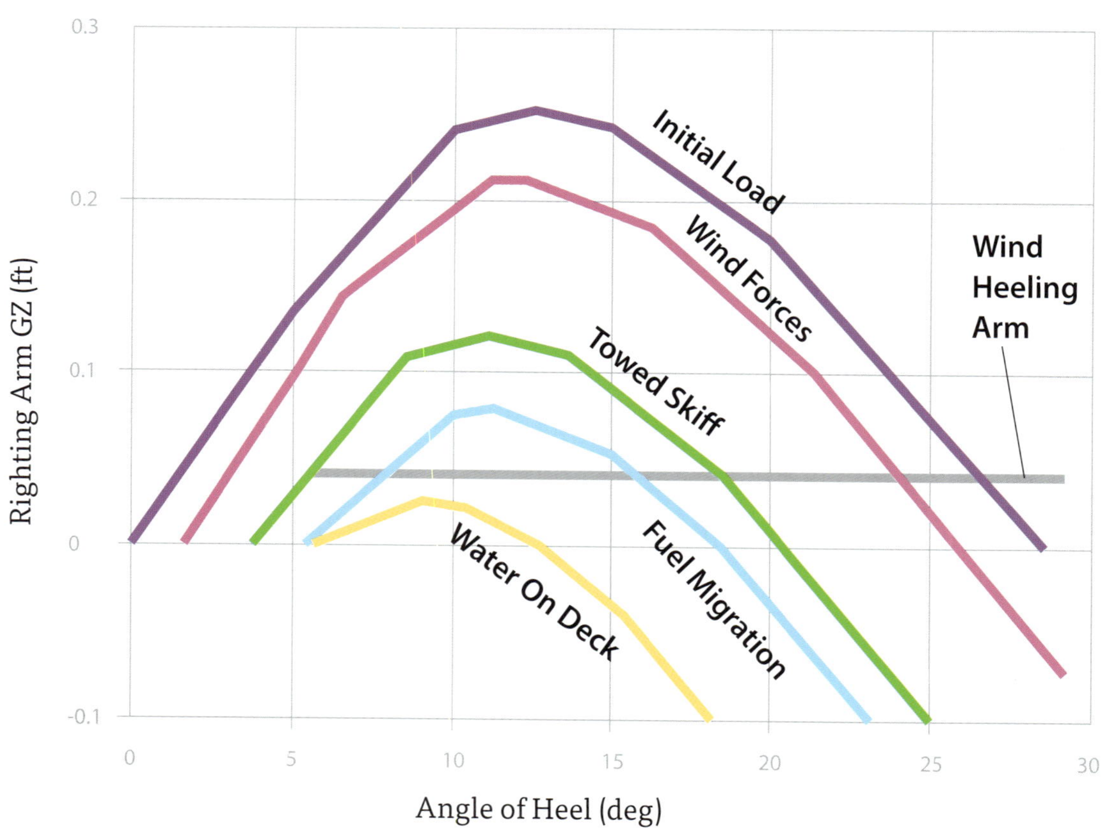

of the small fishing vessel *Cap Rouge*, with the attendant fatalities of many members of one family.

We did a fairly exhaustive analysis of the vessel and its conditions, demonstrating that the review done by the Transportation Safety Board of Canada was rather cursory and had made early and false assumptions. After the Coroner's Enquiry here in BC, a detailed paper on the scope of the analysis and findings, "*Cap Rouge II*: Lessons Learned from a Fishing Vessel Tragedy," was written by R.G. Allan for a Royal Institution of Naval Architects (RINA) Conference on Fishing Vessel Safety held in London in 2005. The paper demonstrated how the incremental accumulation of many small factors could lead to a catastrophe and was lauded (and is still being used by university lecturers) as an example of the textbook methodology to use in the analysis of similar incidents.

After the Coroner's Enquiry into this incident, the WCB worked very hard to promote education and safety in the fishing community of BC and generated the FishSafe program, which undoubtedly saved more than a few lives in what has always been a rather risk-prone industry. We are proud to have played a small part in advancing that cause.

POLICE
POLICE

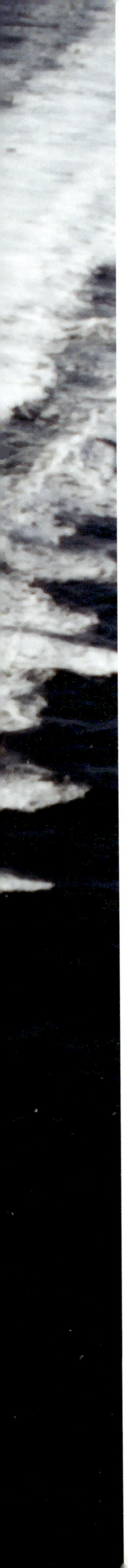

5

Coastal Patrol Vessels

Patrol boats are a vessel type that seem to come in waves, about a generation apart. It is a genre of boats for which the company is not widely recognized, and yet over the decades there have been a significant number of such craft developed under the Allan name.

In the 1920s and 1930s emerging departments of both the provincial and federal governments with coastal mandates of various kinds were each busy developing their own fleets of coastal patrol vessels to enable them to exercise their duties. This served Robert Allan well, as it certainly appears from the records that there was little else to keep him busy in the early years of the "Dirty Thirties." A wide variety of wooden patrol boats were commissioned, nearly all built by one or the other of the numerous boatyards that existed in Vancouver's Coal Harbour in that period.

One of the earliest of his coastal patrol boat designs was the elegant 62-foot *Salucan v*, built in 1932 at Walkem's Shipyard in Vancouver as the William's Head quarantine launch for the federal minister of Pensions and National Health. The drawings for this boat, though few in number, are lovely examples of the sort of design work being done in those days. Skilled shipwrights didn't need a great amount of detail; they worked typically from the Lines Plan, General Arrangement and some basic structural drawings. The rest were resolved on the loft floor and in the yard. The three drawings shown below indicate everything a 1930s yard needed to know to both bid and build the vessel.

Opposite: The RCMP patrol vessel *Nadon*, later used to recreate the voyage of the *St. Roch*.

At about the same time (1929–33), the federal Department of Fisheries built a series of Allan-designed patrol vessels in the 45-to-52-foot range. The *Swantail II*, *Cloyah II*, *Clupea*, *Onerka*, *Egret Plume 2*, *Black Raven 2*, *Onerka 2* (the first *Onerka* burned in 1930), *Merry Sea 2* and *Bonilla Rock* are all credited to his name, but unfortunately no drawing record for this significant set of

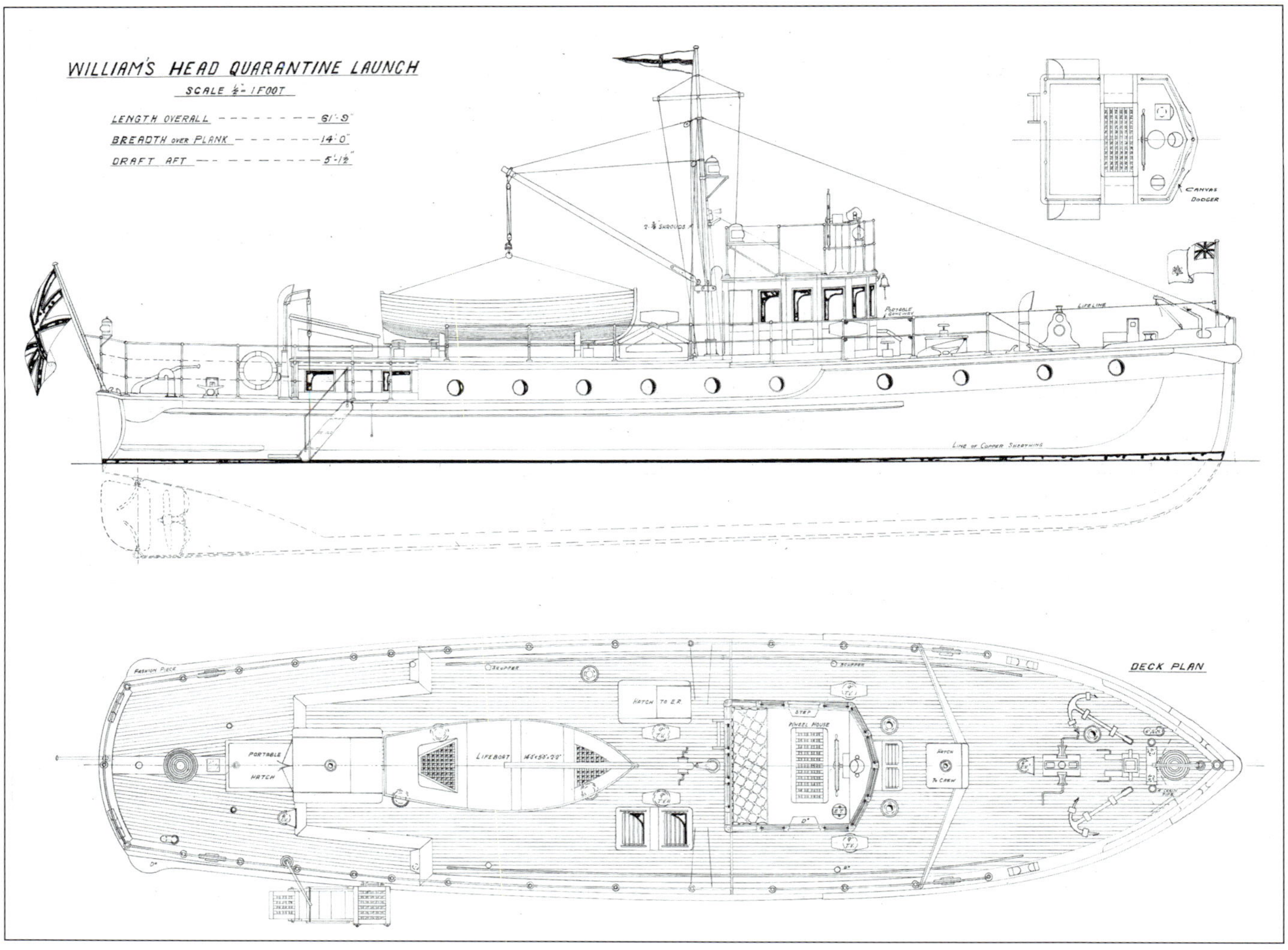

Top: Quarantine launch *Salucan v* on trials in English Bay; **above:** Outboard Profile and Main Deck Plan for *Salucan v*; **opposite top:** Inboard Profile, Lower Deck Plan and Structural Arrangement for *Salucan v*; **opposite bottom:** Typical Structural Sections and Scantlings List for *Salucan v*.

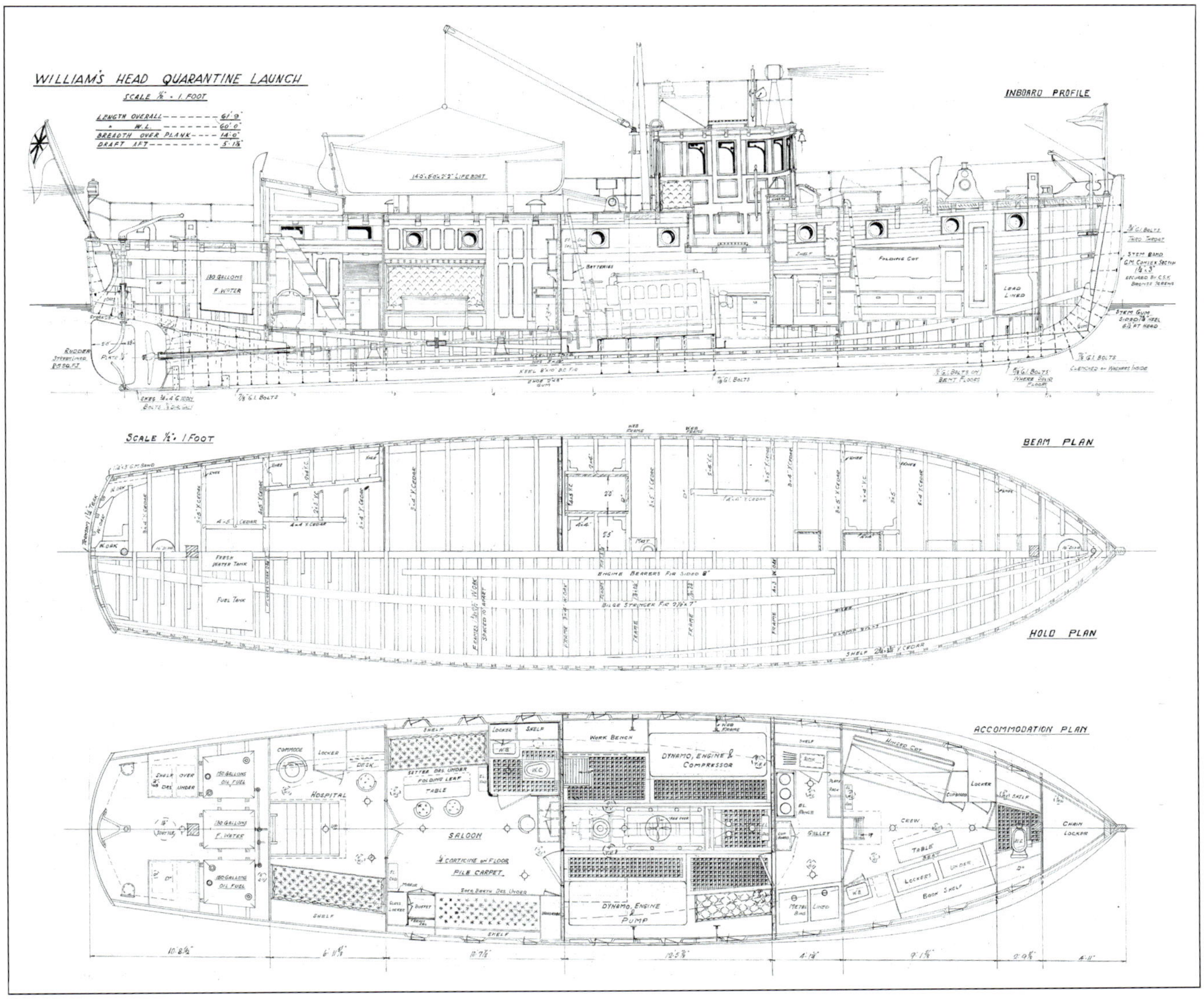
WILLIAM'S HEAD QUARANTINE LAUNCH
SCALE ½" = 1 FOOT
INBOARD PROFILE
SCALE ½" = 1 FOOT
BEAM PLAN
HOLD PLAN
ACCOMMODATION PLAN
HOSPITAL
SALOON
DYNAMO, ENGINE & COMPRESSOR
DYNAMO, ENGINE PUMP
GALLEY
CREW
CHAIN LOCKER

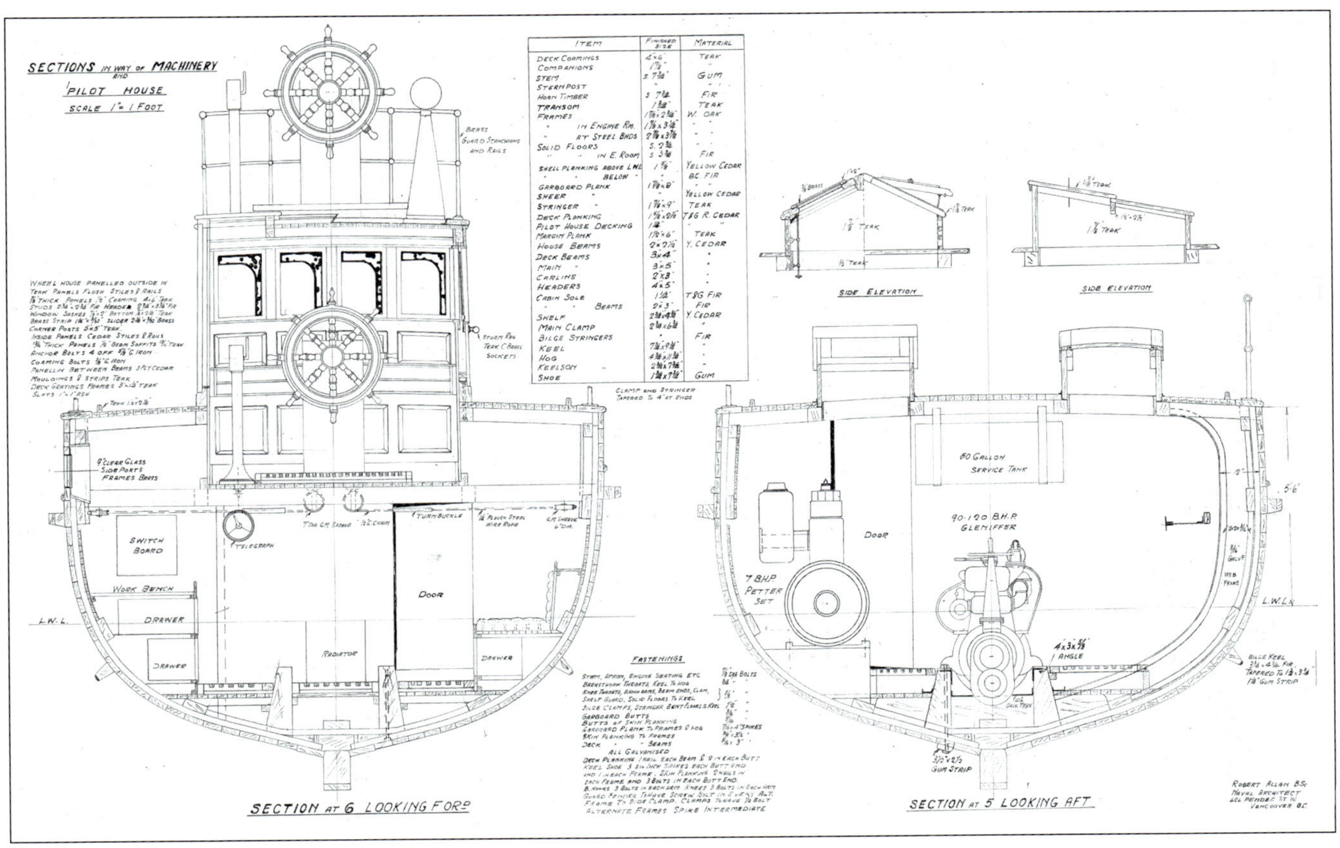
SECTIONS IN WAY OF MACHINERY AND PILOT HOUSE
SCALE 1" = 1 FOOT
SIDE ELEVATION
SIDE ELEVATION
SWITCH BOARD
WORK BENCH
DRAWER
DOOR
80 GALLON SERVICE TANK
90-120 B.H.P. GLEMIFFER
7 B.H.P. PETTER SET
FASTENINGS
SECTION AT 6 LOOKING FOR'D
SECTION AT 5 LOOKING AFT
ROBERT ALLAN B.Sc.
NAVAL ARCHITECT
VANCOUVER B.C.

PGD #2 in original government livery.

Below: PGD #1, reborn as *Townley Isle*, still cruising the BC coast.

boats can be located today. Most likely they migrated into later projects. In this same time-frame, the BC Department of Fish and Game had the *PGD #1* (36′) and *PGD #2* (42′ 6″) built at W.R. Menchions & Co. Ltd.

A second wave of coastal patrol boat designs was developed in the 1953–1959 period, when the Department of Fisheries built several wooden patrol boats for fisheries management tasks for specific regional areas from the Fraser River to Bella Coola. These boats were all in the 40-to-50-foot length range and had classic semi-displacement transom-sterned hull forms easily driven with modest power. The *Gavia*, *Brama*, *Seal Rock*, *Star Rock*, *North Rock* and *Temple Rock* were all built to Allan designs for Fisheries in this period, and R.F. (Bob) Allan's influence definitely showed in these designs.

One of the treasures in the company archives is a small 7-by-9-inch notebook in which Bob Allan kept many (but regrettably not all) the basic calculations for his designs. Nearly all of these patrol craft, however, are recorded in this booklet, which also includes some interesting anecdotal tidbits, such as the following:

Gavia; (42' × 10' 10") price $28,800 plus Sales tax. 12.13 knots at 2,150 rpm with a GM 6-71

Indian Affairs launch (*D.M. MacKay*) 1958; Star $48,528; Alberni $46,118; Bel Aire $49,815 (*clearly indicating how competitive these yards were at the time*)

45' Game Commission Protection Cruiser (*PGD #3*) – $43,285 tax incl.

An example from that little workbook shows the detailed displacement and hydrostatic calculations for two notable coastal patrol vessels, the *PGD #3* and the *Star Rock*, both mid-50s designs.

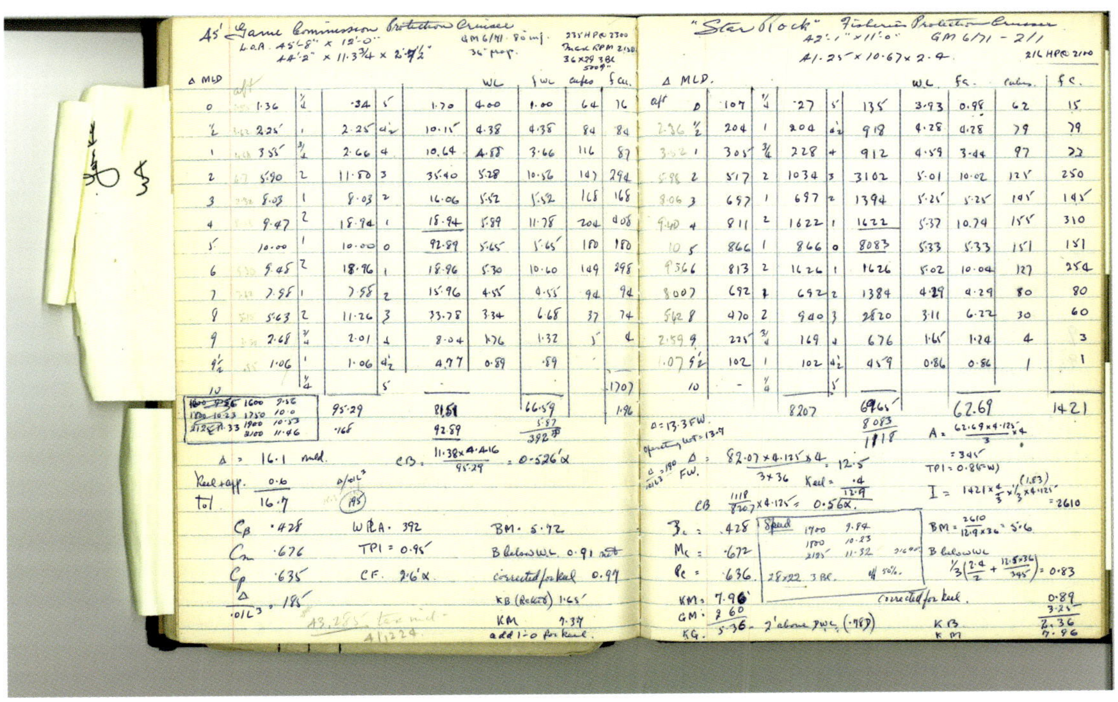

Sample pages from Bob Allan's early design notebook.

Fortunately, many of these well-built wooden boats are still in service as private yachts and are being well maintained.

Of a similar style, the *D.M. MacKay* was built as a Bella Coola agency launch for the Department of Indian Affairs in 1959. In 1980 it was sold by Crown Assets to the Coastal Missions Society and renamed *Coastal Messenger*. The *PGD #3*, built in 1959 for the BC Pacific Game Commission, and the 45-foot *Forest Ranger II*, built in 1953 for the Forestry Department, were of similar style. All are still cruising the coast in private hands. These boats were modestly powered, many with GM 6-71 diesels rated at about 215–230 bhp and achieved speeds in the 11–11.5 knot range, making them ideal boats for casually and economically exploring the BC coast.

The RCMP also commissioned a wide array of vessels across the country, selected according to the local patrol and policing needs. This included some Allan-designed boats built in BC in the 1950s. The *M.L. 1 (Westview)* and *M.L. 2* (*Tahsis*) were 33-foot patrol launches built for local enforcement duties, the former at McKay-Cormack Ltd in 1953 and the latter in 1954 at Foster's Shipyard, both in Victoria. The 55-foot Detachment class vessel *Sidney* was another design from the Allan boards, built in 1959 at A.C. Benson Shipyard in Vancouver.

Inboard Profile and Beam Plan for the Fisheries patrol vessel *Brama*.

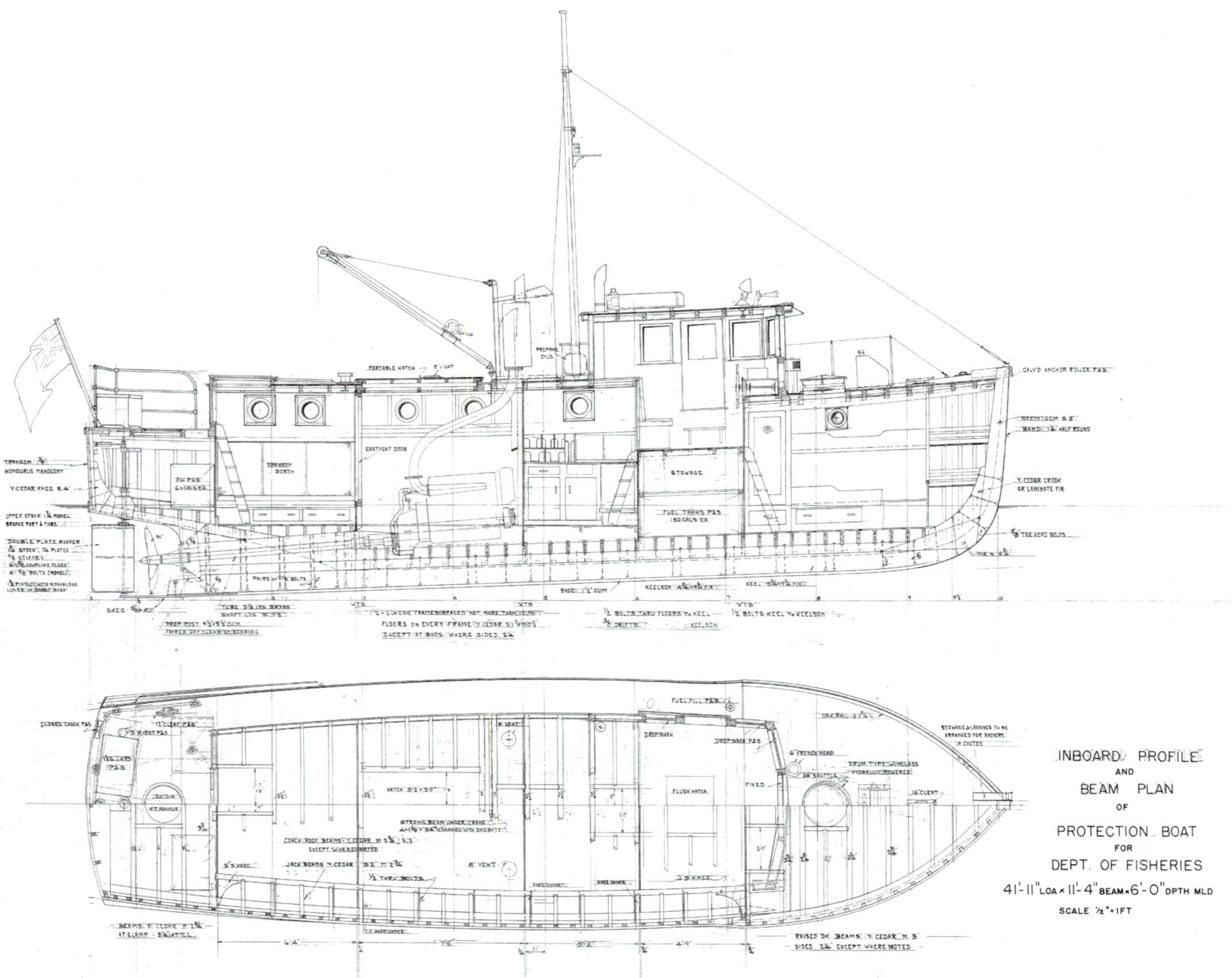

Above and right: Fisheries patrol vessels *Brama* and *Temple Rock*.

RCMP patrol boat *M.L. 1, Westview*.

Star Shipyard seems to have won the lion's share of orders for these boats in the mid-to-late '50s, the last era of extensive wooden boat building in the Lower Mainland, as steel was gradually taking over as the material of choice for new builds. Quality woods were becoming increasingly difficult to source and presumably wooden-boat-building skills were also waning. Metal boats offered more durable construction and less regular maintenance. However, the federal Department of Fisheries had one last crack at a larger wooden patrol vessel, commissioning from Robert Allan Ltd the design for the 105-foot *Golden Bay* in 1972 for patrol duties in Newfoundland. This handsome boat, built in Nova Scotia in 1975, was later transferred to Gaspé and rechristened *Pierre Fortin*. This project was assumed to be an attempt to support some of the last commercial wooden-boat builders in Atlantic Canada, but sadly represents the last wooden vessel designed at Robert Allan Ltd.

Robert Allan Ltd had some overseas success in the late 1970s with a series of 19 aluminium fast patrol boats designed for the Singapore Marine Police. These 36-foot 30-knot PX class boats were one of the first project successes of the SeaConstruct joint venture in Singapore (Chapter 2).

The latest wave of patrol vessel designs from Robert Allan Ltd came in the early '90s with a trio of fast 58-foot 30-knot aluminium catamarans built by Shore Boat Builders for the RCMP: the *Nadon*, *Higgitt* and *Lindsay*. These multi-hull vessels, the first of that type designed by the company, were quite unique, featuring a very low draft for operation in very shallow waters and fitted with surface-piercing Arneson drives.

The *Nadon* earned distinction in 2000 when, renamed as *St. Roch II*, it was used to recreate the 1940 transit of the Northwest Passage and subsequent circumnavigation of North America by her famed wooden namesake.

The 105-foot *Golden Bay* as *Pierre Fortin* after sale to the Province of Quebec.

Top left: *PX 3*, one of a series of fast patrol craft built in 1978 for the Singapore Marine Police; **top right:** *Nadon*, one of a trio of 30-knot catamarans for the RCMP; **bottom:** The RAven class patrol catamarans built at Shore Boatbuilders Ltd for the RCMP.

These vessels were followed a few years later by larger cousins, the *Inkster* and *Simmonds* for the West Coast, and the *Murray*, built in 2003, for the East Coast. Robert Allan Ltd also had commissions from the Port of Vancouver for the 45-foot aluminium harbour patrol boat *Kla-Wichen*, built by Allied Shipbuilders in 1985, and the larger *Takaya* in 1998.

During the writing of this memoir Robert Allan Ltd is completing the design of an all-electric patrol boat for the Vancouver–Fraser Port Authority, indicative of the emerging demand for vessels of all types with lower emissions profiles in coastal environments.

Top: The *Inkster*, built by Allied Shipbuilders for West Coast duties; and **bottom:** The *Murray*, serving the Canadian East Coast.

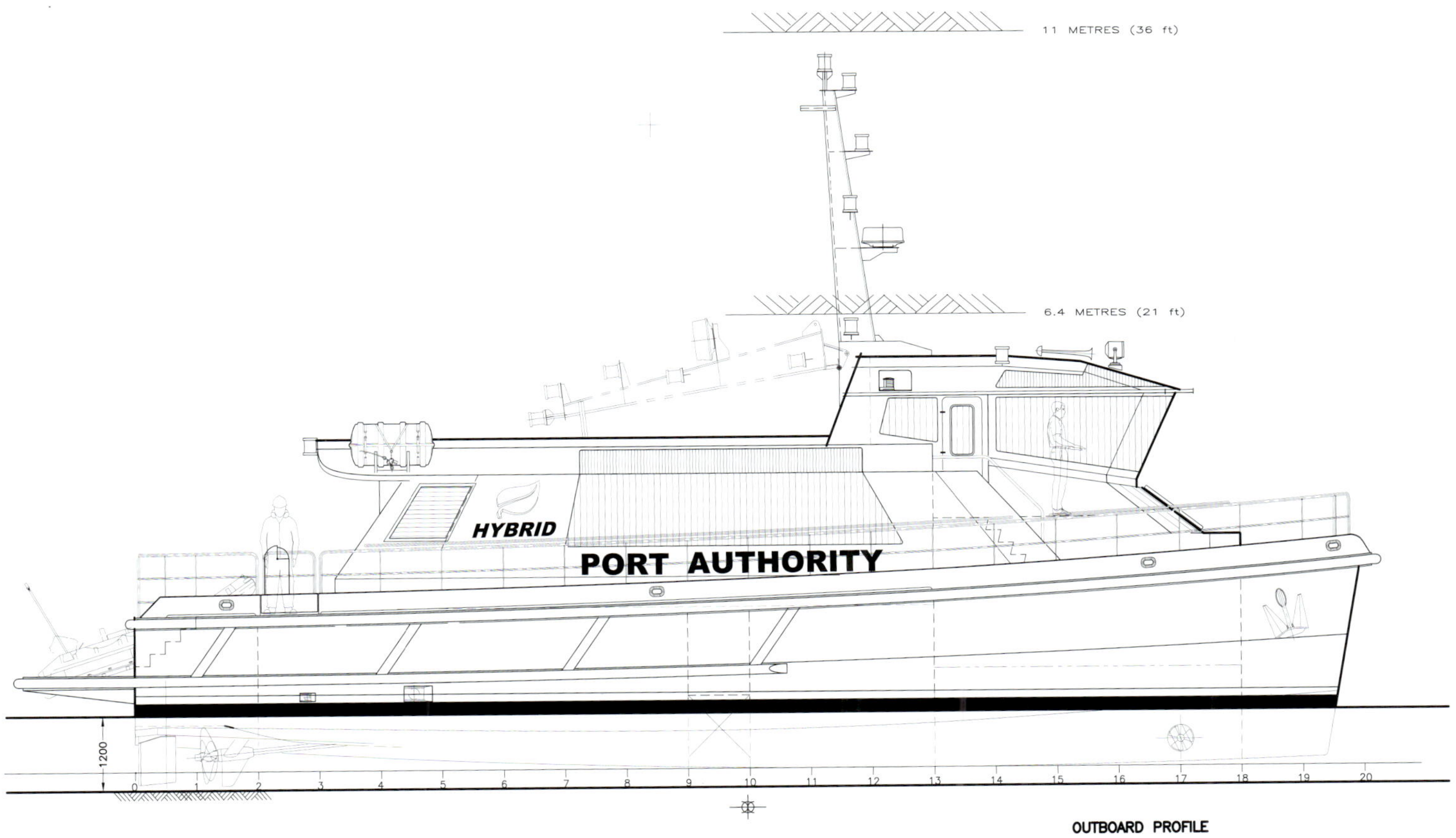

Top left and right: Port of Vancouver patrol boats *Takaya* and *Kla-Wichen*.

Bottom: Arrangement for proposed all-electric patrol boat for the Vancouver Fraser Port Authority.

COLUMBIA III

6

Coastal Mission Vessels

Of all the vessel types designed by Robert Allan, the several mission boats to his credit are perhaps the most unexpected, because the elder Allan was a well-declared atheist, and the more mellow Bob Allan was quite happy to describe himself as agnostic. It is unclear how Robert Allan made a connection with the Columbia Coast Mission organization, but he clearly established a close personal relationship with Canon Alan Greene, superintendent of the Columbia Coast Mission from 1936 to 1959, which resulted in the design of three fine wooden vessels for this demanding humanitarian service. It is notable that all three vessels are maintained in excellent condition still today (2022) and continue to ply the coastal waters of BC, albeit in quite different roles from their original intent.

Coastal missions were operated by several of the churches in the province. Their role was obviously to advance the aims of the church, but far more importantly these vessels brought much-needed health care, medical and dental services, and vital communications and news to many remote coastal communities when no other services existed. They served as chapels for weddings, christenings and funerals. They provided birthing services. The Columbia Coast Mission, operated by the Anglican Church of BC, was one of the largest of these missions. It operated on the BC coast from 1904 until the mid-1970s, when floatplanes made the much slower boats redundant.

Opposite: The lovely *Columbia III*, "The best cared for wooden boat on the BC coast."

The first mission boat purpose-designed by Robert Allan early in 1955 was the *Columbia*. This was in fact the third vessel in this service to bear that historic name, but *Columbia* as the registered name was carried over to each subsequent vessel. The services offered by the Columbia Coast Mission, and this boat in particular, are well documented in several books, including *The* Columbia *is Coming!*, by Doris Andersen. The colourful stories of this important era in BC coastal history are certainly worth reading. This handsome 68-foot vessel was a lifeline on the coast and is still celebrated in many communities for her distinguished service. *Columbia* is stoutly built of fir and yellow cedar planking over oak frames, as illustrated, and is powered by a 125 hp 6-cylinder Gardner diesel engine.

She served the BC coast until 1968, when she was retired and was then used in a variety of coastal services. At some stage she was renamed *Columbian* and thus lost her claim to the original name. In a state of some neglect, she was purchased in the late '90s and lovingly restored by Bill McKechnie of Victoria, BC, and renamed *Columbia III*. In 2005 she was purchased by Ross Campbell of Heriot Bay and adapted to the role she still provides today, as the flagship of the eco-tourism trade on the BC coast, sailing under the banner of Mothership Adventures. She is generally considered to be the best-kept wooden vessel on the coast and is beautifully maintained.

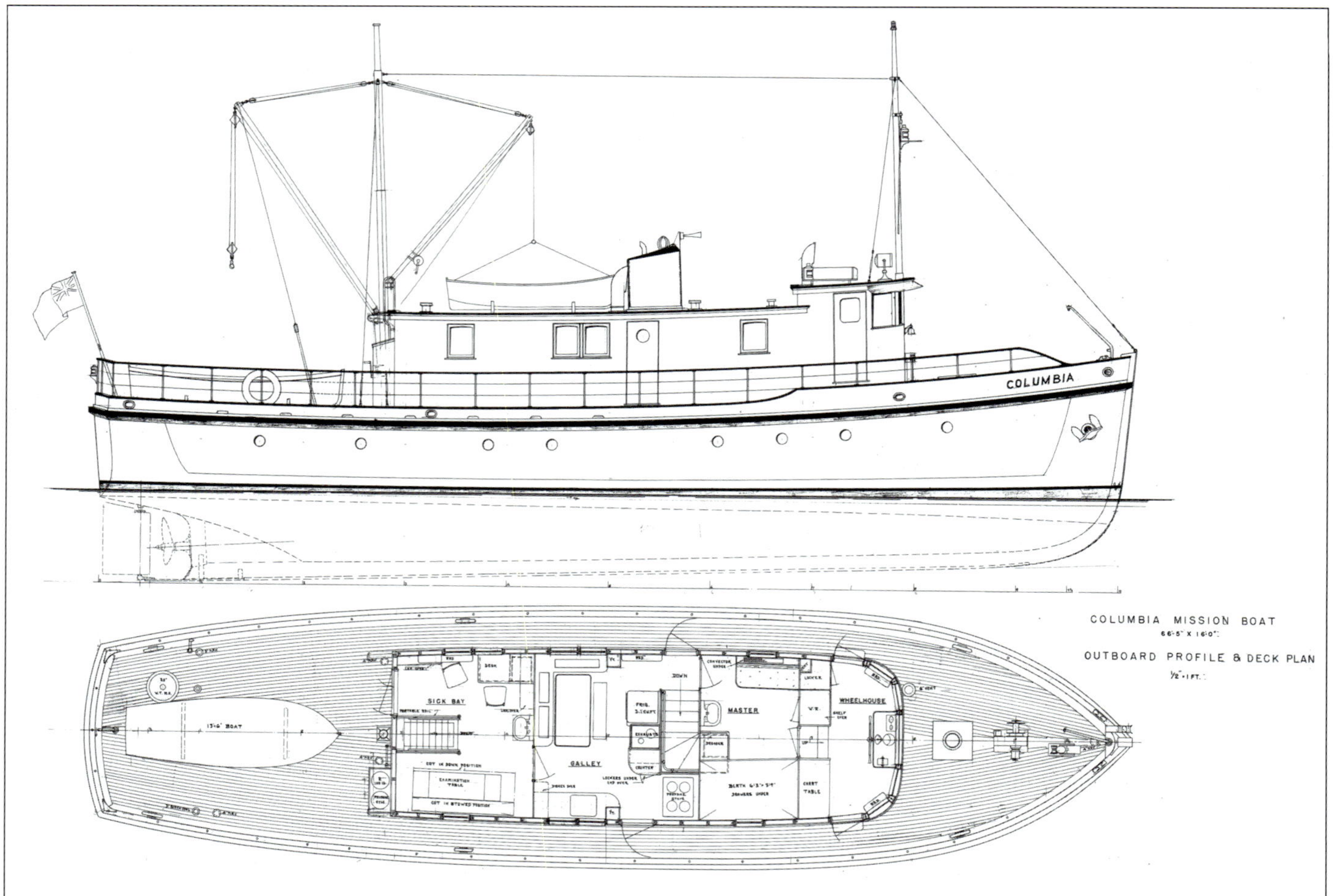

Above and opposite bottom: General Arrangement drawings for the *Columbia*, the first Allan-designed mission boat.

Opposite top: Midship Section drawing for *Columbia*.

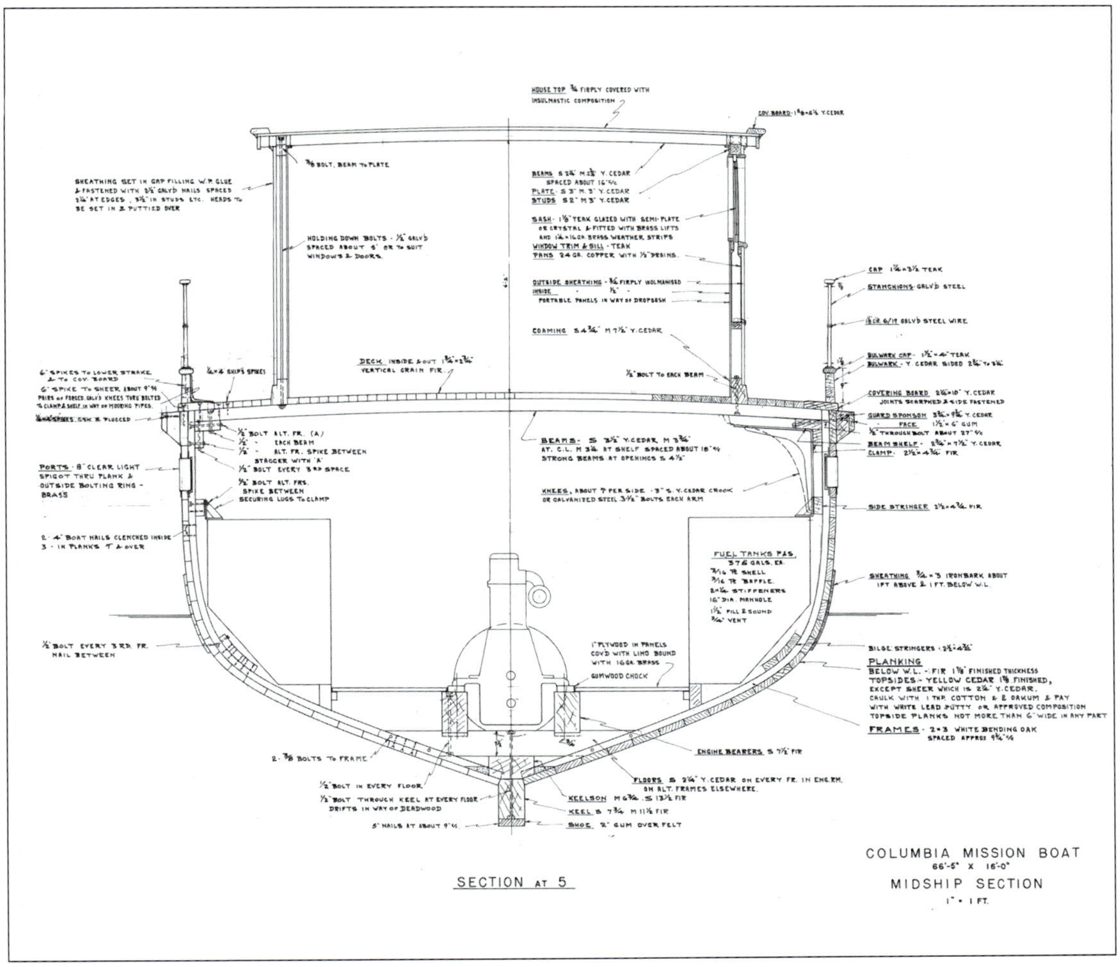
SECTION AT 5
COLUMBIA MISSION BOAT
66'-5" x 16'-0"
MIDSHIP SECTION
1" = 1 FT.

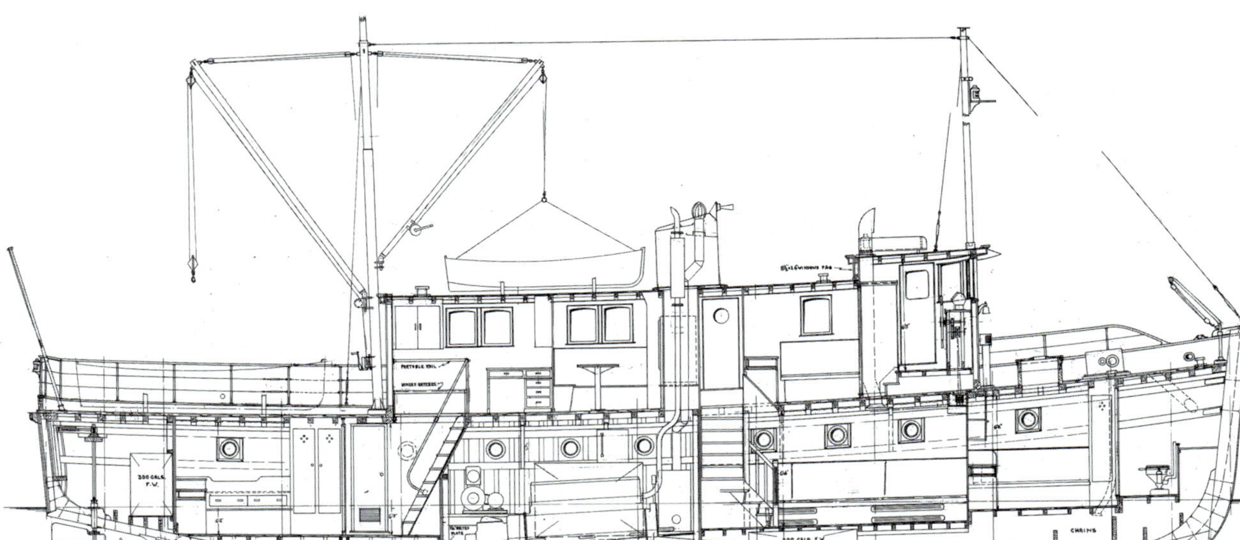

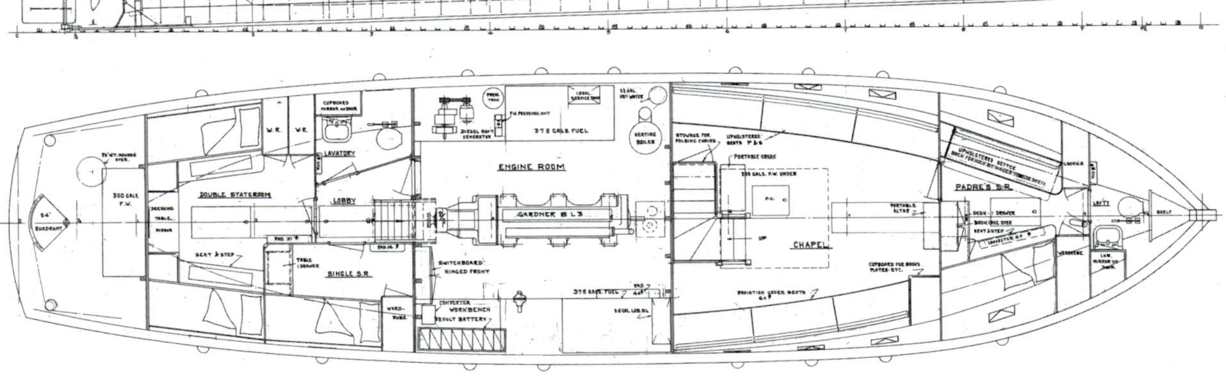
ENGINE ROOM
DOUBLE STATEROOM
LOBBY
SINGLE S.R.
GARDNER 8 L 3
CHAPEL
PADRE'S S.R.

A Mothership Adventure

In 2005 we were approached by Mothership Adventures to provide a drawing of *Columbia III* which could be used to embellish the T-shirts offered to their guests. We gladly provided that, but I was intrigued to find out more about this boat, for which I had only a vague understanding. Ultimately my wife, Enneke, and I signed up for a kayaking trip through the Broughton Archipelago with the Mothership team. It was a truly amazing experience, and we enthusiastically signed up for several subsequent trips, including one charter with our whole family. These trips enabled us to get a really up-close-and-personal view of the wondrous mid and north coasts of BC. We count these local trips as among the most favourite of our many adventures travelling around the world.

—R.G. Allan

The *Columbia* was followed in 1956 by the *John Antle II*, a 46-foot "little sister." Both boats were built by Star Shipyard (Mercer's) Ltd in New Westminster. Named in honour of the Reverend John Antle, the founder of the Columbia Coast Mission, the *John Antle II* replaced the original *John Antle*, described as a "little 25-ton motor yacht," built in France. The Reverend reportedly sailed that vessel across the Atlantic, through the Panama Canal and up to BC in about 1933. The *John Antle II* served the mission until

Opposite and left: Photos of the *Columbia III* in her eco-tourism role with Mothership Adventures.

Built in 1956, the 46-foot *John Antle II* was a "little sister" to the *Columbia*. As of 2021 she was well cared for in private hands as the *Chelsea II*.

General Arrangement drawings for *John Antle II*.

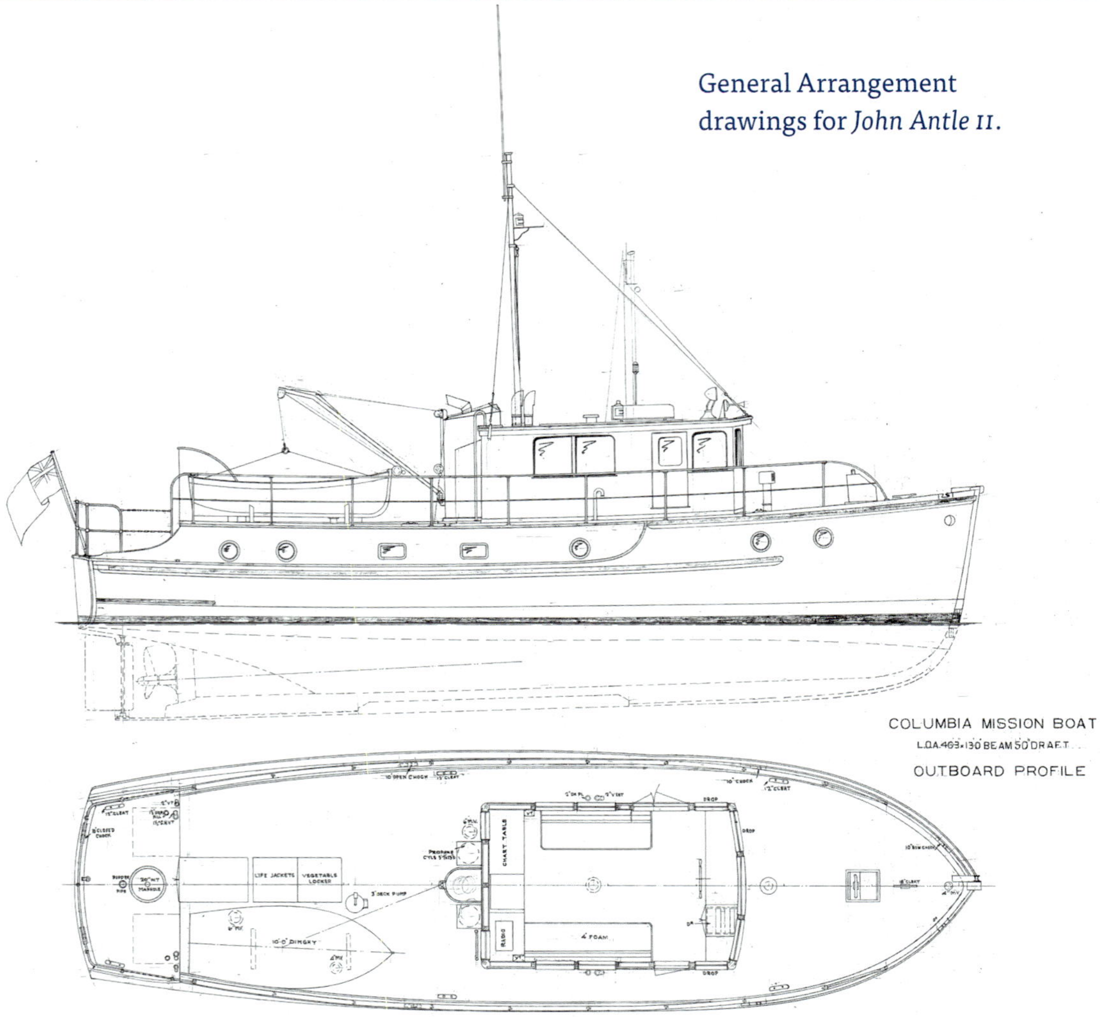

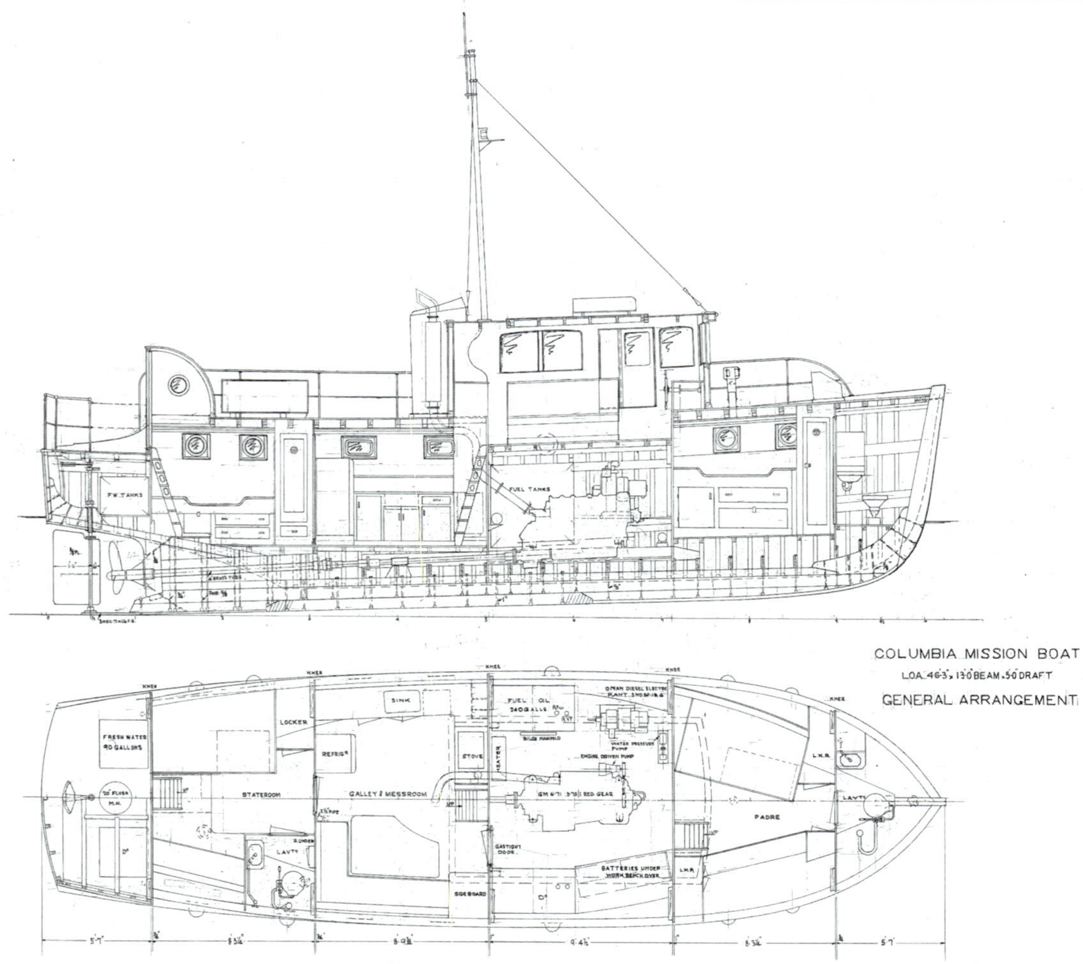

1964, after which she was sold into private hands and converted to a private yacht or charter vessel, in which guise she still plies the coastal waters of BC. She was designed very much in the style of the many coastal patrol boats from Robert Allan's drawing board in the same era.

The last of the trio of boats designed for the Columbia Coast Mission was the much smaller *Alan Greene*, named after Canon Alan Greene himself. The design is identified as the "Kingcome Inlet Launch," but it is unlikely that was the only area in which she worked. This 35-footer was built in 1959 by Bissett and Gilstein of North Vancouver and served the mission until 1967. It too was then sold into private hands and remains so today.

The (un)common thread in all these boats is the relationship between Robert Allan (Sr.) and Canon Alan Greene. The communications between the two men indicate that Allan, the Scotsman, was quite generous in reducing the fees (if any) he charged the mission for this design work. Unfortunately, no financial records from those days exist, but Canon Greene was quite a prolific cartoonist, and he expressed his gratitude in a series of cartoon messages throughout 1966 and 1967. Echoing the Scottish anthem "Scots wha hae (wi' Wallace bled)," one of Greene's cartoons shows Allan (replete with bagpipes!) arriving at the pearly gates to be greeted by St. Peter, who is accompanied by Alan Greene himself. As both men were getting on in years, there was an obvious kinship, and it seems that Canon Greene, while thanking Robert Allan for his generosity, was also perhaps trying to get the "Auld Scot" to accept the church before it was too late!

A multitude of blessings for the *Alan Greene* at the rather utilitarian-looking boat's well-attended commissioning. She was surely the most blessed of any boat on the BC coast given the nature of the assembled guests.

General Arrangement drawings for the *Alan Greene*.

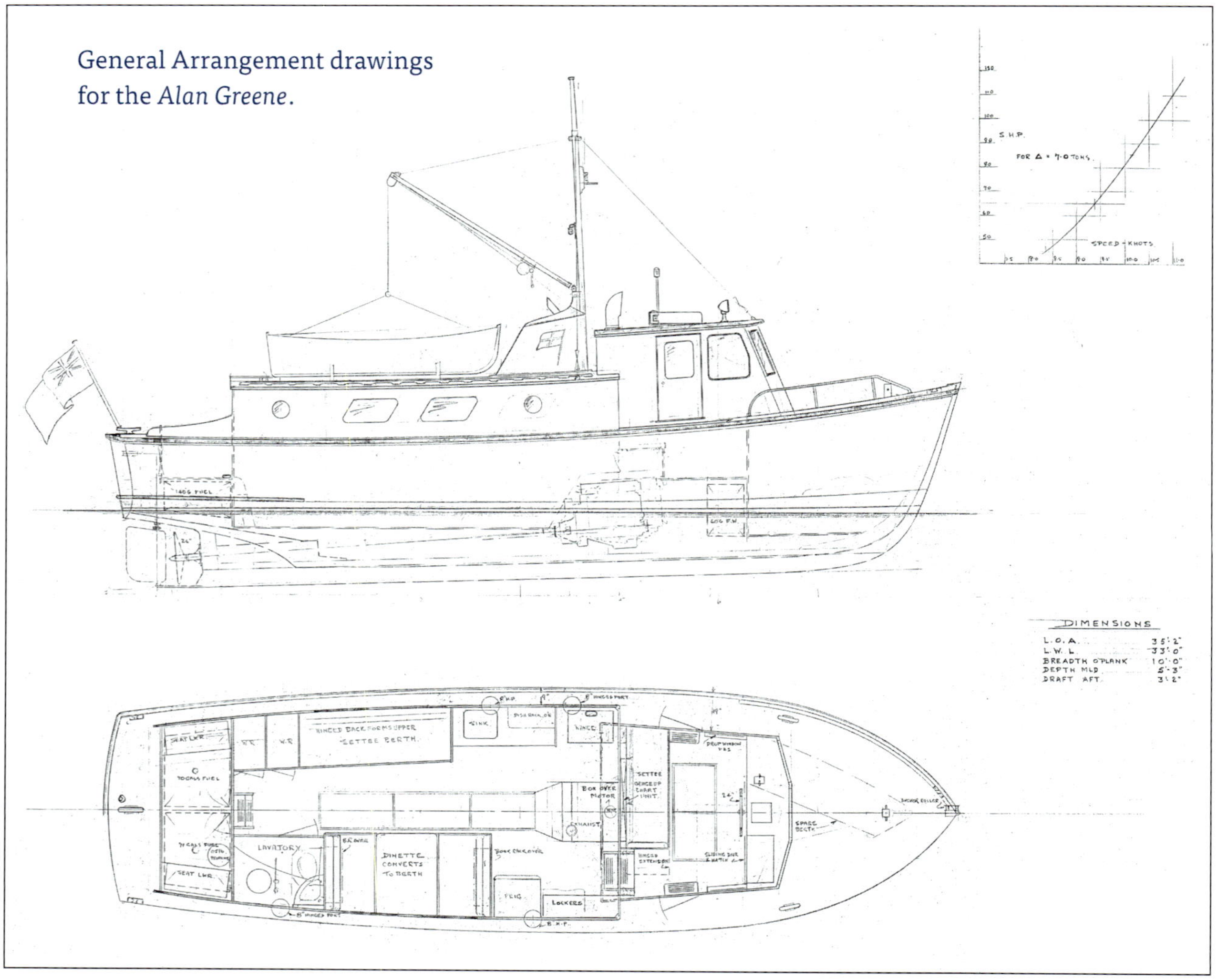

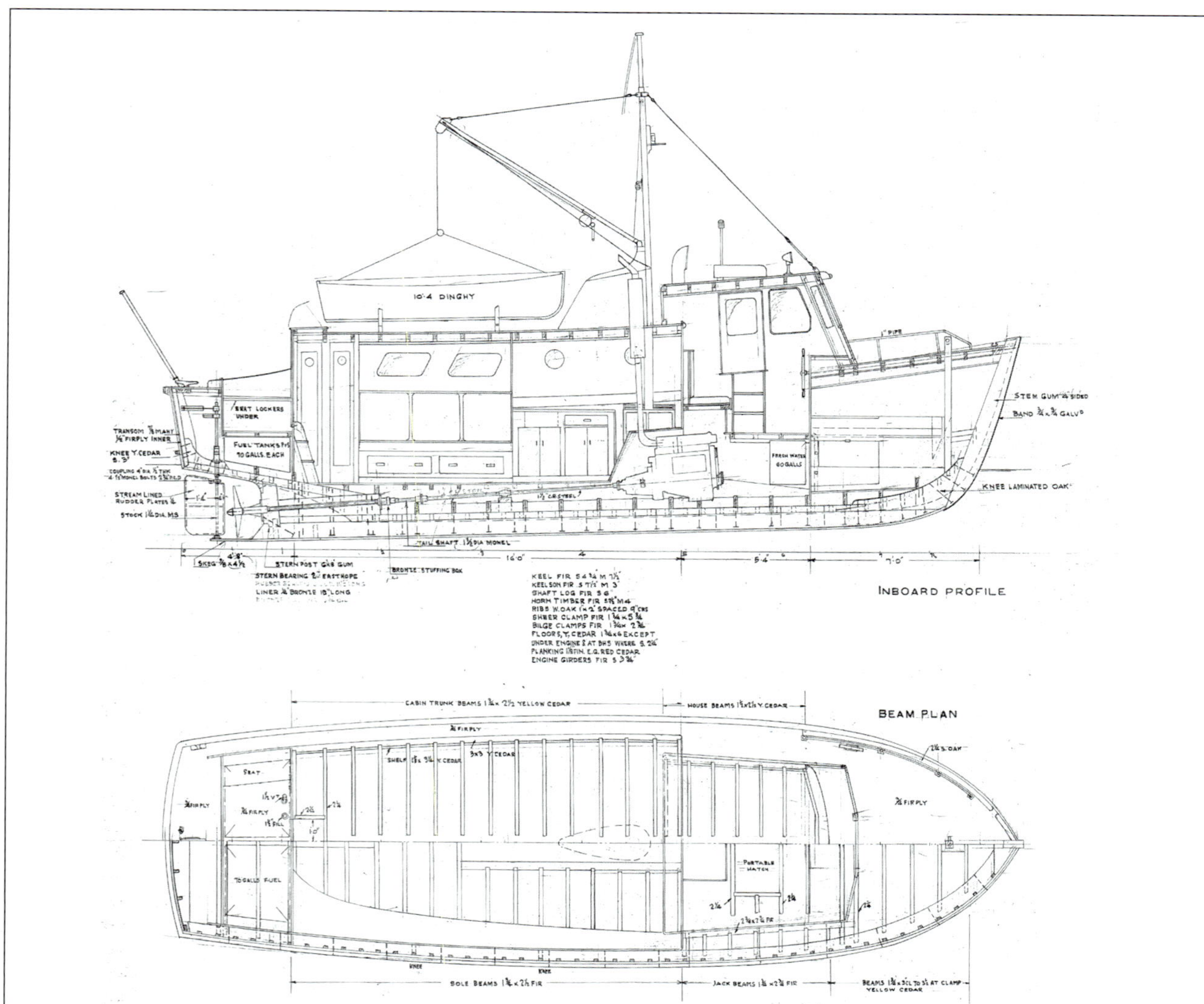

A 1966 cartoon penned by Canon Alan Greene (self-portrayed at bottom left) showing Robert Allan welcomed, "his dollars shed," at the pearly gates.

Two other Allan designs went into coastal mission service, though neither was designed initially for that purpose. The *Meander* (described in detail in Chapter 3) served from 1950 to 1963 as the *Melvin Swartout* for the United Church of Canada's West Coast mission. The *D.M. MacKay*, built in 1959 for the federal Department of Indian Affairs as the Bella Coola Agency launch, was sold to Coastal Missions in 1980 and renamed *Coastal Messenger*. She served that group faithfully until 1993 when replaced by a new steel namesake. The subsequent fate of the wooden boat has not been traceable.

Below: The *D.M. MacKay*, built in 1959 for the federal Department of Indian Affairs, operated for more than 20 years as the *Coastal Messenger*.

DRIVERS AND PASSENGERS
Must remain in their vehicles at all times during ferry crossing. Vehicle engines must be shut off and emergency brakes must be applied.
Thank you for your cooperation
EMPLOYEES ONLY

7

Ferries: More Than You Think

At first glance, the ferry market may seem to have been a rather elusive one for Robert Allan Ltd, especially for larger passenger/vehicle ferries, but within the archives there is a wide array of ferries, from the smallest of river-crossing vessels to large ocean-going ferries. It is a highly specialized market and an inconsistent one, as the demand is also quite sporadic, but they do represent very interesting design challenges.

Opposite: The *George Black*, see page 123.

From the earliest days of the business to the present there has been a constant, if rather thin, thread of ferry designs for all manner of crossings, either for just passengers, passengers and vehicles, or for trailers and railcars. The following examples are illustrative of the many of these hard-working craft from the Allan drawing boards.

Atrevida

The ferry story of Robert Allan begins in the earliest days of his independent practice, with the design of the *Atrevida* in 1931 for Capt. H.C.R. Davis of the Gabriola Ferry Company, to serve the Nanaimo–Gabriola Island run with freight, vehicles (5) and passengers (40). This was a rather unusual-looking vessel for a ferry, being more of a small cargo ship with incidental loading of vehicles rather than dedicated to vehicle service. One can imagine that vehicles may well have been rather rare on Gabriola in the early 1930s. She went on to serve the Texada Island–Westview run from 1955 to 1969.

Sturdily built of wood, she was constructed in 1931 by John Stokkeland, a builder whose name appears often in the quality local work of the 1930s. The vessel has long since been replaced in the ferry service, and she was later converted, initially into a "bakery vessel" and eventually to a

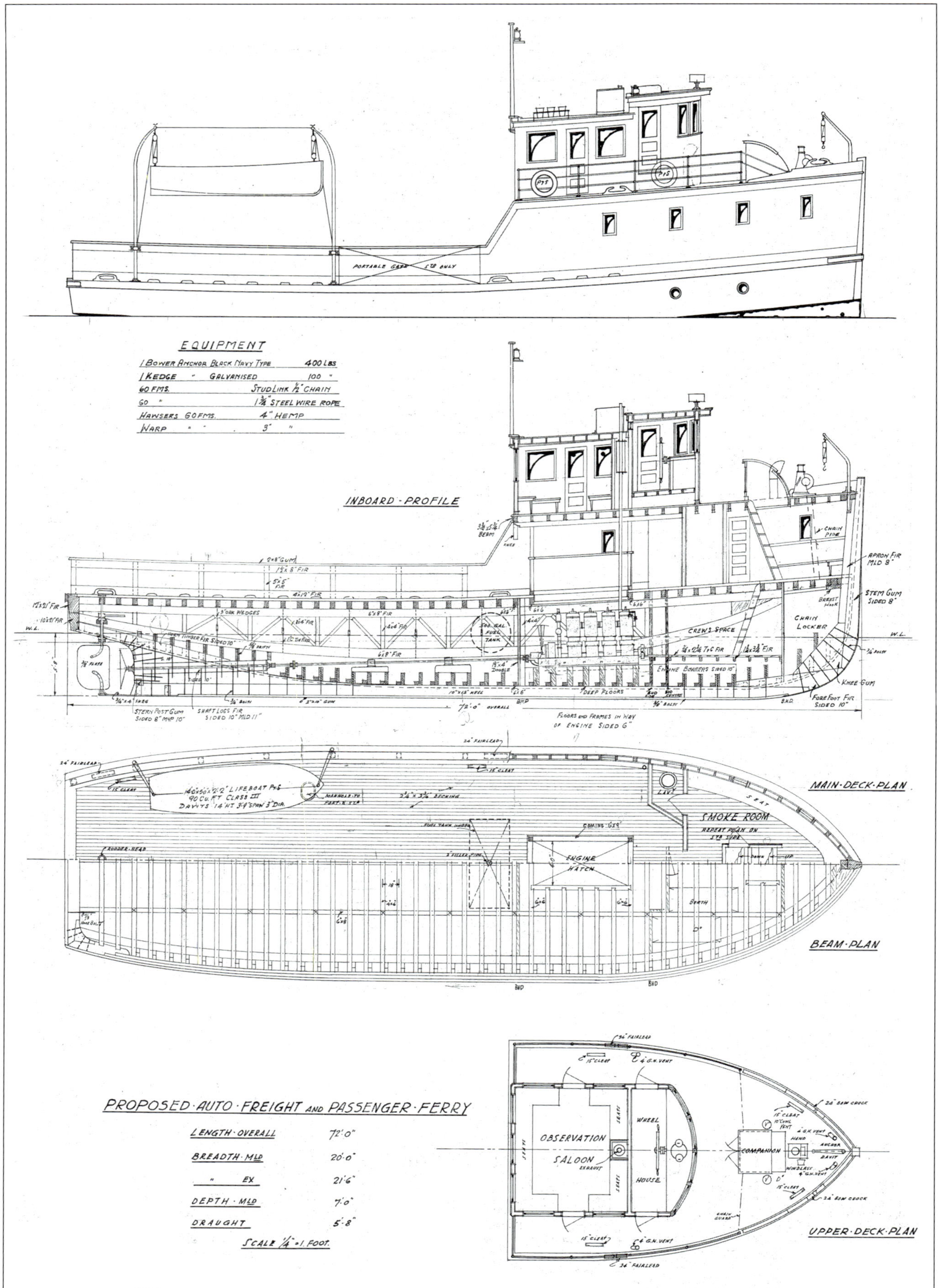
EQUIPMENT
1 Bower Anchor Black Navy Type 400 Lbs
1 Kedge " Galvanised 100 "
60 Fms. Stud Link 1/2" Chain
60 " 1 3/4" Steel Wire Rope
Hawsers 60 Fms. 4" Hemp
Warp " " 3" "
INBOARD · PROFILE
MAIN · DECK · PLAN
SMOKE ROOM
ENGINE HATCH
BEAM · PLAN
OBSERVATION SALOON
WHEEL HOUSE
COMPANION
UPPER · DECK · PLAN
PROPOSED · AUTO · FREIGHT and PASSENGER · FERRY
LENGTH · OVERALL 72'0"
BREADTH · MLD 20'0"
" EX 21'6"
DEPTH · MLD 7'0"
DRAUGHT 5'8"
SCALE 1/4" = 1 FOOT

Opposite: General Arrangement of the Gabriola ferry *Atrevida*.

Right: MV *Atrevida*

"tug-yacht," looking nothing like her original configuration. Remembered fondly by many as "The Cinnamon Bun Boat" out of Montague Harbour, she was ultimately and sadly wrecked in Maple Bay during a winter storm in 2018.

North Vancouver Ferry #5

In late 1939 or early 1940 Robert Allan obtained the contract to design the wistfully named *North Vancouver Ferry #5*. This 150-by-48-foot double-ended passenger-vehicle ferry was built by West Coast Salvage and Contracting Ltd and delivered in 1941 to replace the *North Vancouver Ferry #3* on the cross-harbour route. According to a *Harbour and Shipping* article in April 1941, the contract price was $113,958. Its primary function was to bring large numbers of wartime workers to North Shore shipyards from the City of Vancouver. It had a capacity of 40 cars and trucks and 600 passengers. Built of wood, this vessel served on that route until 1958, when the opening of the Second Narrows (later the Iron Workers Memorial) Bridge made her redundant at a relatively young age. In 1959 she was modified to become the landmark Seven Seas Seafood Restaurant located at the foot of Lonsdale Avenue until that business closed in 2001. The ship was then sadly broken up for scrap.

The set of drawings for this ferry are another wonderful example of the "art" of wooden ship design. The Inboard Profile drawing in particular shows the incredible detail provided by the designer. There is some interesting composite structure incorporated into this design, including a steel engine-room casing and two fore-and-aft steel trusses, partnered with two fore-and-aft wooden trusses, all designed to

cope with the expected hogging strains. There is 36 inches of solid wood, including keelsons, floors and fillers, under the engine bed to cope with the 30-ton weight of the engine. Also notable from this drawing is the fact that power for the shafts at both ends came from a single 8-cylinder 360 hp "double-ended ferry-type" Union diesel fitted with thrust bearings at both ends of the engine.

The Midship Section drawing is encyclopaedic in its content, effectively serving as the specification for the construction of the ship. The construction of the ferry was described in extensive detail in the October 1941 issue of the UK-based *Shipbuilder and Marine Engine-Builder* magazine and featured this fact-packed drawing, the original of which has sadly been lost.

Top: The *North Vancouver Ferry #5* transported wartime workers from the City of Vancouver to the North Shore shipyards; **left:** The *North Vancouver Ferry #5* was reborn as the iconic Seven Seas Seafood Restaurant in North Vancouver; **opposite top:** Outboard Profile and Passenger Deck Arrangement for *North Vancouver Ferry #5*; **opposite bottom:** The incredibly detailed Inboard Profile and Main Deck Structural Plan for *North Vancouver Ferry #5*.

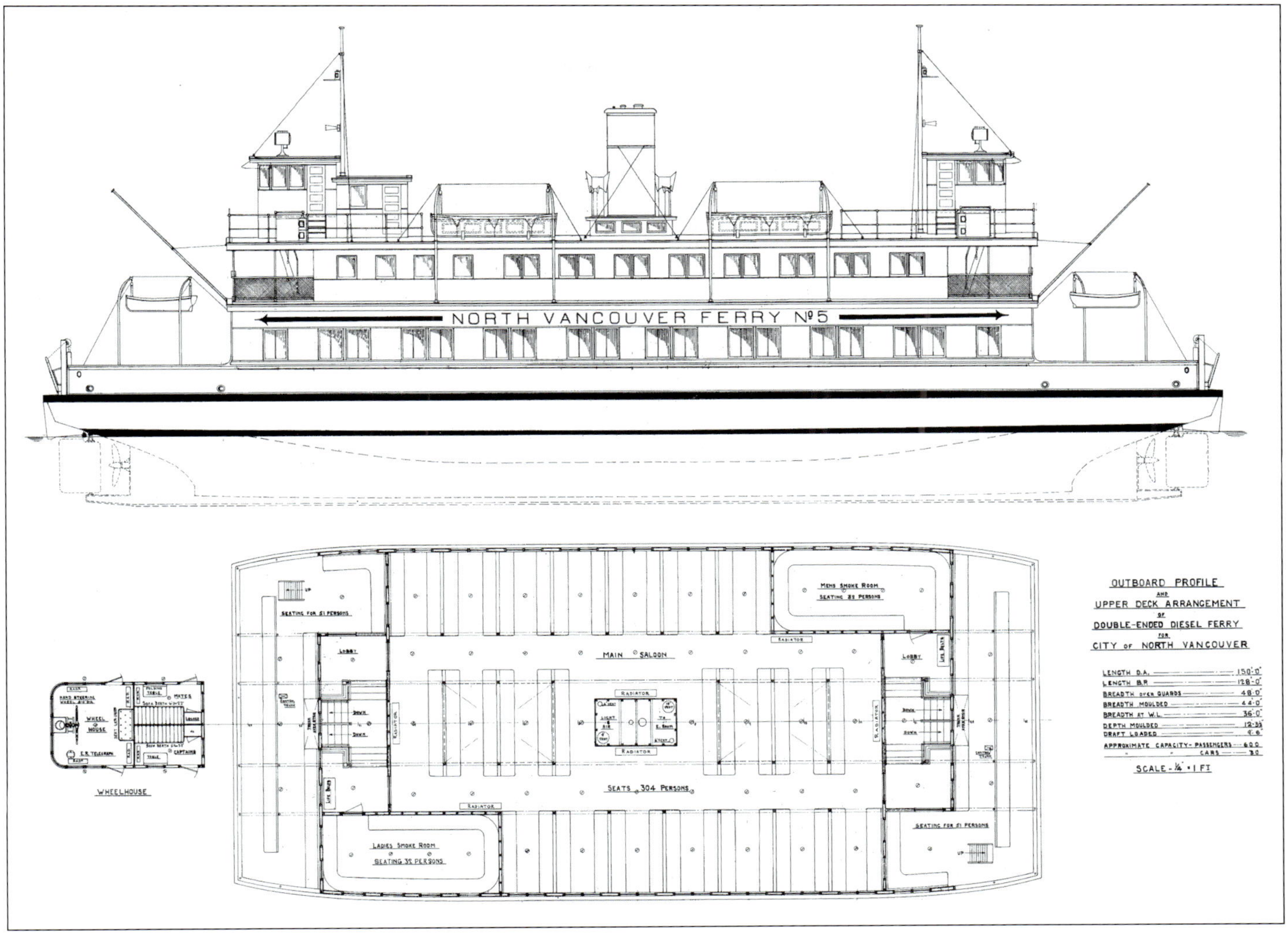
NORTH VANCOUVER FERRY No 5
MAIN SALOON
SEATS 304 PERSONS
MENS SMOKE ROOM
SEATING 31 PERSONS
LADIES SMOKE ROOM
SEATING 31 PERSONS
SEATING FOR 51 PERSONS
WHEELHOUSE
OUTBOARD PROFILE
AND
UPPER DECK ARRANGEMENT
OF
DOUBLE-ENDED DIESEL FERRY
FOR
CITY OF NORTH VANCOUVER
LENGTH O.A. 150'-0"
LENGTH B.P. 128'-0"
BREADTH OVER GUARDS 48'-0"
BREADTH MOULDED 44'-0"
BREADTH AT W.L. 36'-0"
DEPTH MOULDED 12'-3½"
DRAFT LOADED 9'-6"
APPROXIMATE CAPACITY - PASSENGERS 600
CARS 30
SCALE - ¼" = 1 FT

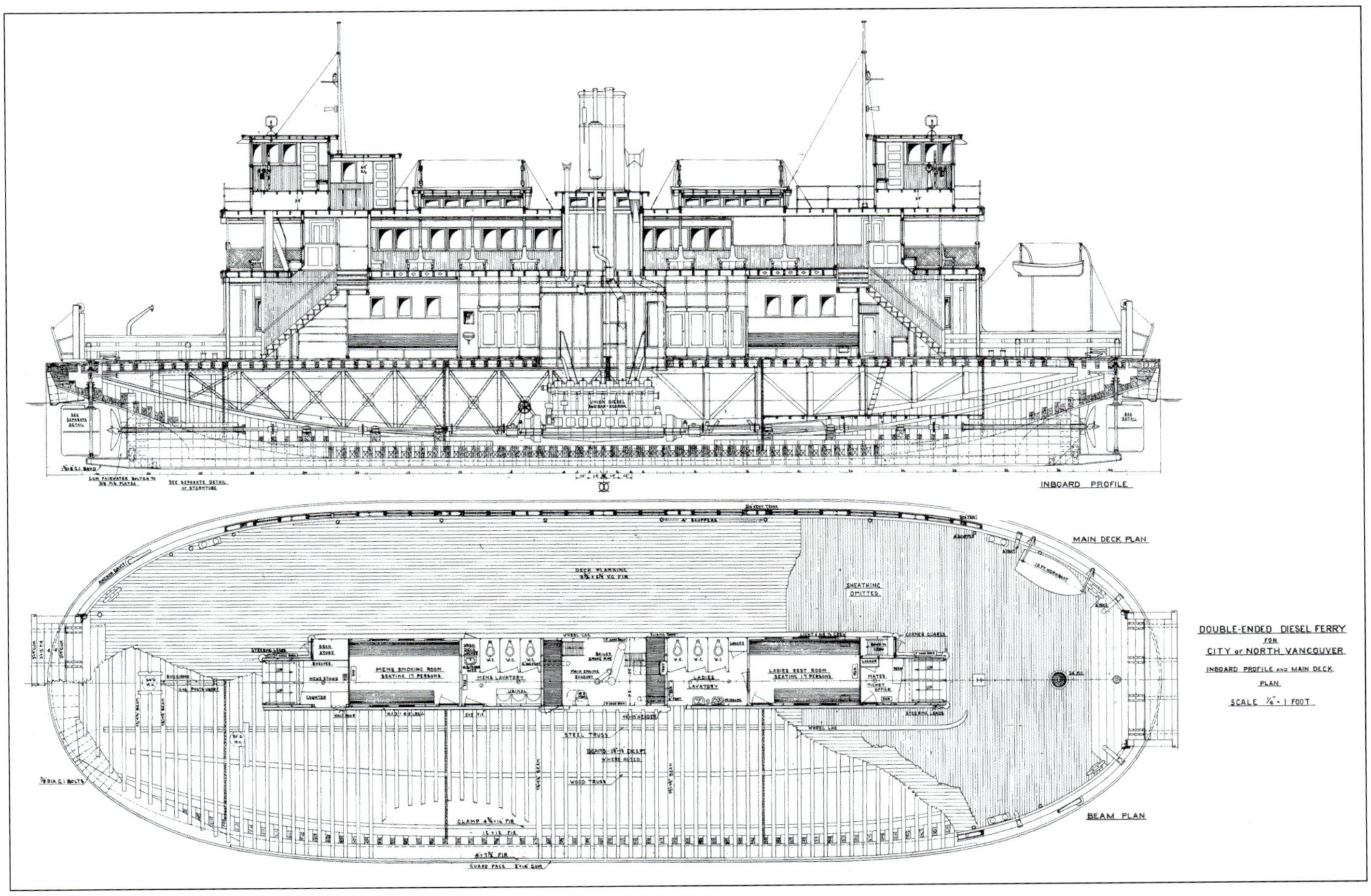
INBOARD PROFILE
MAIN DECK PLAN
MENS SMOKING ROOM
LADIES REST ROOM
BEAM PLAN
DOUBLE-ENDED DIESEL FERRY
FOR
CITY OF NORTH VANCOUVER
INBOARD PROFILE AND MAIN DECK
PLAN
SCALE ¼" = 1 FOOT

THE "NORTH VANCOUVER FERRY NO. 5."

A Double-ended Ferry of Wooden Construction, with Diesel Propelling Machinery.

THE way in which a municipal or state authority can ensure that public money is economically dispensed is not necessarily always that of cutting down utility services or similar encroachments on efficiency in the name of careful spending. An alternative method, and one

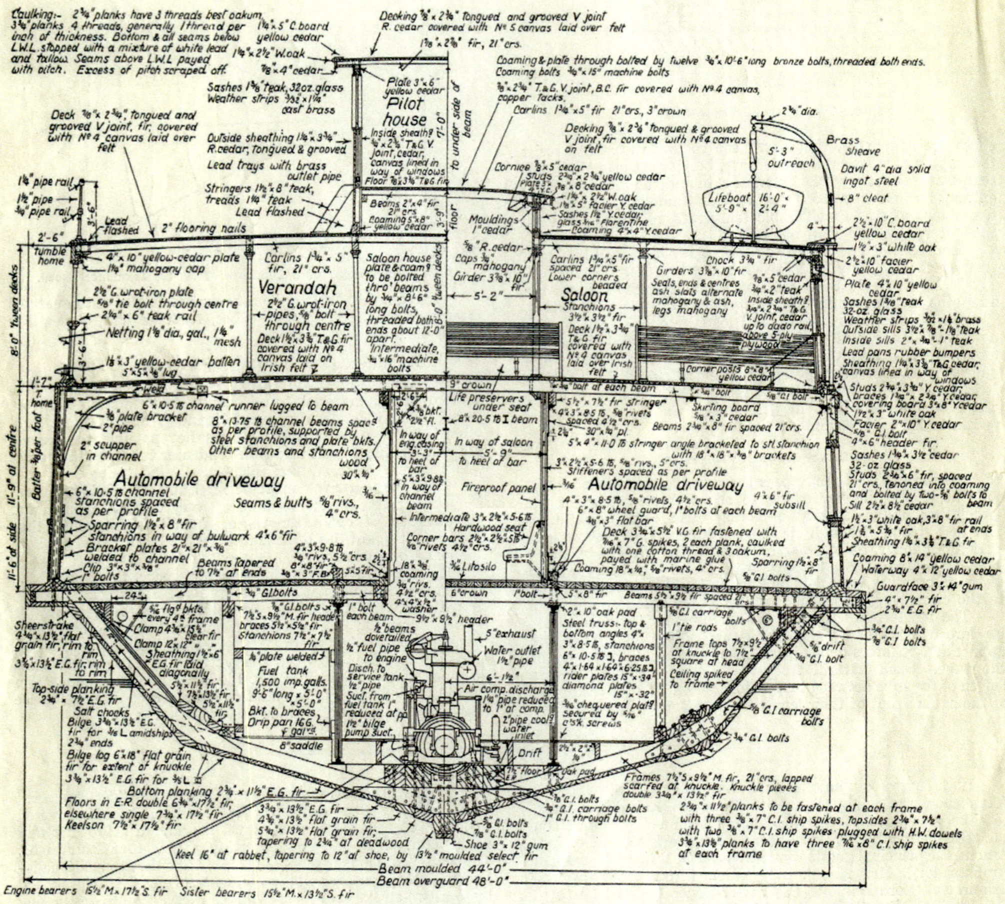

Fig. 1.—Constructional Sections of the "North Vancouver Ferry No. 5."

which apparently is that adopted by the guardians of the money of North Vancouver ratepayers, is by preserving a progressive outlook, to eliminate existing waste accruing from out-of-date equipment and methods.

Backed by the active support of several local organisations, the North Vancouver authorities, during the early part of 1941, examined the possibilities of replacing one of the older units of the ferry service across Burrard Inlet with a modern vessel, designed to meet present-day requirements. The examination into the financial aspect of the proposal brought to light some revealing statistics, which may serve as matter for a useful lesson to many other authorities who rely on equipment which has been left behind in the advancement of industrial progress.

The older vessel in question—*North Vancouver Ferry No. 3*—which was built some 30 years ago, is fitted with steam-driven propelling machinery, while a more recent unit of the service—*North Vancouver Ferry No. 4*—is Diesel-driven. The comparison of running costs of these two vessels showed that the operating cost of *No. 3* exceeded that of *No. 4* by some 23,400 dollars per annum. It was thus seen that, in five years' employment, a more modern replacement of *No. 3* would earn more than the 100,000 dollars which it was estimated would be the approximate initial cost of a new vessel.

Once having established the fact that a new ferry would be

The Midship Section drawing of the *North Vancouver Ferry #5* showing virtually every construction detail.

Jacob Henkel

In 1948 Robert Allan received a significant commission from the BC Department of Highways for a new ferry for Francois Lake, west of Prince George. Steel-hulled and with a wooden deckhouse, the *Jacob Henkel* was a typical "flat-top" ferry, measuring 114 by 32 feet, with capacity for about 15 cars. She was built by Yarrows Ltd and commissioned in 1948. The design package of 21 drawings is comprehensive and, in many areas, of sufficient detail to be considered as construction working drawings. It is assumed that the ferry was built fully in Victoria then sectioned for transport to the depths of the BC Interior for re-assembly on the lakeshore.

Jacob Henkel, built in 1948 for the BC Department of Highways.

General Arrangement and Structural Section drawings for the *Jacob Henkel*.

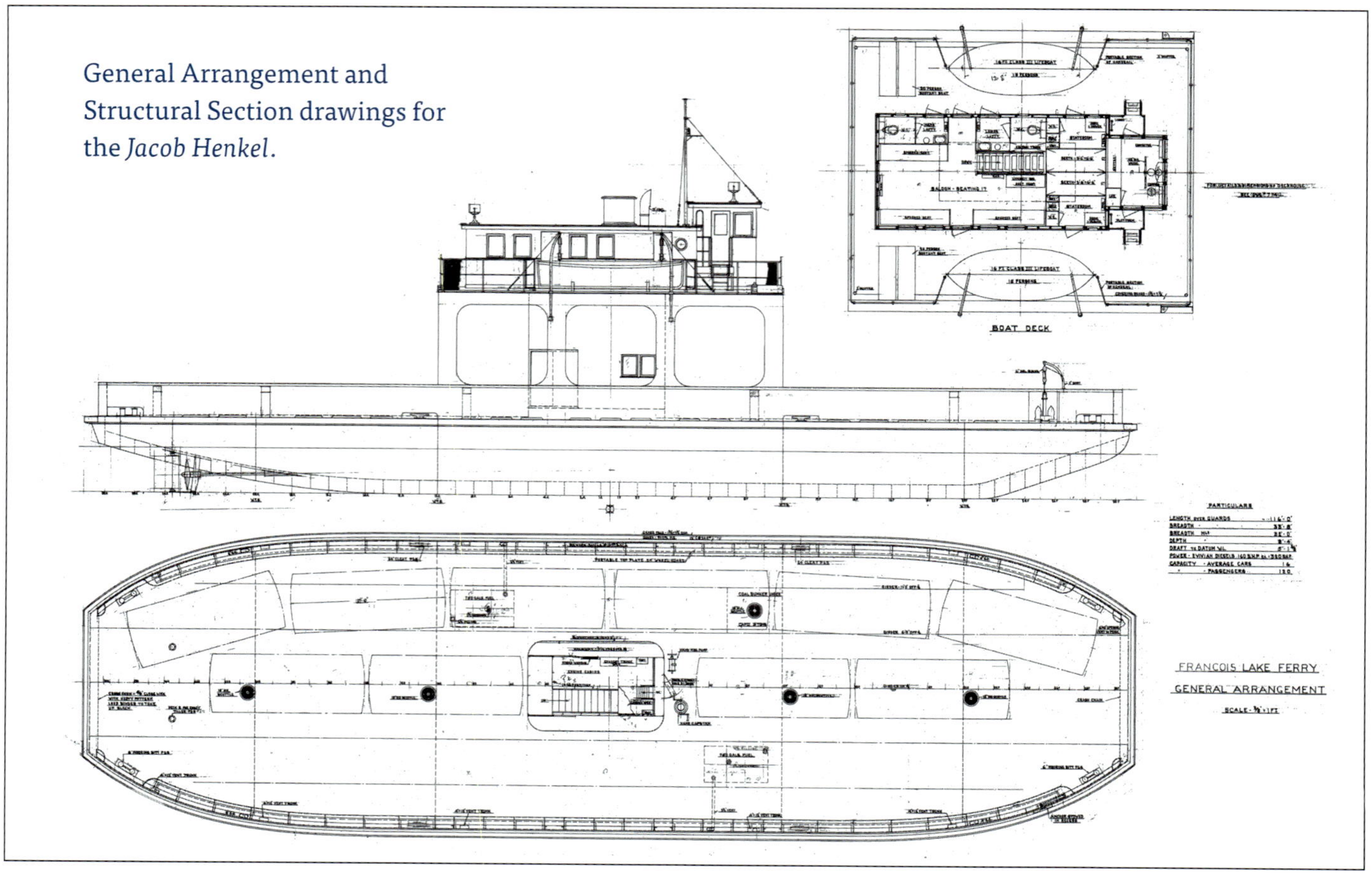

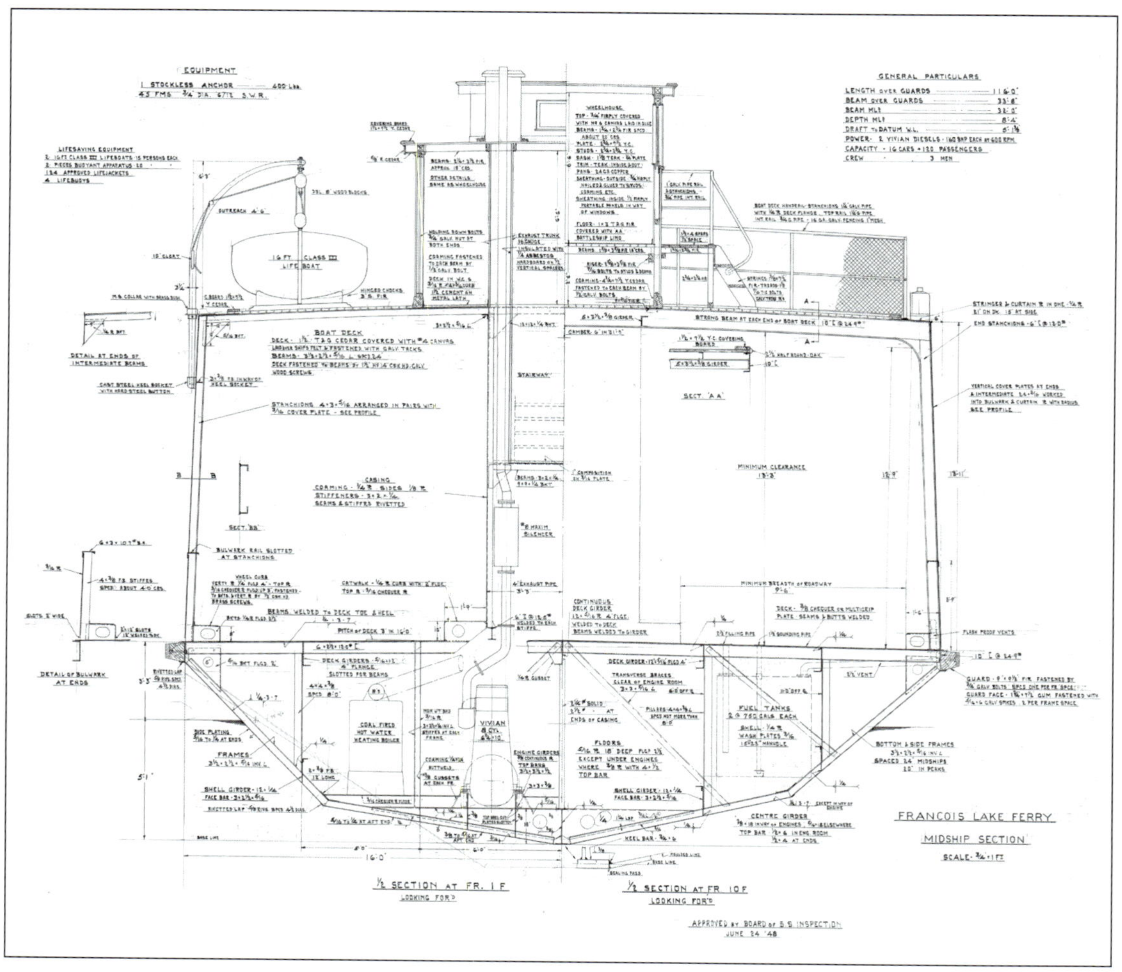

Delta Princess

On the heels of the Francois Lake ferry design, Robert Allan received another commission in 1948 from the Department of Highways, this time for a vessel to serve the Lower Mainland, crossing the south arm of the Fraser River between Ladner and Woodward Landing, in Richmond. Built by Yarrows Ltd in 1949 and christened the *Delta Princess*, she was made redundant when the Deas Island Tunnel (now the George Massey Tunnel) opened in 1959. She was then purchased by the Gulf Island Ferry Co. and moved to Gulf Island routes until the assets of that company were purchased by BC Ferries in 1962. The ship was then renamed the *Saltspring Queen* (and later dethroned to the *Saltspring Princess*!). She was finally retired in 1996.

A relatively straightforward double-ended single-deck ship, she had little in the way of amenities for passengers save a small sitting room and toilets, as the river crossing took just a few minutes.

Delta Princess crossing the Fraser gently.

Particulars of this all-steel ferry were as follows:

Length Overall	136' 7"
Beam, Moulded	36' 0"
Depth, Moulded	8' 0"
Power	480 bhp
Vehicle Capacity	36 cars
Passenger Capacity	187

The design package for this ferry is also very detailed, comprising about 35 drawings, the majority of which are ink-on-linen and thus form ideal archival material. It appears that these were intended to be virtually production-level drawings, as illustrated in the examples shown.

The projects, and presumably also the design fees, were finally getting larger!

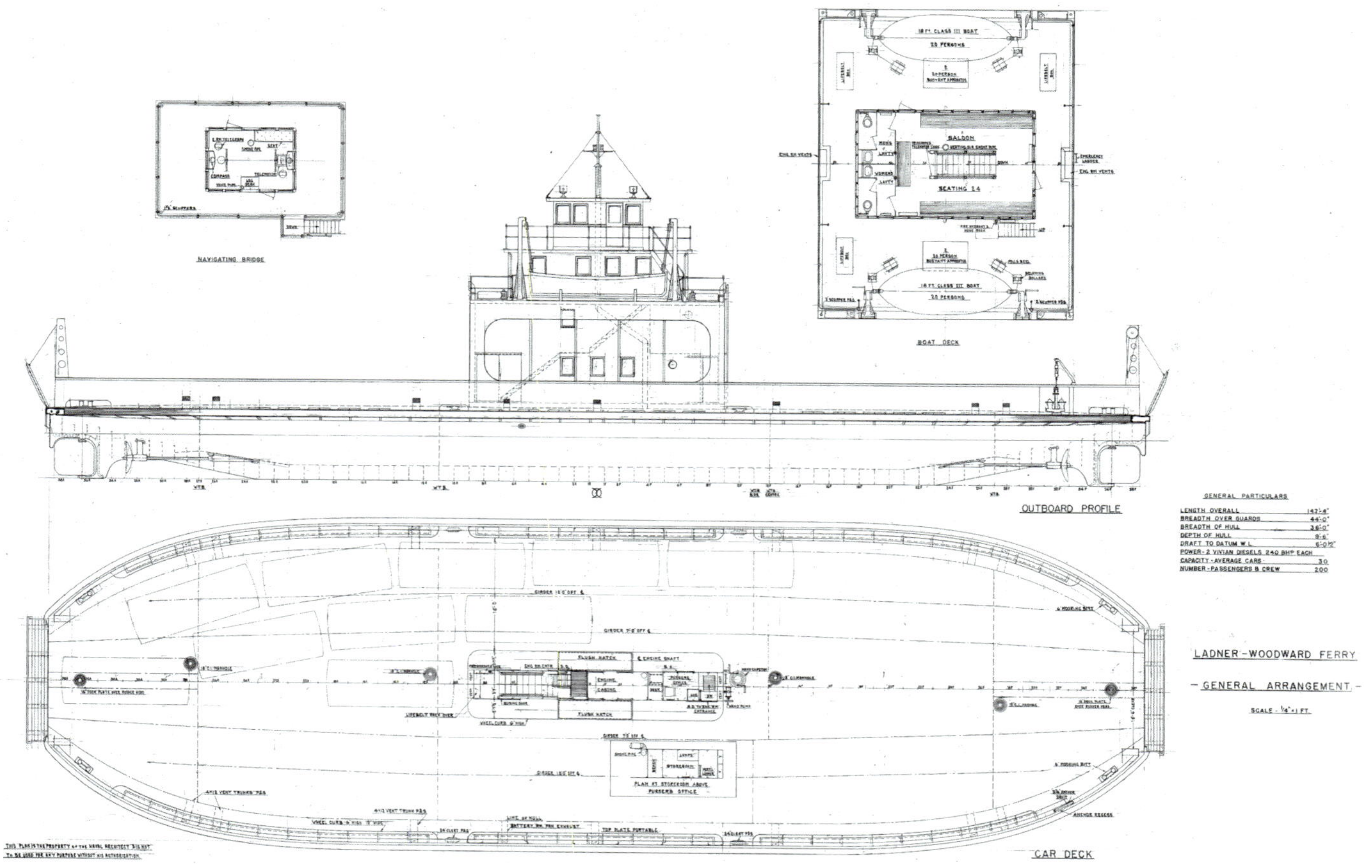

General Arrangement drawing for *Delta Princess* (and for the *Lloyd Jones*).

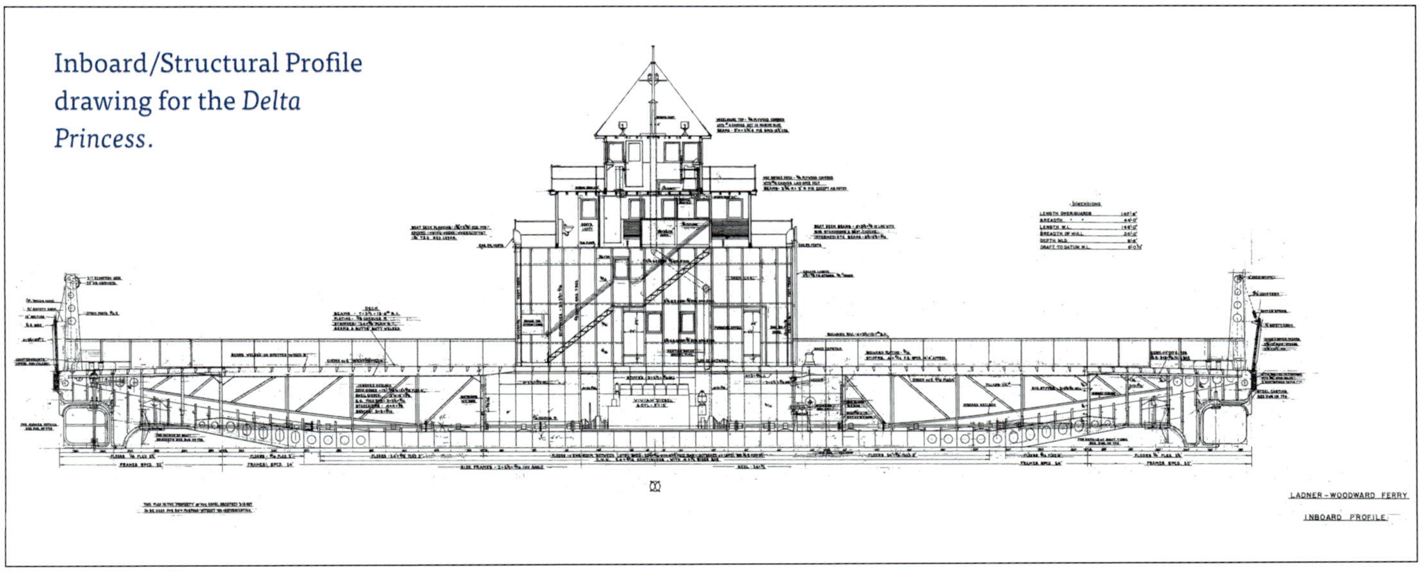

Inboard/Structural Profile drawing for the *Delta Princess.*

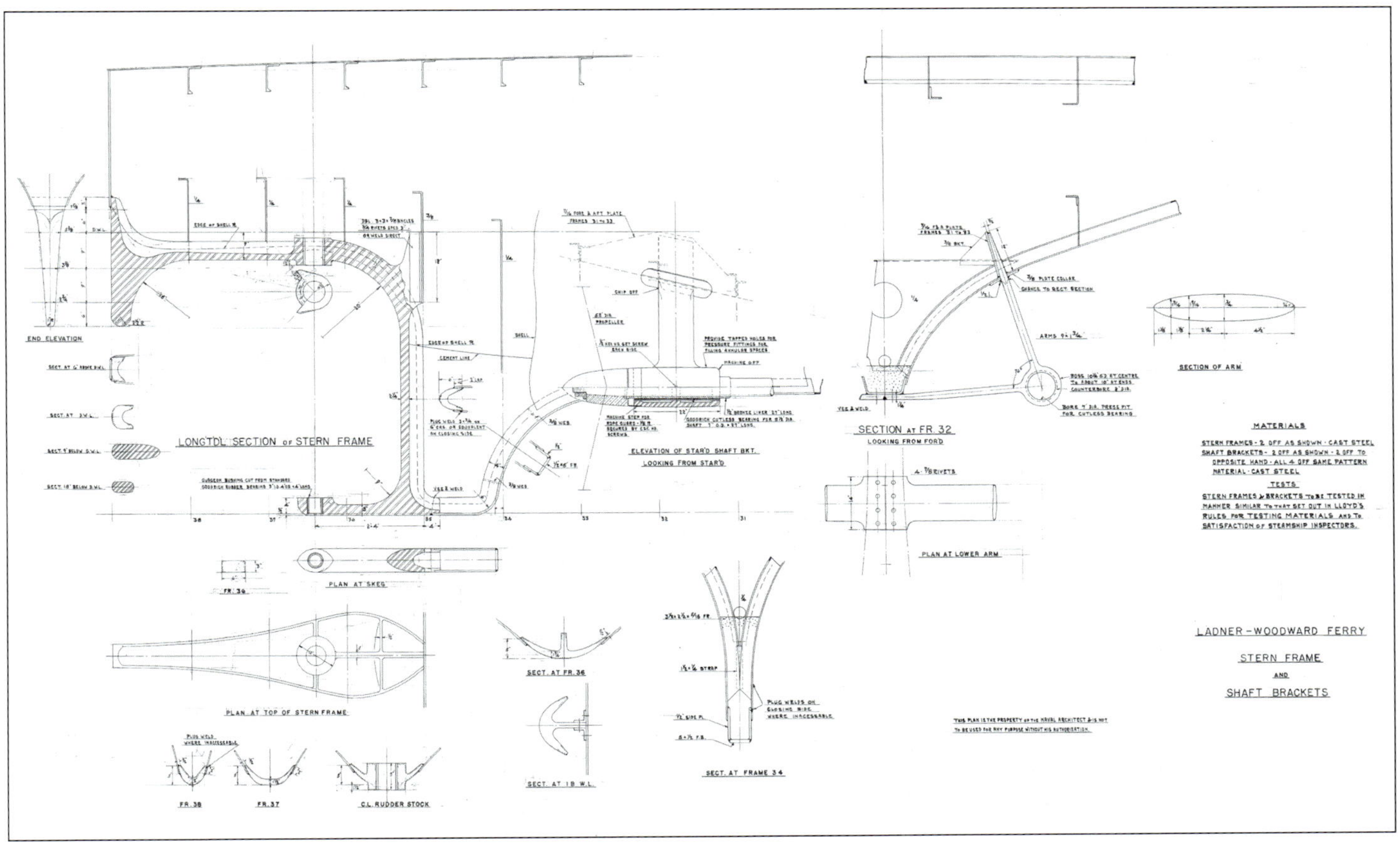

Highly detailed drawing of stern frame and shaft brackets for the *Delta Princess.*

Lloyd Jones

In 1950 the BC Department of Highways used the same design as the *Delta Princess* to call for tenders for a ferry to operate on Okanagan Lake. The *Lloyd Jones* was built by Victoria Machinery Depot in that same year and served the city of Kelowna until a floating bridge was installed across the lake in 1958. The *Lloyd Jones* was still a relatively new vessel when made redundant, and in 1962 it was purchased by BC Ferries, then cut into sections and brought down to the coast, where it was reconstructed to become the *Bowen Queen*, and later the *Vesuvius Queen*. In 1998, having put in 68 years of continuous service in BC, it was sold to the Dominican Republic, where it may well still be operating.

The *Bowen Queen*, formerly the *Lloyd Jones*, in her BC Ferries livery.

West Creston Ferry

One of the more unusual designs from the early Robert Allan drawing board is the *West Creston Ferry*, also designed in 1948 for the BC Department of Highways. (Clearly 1948 was a "ferry good" year!) At 90 feet long with a 28-foot beam, it was basically a steel-hulled barge with a simple wooden deckhouse, built by West Coast Shipbuilders in 1949. This was a cable ferry, with fixed lines overhead to hold position across the current of the Kootenay River and using simple paddle-wheels for motive power, as illustrated.

Sidewheeler/cable ferry *West Creston Ferry* on the Kootenay River during the flood of 1948.

General Arrangement for the *West Creston Ferry*.

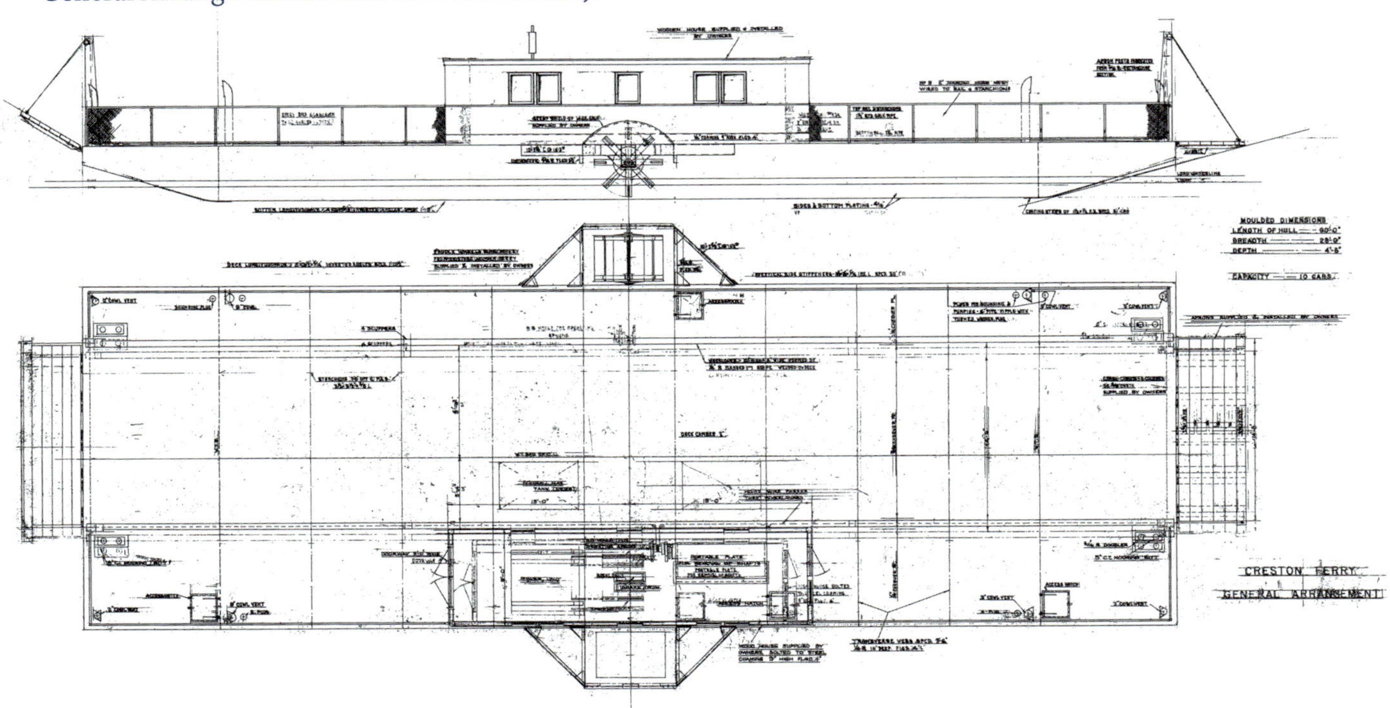

Peace River Pontoon Ferries

In 1950 the first in what became an interesting series of ferries for the Peace River in northwestern Alberta was commissioned. All these ferries essentially served the hamlet of La Crete, and thus the records frequently refer to them as the "La Crete Ferry," but they had various names throughout their existence. The first *Dunvegan Ferry* was a "pontoon ferry" consisting of a pair of boat-shaped hulls supporting a transverse bridge that constituted the vehicle deck. This "barge" was then pushed by a 38-foot shallow-draft tugboat that settled into the notch between the two pontoons. The tugboat portion was built by Allied Shipbuilders Ltd, and the barge was built by Western Bridge & Steel Fabricators, with the assembled units going to work in 1951. The ferry operated at Dunvegan until 1961, when, having been replaced by a bridge, it was moved to Tompkins Landing, where it operated until replaced in 1987. That "vessel" is now rather landlocked as part of the La Crete Mennonite Heritage Museum, as shown.

Top: The 1950-vintage *Dunvegan Ferry* on display at the La Crete Mennonite Heritage Museum; **bottom:** the 1951 *Dunvegan Ferry* after it was moved to Tompkins Landing. This image clearly illustrates the need for an elevated wheelhouse on the tug!

In 1962 a very similar (or perhaps identical) ferry was built, known initially as the *Fort Vermillion Ferry*, and operated until a bridge was built there in 1974. The ferry was then moved to the Shaftesbury crossing (just south of the town of Peace River), where it is currently still operating, renamed the *Shaftesbury Ferry*. It is also obvious from available images that the later loads for these ferries had grown to sizes and weights certainly not envisaged in the 1950s or 1960s.

In 1987 Robert Allan Ltd was retained by the Alberta Department of Transportation to design a replacement for the 1962 version of the *La Crete Ferry*. In recognition of the changing nature of the loads using the local highways, this new vessel had a significantly higher payload than its predecessors, consisting of two "B-Train" semi-trailer units with a combined weight of 95 tonnes. This ferry was designed as a self-propelled vessel but configured in essentially the same manner as the tug-barge units in order to best cope with the strong river currents. However, in lieu of an independent pusher tug for motive power it was propelled by two deck-mounted, diesel-driven Z-drive propulsion units. This vessel thus has the distinction of being the only vessel in the Allan portfolio that is effectively as wide (32.0 metres) as it is long (33.4 metres)

All these Peace River ferries were uniquely configured to cope with the swift currents in which they operate, essentially stemming the current when landing in order to align the cross-deck with the roadway landings.

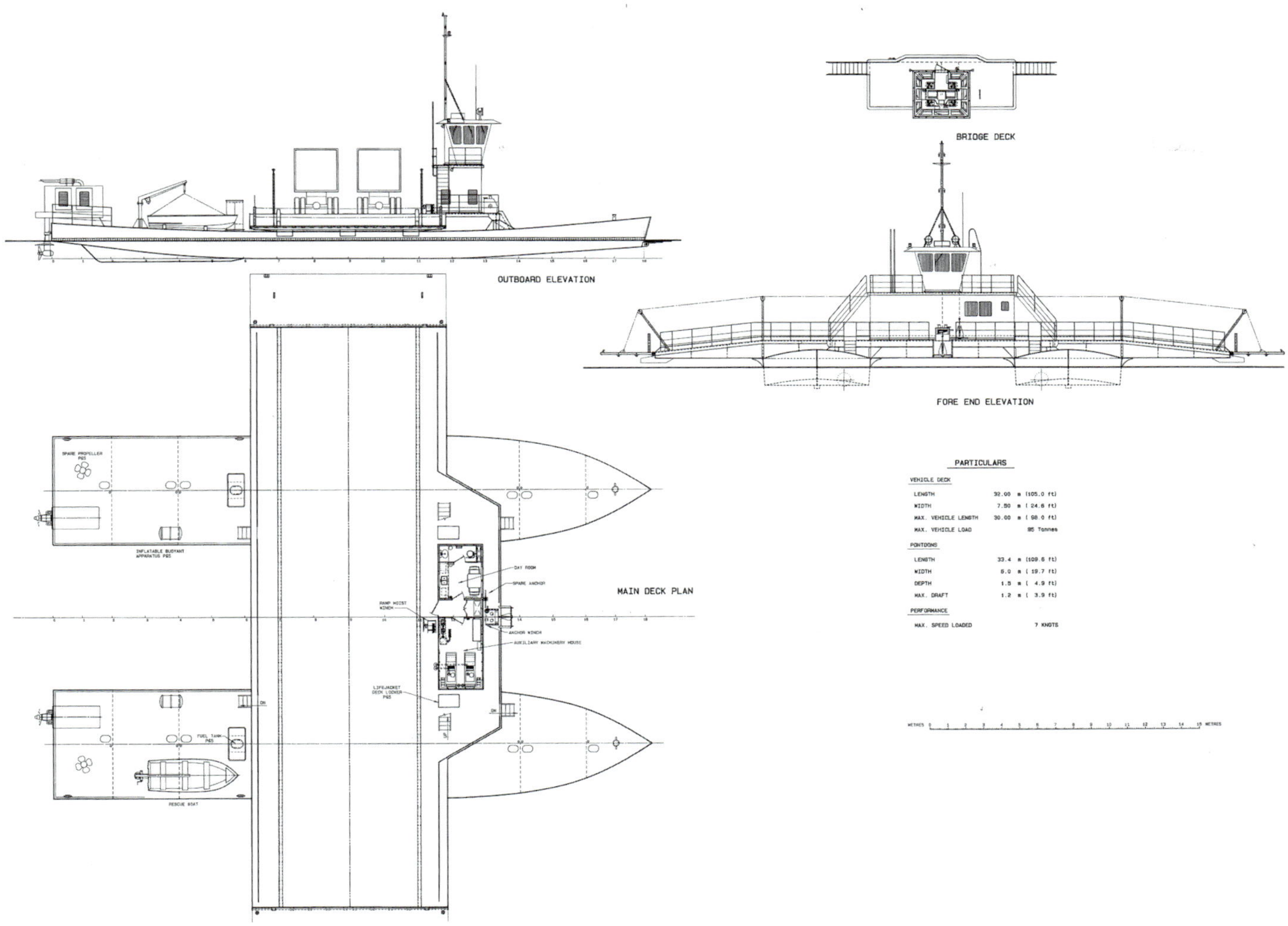

The 1987 version of the *La Crete Ferry*, a self-propelled Z-drive-powered pontoon ferry replacing the previous tug-barge combinations.

La Crete Ferry traversing the Peace River.

New Brunswick Ferries

In 1952 Robert Allan was awarded a contract by the Government of New Brunswick to design a new ferry for the Deer Island–Letete route in the Bay of Fundy. At 92 feet by 34 feet 4 ½ inches and powered by two 190 hp diesels, this was a small all-wooden flat-top ferry designed for 14 cars. It was built by Saint John Shipbuilding and Drydock Co. in Saint John, NB, in 1952.

This vessel was thoroughly described in the April 1953 edition of *Harbour and Shipping* magazine, including its perhaps most unique feature: "...a double-ended ferry capable of making landings directly on the gravel beaches at Deer Island and Letete, thereby avoiding the necessity of constructing wharves and floats costing upwards of $200,000 and at the same time having seaworthiness and speed to combat vicious tide rips encountered in a tortuous route through rocky islets."

This vessel was replaced in 1974 by another Robert Allan Ltd ferry design of significantly greater capacity, and this time intended to be served by "wharves and floats." The new ferry was built by Saint John Shipbuilding and Drydock Co. Ltd. It measured 129 feet by 52 feet and was powered by four 350 hp engines, with two propellers at each end.

Postcard showing the 1952 Deer Island–Letete ferry *Abnaki* and its unique beach landing arrangement.

General Arrangement of first Deer Island to Letete ferry, 1952.

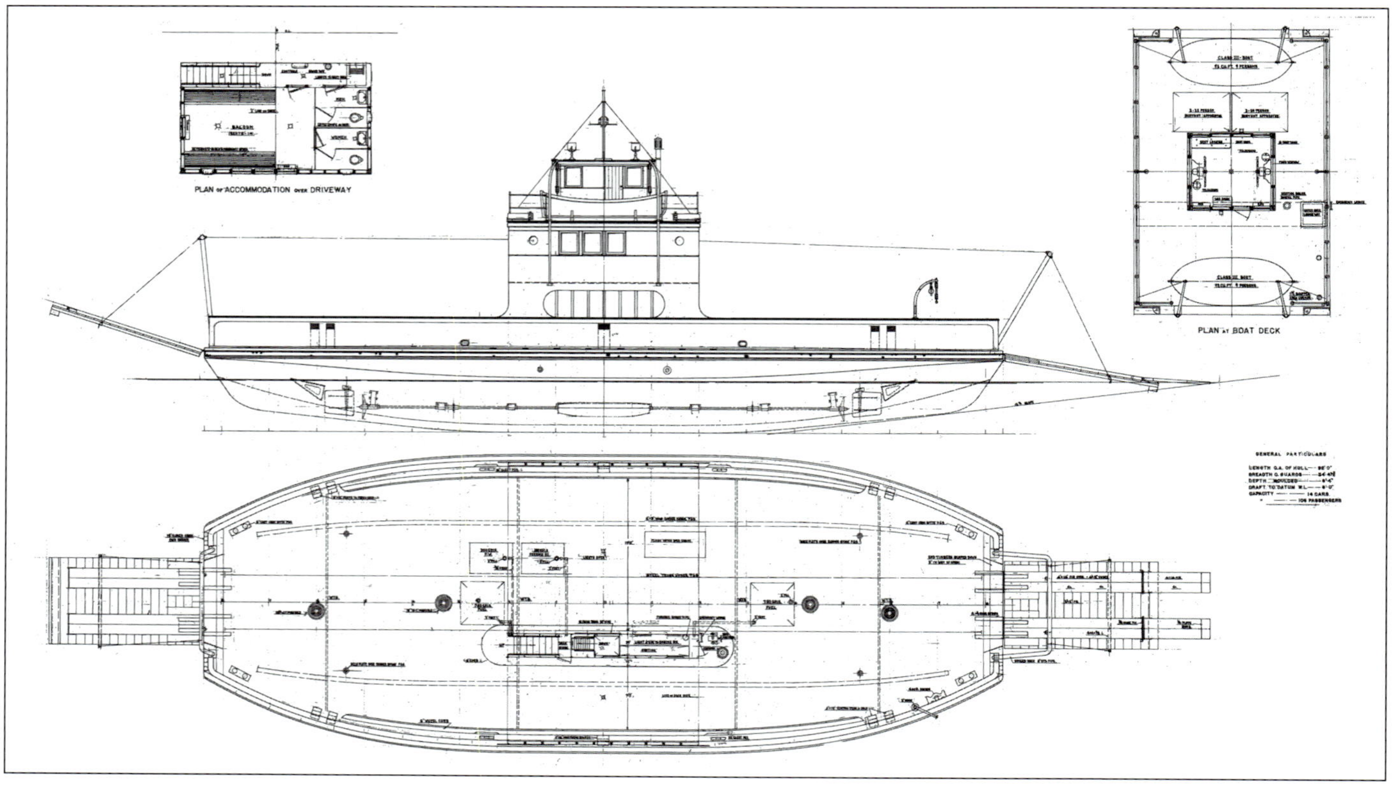

General Arrangement for 1973 Deer Island to Letete ferry.

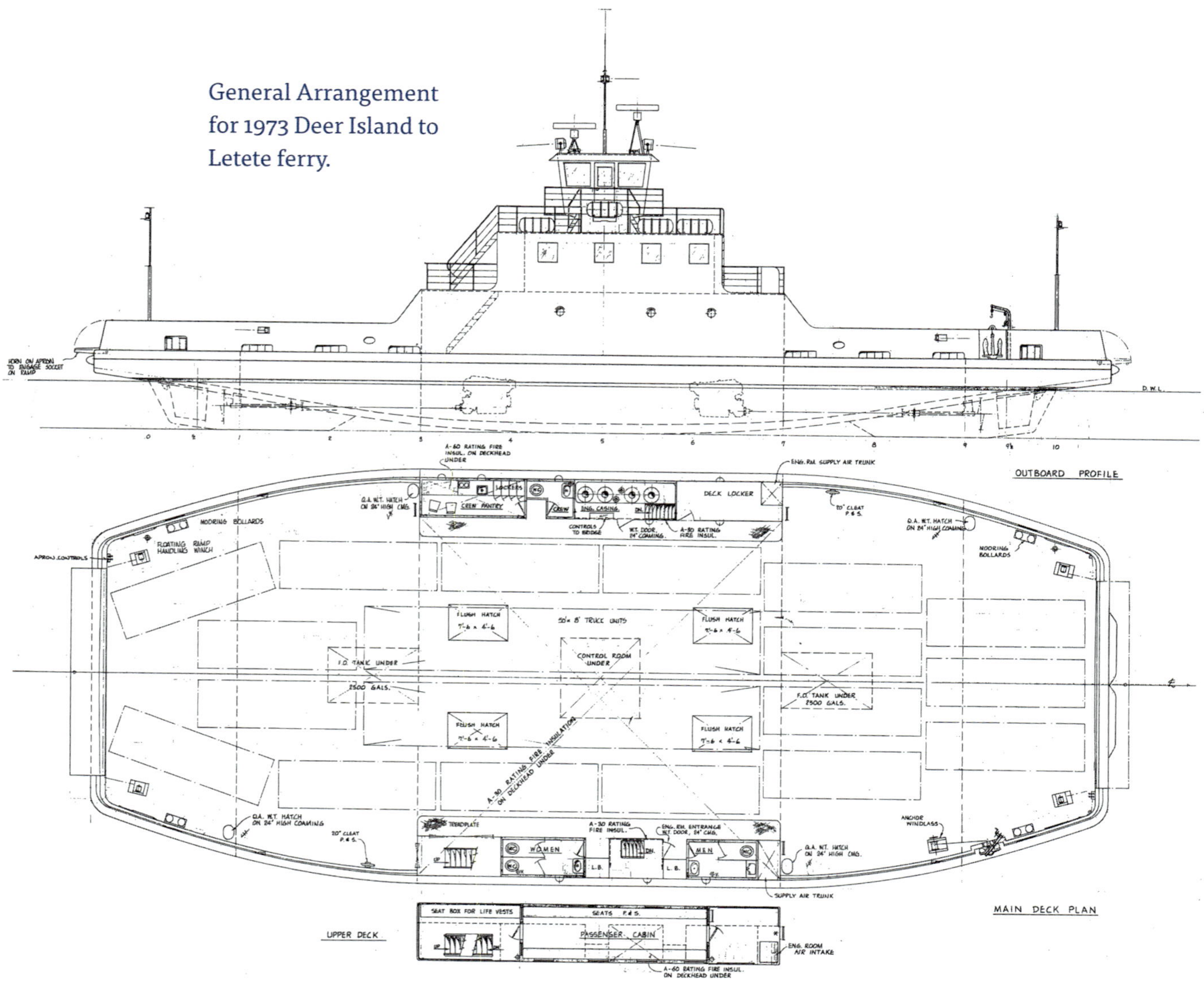

Gulf Island Ferries

In addition to *Delta Princess* and *Lloyd Jones*, mentioned above, which migrated into service in the Gulf Islands of BC, a couple of ferries were specially designed for Gulf Island service when Coast Ferries Ltd, owned by Oswald "Sparky" New (see *Hyak*, Chapter 3), operated these inter-island routes. The first of these was the simple 20-car ferry *Mill Bay*, designed in 1955 for the Brentwood–Mill Bay crossing. This 122-foot by 32-foot-6-inch ferry was built in 1956 by Victoria Machinery Depot Co. Ltd and operated on this route continuously for 55 years, until 2011, and since 1969 under the BC Ferries banner.

In 1957 Gulf Islands Navigation Ltd (a subsidiary company of Coast Ferries Ltd) commissioned the design of a larger ferry to serve the Steveston to southern Gulf Islands route. This vessel, named *Island Princess,* measured 131 feet 3 inches by 34 feet 6 inches and was powered by two 256 hp diesels. Built by Allied Shipbuilders, it entered service in 1958. Unlike more typical double-ended ferries, the *Island Princess* was designed for single-direction running operation with twin screws aft, but still configured for regular through-deck roll-on/roll-off operations, equipped with short loading ramps fore and aft. This design became the model for a number of similar ferries designed by others and built over the ensuing years.

After serving the southern route until 1961, the *Island Princess* was used on several other coastal routes, including Kelsey Bay to the north end of Vancouver Island, until she was purchased by BC Ferries in 1969. In 1970 John Case (a former employee of Robert Allan Ltd) got the job to "catamarize" this ferry, splitting the hull then widening and lengthening it for a significant increase in capacity. In this new and rather unique configuration, now 173 feet 6 inches by 57 feet, she was renamed *North Island Princess*, and among other routes served the Powell River to Texada Island trip for many years, eventually being retired in 2020.

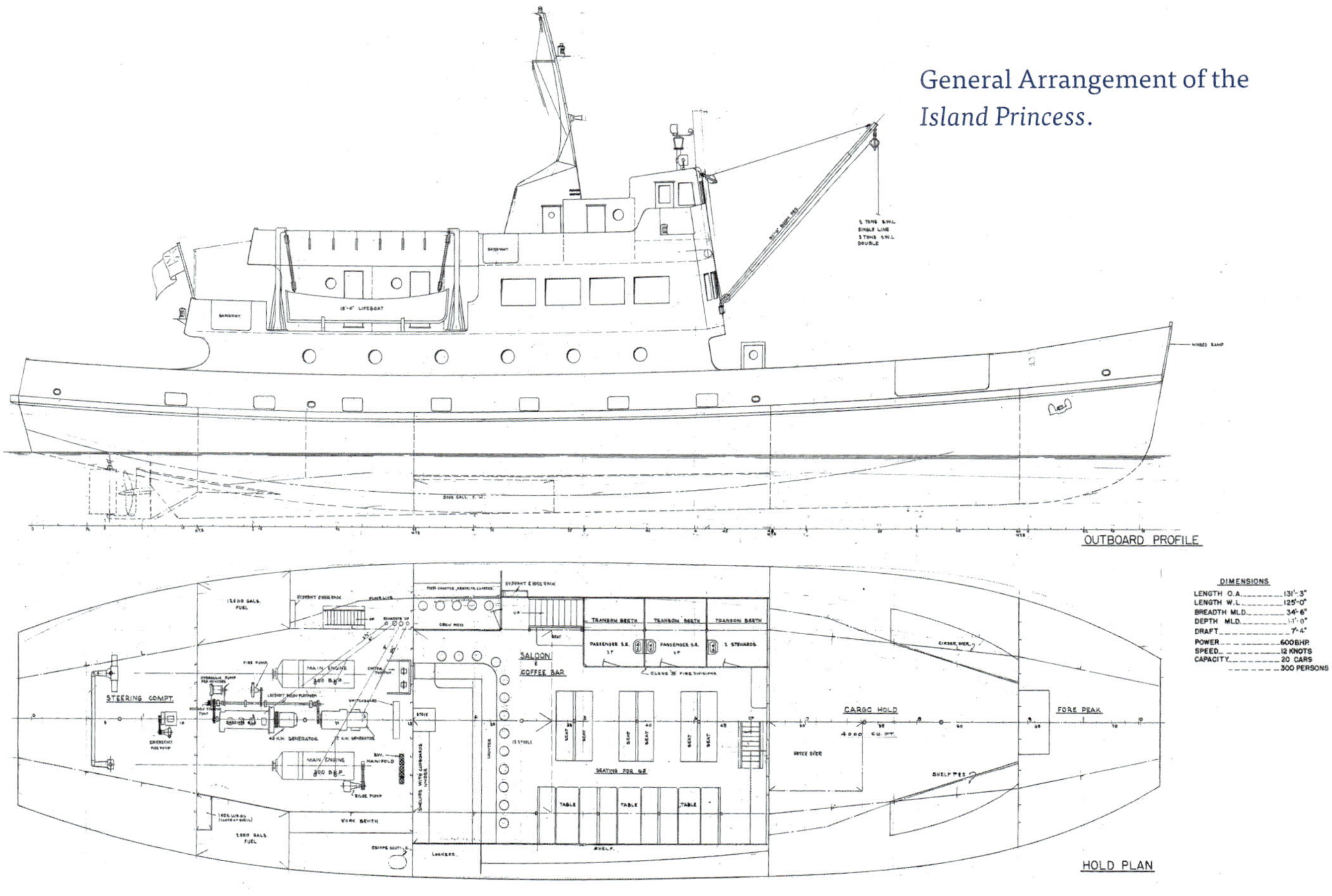

General Arrangement of the *Island Princess*.

Top: The *Mill Bay*, probably in the early 1970s.

Right: *Island Princess* before her conversion.

Bottom: The *Island Princess* was "catamarized" into the *North Island Princess* by BC Ferries.

Northern Ferries

As will be covered further in Chapter 8, the Allan connection with the federal Department of Public Works for a significant number of specialized workboats seems to have also worked in their favour, leading to designs in the 1960s for a variety of river-crossing ferries for the Yukon and Northwest Territories.

In 1960 the design for the *Johnny Berens* was completed to initiate the Fort Providence crossing of the Mackenzie River. This simple little 80-foot ferry was built by Rimmer and Sons Construction Co. (probably of Edmonton) in that same year.

In 1966 a very similar ferry was designed for the Sixty Mile crossing of the Yukon River. Named *George Black*, that 83-foot vessel, with a nominal capacity of 14 cars or a maximum payload of 65 tons, was built in 1967 by Allied Shipbuilders and is still in service near Dawson City to this day. She operates 24/7 taking passengers across the Yukon River, heading for the magnificent Top of the World highway to Alaska. (A highly recommended trip!) The props at each end are located within tunnels to their full diameter to ensure protection when the ferries ground for loading.

In the late '70s the self-propelled work barge *Kap III*, built for Kaps Transport Ltd during the oil exploration boom on the Mackenzie River in the late '60s and early '70s, was purchased by the Canadian Department of Public Works for use as a simple ferry and renamed *Louis Cardinal*. However, the original single-ended configuration wasn't well suited to the intended roll-on/roll-off operation, and in 1985 Robert Allan Ltd was commissioned to undertake the significant transformation of this vessel to a double-ended ferry.

This involved cutting off the two ends of the basic barge-shaped hull, designing new ends, new propulsion machinery, deckhouse, engine casings, ramps, etc. The end result, which uses only the parallel body of the original hull, is now a quite useful 125-foot ferry which operates on the Arctic Red River and the Mackenzie River at Tsiigehtchic.

The first ferry at Fort Providence made its inaugural crossing in 1961. It was named after Johnny Berens, a skillful riverboat pilot who commanded the SS *Distributor* along the lengths of the Mackenzie River for 26 years.

Chapter 7

The *George Black* on the Yukon River, 2019.

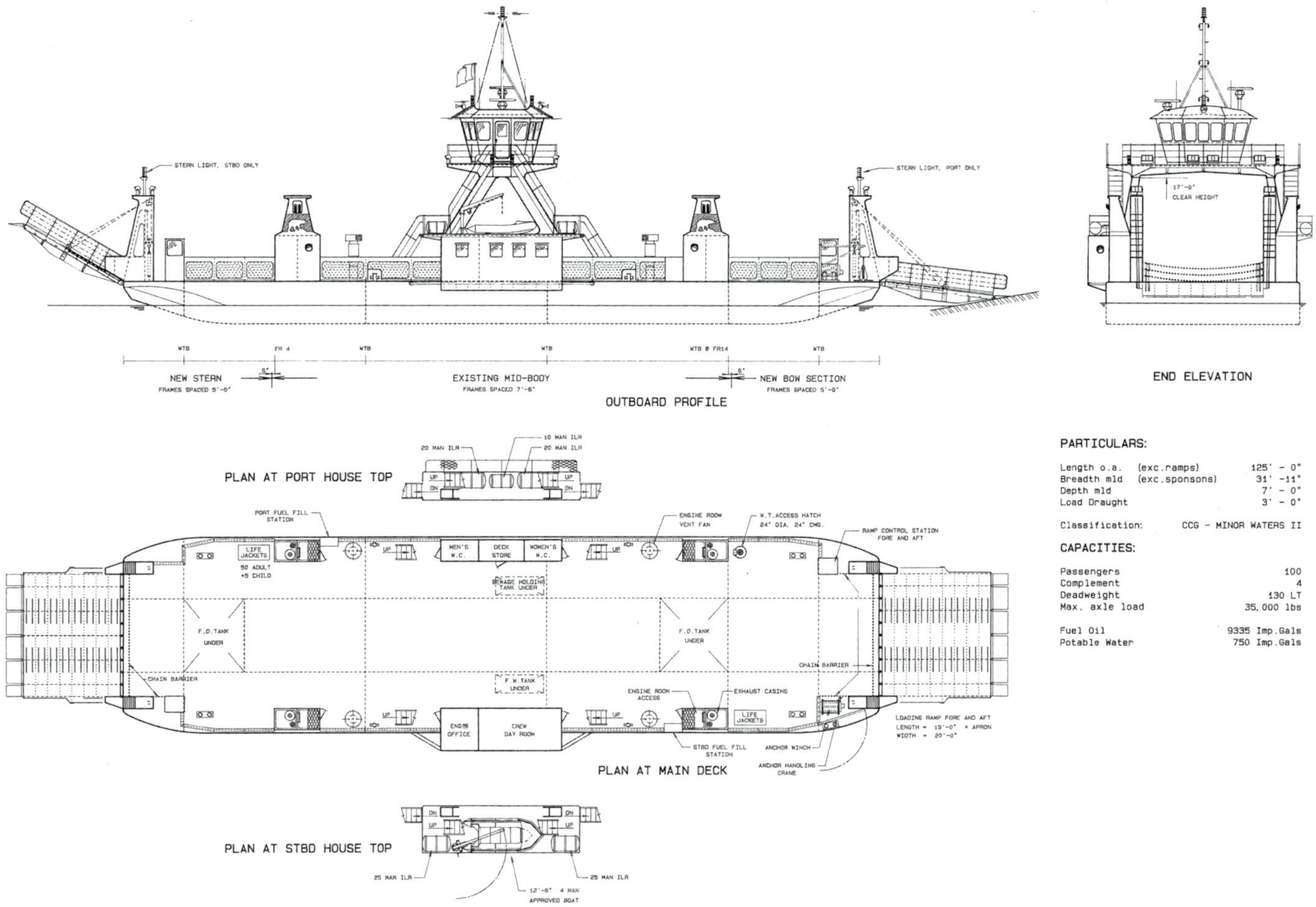

Top: General Arrangement for the transformation of the *Louis Cardinal* (ex. *Kap III*) to a Ro-Ro ferry.

Left: The *Louis Cardinal* reincarnate.

The Yorke Trailer Ferries

Not every ferry designed by this office is a small flat-top. In 1963 Bob Allan designed a major trailer and railcar ferry for F.M. Yorke and Sons, at the time a major client. This was the first self-propelled vessel in commercial service between the Lower Mainland of BC and Vancouver Island. The *Greg Yorke* was built by Allied Shipbuilders Ltd in 1964. She measures 325 feet by 55 feet, and with twin 1952 hp diesels has a service speed of 12 knots.

This vessel and the railcar/trailer service were taken over in the creation of the Seaspan organization in 1970. She became the *Seaspan Greg* and continues to work on this route at the time of writing.

Shortly after the *Greg* went into service, Lorne Yorke decided to expand their railcar transfer operations and commissioned the design of a double-decked railcar/trailer ferry in 1966. The resulting *Doris Yorke* was built by Victoria Machinery Depot Co. Ltd in 1968.

At 325 feet by 56 feet, her deck plan was not dissimilar to the *Greg Yorke*, as they shared common berthing facilities, but she was designed to carry railcars on the main deck and trailers on the upper deck. The upper deck was accessed by a simple cable-operated elevator. This did not result in speedy loading and off-loading of trailers, but it did serve the purpose quite efficiently without taking up valuable main-deck space with long loading ramps.

The *Doris* was later integrated into the Seaspan fleet as the *Seaspan Doris*, and she was finally retired from this coastal operation in 2018 after 50 years of service.

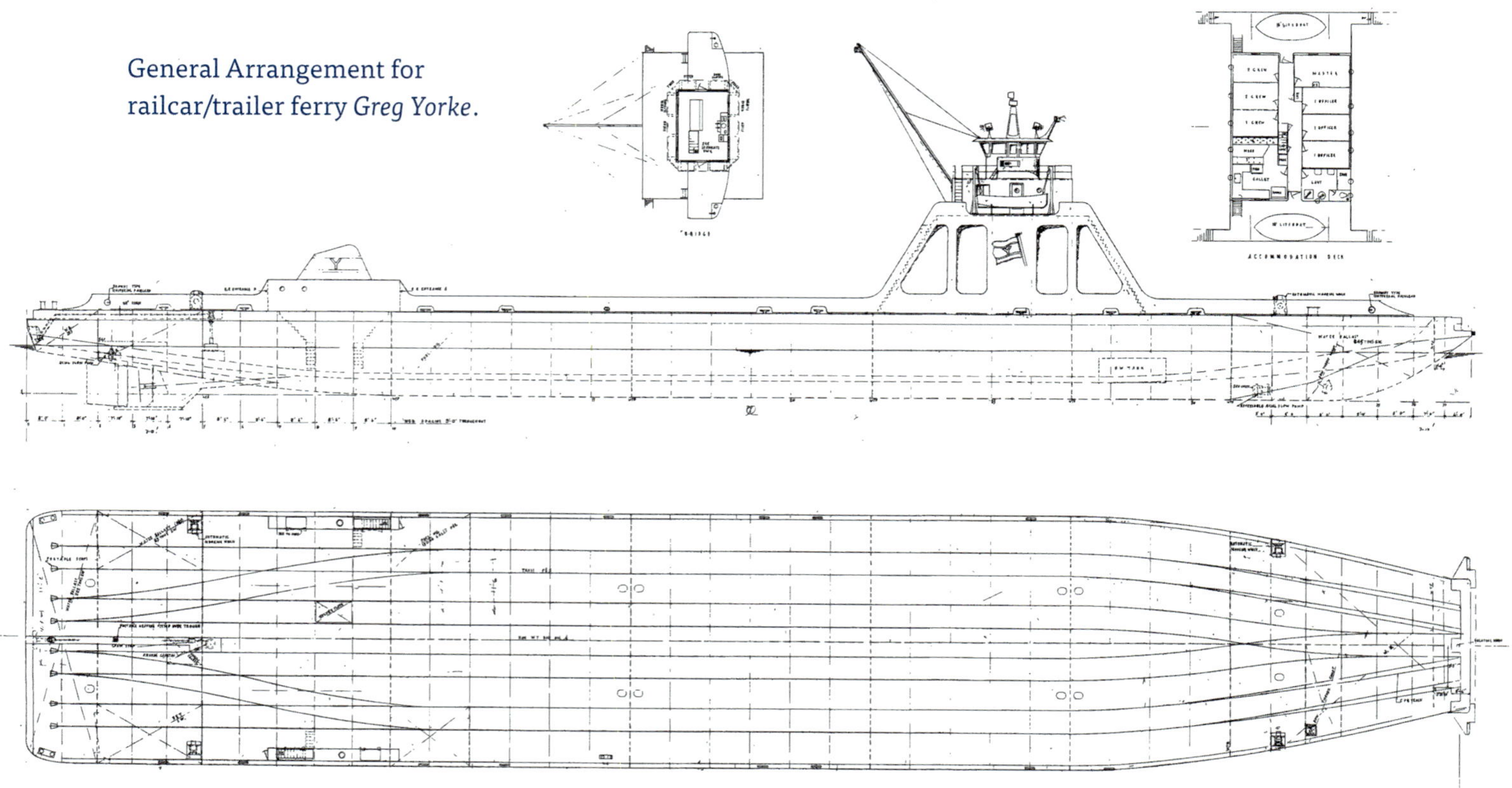

General Arrangement for railcar/trailer ferry *Greg Yorke*.

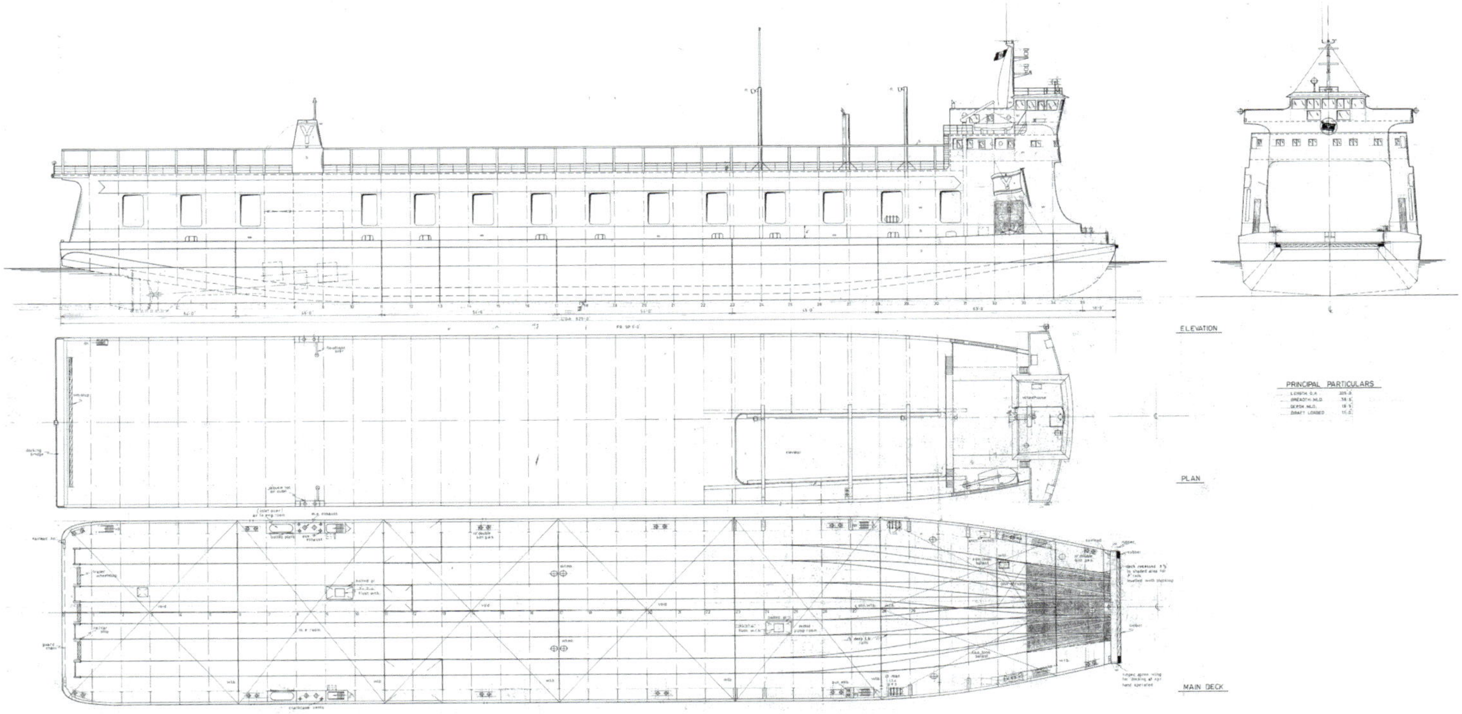

Top: The railcar ferry *Greg Yorke*, built for F.M. Yorke and Sons; **bottom:** General Arrangement of the *Doris Yorke*.

Above: *Doris Yorke*, a double-decked railcar and trailer ferry for Vancouver Island service.

Right: The *Doris Yorke* as the *Seaspan Doris*.

Cable Ferries

Among the least glamorous vessel types from this office are cable ferries, but over the years a significant number of these have been designed, several across BC, but more particularly and most recently across all the prairie provinces. These are essentially simple roll-on/roll-off barges with ramps, but the variety of methods for motive power and the details of making those work are interesting and instructive design challenges for young naval architects.

The typical basic concept for a cable ferry is to have one or two fixed "guide cables" along the route (usually a short river crossing). These can be overhead or near water level but generally need to be quite tight to control the route of the ferry. The drive system can be either paddle-wheels (as seen in the *West Creston Ferry*) or more typically by a "drive cable" wrapped around a large diameter drum on deck and led through roller fairleads at each end of the ferry. The drum rotation, either direct diesel-driven or hydraulic-driven, then moves the ferry back and forth.

Throughout the 1970s and continuing into the 2000s, Robert Allan Ltd had a fairly steady diet of cable ferry designs, particularly for the governments of Manitoba and Alberta. The accompanying table lists these:

Cable Ferry Designs

Proj. No.	Year Built	Vessel Name	Client	Length (ft)	Beam (ft)
7-76	1977	*James Apetagon*	Gov't of Manitoba	50.0	20.0
42-76	1977	*Ingemar Carlson*	Gov't of Manitoba	75.0	30.0
42-76	1977	*Thomas A. McIvor*	Gov't of Manitoba	75.0	30.0
37-81	1981	*Alfred Settee Sr.*	Gov't of Manitoba	75.0	30.0
56-91	1991	*Crowfoot Ferry**	Gov't of Alberta	90.6	34.0
120-94	1994	*Bleriot Ferry* *	Gov't of Alberta	90.6	34.0
200-108	2001	*Finnegan Ferry* *	Gov't of Alberta	73.9	34.0
203-024	2004	*Johnny Paul*	Gov't of Manitoba	90.6	34.0
* = joint design with Brandlmayr Marine Ltd					
12-73		*Fort Vermillion Ferry*			
22-61		*Celgar*			
1948		*Creston Ferry*			

The Brandlmayr Connection

The projects for the Government of Alberta were done in collaboration with Brandlmayr Marine Ltd, which, under the late John Brandlmayr, had designed several of the previous generation of cable ferries for that prairie province. John's son Grant was a talented designer in his own right, and we enjoyed working with him on various projects when he was the lead designer at Shore Boat Builders. Grant had received a request from Alberta to develop a new design to succeed one of his father's designs but felt that he needed more support and approached Robert Allan Ltd to team with him for that contract. It was a very successful collaboration, and we did a couple more in a similar fashion. Eventually Grant was enticed to join the Robert Allan Ltd team, and he has since become a shareholder and a highly valued member of the management team in the office.

—R.G. Allan

The drawing below illustrates one of the more recent Alberta cable ferries. There is a wide mixture of cable configurations used for these ferries, depending upon the amount of other navigation on the waterways, the water depth, length of crossing, water level variations, ice conditions and so forth. Ensuring that the cable drive system works properly in all conditions is the major design challenge.

General Arrangement for the *Finnegan* cable ferry.

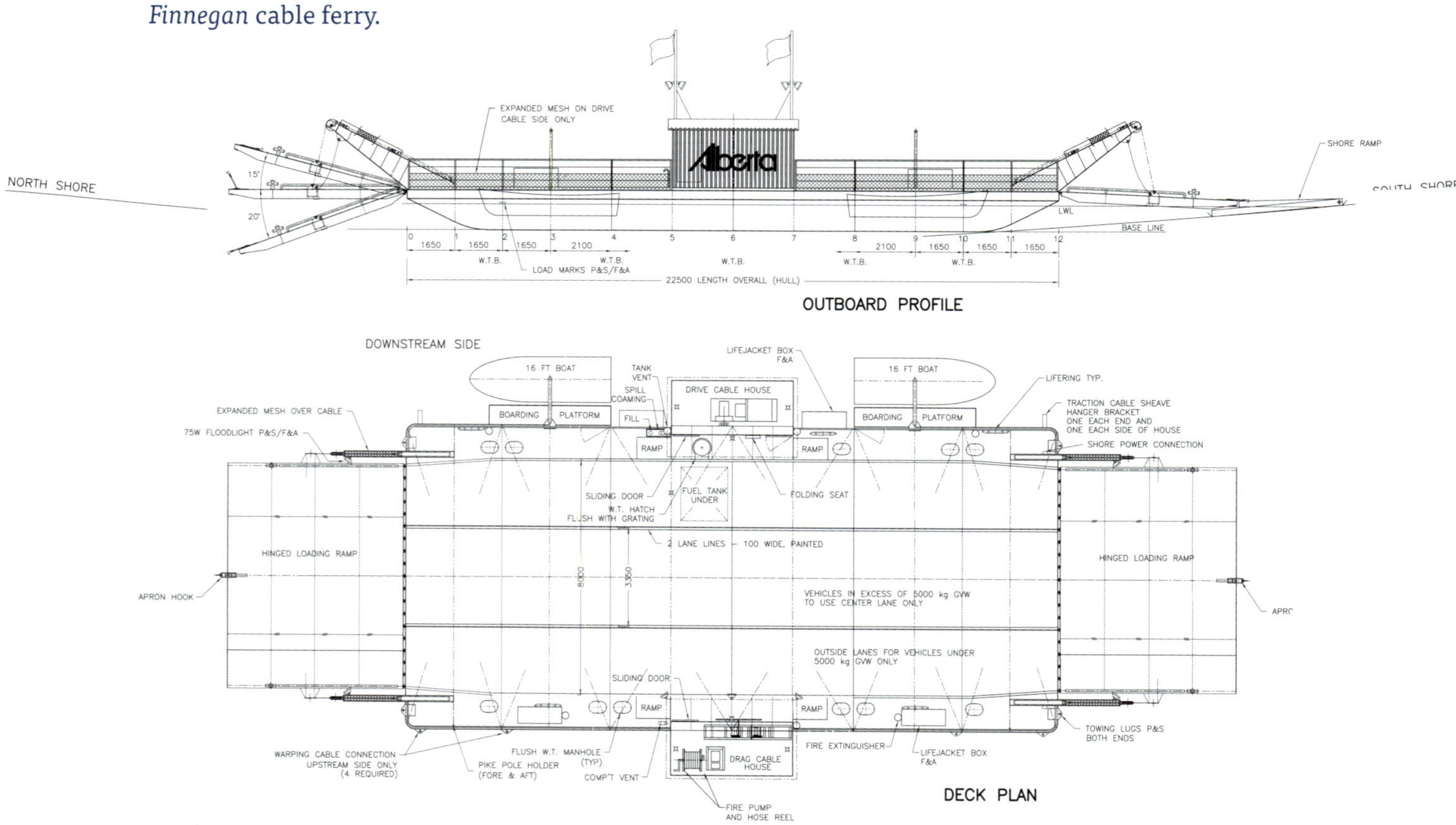

Above: *Bleriot* cable ferry crossing on the Red Deer River in southern Alberta.

Right: *Crowfoot* cable ferry on the Bow River, southern Alberta.

The Manitoba Ferry Fleet

Early in 1976 Robert Allan Ltd won a contract from the Manitoba Department of Renewable Resources and Transportation Services (a name which proved a challenge to our standard drawing title block size!) to design an entire new fleet of ferries to serve a range of routes around that province of a million lakes and rivers, the majority of which served remote Indigenous communities.

The fundamental challenge was that the ferries had to be designed to be assembled and commissioned in some extremely remote areas without much in the way of services and amenities. Designing for simplicity and reliability of construction, maintenance and operation was of paramount importance.

Some of the fleet were the 50- and 75-foot cable ferries described previously, but the greater value portion of the program was for two classes of self-propelled ferries to serve longer routes with higher vehicle and passenger capacities. The two classes were designated as the Lake Winnipeg (LW) class and the Northern (N) class. The designs were identical except that the N class had two deck-mounted Z-drive propulsion units mounted diagonally opposite to each other, as shown below, and the LW class had four Z-drives, one mounted on each quarter.

Originally five vessels were planned, two LW class and three N class; however, the records indicate that probably only one of each class was built, the *Charlie Sinclair* (N), and the *Joe Keeper* (LW). The hulls measured 112 feet overall length, with a beam of 34 feet. Their operating draft was 3 feet 6 inches. The hull construction was designed to facilitate prefabrication in the south of the province, with road or rail transportation of units and panels to the remote construction sites. The structural section illustrates how this was accomplished, with a simple rectangular central section supported by two longitudinal bulkheads, delivered in simple panels, and the outer "triangular" side sections prefabricated in the south. All the deckhouse and ramp units were similarly designed to be road or rail transportable in prefabricated sections.

In addition to the design contract, Robert Allan Ltd was awarded the contract to oversee the construction of these vessels on behalf of the client in 1977. Ex-Canadian Navy marine engineer Claire Johnston had been hired by Robert Allan Ltd as part of the project supervision team to work on a major ice-breaking supply vessel construction project in 1974. As Claire was originally from Manitoba, it was only logical that he be given command of this inland fleet program, which he managed exceedingly well and with good humour over many months and in very challenging conditions.

The photos below illustrate the conditions in which the builders' crews from Purvis-Navcon Shipyard and Selkirk Machine Works had to operate, essentially carving a small shipyard, complete with launch ways, out of the bush.

General Arrangement of Manitoba's "N" class ferries. The "LW" class were virtually identical but with four drive units.

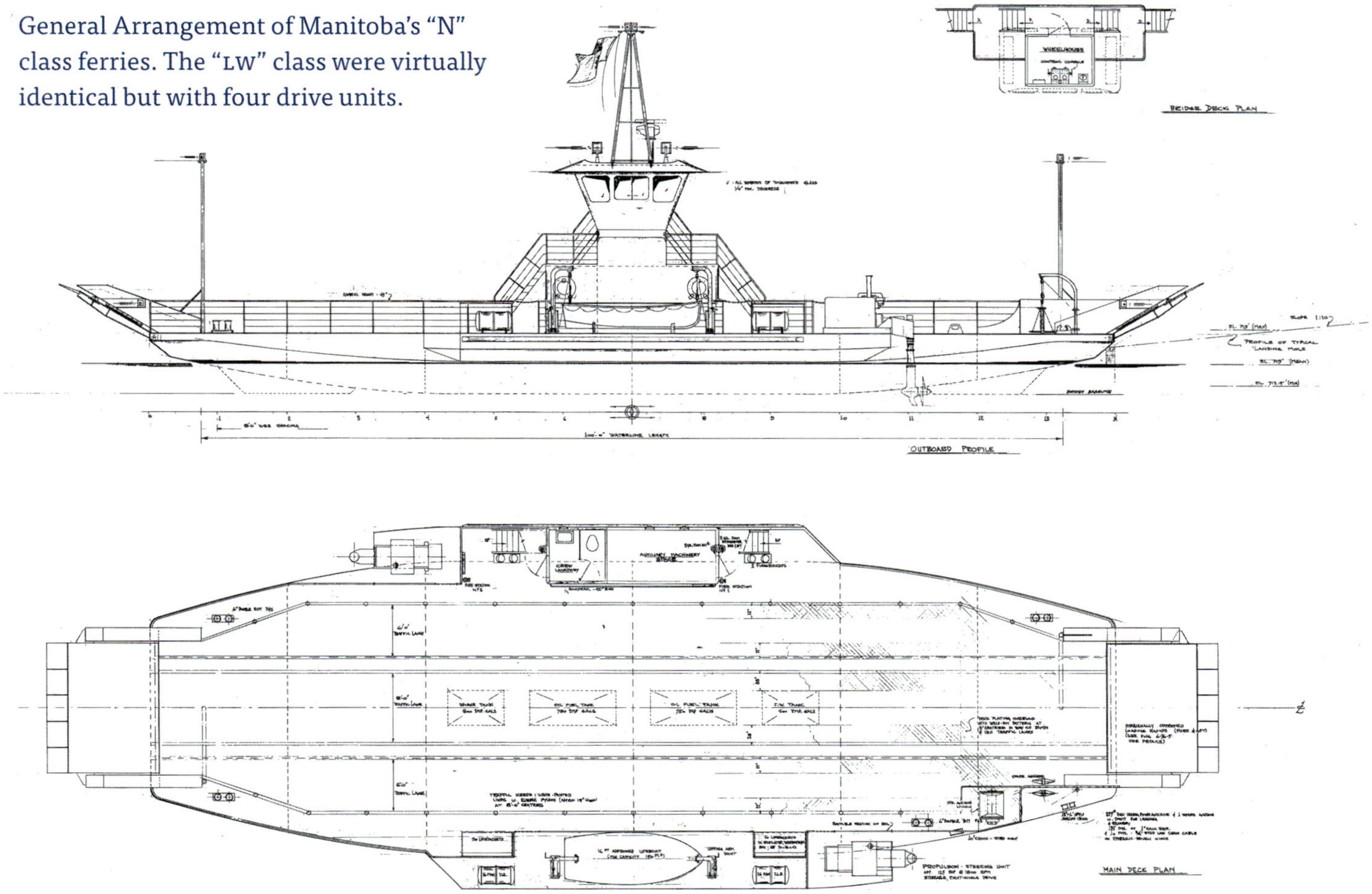

Typical Structural Section of Lake Winnipeg and Northern class ferries.

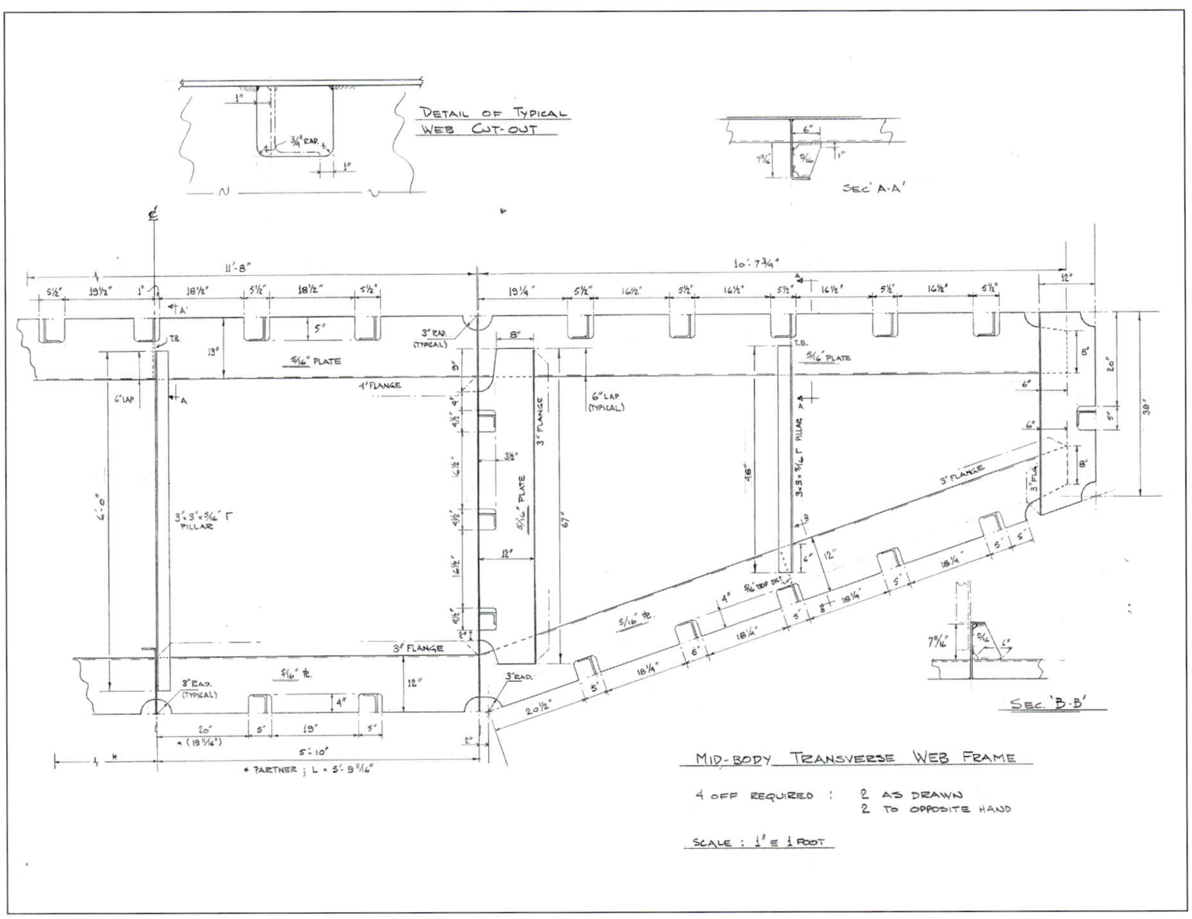

Left: *Joe Keeper*, a Lake Winnipeg class ferry for Manitoba; **below:** rudimentary launch ways for the Manitoba ferries; **bottom:** Launching one of the N class ferries.

Singapore Commuter Ferries

A duo of interesting passenger-only ferries was designed as part of the company's Singapore operations in 1978. The *Sea Breeze* and *Sea Coaster* were commuter ferries built in steel with an aluminium deckhouse and designed to carry 220 passengers each at a speed of 12 knots. Built by Singapore Shipbuilding and Engineering Pte. Ltd, these 31.77-metre-long vessels were in regular service on the Singapore waterfront until about 2010.

Sea Breeze, one of a pair of commuter ferries for Singapore.

Prince Rupert Airport Ferry

Another long-serving BC ferry is the descriptively named *Prince Rupert Airport Ferry*, which serves that city's airport, located on Digby Island, making it a vital link in every air commuter's day. This 113-by-34-foot ferry was built in 1969 by Vancouver Shipyards and continues to serve this route to this day.

Due to the high tidal range in this northern BC town, the ferry docks to a floating pontoon, from which a long ramp leads onto the shore, thus keeping the onboard ferry ramps conveniently short and lightweight.

General Arrangement of the *Prince Rupert Airport Ferry*.

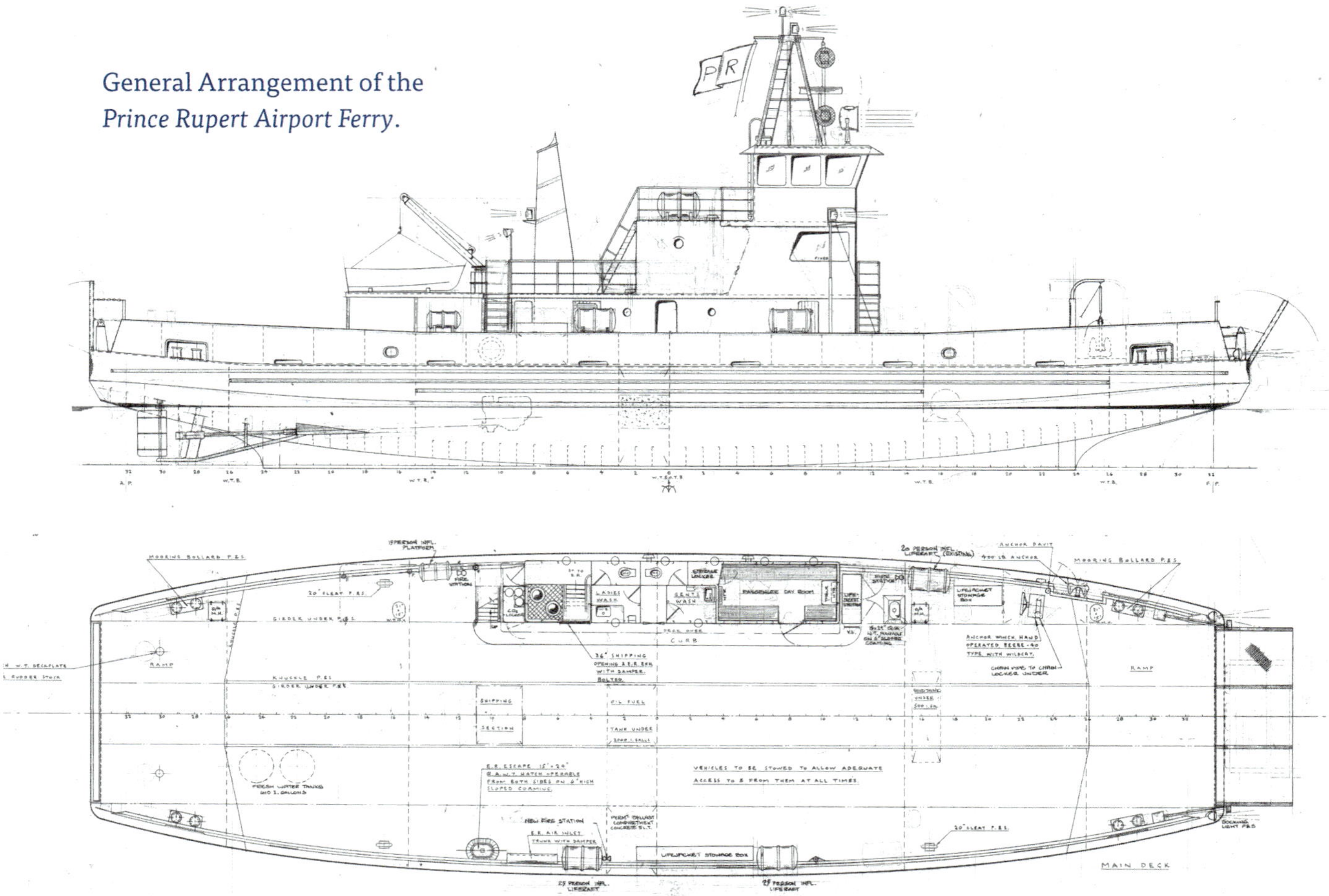

Prince Rupert Airport Ferry

Britannia III

This passenger ferry, designed by Rob Allan in 1979, had an interesting genesis. The owners, Harbour Ferries Ltd, obtained a contract to move 500 workers daily to a planned pulp mill construction site in northern Howe Sound. That project failed to come to fruition, however, and the ship was modified to become a significant feature on Vancouver's waterfront harbour cruise scene. This tourist configuration had always been envisaged as part of the vessel's long-term operations however, so the modifications were not difficult to implement. All steel and measuring 135 feet 6 inches by 31 feet 8 inches, she was built by Allied Shipbuilders in 1981. At some stage Harbour Ferries was able to drop the "III" designation and have her renamed *Britannia*, most likely when the Royal Yacht of the same name was retired.

Britannia in False Creek, Vancouver.

Sir Robert Bond

At 444 feet in length and with a 70-foot beam, the *Sir Robert Bond* is the largest self-propelled vessel of any type ever designed by Robert Allan Ltd. Designed in 1972 for the Canadian Ministry of Transport and built in 1975 by Port Weller Drydock Co. Ltd as a dedicated railcar ferry, she operated initially under the CN Marine banner. As described on the *Marine Atlantic* home page: "Named after the first prime minister of the Dominion of Newfoundland, the vessel originally served the Port aux Basque–North Sydney crossing before moving to the Labrador service in 1977. Originally designed as a freight/rail-car ferry, it underwent a significant refit to add accommodations and passenger amenities to better suit the needs of Labrador customers. The 10,500-ton ship transported significant volumes of freight and passengers destined for Labrador communities during its years of service with CN Marine and Marine Atlantic."

CN Marine ro-ro (roll on/roll off) train ferry *Sir Robert Bond*.

The Fast Ferry Project

Sometimes, no matter how good our design and engineering work, projects just don't work out. Such was the ill-fated "Fast Cat" project for BC Ferries. When the BC Ferry Corporation decided in 1994 to join what was then a worldwide trend to fast multi-hulled ferries, the plan was to replace the two conventional ferries on the Horseshoe Bay to Nanaimo run with three smaller fast catamaran ferries. This would shorten trip times, increase traffic throughput, and enable the fleet to be more closely tailored to the seasonal demands on this route, using just two ships in the winter season. One of the major cost savings was expected to be in crew size, as the smaller ships would require fewer crew overall. While this feature was never widely proclaimed, it certainly got the attention of the ferry workers' unions, and they went into action to disrupt the project. The project was almost doomed from the start, as numerous so-called experts made their unsubstantiated opinions public and the press found it much more interesting to feature detractors than to examine the significant technical achievements of this project.

For the design of these ferries Robert Allan Ltd had teamed with Incat of Australia, at the time the leading fast-ferry designers in the world, led by Phil Hercus. Incat limited their scope of work to the hull geometry and arrangement, powering and basic structure, and were used to leaving all the machinery, outfitting and electrical system design to shipyards. Robert Allan Ltd was tasked with doing all the design work for those critical systems. The teaming worked very well. For all his international success in this field, Phil was a modest and unassuming man and a real pleasure to work with.

The major challenge of the design, however, was to make a wide-bodied multi-hull fit into berths that had served several generations of conventional mono-hull ferries, with dual-level loading and unloading. Phil had a brainwave which involved creating a large, hinged bow door that when dropped would emulate the shape of the main deck of the conventional ferries. While in some respects this was a very elegant solution, these bow doors were large and quite complex structures and became quite heavy.

With a capacity for 250 vehicles and 1,000 passengers each, the "Pacificats" had the following particulars: length 122.5 metres; beam 25.8 metres; draught 3.76 metres. Propulsion was delivered by four MTU 20V1163 TB3 diesel engines, each rated at 6500 kW, driving four KaMeWa 112 SII water jets, achieving an operating speed of 34 knots.

Although model-testing had been done, the generally unanticipated problem with these ferries was the wake generated by a relatively heavy ship travelling at high speed. The wave impacts on Bowen and Passage Islands were quite dramatic, and the waterfront-owning public howled, with some justification.

Ultimately the ferries were declared a failed experiment and were sold off to private interests. This was a sad ending, as we had been proud of the comprehensive work done in the Robert Allan Ltd office.

One of the Pacificat trio on trials in English Bay.

COVE ISLE
Canada
COVE ISLE
Coast Guard
Garde côtière

Opposite: *Cove Isle*, see page 158.

8

Workboats for Every Whim

One of the many great pleasures of naval architecture is being challenged to solve unique operational problems or restrictions for a client and then seeing those results come to fruition in a new hard-working boat that hopefully continues serving in that intended role for many years. While many vessel types end up being somewhat similar, or even repetitive, the catch-all "workboat" category is one from which many truly unique designs emerge, and this creative custom-designing is an area in which Robert Allan Ltd has excelled for many years. These diverse workboats may or may not be somewhat similar to other vessel types, but each has its own individual character and design constraints. Some of the many unique workboat designs developed throughout the history of the company are described in this chapter.

Hilunga

Hilunga was a coastal survey boat designed in 1949 for the Canadian Department of Public Works (DPW). At the time DPW had a mandate to build, operate and maintain numerous dock facilities along the BC coast, and they had a fleet of vessels engaged in the various activities associated with those responsibilities. *Hilunga* was an elegant wooden vessel, 86 feet 10 inches long by 17-foot beam, equipped with a 400 hp Vivian diesel engine. Her role was described in *Harbour and Shipping* in May 1950 as "transporting engineers on inspection trips and for the surveying of harbour and dock installations."

It is noteworthy that this design was model-tested at the National Research Council model basin in Ottawa to verify that a design speed of 12 knots could be attained. Such model-testing for other than warships was a rare event in those days, and particularly for a relatively straightforward hull type. This is the first record of an Allan design being model-tested.

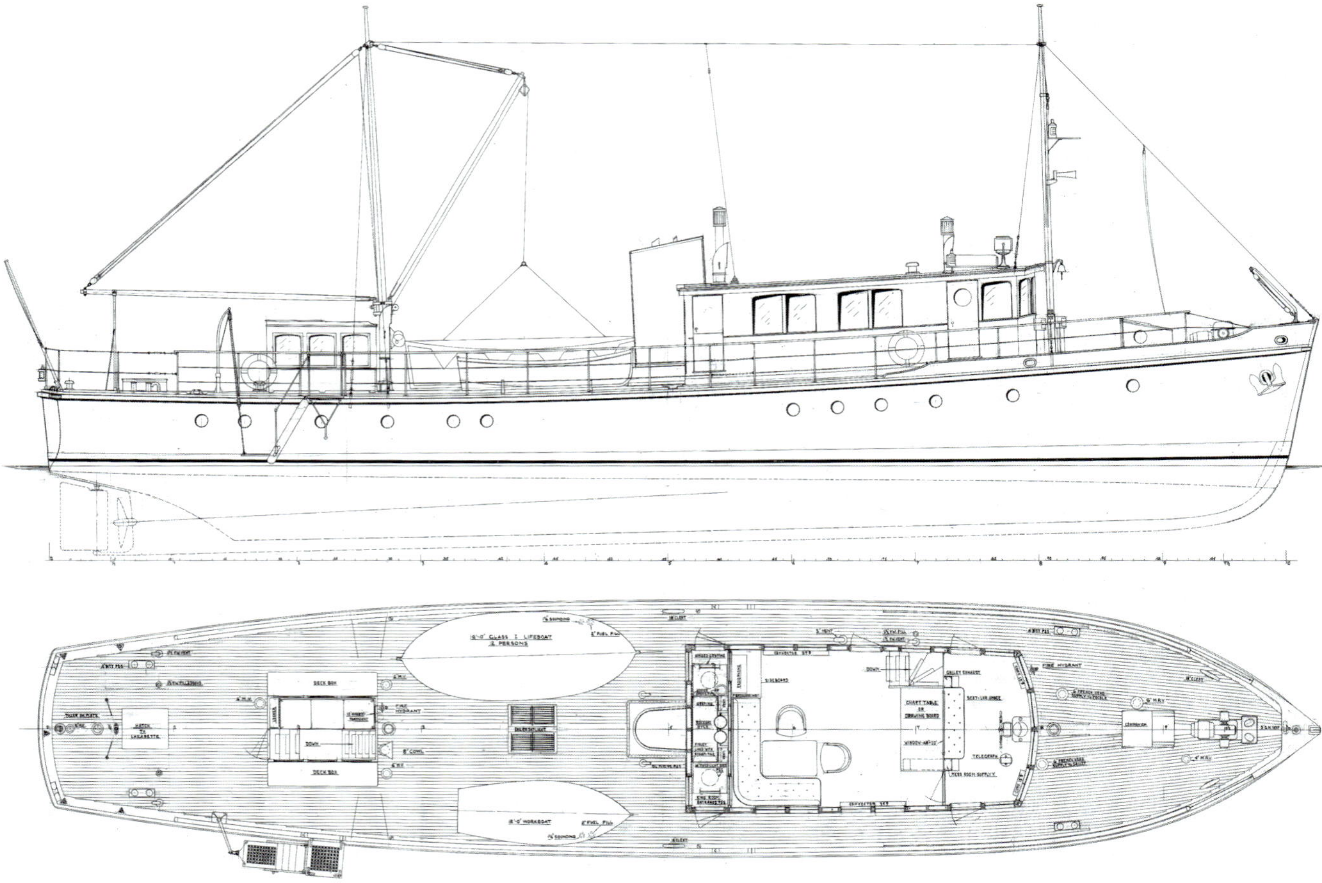

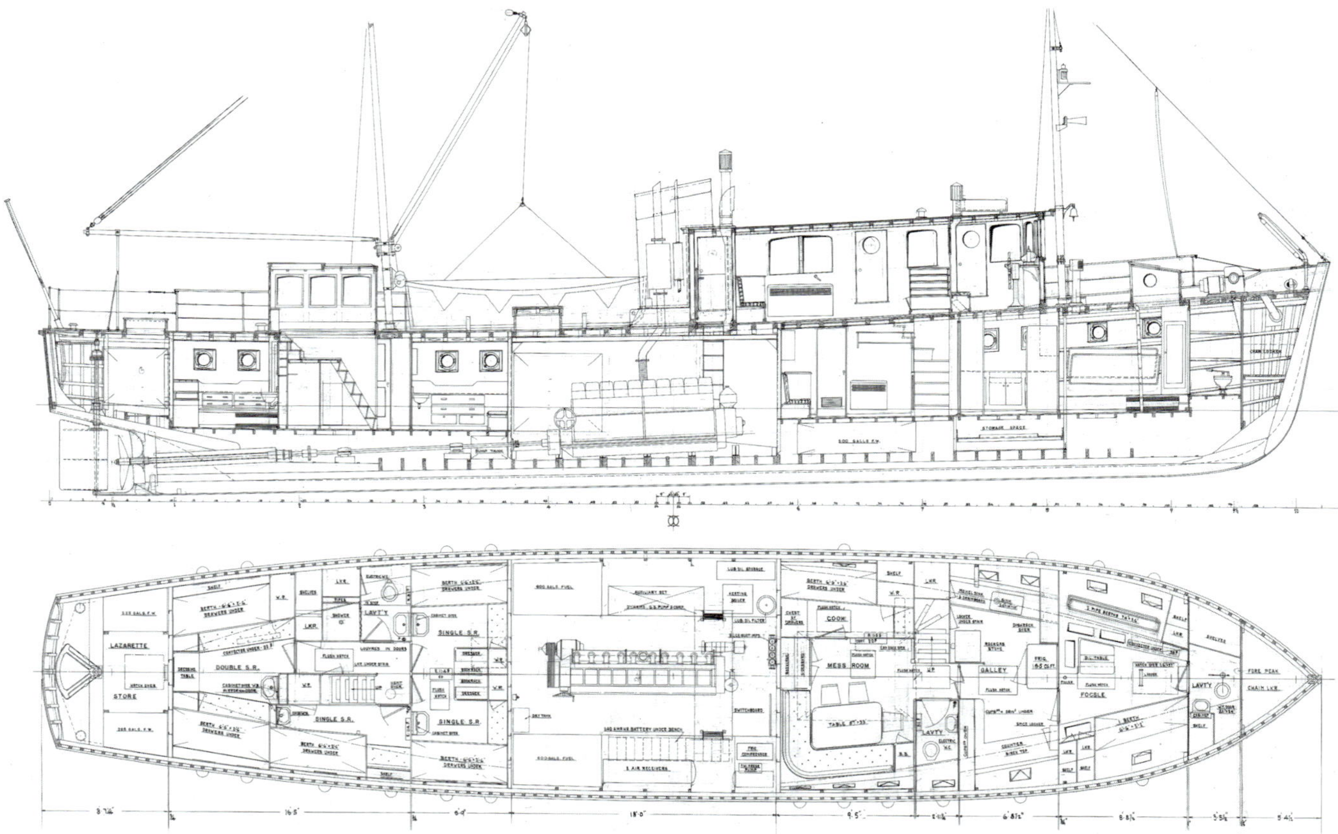

Top and above: General Arrangement drawings for the Department of Public Works survey launch *Hilunga*; **opposite top:** Structural Arrangement of *Hilunga*; **opposite bottom:** The elegant *Hilunga*.

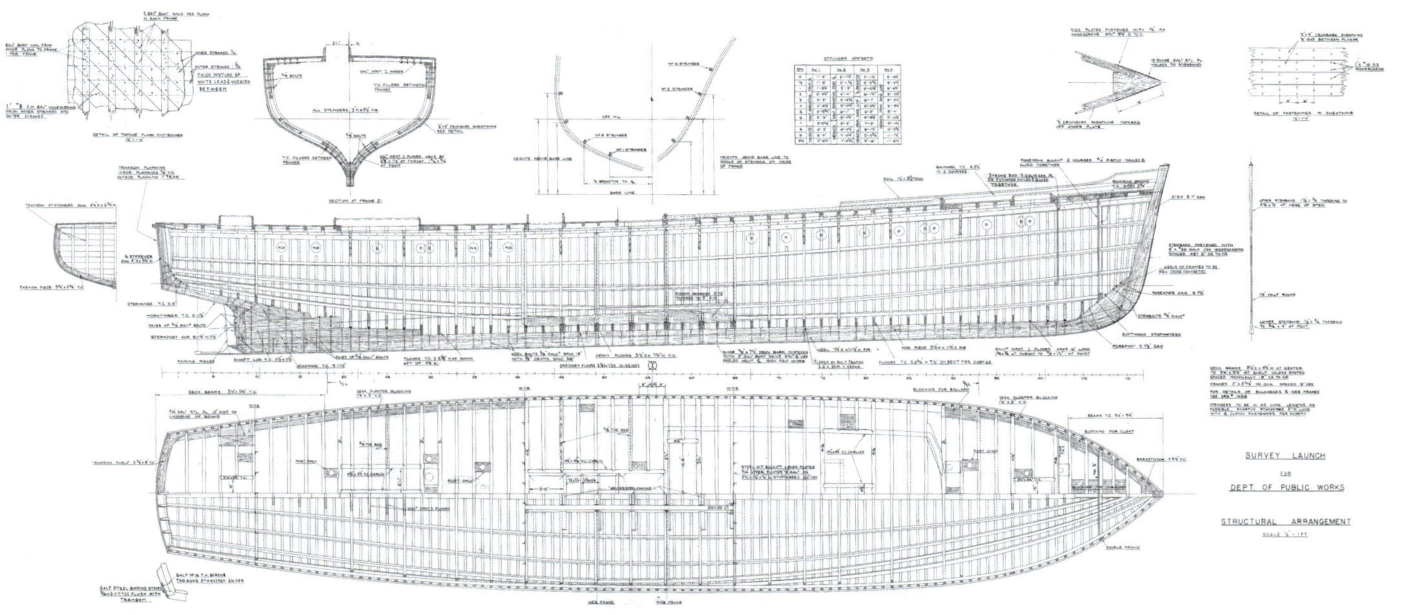
SURVEY LAUNCH
FOR
DEPT. OF PUBLIC WORKS
STRUCTURAL ARRANGEMENT

HILUNGA

Hilunga was built at Star Shipyard (Mercer's) Ltd in New Westminster and launched in 1950. At the time, this was described as the largest wooden shipbuilding contract since the war. The construction is an interesting hybrid combination of materials and configurations. Designed for strength with minimal weight, the framing is a mixture of both transverse and longitudinal frames. Closely spaced bent-oak transverse ribs are employed in association with longitudinal stringers. At every 36 inches in the midship region, and at 54-inch centres towards the ends, there is a "strong frame" made up of a normal rib with an additional bent oak rib inside the longitudinals and over top of the floor timbers, extending from gunwale to gunwale, as illustrated in the Structural Arrangement drawing.

In addition, there are three watertight steel bulkheads and two steel web frames fitted port and starboard within the engine room. The deck planking is teak, while hull planking is a combination of Douglas fir and yellow cedar.

Hilunga met with catastrophe in February 1956, an event reported nationwide. It was described in *The Daily News* of St. John's, Newfoundland, as follows: "The five man crew of the government vessel *Hilunga* was found safe and well Friday, 100 miles from where they thought their ship had been Thursday. Survival of the men in one of northern British Columbia's worst storms in many years was termed 'almost a miracle' by search officials. With their 82-foot Canadian survey vessel *Hilunga* breaking up beneath them, they sent out a position report fixing them at Cape St. James, 100 miles away from where they were found. Then they quit the ship in a lifeboat...The airmen spotted a broken lifeboat and then five figures waving from the beach. The wrecked *Hilunga* was seen on the rocks offshore."

Apparently a very smart salvor had calculated that *Hilunga* could not have reached Cape St. James, at the southern end of the Queen Charlotte Islands (now Haida Gwaii), and his wise analysis of the prevailing conditions led them to the area far to the east (believed to be on Aristazabal Island), where the crew were actually found.

The vessel was thankfully salvaged and repaired after this heavy grounding. The records indicate that she continued in DPW service until sold by Crown Assets in 1984 and has served since as a private workboat or yacht.

The design package prepared for *Hilunga* was comprehensive. There are a total of 17 drawings, which in addition to the typical Lines Plan, Arrangement and Structural drawings, includes shafting and rudder details, tank construction details, and numerous piping and wiring drawings that for other vessels at this time had typically been left to the shipyards to develop. From an internal perspective, another notable aspect of this design is that she was project No. 1-49, marking the initiation (finally) of a sensible project numbering system (project number-year), which persists to this day (though in anticipation of problems after the year 2000 we inverted the order to put the year first). Prior to this there were either no project numbers at all or a consecutive-drawing numbering system without distinction between projects, for which no complete record of entries exists and in which there are several large gaps. (The last existing numbered drawing of that previous system is #893, but fewer than 100 drawings from that era remain.)

Tsekoa

Another DPW vessel, this one designed in 1951, was the *Tsekoa*. Built at Star Shipyard (Mercer's) Ltd in 1953, she was 65 feet long and 17 feet 6 inches in beam. Her role was similar to that of *Hilunga*, but more focused on actual repair and maintenance work on the docks. The design drawings show a robust and workmanlike vessel, part tug and part seiner.

Tsekoa remained in government service until 1987, when she too was retired and sold into private hands. However, DPW's duties on the coast still existed at that time, and another generation at Robert Allan Ltd was ultimately retained to design her replacement.

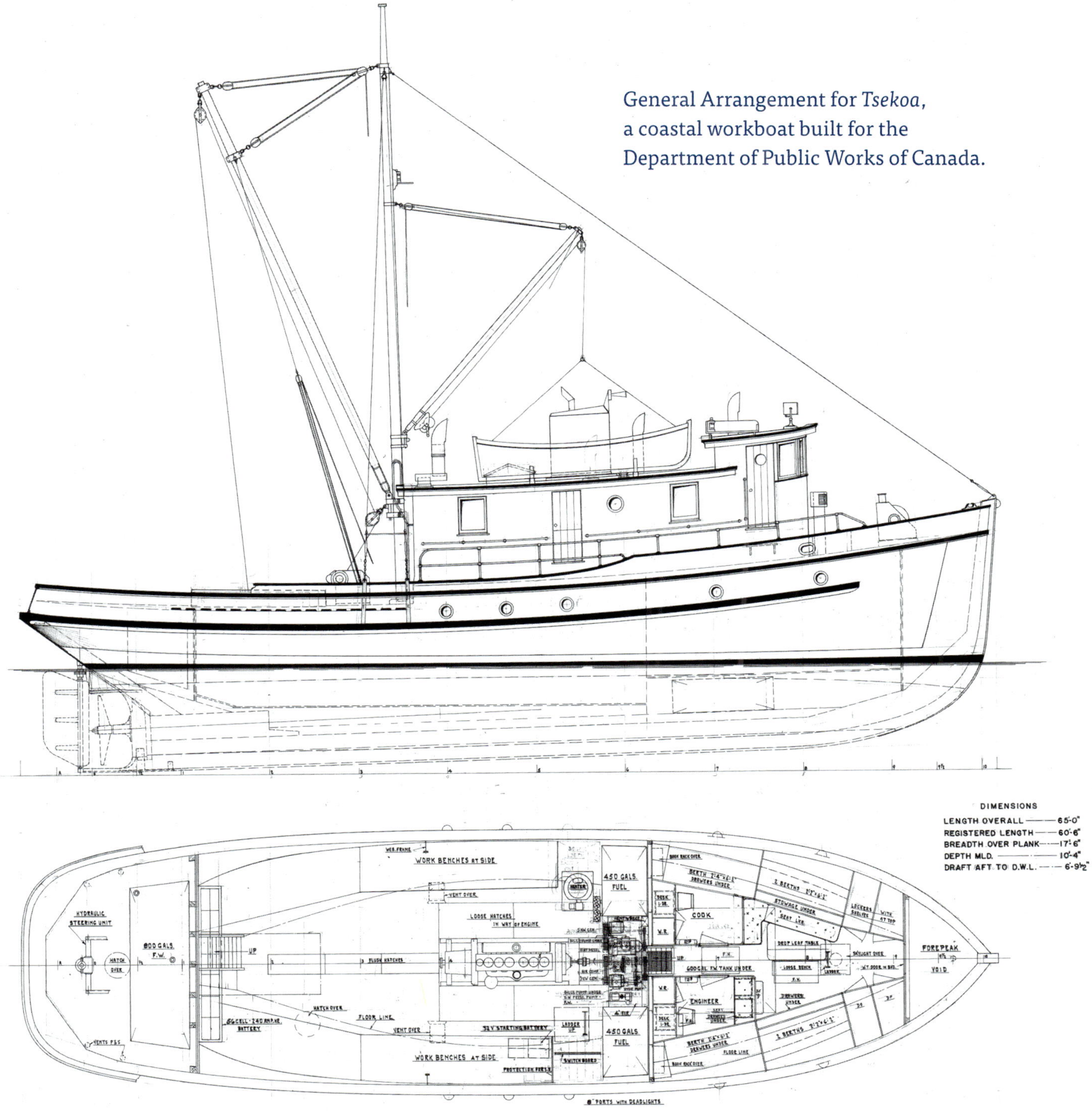

General Arrangement for *Tsekoa*, a coastal workboat built for the Department of Public Works of Canada.

Tsekoa, in private hands after her government service. The extended aft cabin doesn't particularly flatter her appearance.

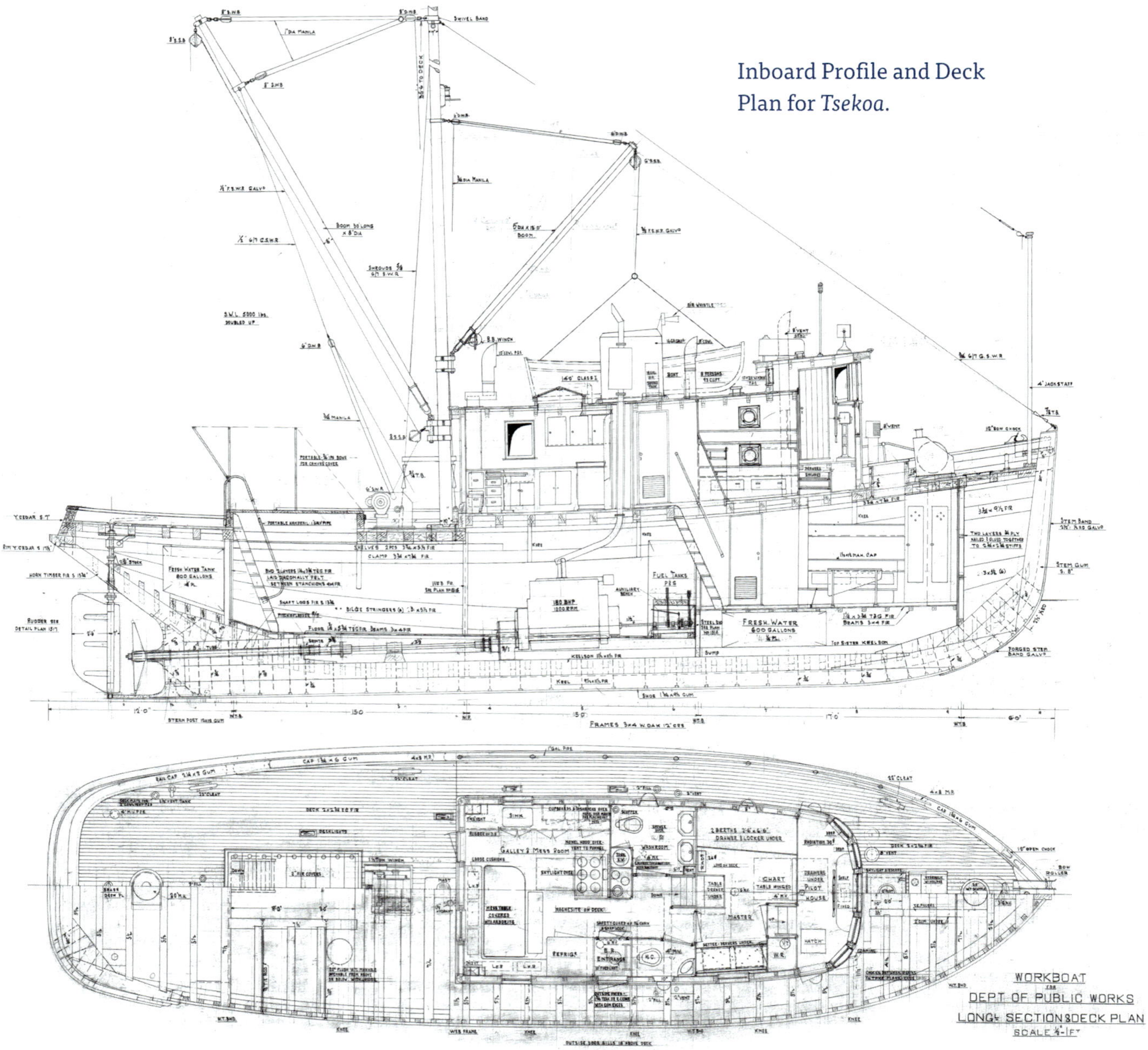

Inboard Profile and Deck Plan for *Tsekoa*.

Tsekoa II

This workboat was designed by Rob Allan in 1978, but her construction by Allied Shipbuilders was not initiated until 1984. Her mission, like that of her predecessor, was to support the maintenance and repair of the many coastal public wharf installations, operating in virtually all BC coastal waters, including the west coast of Vancouver Island. The vessel worked on a two-week round-trip basis, carrying lumber, concrete anchors, chain, and fully fabricated 40-foot access ramps for floats, etc. For handling this mixed cargo, she was equipped with a 10-ton hydraulic knuckle-boom crane on the aft deck. Built in steel with an aluminium deckhouse, *Tsekoa II* measures 87 feet 7 inches overall length by 23 feet 9 inches beam. With twin Cat 3408 diesels, she managed a 12.2-knot top speed. She was handy-sized and quite a rugged vessel for this demanding service.

The Department of Public Works ceased their coastal operations in about 2004, and *Tsekoa II*, among other assets, was sold off. This boat ended up in the hands of the Department of Fisheries and Oceans until 2009, then subsequently was sold into private ownership, where she functions to this day as a survey vessel for commercial owners.

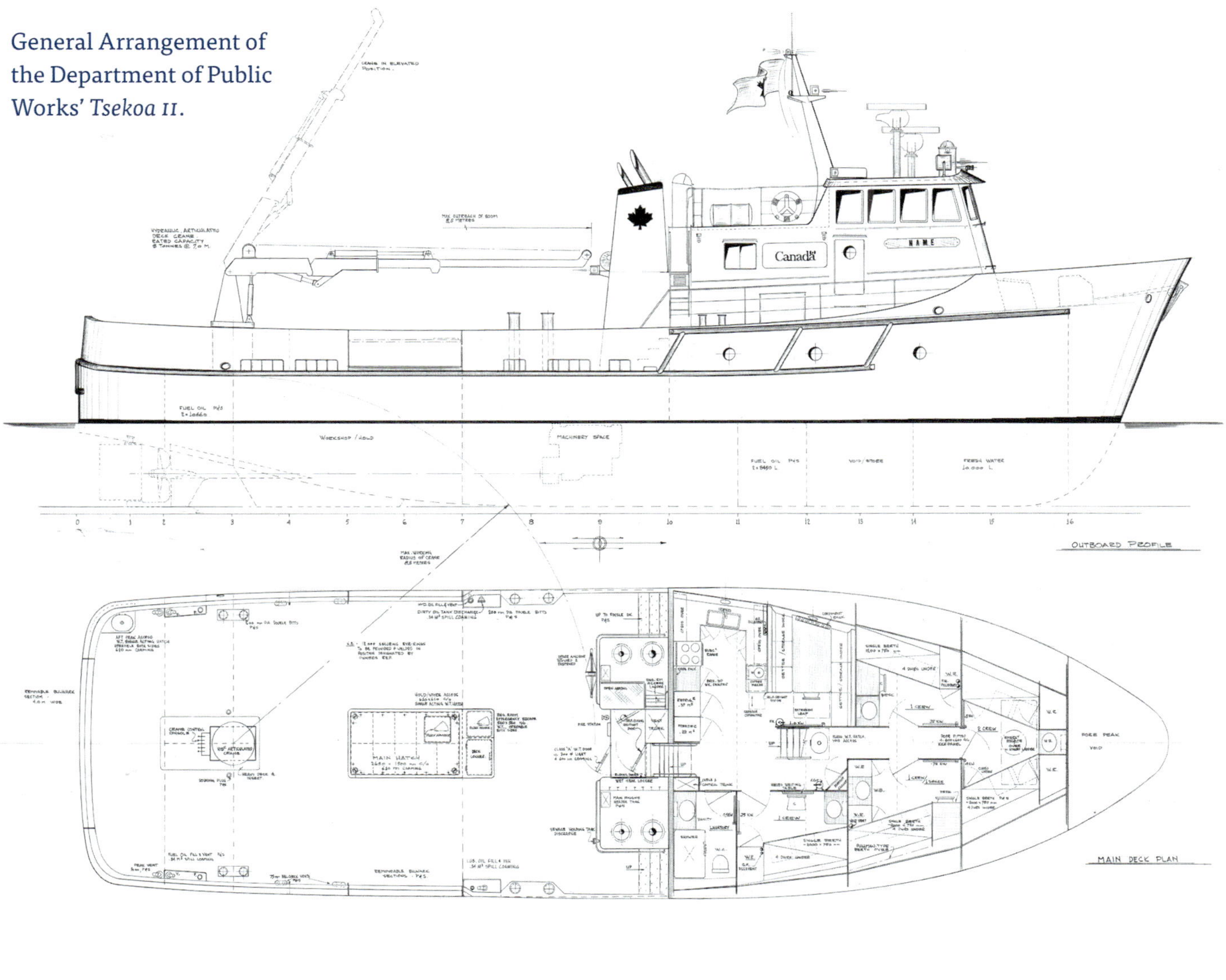

General Arrangement of the Department of Public Works' *Tsekoa II*.

Tsekoa II on trials in Vancouver harbour.

A Memorable Photo Shoot

The trials for *Tsekoa II* were conducted primarily within the confines of Vancouver's inner harbour. It is customary that the shipyard hires a photographer to take good action shots so that the publicity surrounding a new delivery shows the boat off to advantage. Murray McLellan of *Harbour and Shipping* magazine was the photographer of record on that day. He was stationed dockside at Ballantyne pier, while the boat, full of owners' representatives from various government departments, shipyard personnel (including Allied Shipbuilders' Arthur McLaren), equipment reps, and me, the naval architect, ran through a variety of performance tests in mid-harbour. Murray wanted the boat to get closer to his position so he could get much more focused shots, so he called the boat to make full-speed circling approaches toward the pier. The first couple of passes still weren't close enough to satisfy Murray, so he called for one more fast approach. The skipper, handling a new boat, wasn't yet sure of her manoeuvrability and he misjudged the approach. The boat crashed heavily into the concrete pier on her starboard bow, and many of those aboard, including Arthur, ended up in a heap on the wheelhouse floor! Fortunately, no one was hurt, but the boat certainly needed some rework, the cost of which was more painful to Arthur than the bump on his head.

—R.G. Allan

Gikumi

Gikumi was designed in 1954 for Mr. Fred Wastell, to service the Broughton Lumber and Trading Company sawmill in Telegraph Cove, on northern Vancouver Island. She was built by the Bissett and Gilstein boatyard in North Vancouver and launched in 1956. Her story is well told on the website of the current owners, Seabridge Ventures Ltd (www.gikumi.ca/history): "The name 'Gikumi' means 'Chief' in the tongue of the Namgis First Nation (Alert Bay) in the dialect known as Kwakwala. From her launch on Labour Day 1955 to the early eighties she worked delivering freight and towed logs and scows out of her home port of Telegraph Cove. She even did some ship docking and worked as a pilot boat over her long career. When the mill closed in Telegraph Cove, she became BC's first whale-watching vessel and remained in Telegraph Cove providing tours to the spectacular Broughton Archipelago until 2017."

On her 50th anniversary, in 2006, *Gikumi* was the subject of a feature article in *Western Mariner* magazine entitled "M V Gikumi: Three Men and a Boat," written by Mary Borrowman. The article describes the vessel's history and also honours the 25th anniversary of Stubbs Island Charters Ltd, who were *Gikumi*'s second owners, operating her in that pioneering whale-watching enterprise. Described as a much-loved local boat, the article focuses mostly on "...three mariners who have loved *Gikumi* the most—Fred Wastell, Jimmy Burton and Jim Borrowman." Jim and his business partner Bill Mackay had purchased *Gikumi* from Fred Wastell in 1983.

The vessel was sold again, in 2017, to John Turpin of Seabridge Ventures Ltd, who invested in a major restoration and refit, and this historic utility workboat turned passenger vessel looks splendid and fit for many more years of service to the coast of BC.

A Visit with *Gikumi*

Gikumi looking very distinctive in her latest colours.

In about 2013, my wife, Enneke, and I were on our way "up island" to participate in a North Coast kayaking trip. We took a little side-trip into Telegraph Cove, a place we had not previously visited. After a couple of hours of taking in this charming village we found ourselves on a dock overlooking the float where *Gikumi* was moored and taking on supplies. We were no different than any other tourists in the area, just watching the waterfront activities, when I heard a voice calling, "Are you Robert Allan?" Jim Borrowman, who I had never previously met, had somehow identified me from 100 feet away and called us down to have a look at his pride and joy. Unfortunately, we were running short on time and couldn't stay long enough to take her all in, but we had a good but too brief chat. It was abundantly clear that this man and his lovely boat were truly connected.

—R.G. Allan

Essington II

The *Essington II*, heir to the name of a famous wooden sternwheeler that worked the Fraser River for many years, was designed by R.F. (Bob) Allan in 1956 for the Canadian Department of Public Works (DPW), an agency that clearly had established a liking for the functional design work of the Robert Allan office. This 120-by-34-foot "snag boat" was fitted with a large slewing/luffing crane on the fore deck and was intended primarily to keep the lower Fraser River safe for navigation by removing the numerous "snags" which became embedded in the river bottom. Other duties included buoy work and light pile-driving.

The *Essington II* was built by Victoria Machinery Depot in Victoria and delivered in 1958. The ship, like most of the DPW's coastal assets, was sold to private interests in 1993 and continues to serve the coast to the present day in similar service.

Essington II when new.

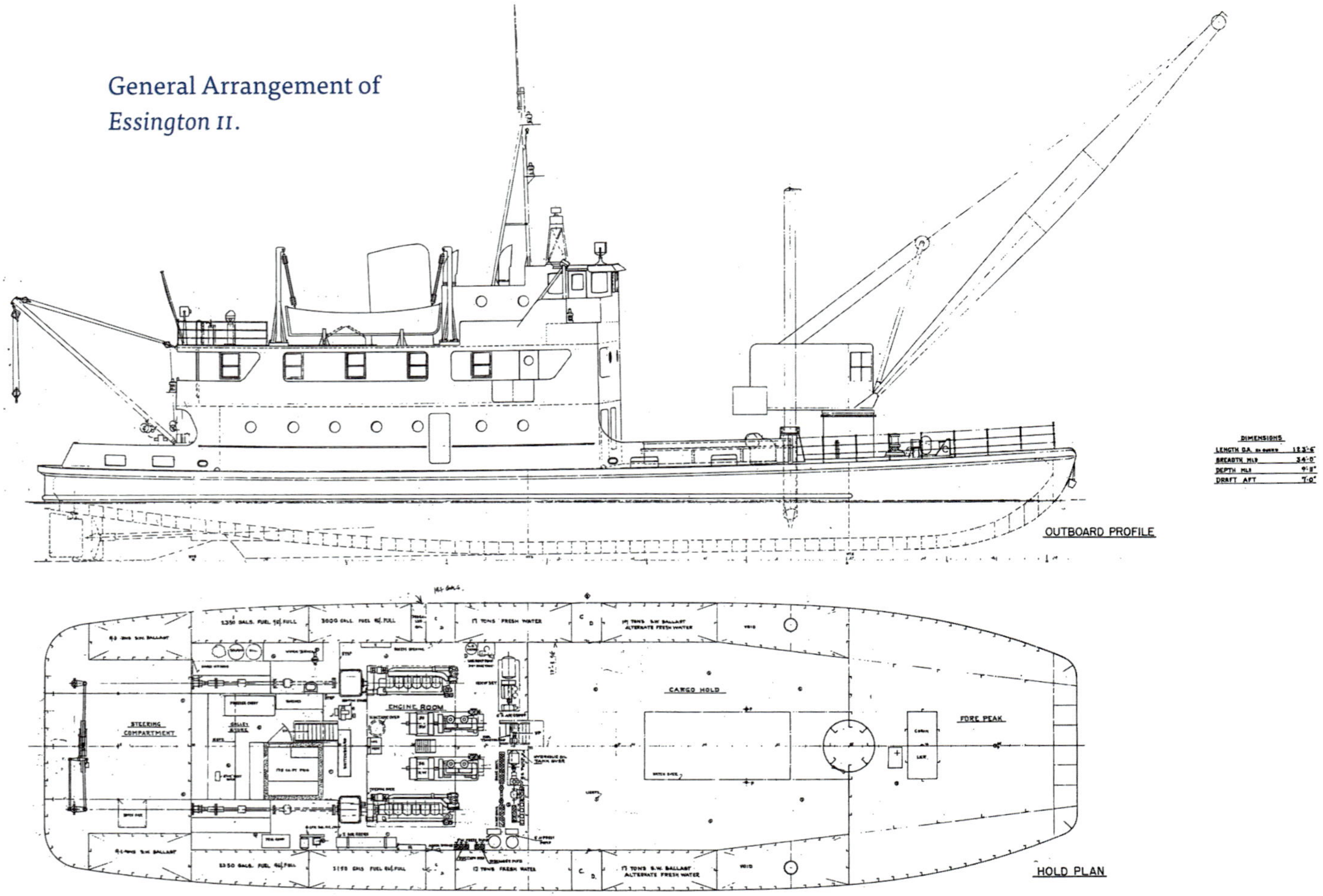

General Arrangement of *Essington II*.

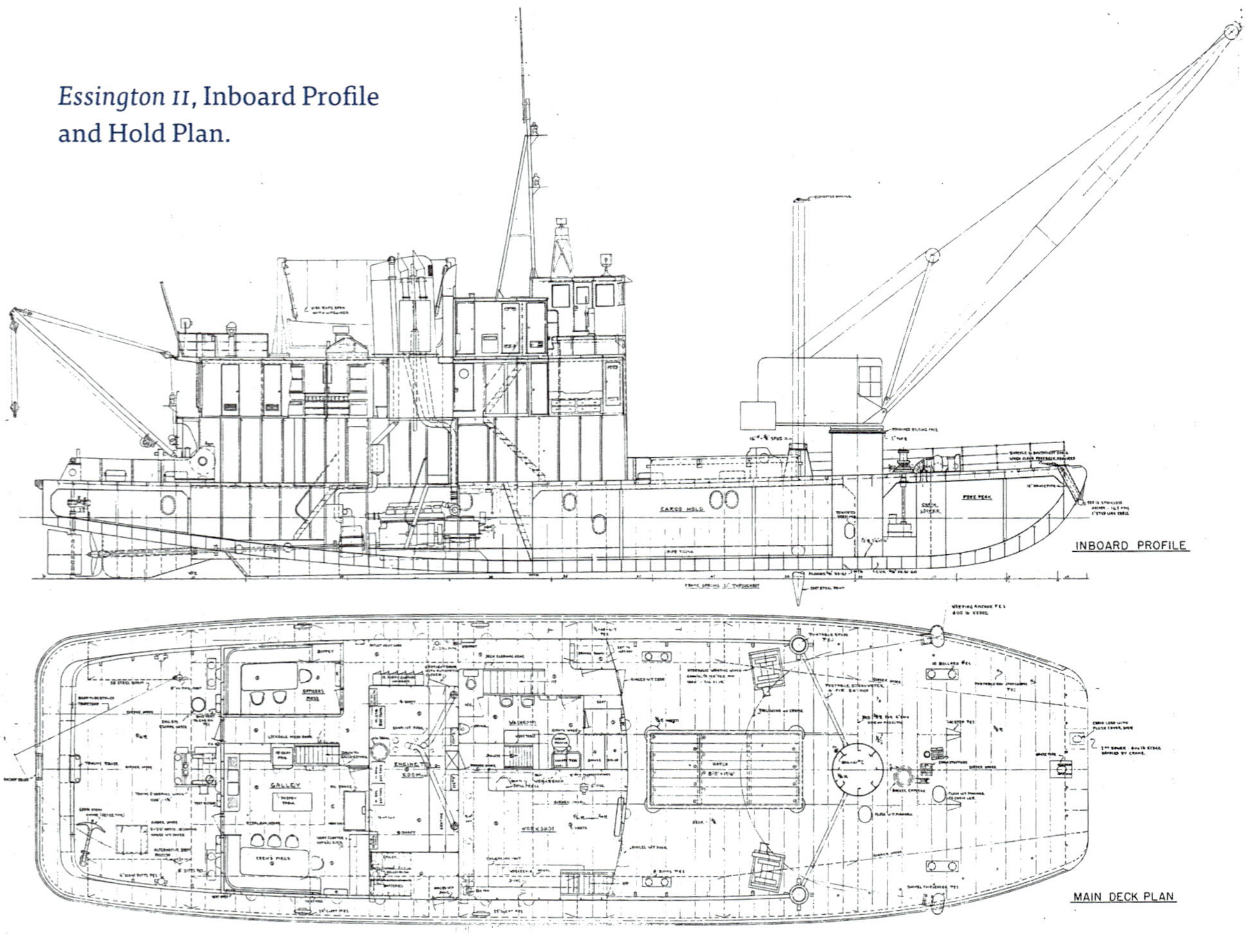

Essington II, Inboard Profile and Hold Plan.

Hugh A. Young

By the late 1950s Bob Allan was clearly making a name for himself as a highly capable designer, with his father effectively retired and the business starting to grow significantly. Several vessels had been designed by him for the Department of Public Works on the BC Coast (*Hilunga*, *Tsekoa*, *Essington II*), and in 1960 DPW returned to the Allan office with a commission for a survey vessel for the Mackenzie River system.

The *Hugh A. Young* measured 87 by 22 feet and had a 3-foot 8-inch draft. She was built by Allied Shipbuilders in 1962 and assembled in Hay River on the shores of Great Slave Lake, her long-time home base.

The role of the *Hugh A. Young* was to take a survey crew up the entire length of the Mackenzie River system every year in advance of the fleet of dredges that DPW operated throughout the 1960s and '70s. The crew would take bathymetry readings in all the small hamlets where vessels would call, and then set up sighting markers for the dredge crew to use. As this large, shallow and fairly silty river would freeze almost completely during the winter, every "port" would silt up and require dredging to enable essential supplies to be offloaded in the following summer season.

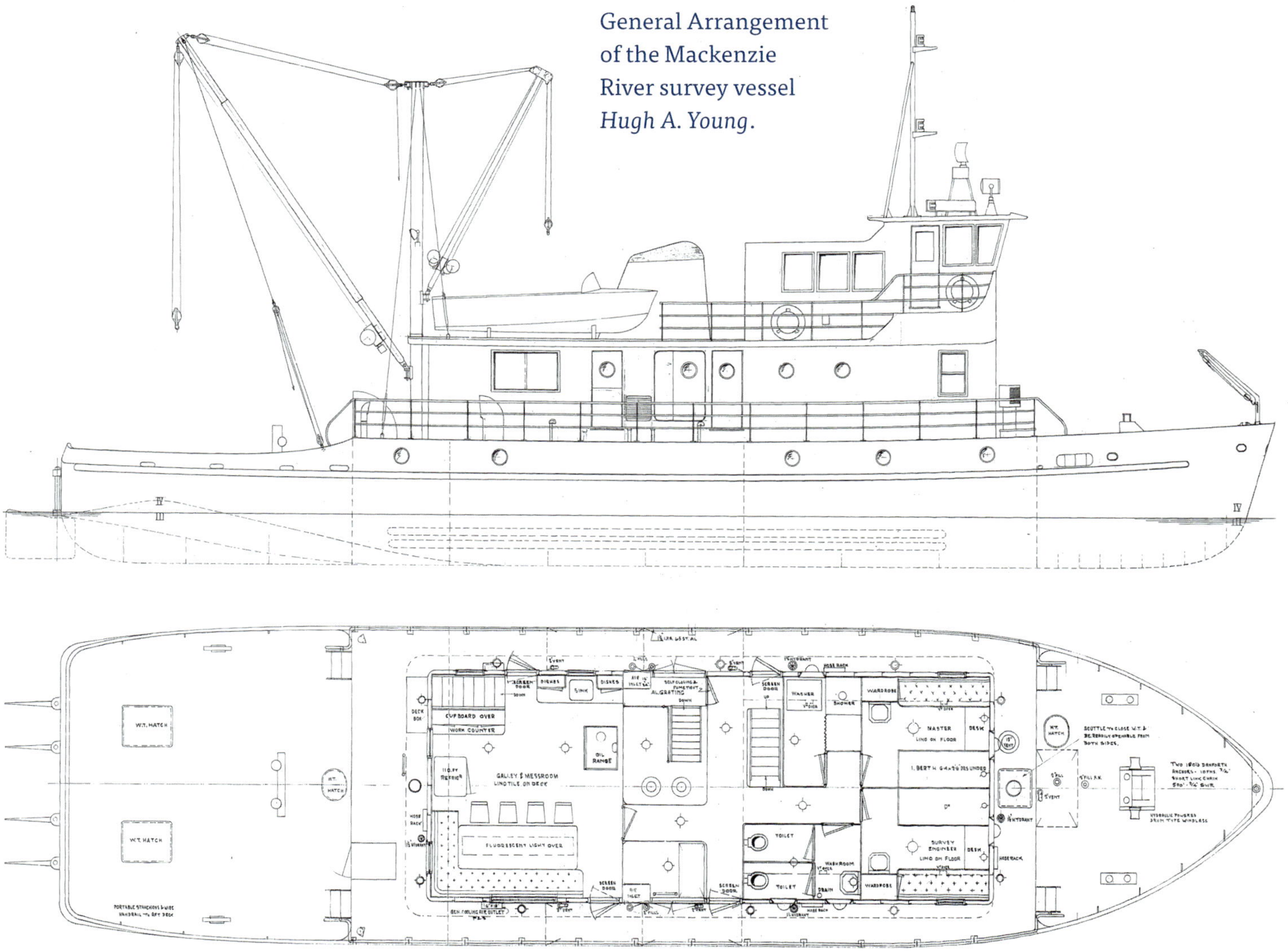

General Arrangement of the Mackenzie River survey vessel *Hugh A. Young*.

Hugh A. Young in Hay River, NWT.

A Summer on the Mackenzie

The *Hugh A. Young* has a special place in my memory, as I spent the entire summer of 1966 working on her as a deckhand. I was attending the University of British Columbia prior to transferring to the University of Glasgow, and from early May to late August we travelled the entire 1,080-mile length of the Mackenzie River, from Hay River to Inuvik and back (I left the ship at Fort Simpson on the return voyage in order to resume my studies in time). It was a tremendous experience to see the whole length of the river at that time, before much commercialization had taken place, with the notable exception of the development at Norman Wells for the Canol Pipeline during WWII.

The crew was interesting. It consisted of master, mate, chief engineer, the cook and two deckhands. They were all fairly easy to get along with, save the other deckhand, with whom I had virtually nothing in common but with whom I shared a rather compact cabin. The survey crew consisted of the chief surveyor, who had spent many years in the North and who was a font of information about the river, and two engineering students as his assistants. I got on very well with those two, but the "ship's rules" frowned on mere boat crew fraternizing with the more elite survey crew.

Rob Allan (top right) and the crew of the survey vessel *Hugh A. Young*, 1966.

My main memory of that summer is learning to cope with millions of mosquitoes and blackflies. When we had to go ashore to cut line for the dredges, we would be completely covered and have mosquito nets over our heads and necks, spray ourselves and our clothing liberally with bug repellent, and still get hundreds of bites. We were working in the marshy side banks of the river and its tributaries, cutting through dense willow groves with machetes in order to establish sight-lines to markers placed for the dredge to follow. It became our coffee-break challenge to see how many blackflies one could kill with one swat—I think the record was more than 30!

It was a great experience. The river is awesome in spots, but also often quite boring as it wends its way north through the boreal forests and ultimately through tundra. Although the pay wasn't exceptional, there was no place to spend anything, so I returned to Vancouver with all my summer wages intact and enough in the bank to buy my first car, pay for another year of university tuition and still have something reasonable left over.

—R.G. Allan

Multi-Purpose Workboat for Turkey

In the character of *Tsekoa II*, the *Salim Düzgit* is a recent RAlly 2500 TS class vessel designed by Robert Allan Ltd and built in 2019 by Sanmar Shipyards for Gemtac, part of the extensive Turkish Duzgit Group. The now 24.7-by-7.0-metre twin-screw utility boat is capable of many different tasks, including cargo transportation, crew boat service, construction and maintenance work, and light towing. The 90-square-metre cargo deck is rated to carry up to 32 tonnes. The boat is outfitted with seating for up to 12 passengers. This little workhorse notably started life in 2018 as a 19-metre-long boat and shortly after delivery was lengthened to its current almost 25-metre length. Powered by a pair of 404 kW Volvo diesels, she has a top speed of 12.5 knots.

Salim Düzgit, a multi-purpose "Sub-Agency Boat" serving the Turkish Straits.

Hudson Handler

Some designs are truly unique one-offs developed for a single, very specific purpose, and in some cases even for a single mission. Such a vessel is the *Hudson Handler*, designed in 1970 for Al Trice of International Hydrodynamics Ltd of North Vancouver.

Designing the *Hudson Handler*

One of my first projects after joining Robert Allan Ltd in 1970 was a design for the *Hudson Handler*, a tender for HYCO. HYCO's Al Trice came in and talked to Mr. Allan. He wanted a vessel that could launch and retrieve a PISCES submersible. It would have to be modular, portable, self-contained and self-powered. Initially, this was to service a contract in Hudson Bay for a petroleum company, Aquitaine. Al's idea was to build the tender, then take it up in pieces on the railway to Churchill, Manitoba, and put it together on the beach. This meant that each module had to be small enough to fit on a railway car. The vessel overall was to be 90 feet long and 40 feet wide, with two GM diesels driving deck-mounted Z-drive units. The hull was designed with six pontoon elements and then a hanger and launch and recovery ramp. It had a pilothouse and accommodation for about 10 crew. A very good Dutch engineer, Daan Hengeveld, was working for us at the time. Our biggest concern was how to assemble these modules—how we were going to hold them together, especially as it might not be calm weather when they got there. We didn't want them to all come apart. We looked at the Bailey Bridge pontoon system and how they made the joints and how they handled the stresses and everything. But then Daan came up with a better idea. It was sort of a clamp design. One of the advantages of that was that you could launch the pontoons and then clamp them together in the water without divers.

The *Hudson Handler* was composed of six pontoons, a hanger, and a launch and recovery ramp, with accommodation for about 10 crew.

I was supposed to do most of the drawings and design this clamp thing. It was a bit scary for me because we didn't have access to finite element analyses and other computer methods at that time. Then I visualized that if this thing came apart, it would come apart like a zipper, all these clamps would fly off. That was a little bit of a stressful thing. Mr. Allan looked at the design, and then Daan looked at it, and it seemed to be okay. So they built it, and Al Trice took all these parts up there and put them together on the beach, launched it and put a submersible on it. I believe their first job, it was to look for a piece of expensive equipment that was lost overboard, maybe by Aquitaine. The story that came back was that they found it on the first trip, to everyone's pleasant surprise. I later heard from Al Trice that after that job the boat was taken apart, transported to the East Coast, and reassembled for underwater pipeline and cable surveys. They were able to do this in quite severe weather, which was surprising, considering it was just an assembly of pontoons held together by clamps. We subsequently used the same clamp concept for at least half a dozen other modular pontoon barge projects, and all were actually quite successful.

—Hans F. Muhlert

Navigation-Aids Tenders

The Canadian Coast Guard (CCG) has jurisdiction over all of Canada's waterways and is responsible for maintaining all the floating (and many fixed) aids to navigation within them. This demands an array of vessels that are best suited to the specific regions in which they operate. The CCG fleet comprises everything from very shallow vessels working on the Mackenzie River to heavy icebreakers working in the Gulf of St. Lawrence and its Seaway system.

The vessels performing these tasks are variously known as "Navigation-Aids Tenders" or "Buoy Tenders." These are interesting and rather challenging designs, with very mission-specific cargo-handling gear. As well, the ships need to be able to carry out all the many secondary mission objectives required by CCG. The following describes the various navigation-aids tenders that Robert Allan Ltd has designed over the years.

Cove Isle, the first of two 20-metre nav-aids tenders for CCG in Parry Sound.

Isle Class Nav-Aids Tenders

In 1977 we won a contract to design a small nav-aids tender for the Parry Sound region of Lake Huron. It was a relatively simple little boat, but also was an interesting design challenge, fitting a lot of demands into a 65-foot shallow-draft hull. The Coast Guard had two of these vessels built at Canadian Dredge & Dock in Kingston, Ontario: the *Cove Isle* and *Gull Isle*. These boats were 20.0 metres long by 6.0 metres beam, with a 1.35-metre draft. They were powered by twin 250 bhp diesels, and the working deck and the cargo hold was served by a 7.0 T-m knuckle-boom crane.

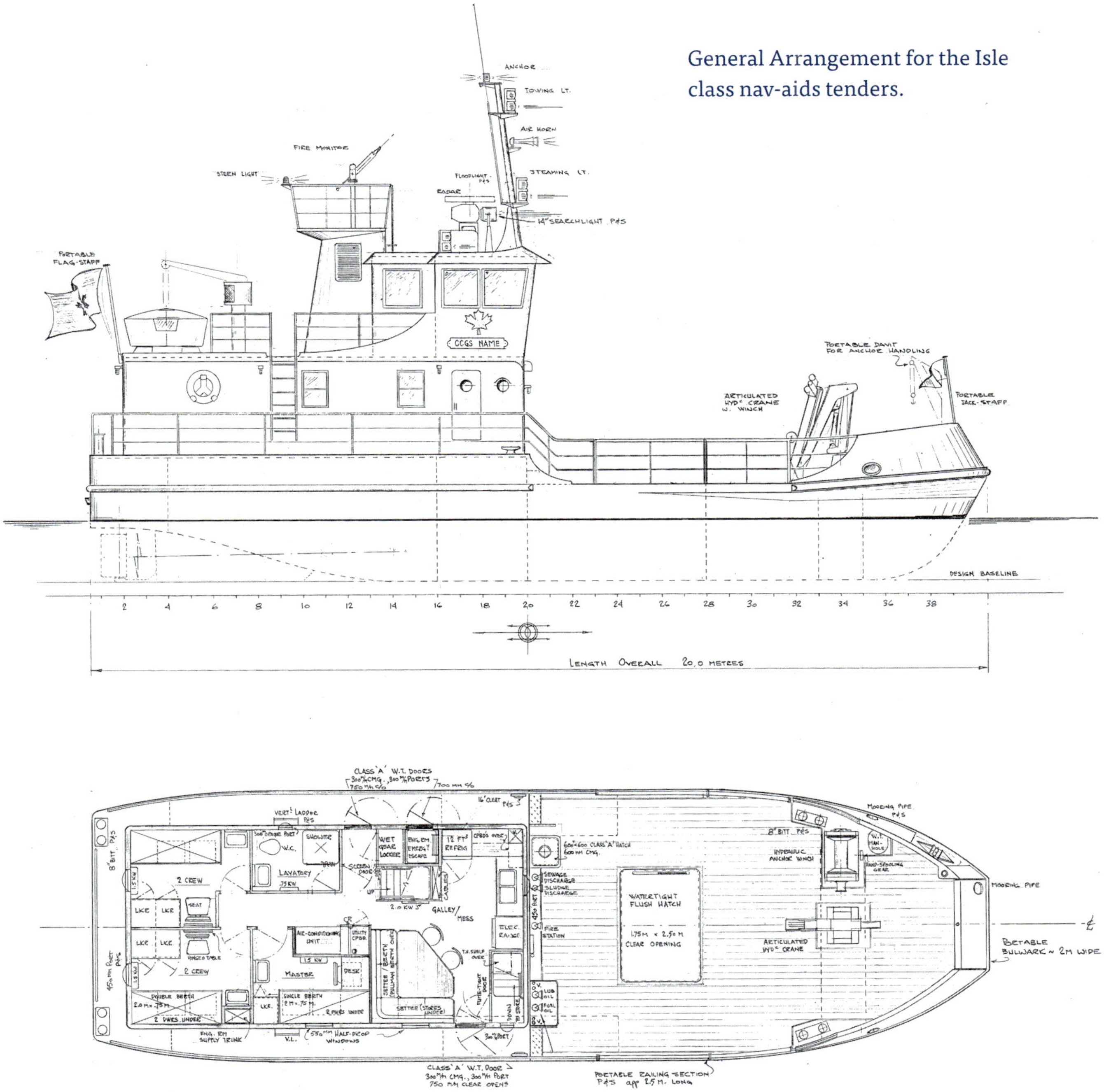

General Arrangement for the Isle class nav-aids tenders.

Type 800 Class Nav-Aids Tenders

In 1983 the company was awarded a contract for a further evolution of these inshore nav-aids tenders, which became designated as the Type 800 class. CCG had four of this 23-metre-long class built throughout 1985–86, all at Breton Marine Industries in Nova Scotia. Named *Caribou Isle, Partridge Island, Ile Saint Ours* and *Ile des Barques,* these handy-sized buoy tenders currently still operate throughout Eastern Canada. They are described by Mac Mackay's *Shipfax* website as "...miniature buoy vessels, besides being endearingly cute...are self-contained units [that] could go places that large ships could not."

The Type 800 class vessels measure 23.0 metres length overall and 6.0 metres in breadth. They draw only 1.35 metres on a hull depth of 2.25 metres. Propulsion is by a pair of Detroit Diesel 6V-92 main engines delivering 636 bhp to two controllable-pitch propellers. There is a knuckle-boom deck crane for handling small buoys and other equipment. What looks like a funnel is actually the base for a firefighting platform, with the engine wet exhausts exiting through the transom.

General Arrangement for the CCG Type 800 nav-aids tenders. This was one of the earliest projects for which our first CADD system was used.

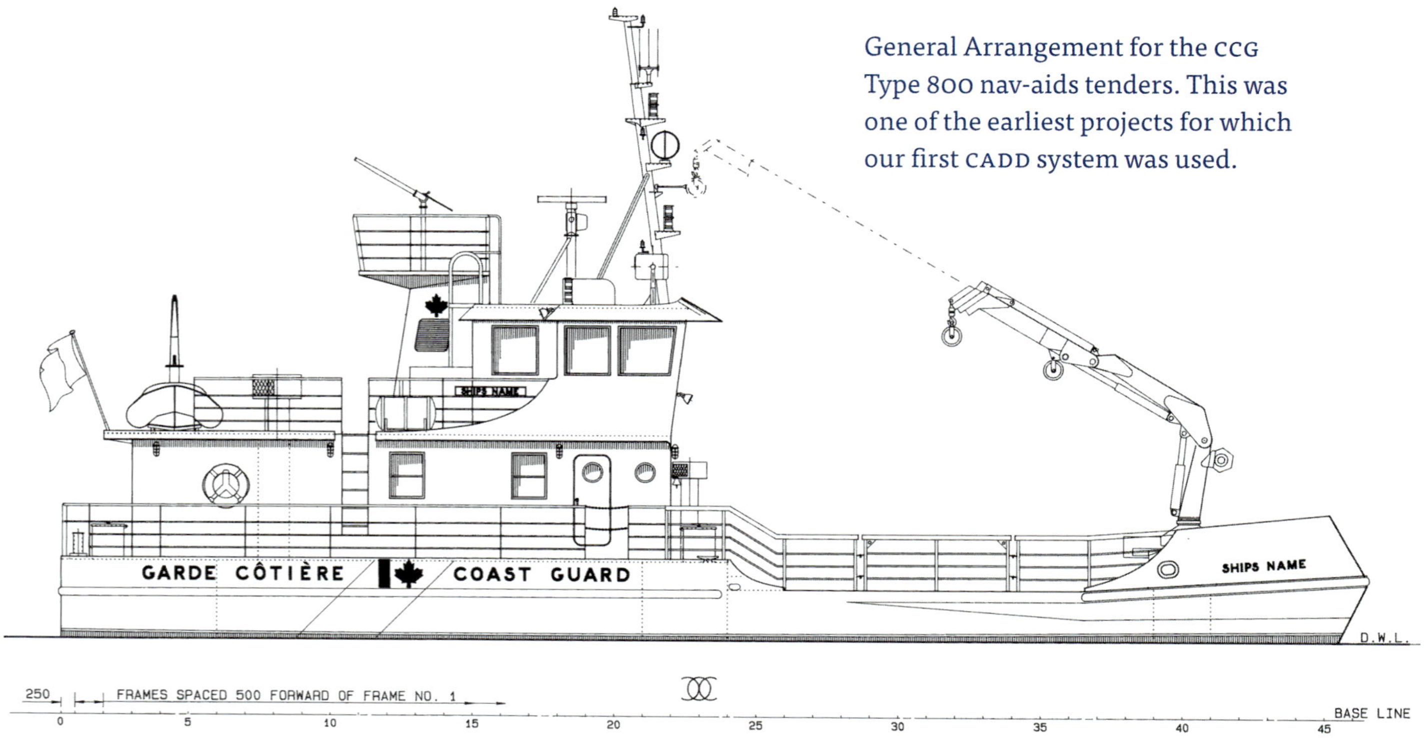

PROFILE

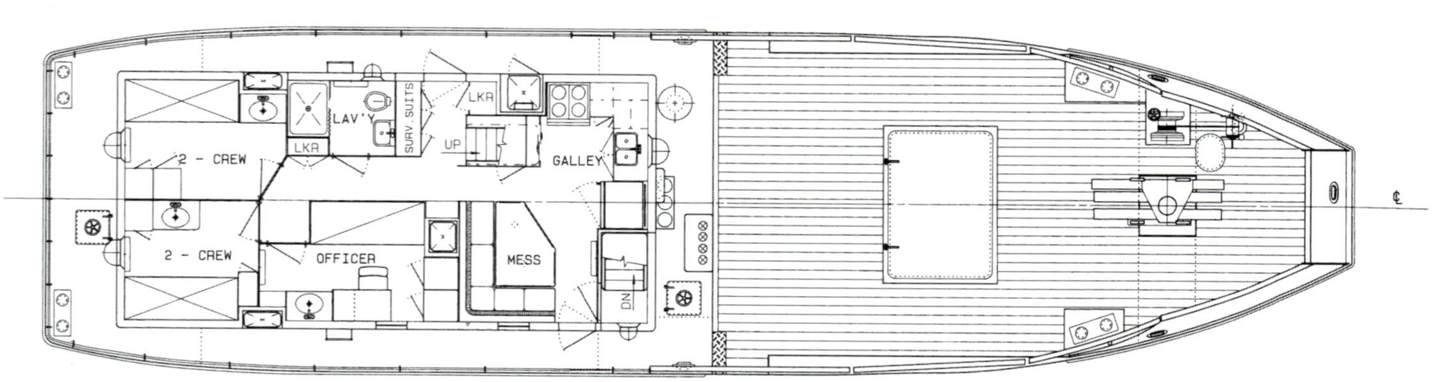

MAIN DECK

Type 800 inshore nav-aids tender *Ile Saint Ours*.

Type 1050 Nav-Aids Tenders

One of the more interesting large design projects that Robert Allan Ltd completed during the lean 1980s was for a new style of ice-breaking nav-aids tender for CCG, based on the general layout of an offshore supply vessel. This was a significant departure from the traditional buoy tender configuration, which has the deckhouse aft and usually twin masts close to the house front with booms extending forward. That set-up had been used on every major CCG nav-aids tender for many decades. The crane-aft concept also didn't quite fit into the recently announced fleet standardization system that CCG had been promoting for several years, with small Search and Rescue cutters identified as Type 100 all the way up to the Type 1200 heavy icebreakers. Accordingly, this new odd-ball design fit between the Type 1000 and Type 1100 ice-breaking nav-aids tenders and was designated as the Type 1050, one of the first vessels built under the new standardized numbering system.

Two vessels of this type were built as part of the Canadian Government Shipbuilding Industry Assistance Program (SIAP) set up in 1976 and which lasted until about 1985, implemented to support struggling Canadian shipyards. Designed in 1981, the *Earl Grey* was built in Nova Scotia at Pictou Shipyards Ltd (the successor to Breton Marine Industries) in 1985, and the *Samuel Risley* was built by Vito Steel Boat and Barge Ltd in Delta, BC, in 1984. The vessels now serve in the Great Lakes and on the eastern seaboard.

All reports concluded that the buoy-handling system with the crane aft worked very well, but the concept has not yet been adopted in any subsequent vessels, mainly because the CCG have not built any additional nav-aids tenders since the Type 1050, Type 1100 and Type 1200 class ships, which were all built in the 1980s.

The particulars of the 1050 class ships are as follows:

Length Overall	69.73 m
Beam, Moulded	13.70 m
Depth, Moulded	5.80 m
Load Draft	5.2 m
Power	4 × 1,625 kW diesels in pairs with combining gearboxes to serve two propellers in nozzles
Ice Class	CASPPR ice class II
Crane	Motion-compensating type: 15 tonnes at 13.5 metres capacity

CCGS *Samuel Risley* in the ice.

Below: General Arrangement for Type 1050 ice-breaking nav-aids tenders.

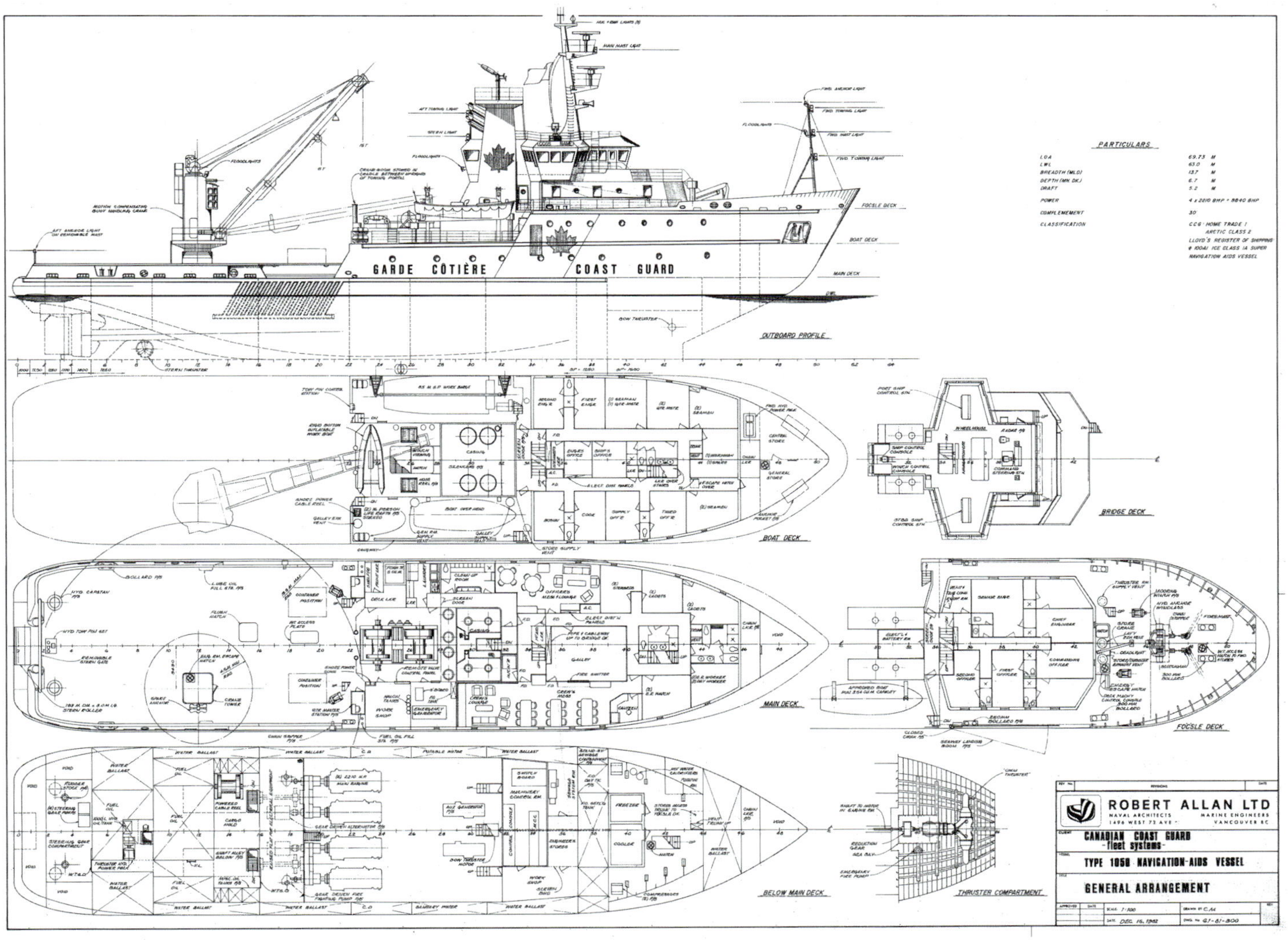

9

Research Vessels

Although relatively few in numbers, research vessels of various types have been present in the Robert Allan Ltd design repertoire for many years and have become increasingly large and more complex over time. These vessels present very interesting design challenges, as one must marry the basic ship design and performance requirements with the on-board science needs of the crews and their specialized equipment and facilities. Sea-keeping characteristics are also critically important, as the science crews are generally not mariners, and their ability to work effectively is impaired by excessive motion.

Opposite: *R.B. Young*, see page 170.

This chapter describes the fleet of research vessels designed by the Robert Allan office from the late 1950s to the present day.

Salvelinus

The MV *Salvelinus*, designed early in 1959, is a rear/side trawler built that same year by Star Shipyard (Mercer's) Ltd of New Westminster, BC, for the Arctic Biological Station of the Fisheries Research Board of Canada. Her working career has been spent entirely in the northern oceans of Canada and at least part of that is nicely described in a web posting entitled "Five summers on board the MV *Salvelinus* in the Western Arctic, 1964–68" by Dr. Jean Boulva of Quebec.

According to Dr. Boulva's notes, the *Salvelinus* was transported to Tuktoyaktuk by the Arctic Sea Lift (an annual movement of gear up into the Arctic, mobilized mostly to support US oil companies working on the North Slope) and was involved in marine research across the breadth of the Beaufort Sea, between Herschel Island and Cambridge Bay, from 1959 to approximately 1975. The nature of that work, described as the MV *Salvelinus* Program, was "...headed by Dr. Gerry Hunter and aimed to study the marine fauna and oceanography of the Western Arctic. Some occasional freshwater work was also done. Oceanography was carried out, on occasions from the ice before breakup, and mostly from the ship afterwards. Navigation season generally lasted from early July to early September."

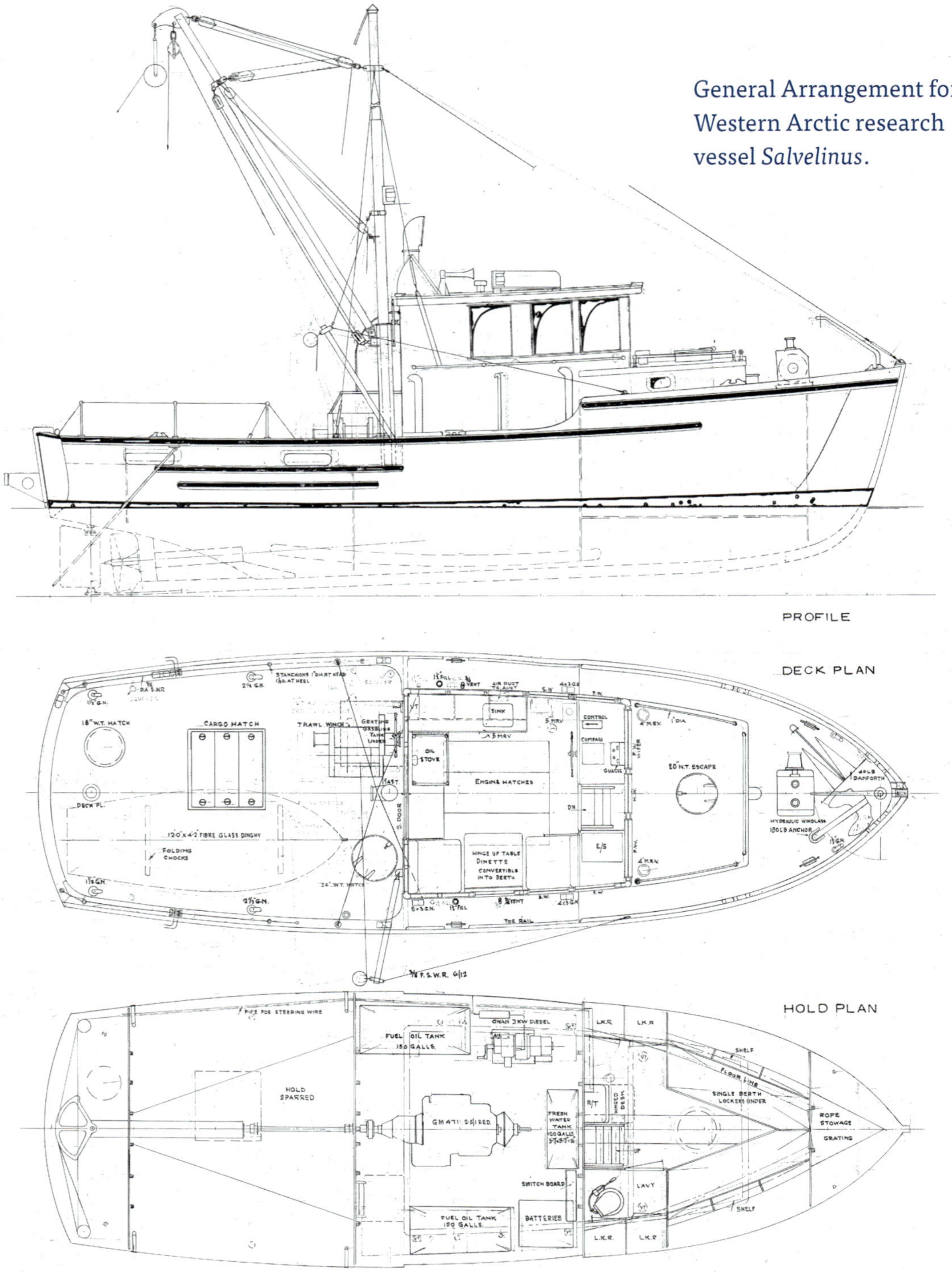

General Arrangement for Western Arctic research vessel *Salvelinus*.

Salvelinus at home in the western Arctic Ocean.

Salvelinus is a small ship by any standards, at 37 feet 2 inches overall length by 12-foot beam, and must have been quite a cozy home for her crew of two or three over the summer research season. In a rather unusual trait, her design includes a sail plan, the purpose of which is not clear, but the two loose-footed sails are small and likely just used as steadying sails during various operations.

Her wooden construction, with primarily 1 ¼-inch edge-grain fir planking supported over very closely spaced oak frames, was obviously sturdy enough to cope with the ice conditions encountered in the early stages of the operating season for many years. For this purpose, the Structural Sections drawing also indicates she was fitted with "Sheathing of ½" gum[wood] laid over tar and felt [in] widths of 1-¾" spiked on to extend all fore and aft 9" above and 9" below LWL also to extend down to forefoot and up in way of anchor chafing." Registry records indicate she was sold to private interests in 1990 and was still in service as of 2019.

Salvelinus in the ice of the western Arctic Ocean.

Caligus

The near-shore fisheries science vessel *Caligus* was designed in 1966 for the Pacific Biological Station of the Department of Fisheries in Nanaimo, BC. An all-steel vessel built by John Manly Ltd in 1967, she measured 56 feet in length with a 15-foot beam and was powered by a single 230 hp GM diesel engine. She was configured as a dragger/trawler. At some stage in her career her initially conventional bow shape was modified to a rather ugly cantilevered form for purposes unknown.

Registry information indicates that she too was sold out of government service in about 2002 and is still fishing in private hands.

Caligus in her original configuration (top) and (bottom) after her "nose job."

Discovery

In 1978 Robert Allan Ltd's Singapore office, in full stride at the time, obtained a contract from the Port of Singapore Authority (PSA) for the design of two 50-foot hydrographic survey launches. Designed for a speed of 16 knots, they were built of aluminium, which was an uncommon material in Southeast Asia at the time, primarily because there were few qualified welders or welding shops for that type of work in the late '70s. These were fairly basically outfitted, equipped as day boats to take soundings in the territorial waters around Singapore. They were, however, an important project in establishing our credibility as a professional design office in Southeast Asia, and over the years the Port of Singapore Authority has become a very good client.

Port of Singapore Authority hydrographic survey launch *Discovery*.

R.B. Young

The next research vessel assignment for Robert Allan Ltd came in 1984, when the company was contracted by the Canadian Department of Fisheries and Oceans to design a coastal hydrographic survey vessel.

Built by Allied Shipbuilders Ltd in 1990, the *R.B. Young* was designed to conduct oceanographic and hydrographic research on BC's west coast. At 30.35 metres long, she had sufficient length to be a handsome little ship, unlike many of her length-constrained counterparts. She has accommodation for five ship's crew and six scientists, a large science lab on the main deck and a fully equipped hydrographic centre aft of the wheelhouse.

In 2003, while still one of the newer assets of the DFO fleet, she was sold to the State of Alaska, where she continues to operate as the research vessel *Kestrel* for the Alaska Department of Fish and Game.

Outboard Profile and Main Deck Plan of the hydrographic survey vessel *R.B. Young*.

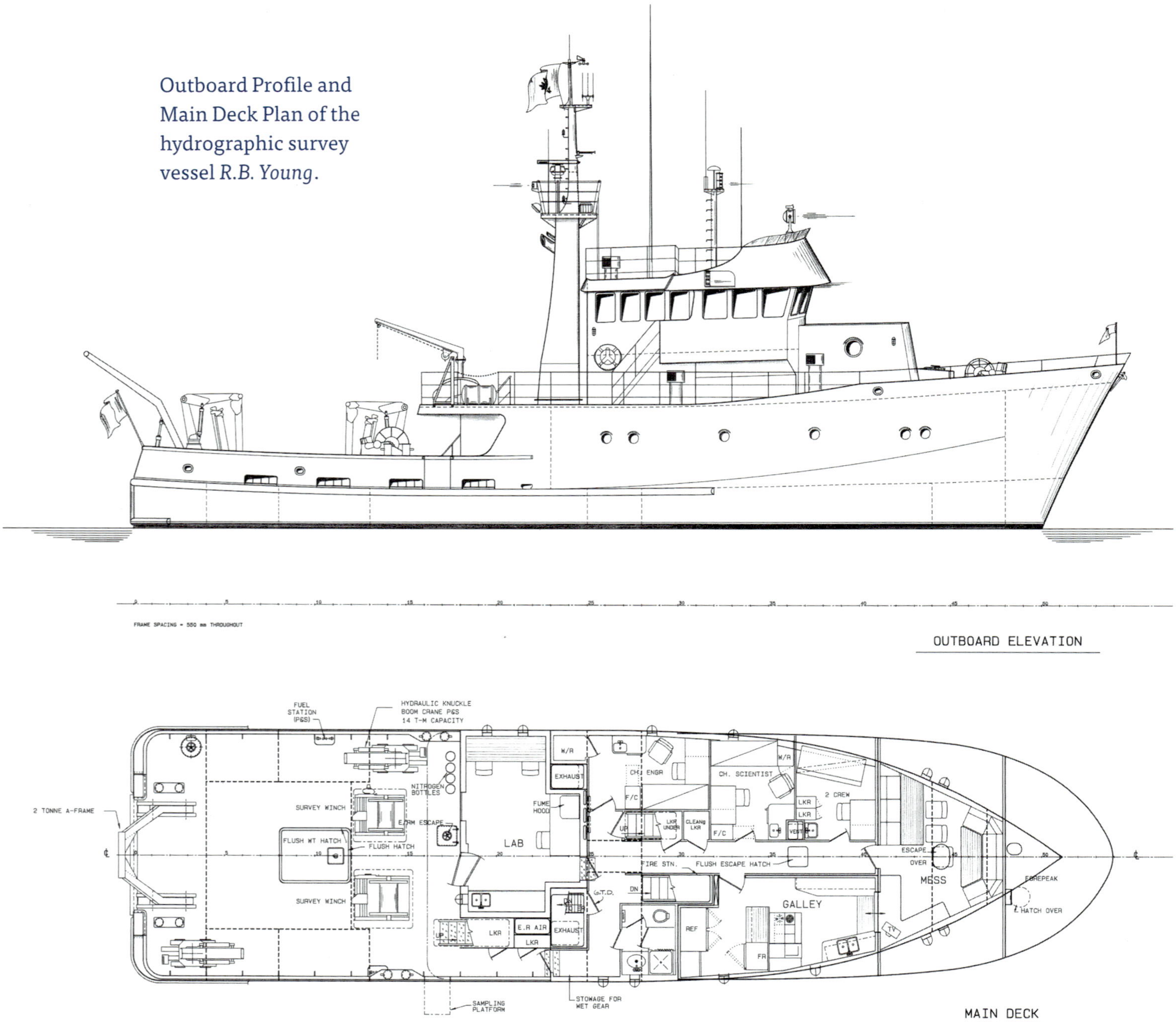

Top left and right: *R.B. Young* on trials.

Bottom: *Kestrel*, ex. *R.B. Young*, now operating in Alaskan waters.

FARV "A"

The files of Robert Allan Ltd contain many "paper ships," vessels designed to varying degrees of detail but never built. Many of these were concept designs for proposal purposes or just for verifying if a design idea was viable or not. Generally, these wouldn't warrant mention, but certainly the largest and most complex of these unbuilt designs is the Fisheries Assessment Research Vessel (FARV) Type "A" designed in 1986 for Fisheries and Oceans Canada. This was a major project, coming as it did in the midst of the lean 1980s, but it was also a really interesting and challenging design.

The FARV "A" was a complex ice-classed ship, 72.1 metres long by 14 metres beam, outfitted for a crew of 20 and 12 scientists. It was twin-engine/single-screw with a CP propeller in a thrust-augmenting nozzle. The basic configuration was that of a deep-sea trawler, with associated fish-sampling rooms and wet labs instead of a fish-hold, plus a full oceanographic sampling winch system and labs.

It is not uncommon to develop a design, usually just to a "proof of concept" stage, in an attempt to win a design competition, but it is very disappointing when a very comprehensive and complete design package such as that which was performed for the FARV "A" is completed and used to obtain construction bids from shipyards, only to be cancelled before the construction phase, for reasons never well explained.

Outboard Profile of the FARV "A" ice-classed fisheries research vessel.

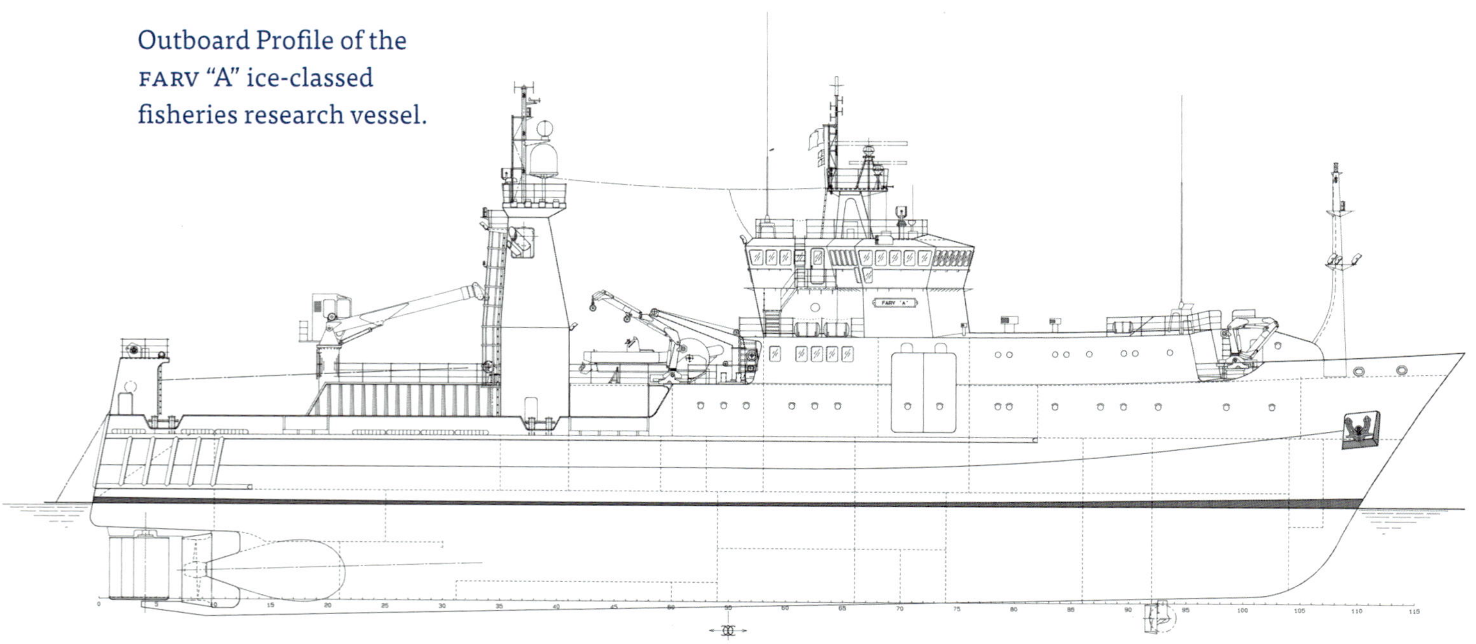

Inshore Fishery Research Vessels

In 2009 Robert Allan Ltd was commissioned by the Canadian Coast Guard (which by this time had integrated the Department of Fisheries and Oceans' fleet into the CCG fleet) to develop a design for a small Inshore Fishery Research Vessel (IFRV). The design was constrained to be only 22 metres long, but the operational requirements were better suited to a vessel closer to 30 metres in length. The rationale behind the length constraint had its roots in local concerns that any boat larger than 22 metres would not be able to manoeuvre well in the small local harbours, not recognizing the enhanced capability offered by the multiple high-lift rudders and bow thruster incorporated into a modern design.

CCG described the mission of these vessels as "a coastal research vessel. As such, it can deploy many kinds of equipment including various types of trawls, scallop dredges, remote-operated vehicles, camera sleds, plankton nets, acoustic and geophysical survey equipment, and water sampling arrays." Secondary missions include search and rescue, environmental response, and the support of marine programs for other government agencies.

The first of the 22-metre IFRV class were the CCGS *Leim* and the *M. Perley*, both built in 2012 by Méridien Maritime Réparation Inc. in Quebec. The *Leim* is based in the Gaspé region and the *M. Perley* in Nova Scotia. At 22 metres by 9.2 metres, this vessel type is unlikely to be described as elegant, but they have proven to be effective and capable small vessels.

Recent 22-metre Inshore Research Vessels for Canadian Coast Guard: CCGS *Leim* (left) *and* CCGS *M. Perley* (right).

Shortly after the construction of the *Leim*, Robert Allan Ltd was contracted to develop a slightly longer version of this Inshore Research Vessel type when DFO undoubtedly recognized that more length would benefit every aspect of performance and operations. The first of this 25-metre class of research vessels was the CCGS *Vladykov* based in St. John's, Newfoundland. This vessel was also built by Méridien Maritime Réparation Inc. in 2012. The *Vladykov* is named after an internationally known ichthyologist, the late Professor Vadim Dimitrievitch Vladykov, who contributed significantly to the study of fish biology in Canada.

At the time of writing an even larger 32-metre version of this same type of vessel is under design development for Fisheries and Oceans Canada.

CCGS *Vladykov*

Offshore Fisheries Science Vessels

The Canadian government introduced a program in 2010 to rebuild aging government fleets and to simultaneously revitalize the Canadian shipbuilding industry. The program, ultimately named the National Shipbuilding Strategy (NSS), resulted in the awarding of two significant contracts: one to Irving Shipbuilding of New Brunswick for all surface combatant vessel requirements, and one to Seaspan's Vancouver Shipyard for all non-combatant type vessels, principally for the Canadian Coast Guard and Fisheries and Oceans.

The first ships designated for construction under this program were three Offshore Fisheries Science Vessels (OFSV). Robert Allan Ltd, in partnership with Alion Canada of Ottawa (teamed as "RALion" for this project), won the contract in 2010 to design these complex multi-tasked research vessels. Alion is a major US firm of naval architects with a lot of experience designing specialized research vessels for the US Navy and other government departments. Their expertise in the underwater acoustics area of design was critically important. We made a good team.

The initial design was developed to address a comprehensive statement of operational requirements (SOR), but without being informed of any capital budget constraint. At the completion of the design, a detailed construction cost estimate was prepared. DFO then advised that this cost was greater than their budget allowed, and that the design had to be "shrunk" in order to meet the previously undisclosed budget. That necessitated virtually a complete redesign, which was then bid by the shipyard charged with its construction under the NSS program. In the meantime, the scientists and operational personnel who would have to spend substantial portions of their professional careers aboard these ships expressed concern (to the shipyard!) about the new constrained size of the ship and the onboard spaces. The shipyard was then charged with redesigning the ship, essentially back to the Phase I design dimensions and configuration. The three vessels, christened CCGS *Sir John Franklin*, CCGS *Capt. Jacques Cartier* and CCGS *John Cabot*, were completed and delivered in 2019/2020.

Due to the convoluted and prolonged evolution of this design, the RALion team was ultimately not given due credit for its work in developing the complex design for these 63.4-metre ships. The program certainly highlighted some shortcomings in the government's vessel procurement processes.

The OFSV CCGS *John Cabot* during sea trials.

RV *Investigator*

Shortly after completing the first round of the design for the Canadian OFSV project, the RALion team collaborated with Sembawang Shipyard in Singapore to win an internationally-bid "design-build" contract to deliver a major ice-classed oceanographic research vessel to the Commonwealth Scientific and Industrial Research Organisation (CSIRO), Australia's National Science Agency.

Commissioned in 2014 (five years before the smaller but similar Canadian ships were finished) and based in Hobart, Tasmania, RV *Investigator* is Australia's flagship research vessel. At 94 metres in length, she is capable of delivering up to 300 research days each year. The vessel has impressive scientific capabilities and possesses a wide range of on-board and modular laboratories and facilities.

According to the CSIRO website: "The vessel has the capacity to undertake biological, oceanographic, geological and atmospheric research, as well as maritime training and education, and related outreach activities. It accommodates 40 researchers and technicians and 20 crew and has an endurance of 60 days and 10,000 nautical miles without resupply. The *Investigator* is the first Australian vessel equipped with dedicated atmospheric laboratories. These world-class atmospheric facilities host a suite of instrumentation measuring meteorological packages, greenhouse gases, radon, ozone and a range of aerosol parameters."

The short time frame and the competitive price at which this contract was completed, and for which a highly successful ship was delivered, represent an interesting contrast to the Canadian government program for similar vessels.

3D images of *Investigator*, a major ice-classed oceanographic research vessel for Australia.

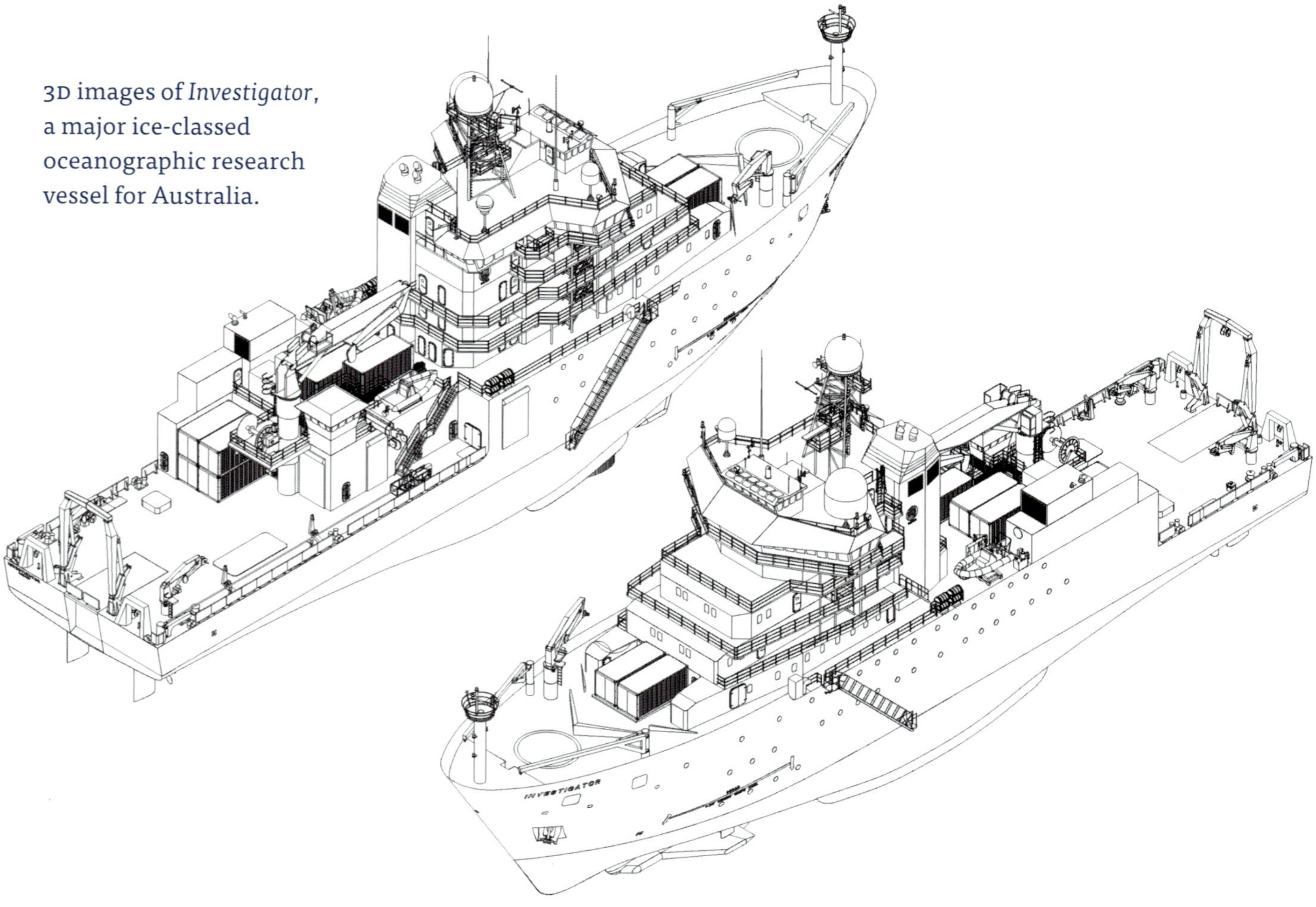

RV *Investigator* working in Antarctica.

Research Ship for Australia

This memorable project started with a design that Alion had for a smaller American research vessel, but Rob said he didn't like the lines of the thing. He said it didn't look like a Robert Allan design, so one of the steps in the process was to reconfigure the outside to make it look better, and Rob did that. The *Investigator* is a classic example of a Robert Allan design. Mathematically, it was a 50:50 allocation, but practically, Alion was far better at certain things, and we were far better in others. They did all the science equipment and most of the electrical design, as well as the deck arrangement and how all the science equipment would be interrelated and co-located. That was one of their many skill sets. We did all the structure and stability, the mechanical bits and pieces and things like that. We pulled it all together and it's a great ship. Our relationship with Alion (now owned by and known as Serco) still goes on, and we still do things together when they come around—largely around research ships.

—Rollie Webb

L.A. FIREBOAT 2

10

Fireboats

Of all the vessel types on which we have had the privilege to work over the years, fireboats represent one of the most interesting and often most challenging. This is not because the technical challenges are so very difficult to meet (though they certainly can be), but there are often competing political influences behind the project, different geographic and operational constraints that must be met, and the client (very often a city government or a port authority) may have restrictions on their purchasing processes which can limit the ability of a designer to develop the design in the best interests of the user department.

Fireboats are nearly always extremely equipment intensive, so the layout of main machinery and the large sizes of pumps and piping require very careful planning to ensure sensible construction, operation and maintenance. Also, because equipment choices will influence both layout and the design of supporting structure, one must make quite definitive choices for main machinery at a relatively early stage of the design. Without this pre-selection, the rest of the design can at best be described as generic, leaving many fundamental design decisions until later in the process. This pre-selection can present challenges in the procurement process, as clearly basing the design on a defined make/model doesn't give the supplier an incentive to sharpen their price. We have had considerable success, however, in pre-bidding the major components competitively before the design is fully detailed, thus bringing the main equipment suppliers into the design process at an early stage, leading usually to good teamwork among all the players.

Opposite: The Los Angeles fireboat *Warner L. Lawrence* led off an extensive series of new fireboat designs for major American cities.

The fundamental performance parameter for most fireboats is pumping capacity and water stream throw distance. Because the units (metric vs. imperial vs. US units) in which the various boats described here have been designed vary according to the year of design and the country of the owner, all the pump capacities in this section have been converted to a common standard measure of cubic metres per hour (m^3/hr) to make comparisons easier. Firefighting (Fi-Fi) capacity for all types of vessels (not only fireboats) is defined by the classification societies, who use the following standard designations:

Fire-Fighting I	2400 m^3/hr	(40,000 lpm)
Fire-Fighting II	7200 m^3/hr	(120,000 lpm)
Fire-Fighting III	9600 m^3/hr	(160,000 lpm)

These standards define much more than just pumping capacity, including the number of pumps, number and throw range of monitors, etc., but for present purposes pump capacity is the governing performance measure. It is important to note that not every fireboat uses these published standards, however, as the overall firefighting performance must be set according to the individual, site-specific needs.

Working with firefighters in developing fireboat designs is a very rewarding activity. These are people who put their lives on the line every time they respond to a major incident, and although those incidents are thankfully quite rare, the boat must be capable of performing in extremis when called upon, providing the safest possible platform for those aboard. We have learned to appreciate how serious these professionals are about their work, and we do our utmost to meet all their sometimes competing requirements, a process both challenging and satisfying.

Finally, a major attraction of these design projects is that almost every port has a quite different set of operational and performance requirements for fireboats: different types and sizes of ships using the port; different geographic or manmade constraints (bridges, canals, locks, etc.); different cargoes being handled in port and thus quite different risks; and of course differing budgets. Every project then starts (or rather should start) with a mission-defining "Needs Assessment," which drives the final vessel definition. Working through this process establishes a firm set of operational requirements for the new design that is usually expressed in a formal "Statement of Operational Requirements" (SOR).

Fireboat contracts don't appear that frequently, as these boats are generally quite long-lived, but there has been a very interesting series of new designs since the early '90s. The tragic events of the World Trade Center attacks on September 11, 2001, gave rise to federal funding in the US that enabled many of their cities to upgrade their often quite aged emergency response equipment, including many fireboats. At about the same time, the emergence of major new ports in China and elsewhere created demand for similar high-performance firefighting vessels for improved port protection. Robert Allan Ltd was well placed to respond to these emerging demands for new and powerful fireboats, and soon established a significant reputation for producing quality, high-performance designs in this unique niche market.

Fireboats in Canada

The Ones That Got Away

The elder Robert Allan had no fireboats to his credit, but he certainly tried hard to obtain the few contracts that appeared in his early days as an independent consultant. There exists in our files a sketch drawing of a proposed fireboat for the Port of Vancouver done probably in 1928. This would have been a proposal for the boat that became the *J.H. Carlisle* (built in 1928), which was Vancouver's only fireboat for many decades, but unfortunately was designed by others.

In 1951 Vancouver called for tenders for the construction of a second fireboat, excitingly named *Fireboat #2*. However, the design contract for this boat was awarded to the eastern Canadian firm of Gilmore, German and Milne, which prompted Robert Allan ("Rate-Payer" in his letter) to write city bureaucrats one of the many scathing letters for which he was quite notorious.

Proposal (unsuccessful) for a 133-foot fireboat for the City of Vancouver, ca. 1928.

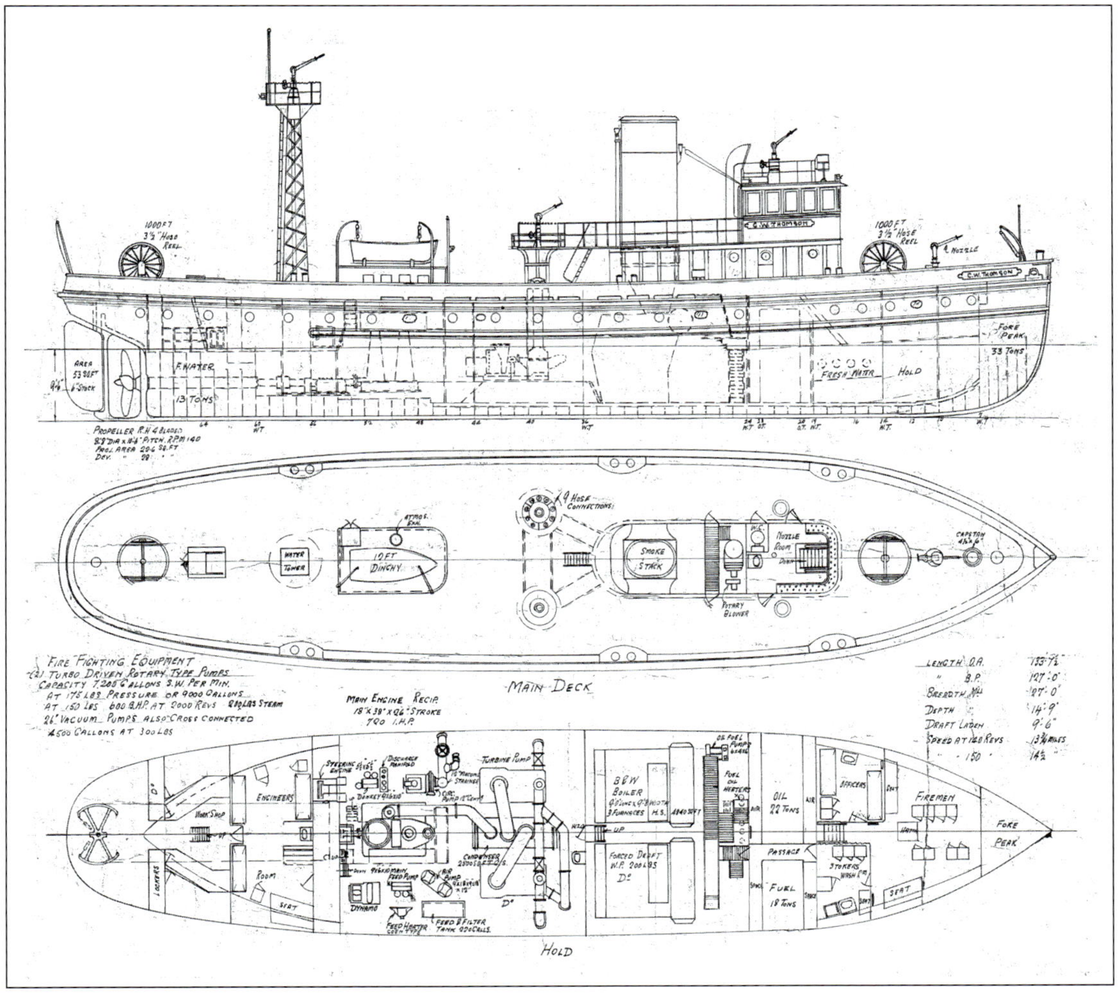

Canadian Navy Fireboats

The first completed fireboat project for the company came in 1972 when Robert Allan Ltd was awarded a contract by the Department of National Defence of Canada to design two fireboats, one each intended for the naval ports of Esquimalt and Halifax. These were interesting boats, embodying one of the earliest applications of azimuthing stern-drive units (Z-drives) in Canada. The manoeuvrability and positioning control afforded by these Z-drives represented considerable operational advantages. The fireboats, built in 1978 at Vancouver Shipyards Ltd in North Vancouver, were 76 feet in length and had a modest pumping capacity of 1,364 m^3/hr. The two fire pumps were each driven by an independent diesel engine that was the same make and model as the propulsion engines. The *Firebird* was stationed in Halifax (East Coast) and *Firebrand* in Esquimalt (West Coast).

These early Z-drive units, built locally by Maritime Industries Ltd, were somewhat problematic and gave rise to vibration problems, which plagued the boats for a few of their early service years until finally resolved. Thereafter, their record is understood to have been very good, though obviously their actual annual operating hours are quite low. *Firebird* was retired from naval duties in 2014, but *Firebrand* continues to serve the West Coast as of 2021.

CFAV *Firebird* in Halifax harbour.

A Lesson in Weight Estimating

This Navy fireboat project was the very first one I worked on after joining Robert Allan Ltd in January 1973. I was given the task of preparing a detailed weight estimate for the project, which was one of the documents required as part of our contract deliverables. The project design work had proceeded within our office based on a one-page handwritten weight estimate that my father had prepared. After several weeks and many pages of detailed calculations, I arrived at a weight value that was within about two percent of the number that my father had estimated. That was a vital lesson to me in the importance of experience. And in the end the weight of the boat was pretty close to both our estimates. I conveniently forget who was closest!

—R.G. Allan

Port of Vancouver Fast-Response Fireboats

The various cities that border Vancouver harbour share the responsibility for fire protection of this extensive and highly developed waterfront, in conjunction with what is now the Vancouver Fraser Port Authority. There has not always been a happy relationship between the parties involved, as the costs to provide and maintain fireboats are considered high for what some politicians see as little return. They perhaps have never seen the result of a major waterfront fire that could not be properly responded to.

In 1990 all the parties managed to put aside their differences and agreed to build a fleet of five fireboats. The concept was for multiple small fast-response boats that would be strategically located around the waterfront, enabling a very short response time to any incident, in which case the pumping capacity of each boat needn't be very large. The first boat on-scene would be there ideally within 10–15 minutes and some of the other boats would arrive as quickly as possible, depending on their home-base locations. The concept design contract for these boats was awarded to Peter Hatfield Ltd, a competing local naval architecture firm, who defined the basic performance parameters and the general concept for these boats. Robert Allan Ltd teamed with Celtic Shipyards Ltd of Vancouver for the detail design, production design and construction of the five boats, and we were ultimately the successful bidder.

This was a very interesting project for our office, as we don't often get to work closely with the production department of a local shipyard. Celtic was a small but well-managed shipyard occupying the former BC Forest Service shipyard facilities at the south foot of Blenheim Street on the Fraser River. The manager at Celtic was Al Dickey, who had previously managed West Coast Salvage. He was a seasoned professional and a pleasure to work with. Unfortunately, Al was diagnosed with cancer before the project was finished and didn't survive to see its completion.

The five fireboats were all delivered in 1992. They are of all-welded aluminium construction, 40 feet in length, with a simple planing hull geometry. They were powered by twin 450 hp diesel engines driving Hamilton waterjets to obtain the design speed of 25 knots. The fire-pumping arrangement, which had been defined by the owners, was unusual and proved to be an ongoing challenge in the operation of these boats. The waterjets were fitted with a diverter valve that could be operated from the wheelhouse, driving water to the fire monitors rather than through the jet outlets. This worked well in theory but required some careful tuning of the flow so that the boats could still manoeuvre while fighting fires. The nominal individual fire pump output was 546 m^3/hr.

Left: The first of the five Vancouver fast-response fireboats; **right:** Port of Vancouver Fast-Response Fleet: *Fireboat #1* through *#5*.

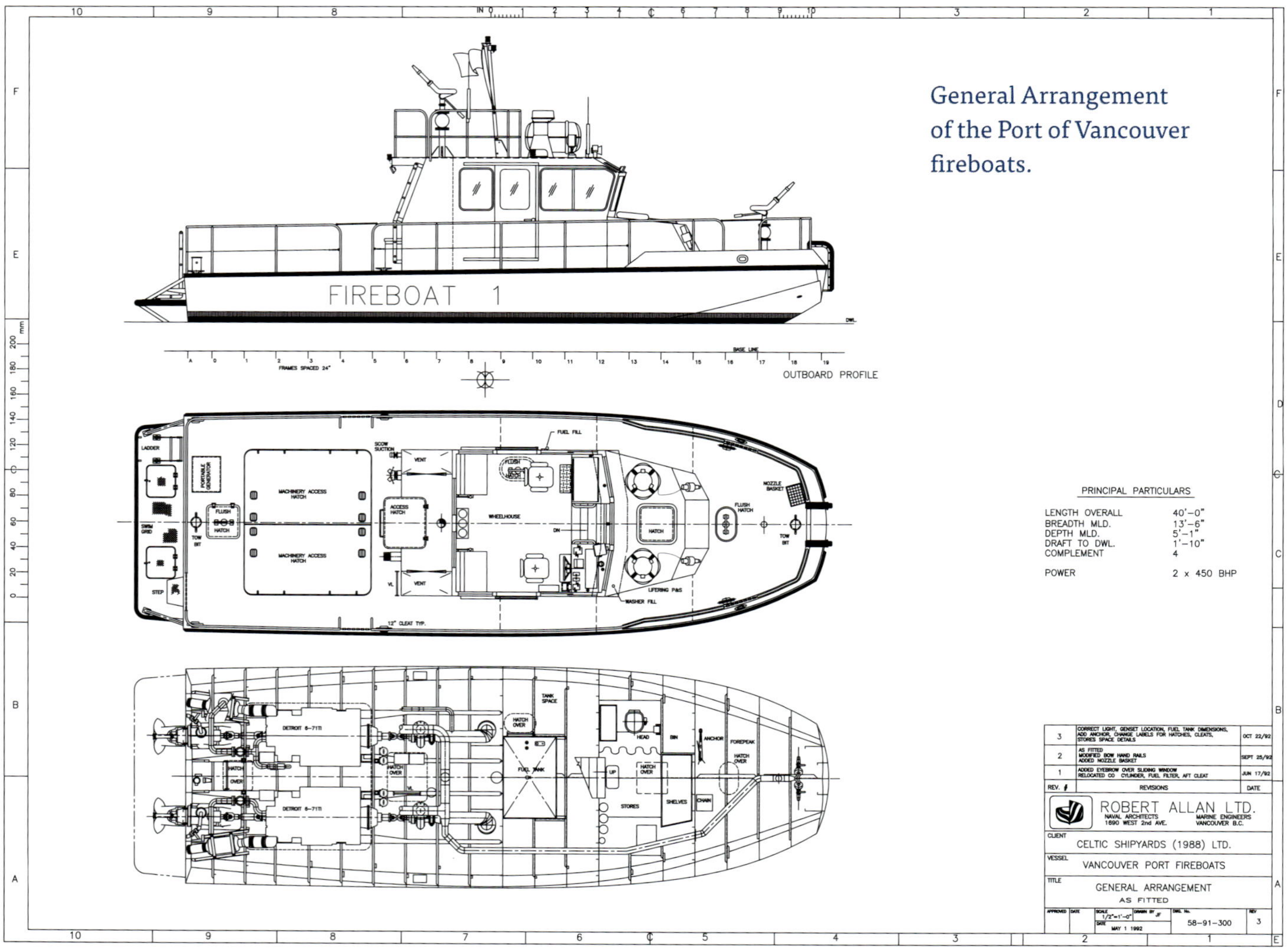

General Arrangement of the Port of Vancouver fireboats.

America Rebuilds Its Fireboat Fleet

In 1999 we were made aware that the Port of Los Angeles was looking to replace its then nearly 75-year-old fireboat, the *Ralph J. Scott*. This began a quite remarkable run of fireboat designs from Robert Allan Ltd for many major US cities, a run which lasted for more than a decade, and which put the company in a very prominent position worldwide as successful designers of modern high-performance fireboats.

Los Angeles Fireboat

The genesis of a new fireboat for Los Angeles had begun nearly a decade before we became involved in 2000. At that time the city had approved the acquisition of a new boat and with it an elaborate new boathouse. Construction of the boathouse was approved, and sometime in the mid-1990s it was completed. The fireboat design was awarded to an obscure Los Angeles naval architect whose idea for the new boat was to essentially replicate the *Ralph J. Scott*, which was of an earlier generation of long, slim boats and which had a modest firefighting capacity by 21st-century standards. When tendered to shipyards, that poorly conceived design was challenged (and proven) as being unable to meet the specified performance requirements, and the whole project ground to a halt—except for the shiny new 30-foot-wide boathouse, which had already been built to accommodate an unfortunately too-narrow boat concept!

The revamped 1999 attempt to create a modern fireboat was thus faced with its most fundamental design constraint, namely the vessel beam. There were more than a few politicians in the background to this project who had serious egg on their faces from the original boat replacement effort, and the new project was put in the hands of the engineer from the *Scott*, Doug Moore. Doug knew very well that he had been set up as the sacrificial lamb if anything went wrong with round two of the fireboat project, but he was also smart enough to know how to work the system to get what he knew the city really needed. He was absolutely the right person for this very challenging project.

The process began with a qualifications-based selection of the naval architect for the project. Several consulting firms, including Robert Allan Ltd, were invited to Los Angeles to present our credentials for being selected for this prestigious project. This was such a positive process for choosing a naval architect, in comparison to many government projects, which are largely driven by a least-cost-fee proposal, which almost by definition implies that all firms are considered by the purchaser as equal in capability, efficiency and experience—which is clearly not the case.

The Right Man for the Job

I recall clearly going to Los Angeles to be interviewed for this project and meeting Doug Moore for the first time. Sometimes it is very easy to know whether you are going to be able to work well with people, and Doug was the kind of person that I knew instantly we would enjoy working with. He was a marine engineer, a firefighter, a commercial fisherman, and a race-car driver and builder. He knew boats and machinery inside out. We had a great discussion about his vision for the new boat, how we could work with him to get to that point, and that became the start of a very positive working relationship and friendship.

—R.G. Allan

Doug Moore, chief engineer on the LA fireboats, was the perfect person to manage the challenging LA fireboat project to a most successful conclusion.

The design contract was awarded to Robert Allan Ltd early in 2000, and we commenced the rather cumbersome contractual process involved in dealing with one of the world's largest cities. That was soon sorted out, however, and we began the design work. The boathouse dictated a maximum length of 105 feet and a beam of 29 feet. Doug had been quite firm that for their operational needs this boat should have Voith Schneider cycloidal propulsion (known as VSP drives). These devices, though large, heavy and less efficient in developing thrust than a more conventional screw propeller, do offer the flexibility of fully controllable pitch and fast directional control—attributes hard to match in any other propulsion type, and highly desirable for vital position-keeping in a fireboat. To a large degree, that propulsion choice defined much of the configuration of the boat, as these VSP drives have to be situated about one third the boat length from the bow and in a relatively flat bottom. With the large size of these units, we were already bumping into the beam constraint.

Showing Off the Firefighting System

Most conventional fireboat designs go to great pains to disguise the fire piping systems within the deckhouse of the boat. As the Los Angeles boat was to have a total output in excess of 7000 m^3/hr, the main pipes were going to be large (>400 mm diameter) and there were multiple monitors to accommodate. I had the idea of making the firefighting system the showpiece of the design, right in your face, where it would not only be so much easier to install and maintain, but also make it abundantly obvious what this boat was intended to do. A quick sketch of a big "ring main" surrounding the top of the deckhouse, topped by the main monitors, showed immediately that this was a first-rate approach to the overall design. It also had the advantage of keeping the interior of the deckhouse totally clear of big vertical pipes and more easily laid out for the other space and equipment requirements. (And any leaks would be "outside"!) My initial pencil sketch of this piping concept and the associated overall arrangement of the boat was not so very different from the end result. That concept design sketch was framed and presented to Doug Moore at the end of this most successful project.

—R.G. Allan

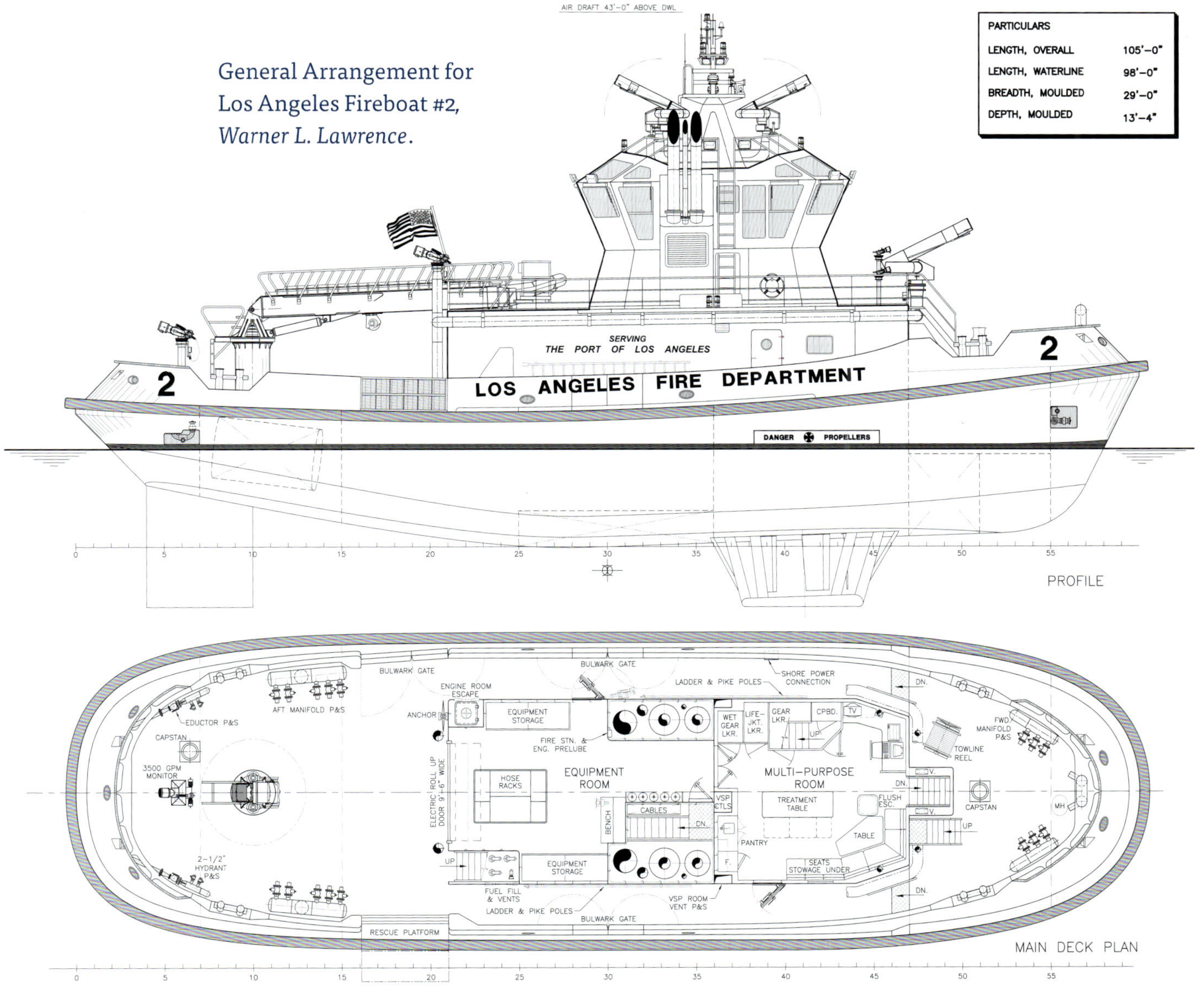

General Arrangement for Los Angeles Fireboat #2, *Warner L. Lawrence*.

The total pumping system was quite elaborate and provided flexibility in both how and how much water and/or foam was delivered to the 12 different monitors. Six different fire pumps provided a total combined output of 7040 m^3/hr. The largest monitor had the prodigious output of 2500 m^3/hr.

When the design was completed for this high-capacity fireboat, the construction project was tendered to qualified shipyards around the US. The contract was awarded to Nichols Brothers Boat Builders on Whidbey Island in Washington State, a nearby location which proved convenient as Robert Allan Ltd was also awarded the contract to provide construction supervision services in the shipyard on behalf of the City of Los Angeles. One of the early project review meetings with owner, shipyard and the design team was scheduled on September 11, 2001. The events of that historic day certainly overtook the daily business. The firefighters involved were all acutely aware of the impact of the terrorist attacks on their East Coast colleagues.

After completion of all successful trials, Fireboat #2, named the *Warner L. Lawrence*, sailed south and was a great hit on its arrival in the city. It managed to just fit into that cursed boathouse, and Doug Moore was heralded for his fine work in managing the project to a highly successful conclusion. The previous "ducking for cover" politicians emerged in great numbers to have their photos taken "pseudo-christening" the fireboat repeatedly so they could all get some of the credit that none of them deserved. Robert Allan Ltd was publicly thanked by the City and acknowledged as "one of the leading naval architects in America"! Cynicism aside, this was really an exciting and very gratifying project, and it set the stage for many that followed.

Above: Fireboat #2, *Warner L. Lawrence*, en route to Los Angeles. The large diameter "ring main" is prominent at the upper deck level, supporting many of the large monitors; **right:** Los Angeles fireboats, including Fireboat #2 (left), fighting a harbour fire at the docks in San Pedro on September 22, 2014.

Philadelphia Fireboat

Later in 2000, Robert Allan Ltd was awarded a contract by the City of Philadelphia to design their new fireboat. We soon learned that firefighters talk to firefighters, and it was obvious that our successful work on the Los Angeles project was a big influence in us obtaining this and many more new US fireboat contracts.

The Philadelphia boat, named *Independence*, was much more modest in its specifications, but this city has a lot of waterfront to cover, and so speed was a primary driver in this design. Robert Allan Ltd worked closely with the City of Philadelphia Fire Department and Office of Fleet Management to develop the final design details and specifications. We also provided construction overseeing services throughout the build period at Derecktor Shipyards in Bridgeport, Connecticut.

General Arrangement for Philadelphia fireboat *Independence*.

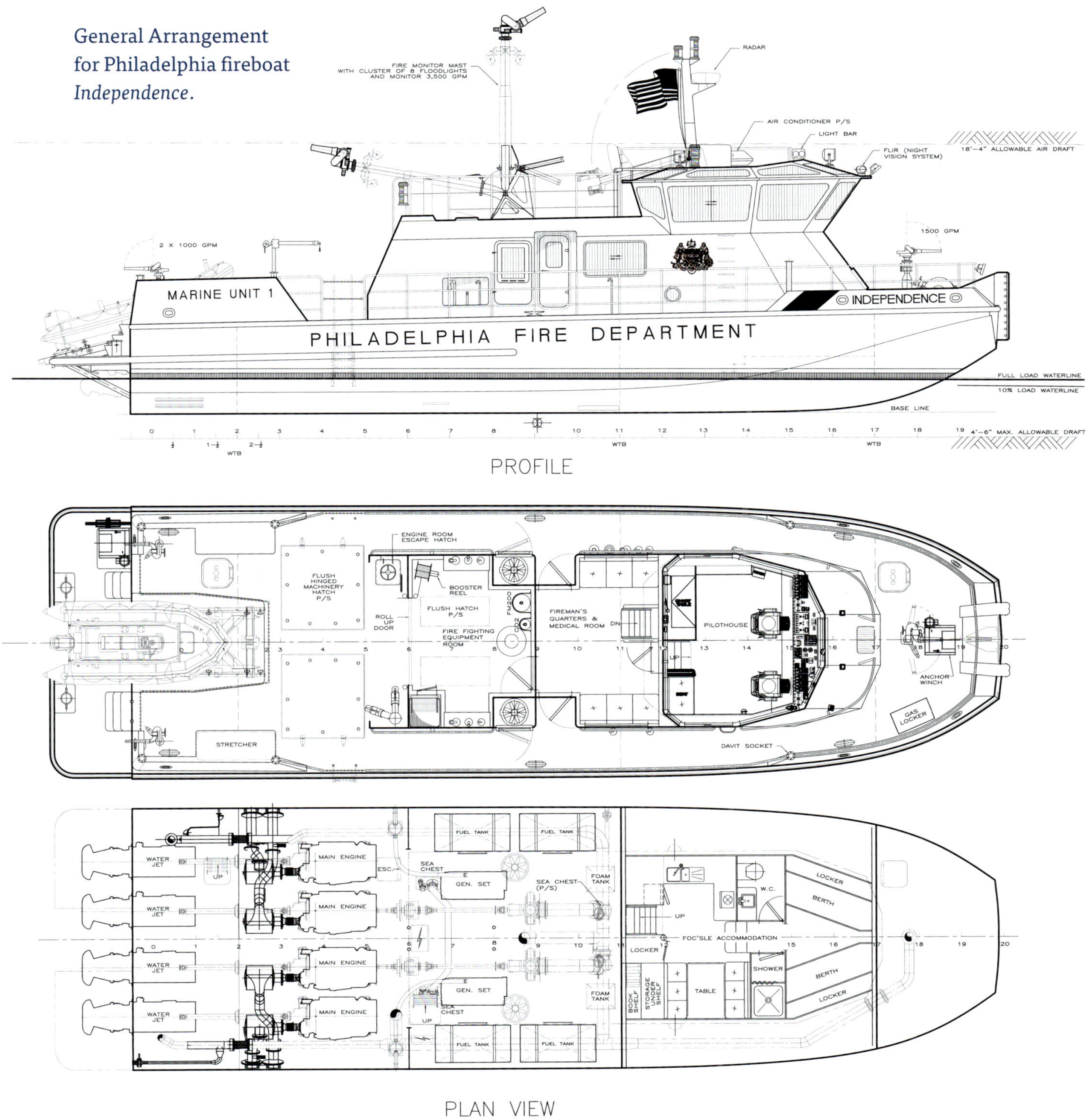

This 66-foot fast-response shallow-draft fireboat is of all-welded aluminium construction. It is equipped to provide emergency services along the expansive waterfront of the Delaware and Schuylkill Rivers and has a regional responsibility from as far north as Trenton, New Jersey, and south to the Delaware Bay. Consequently, the vessel has extreme water depth limits and an air-draft constraint of 18 feet to navigate under city bridges. The hull is strengthened to operate in the thin ice that forms in this region.

Power is supplied by four main diesel engines, each rated at 615 kW and each driving a Hamilton waterjet. This package enabled the fully loaded vessel to achieve an impressive 36 knots during trials. The two inboard engines also drive the two main fire pumps from front-end power take-offs (PTOs). Total fire pump output is 1250 m^3/hr serving four monitors, the largest of which is on a telescoping mast. All four engines are utilised to achieve top speed in an emergency response, then the two inboard engines are declutched from the jets and drive the fire pumps while the outboard waterjets are used to control boat position and heading.

Philadelphia's fast-response fireboat *Independence*, delivered in 2007.

New York City Fireboats

In the aftermath of the 9/11 attacks, the Fire Department of New York (FDNY) recognized the need to significantly upgrade their waterfront firefighting capabilities. Their existing boats were about 60 years old and getting tired. One of the most important lessons learned from the tragedy was the need to have substantial backup pumping capacity from a mobile platform in the event of a failure in the land-based infrastructure. That "pump station" role had been the major function of the existing fireboats during the days following the World Trade Center attack.

New York City followed a process not dissimilar to Los Angeles in selecting their consultant for the design of two large fast-response fireboats, and we now had two recent US fireboat projects under our belt as credentials. Robert Allan Ltd was selected from an international field of consultants for this very significant project. The entire story of designing and delivering the New York fireboats became the subject of a detailed technical paper entitled "Concept Development, Detailed Design and Construction of the Three Forty-Three: North America's Most Powerful Fireboat," co-authored by R. Allan, K. Harford, D. Noon, J. Bjerkeset (all from Robert Allan Ltd), and Chiefs Jim Dalton and Bill Siegel of FDNY. That award-winning paper was presented to the 2010 Society of Naval Architects and Marine Engineers (SNAME) Annual Meeting in Seattle.

The scope of work awarded to Robert Allan Ltd included complete engineering and design of the new boats, from initial concept through detailed engineering design, and included a comprehensive value engineering process. The company's scope also included the support of FDNY through every stage of major component procurement and the shipbuilding contract award, and ultimately providing construction supervision services in the shipyard. It is uncommon for a design consultant to obtain such a broad mandate in any project, but it is professionally most fulfilling to have this type of total responsibility to represent the owner's interest from concept to delivery.

These major fireboats were designed to specifically address the firefighting and rescue needs of the greater New York harbour, including the New Jersey shore. This included the capability within the vessel to respond to any CBRN (Chemical, Biological, Radiological and Nuclear) incident. The complex CBRN system was designed in collaboration with NAVSEA (the research arm of the US Navy) and the FDNY Hazardous Materials Group. This critical feature creates a pressurized clean-air citadel for all ship command operations, with air locks with decontamination facilities for the crew provided at the designated entry to the fireboat superstructure, leading to a hazmat assessment space and then a triage room that is fully equipped as a marine ambulance.

The fireboats are 140 feet in length and have a response speed of 17.4 knots, with a low-wake 12-knot cruising/patrol speed and a total pumping capacity of 8920 m^3/hr. The powering, sea-keeping and wake generation characteristics of the semi-displacement hull form were all verified in an extensive model-testing program.

These were big fireboats, and it is so much easier to make such a size of boat look both purposeful and attractive. There was a lot to accommodate, but in general there was room to make all of that work effectively.

The propulsion configuration of the New York fireboats was determined after extensive evaluation of the cost-effectiveness of numerous options, including Z-drives and waterjets. Given the imposed draft limitations, as well as the speed and fire-pump drive requirements, the final solution was a quadruple-screw controllable-pitch (CP) propeller system. The four main engines are MTU model 12V-4000 M70, each rated

2240 bhp. These each drive a CP propeller through speed-reducing gearboxes, and also drive the four main fire pumps through front-end PTOs. Station-keeping is enhanced by the incorporation of a tunnel-type bow thruster.

The contract for construction was awarded to Eastern Shipbuilding Ltd of Panama City, Florida. Robert Allan Ltd's experienced shipyard supervisory staff, led by Jody Bjerkeset, acted in the capacity of owner's representatives on site in the shipyard throughout the lengthy construction process. The lead boat, named *Three Forty Three* in honour of the fallen firefighters during 9/11, was commissioned in 2010. The second boat, named *Fire Fighter II* after its retiring namesake, was commissioned in 2011.

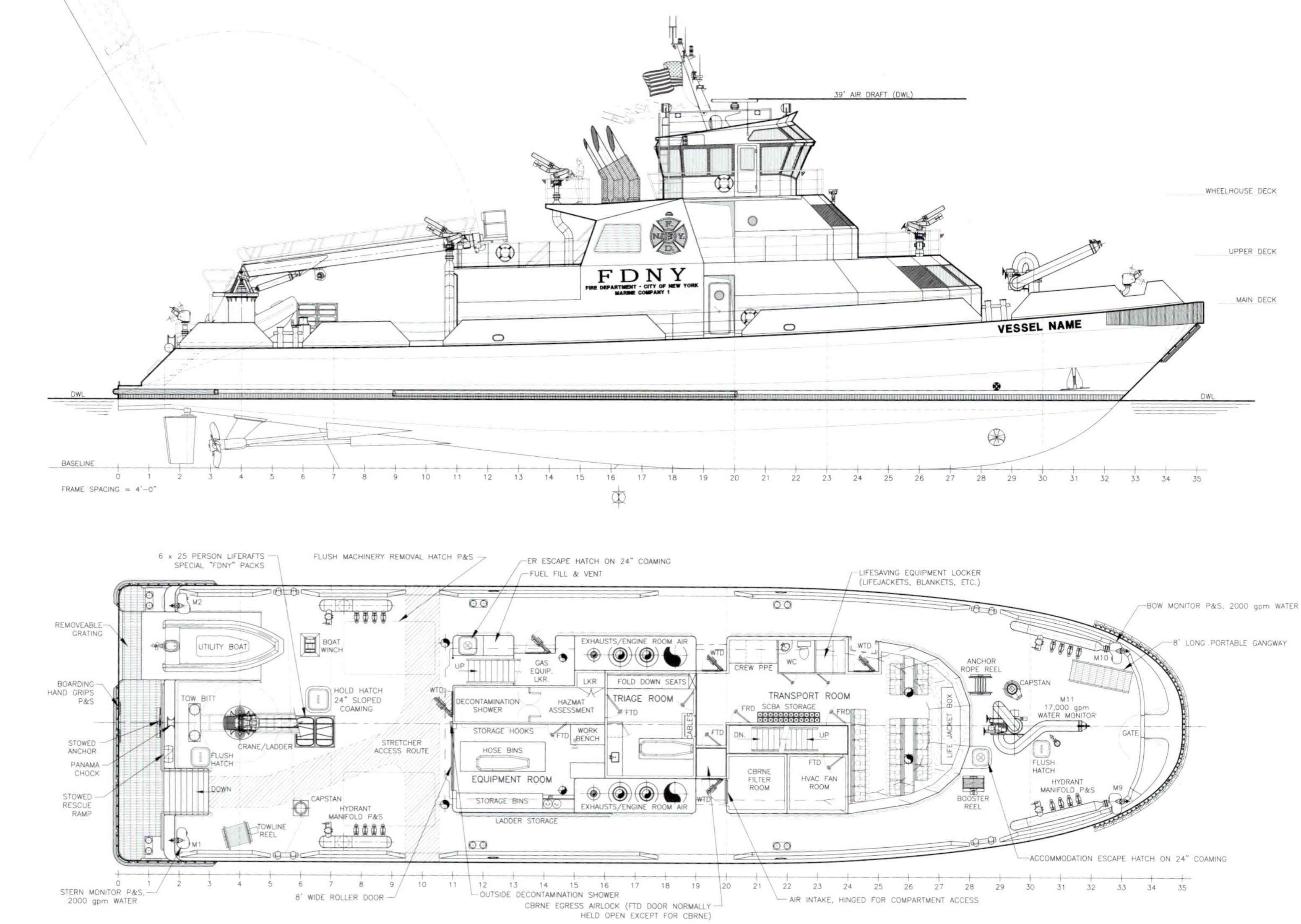

General Arrangement for the FDNY fireboats *Three Forty-Three* and *Fire Fighter II*.

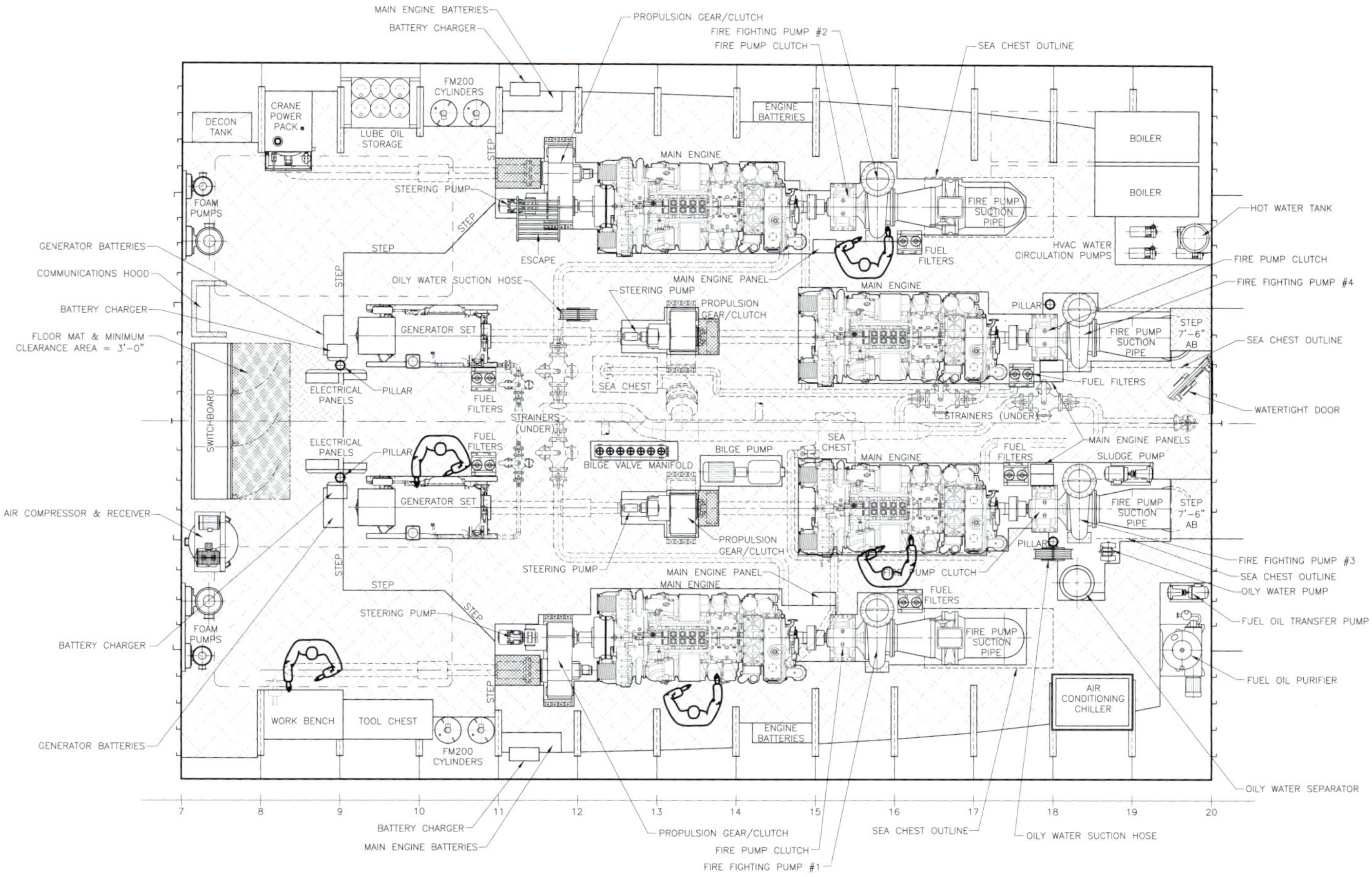

Top: Machinery Arrangement Plan of the New York City fireboats; **bottom:** *Three Forty Three* honours the fallen firefighters of the September 11 tragedy.

Fire Fighter II, looking every inch the serious firefighting machine.

The Fireboat Guy

I was project manager of the New York and Los Angeles fireboats. I was there for the initial discussions and followed the projects through to completion.

When we designed the Los Angeles fireboat in 2000, their previous boat had been built in 1926. One of the New York fireboats being replaced was built in 1938. Similar with other fire departments, the maintenance costs were becoming prohibitive or equipment was obsolete. The people we were dealing with had never built a boat before, but they've operated them, only very old ones. So getting to the definition of a new design was really, really interesting. We had a blank sheet of paper.

The first thing you do with a fireboat is a Needs Assessment: What is it that you do? "We rescue people, we put out fires, we might have to pump out a boat that is sinking." Once these are identified, plus other criteria such as draft limits and response times, then the design begins. Another key is redundancy; the fireboat is responding to an emergency and can't break down. It requires a very different mindset from many other vessel types.

For the Los Angeles and New York fireboats, I was fortunate to visit the vessels under construction all the way through to trials. It was a complete process, and we don't get to do that often. Not only is the process fulfilling, it is also important for continuity of the design to ensure initial concepts or features survive to the finished project. So many of our designs are built overseas, where we can't follow them all the way through construction to completion, and that is the full experience. Any time we get to do that, it is amazing and rewarding.

The most memorable projects for any of us are the ones where you've worked closely with the owner or the operator—ones you've seen from inception to operation.

—Derek Noon

Canadian-Built Fireboats for US Service

When the Conservative Mulroney government negotiated the North American Free-Trade Agreement (NAFTA) in 1992, they neglected to observe that Canada actually had a viable shipbuilding industry, though it was certainly on the wane at the time. The NAFTA eliminated the duties which had previously applied to any US-built vessels imported into Canada, opening the border to allow Canadian owners to import new or used vessels from the US without any tariff. There was almost zero reciprocity, however, as the US retained their *Merchant Marine Act* of 1920, Section 27 of which is known as the *Jones Act*. This act deals with coastwise trade and requires that all goods transported by water between US ports be carried on US-flagged, US-built ships, owned by US citizens and crewed by US citizens. The *Jones Act* almost entirely precludes Canadian shipbuilders from offering their skills to US owners. Fireboats however are *not* engaged in trade and

Left: Baltimore fireboat *John R. Frazier* saluting Lady Liberty; **right:** Emergency response vessel *City of Portland* underway.

are therefore one of a very few vessel types exempt from these provisions. In large part due to the favourable exchange rates at the time, Canadian shipyards had a fair degree of success in meeting the demands of US cities for new firefighting marine assets. This included the following boats, all from Robert Allan Ltd designs.

The 87-foot *John R. Frazier* was delivered to the City of Baltimore Fire Department by Hike Metal Products of Wheatley, Ontario, in 2007. Robert Allan Ltd worked with Hike Metal to win this internationally bid "design and build" contract. The hull is all-steel, topped with an aluminium deckhouse. Propulsion power consists of two 1,650 bhp diesel engines, each driving fixed-pitch propellers through reverse-reduction gearboxes to achieve a full-load trial speed of 17 knots. Two fire pumps, each capable of providing up to 795 m^3/hr, are driven by independent diesel engines.

In July 2009 the fireboat *City of Portland* was delivered to the City of Portland (Maine) Fire Department by another Canadian builder, A.F. Theriault & Son Ltd, of Meteghan River, Nova Scotia. This 65-foot 18-knot boat was somewhat unique in that its role was more that of a paramedic response boat than strictly firefighting. The boat served a widespread maritime community for fire and medical emergency services, with many small villages on the surrounding coastline. The "medical room" layout of the boat was required to emulate the standard land ambulances so that crews would be fully familiar with the location of all emergency equipment. The firefighting equipment includes a single engine-driven 680 m^3/hr pump serving six different monitors.

Outboard Profile of the *City of Portland* fireboat.

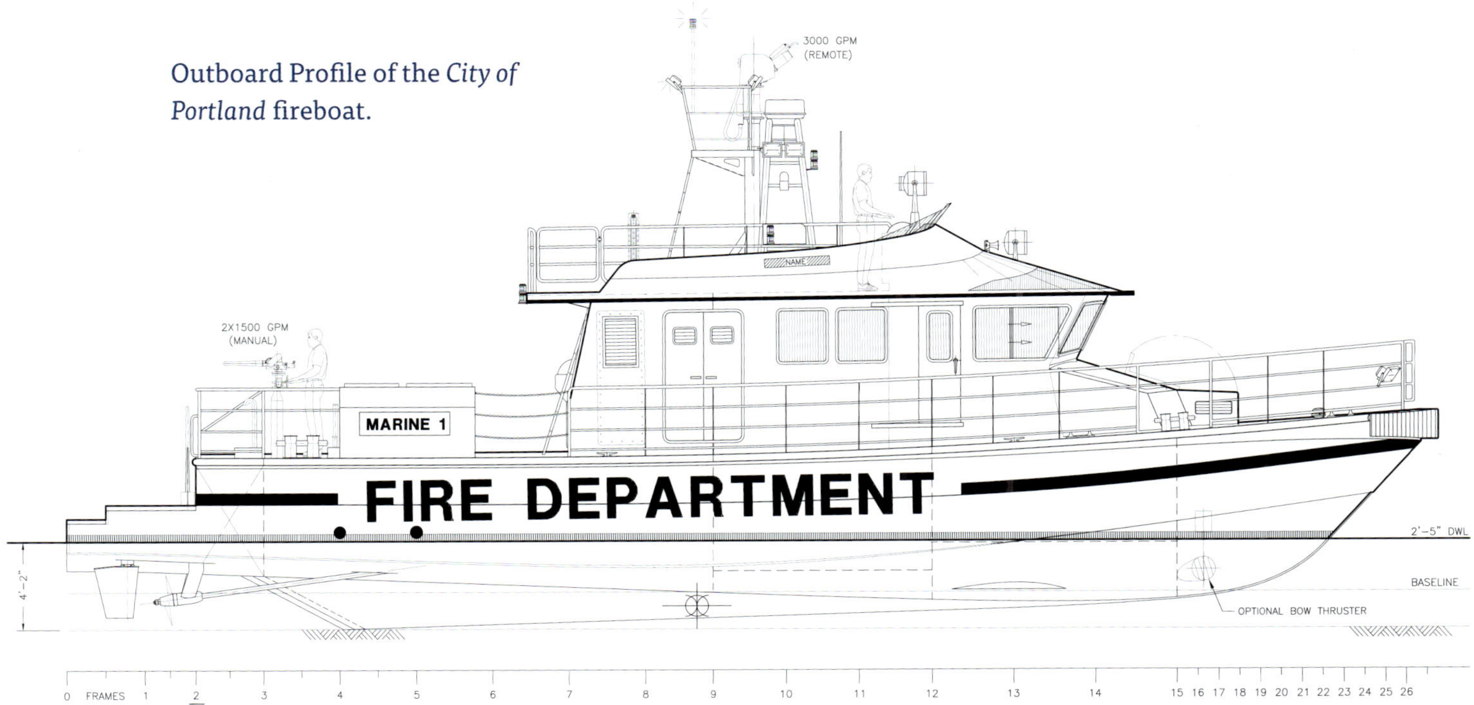

Creative Fireboat Ideas

One of the fireboats we designed was for Portland, Maine. The fire department is also responsible for EMS (ambulance) services, and their jurisdiction includes several island communities in Casco Bay. In case of a fire on one of the islands, what they may do is float a hose to the island and then they'll pump water ashore.

For the new design we worked with a committee comprised of firefighters and EMS personnel. In the initial meeting, we reviewed all the operations the new boat would need to perform. One of their tasks is an ambulance run to one of the island communities, probably once a week or so. I asked one of the EMS guys to walk me through how the ambulance run works. He said that once they get a call, an ambulance drives down the dock in Portland and the paramedics jump on board the fireboat. They get taken to the island, where they stabilize the patient, put the person on the fireboat, bring the person back, transfer them to the ambulance and then to the hospital.

I asked them what they do now with the patient on board. He said, "We just put them on the mess table." I asked if that worked. He said, "No, it doesn't really work. There's pathogens and such." I asked them if I could see the ambulance, and they brought one down. I looked inside and said, "Okay, you know, we could put an ambulance in the back of the boat." They said, "What?" I said we've got enough room here in the back of the boat where we could put all the contents of an ambulance in exactly the same place as in the ambulance. That way, when the paramedics come down, they know everything is where it should be. It was a big success. It's things like that, once you sit down and talk to these guys and you understand what they do, then you can see how best to meet their needs.

—Derek Noon

In October 2011 the fireboat *American United* was delivered to the Massachusetts Port Authority (Massport) of Boston, another build by A.F. Theriault & Son Ltd. The vessel was also named in remembrance of the events of September 11, 2001. Robert Allan Ltd acted in the capacity of owner's representative throughout the construction process.

This 79-foot boat was specifically designed to provide fast emergency response, search and rescue, firefighting operations, capability as an on-scene command post, port security, EMS, and assistance with diving operations and recovery in the vicinity of Boston Harbour and particularly around the Massport airport. The vessel is of all-welded aluminium construction and ice-strengthened to suit year-round operations. It is equipped with a large aft swim platform and 30 liferafts to facilitate rescue of aircraft crash survivors. Also fitted is a FLIR thermal-imaging system and flood-lighting for night rescue operations.

The fireboat was designed for a full-load response speed of 24 knots and a low-wake 12-knot cruising/patrol speed. The propulsion system consists of four main engines, two rated 1,450 bhp and two rated 873 bhp, each driving a fixed-pitch propeller, with the two smaller inboard engines also connected to the two main 680 m^3/hr fire pumps through PTOs at their forward ends.

American United, built in Canada for Massport, Boston.

Chicago Fireboat

The 90-foot *Christopher Wheatley* was yet another Canadian-built fireboat, delivered in 2011 by Hike Metal Products of (serendipitously) Wheatley, Ontario, to the City of Chicago. This was an extremely challenging design as every dimension of the boat was constrained by its riverine operating environment. The beam was constrained by bridge piers and the canal-like river boundaries within the city, draft was limited by the river depth, length was constrained by the tight bends within the boundaries of the river that winds through the city, and most particularly the air draft of the boat was limited by the many street-level bridges that cross over the Chicago River throughout this dramatic Midwest city. In addition, the boat had to be able to break ice!

These limits gave rise to certainly one of the most unusual-looking boats we have ever had to design, but in the final analysis and with some judicious paintwork, it worked. If one wasn't aware of the background to this design, there might be grounds for criticism of its aesthetics.

The propulsion machinery consists of a pair of diesel engines each rated 1,081 kW and driving a fixed-pitch propeller through a reverse-reduction gearbox. The top speed is 13 knots. The firefighting capability is provided by two completely independent pump engines, each rated 745 kW, driving a fire pump rated at 1590 m^3/hr. There are four firefighting monitors.

A few of the many obstacles to normal ship design in downtown Chicago.

The unique appearance of Chicago's fireboat *Christopher Wheatley* reflects a myriad of design constraints.

The profile of the *Christopher Wheatley* illustrates how all the constraints were accommodated in this unique fireboat design.

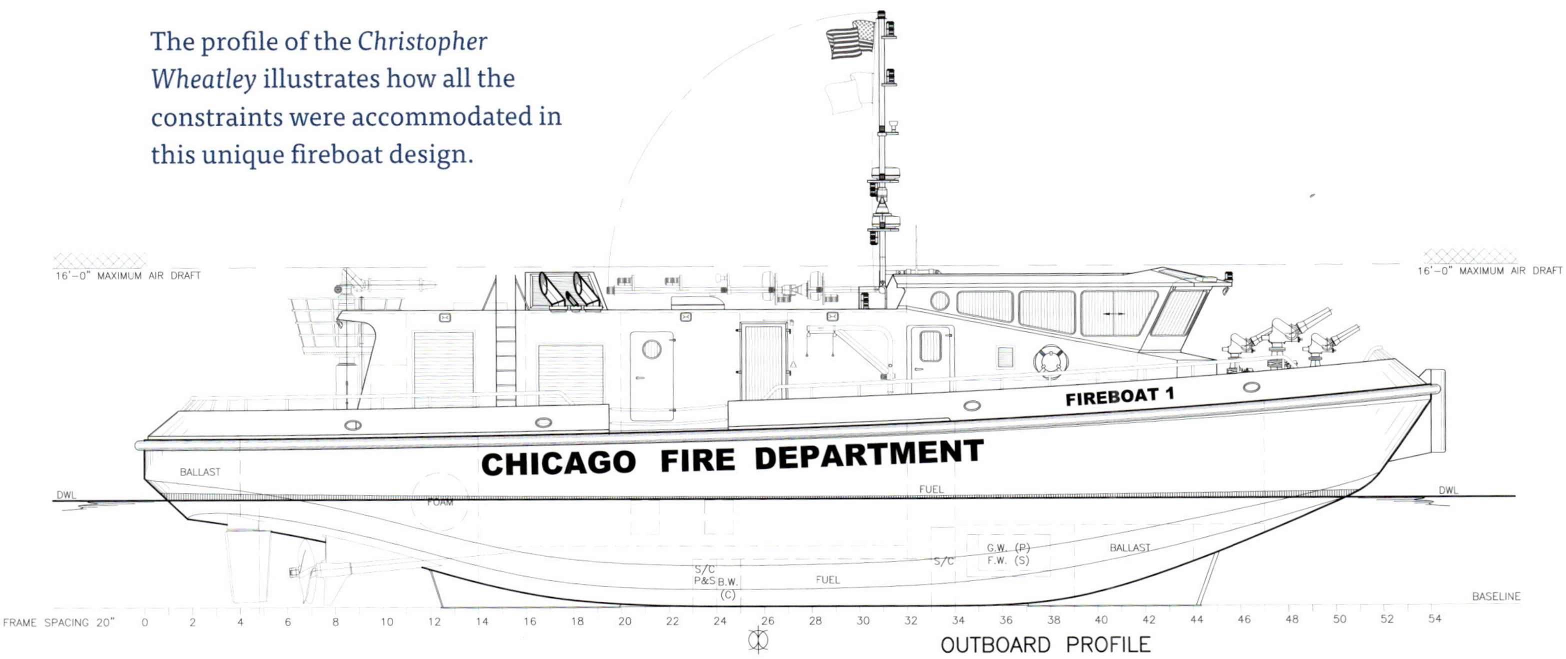

Long Beach Fireboats

At time of writing, the latest fireboat design for the US from the Robert Allan Ltd office was for a pair of high-capacity vessels for the Port of Long Beach. As neighbours to the Port of Los Angeles, the port's vessel requirements were heavily influenced by the nearby presence of Los Angeles' *Warner L. Lawrence*.

As one of the largest ports in America, Long Beach has an administrative set-up larger than many mid-sized cities. Our office was by now well accustomed to dealing with the bureaucracies of civic-level governments, but Long Beach wins the prize for the most complicated. Their systems are well set up for large civil-engineering projects, but as it had been more than 40 years since they had purchased a vessel, it was a struggle to find appropriate pigeon-holes in their system for a custom-designed fireboat.

Eventually, however, Robert Allan Ltd was selected for the design work, which began in 2011 with meetings with the Port of Long Beach and the Long Beach Fire Department to establish the Statement of Requirements for the new vessels. Functional requirements identified in this process included:

- Firefighting of ships, boats and near shore structures
- On-shore support, including shore-side supply of water
- Water rescue, including vessel de-watering, towing and dive support
- Medivac services and EMS/paramedic support
- Harbour security patrol, including detection of hazardous agents
- Disaster response, including a Command Information Centre
- Operation in contaminated (CBRN) areas, including vessel self-protection and decontamination facilities for person

Numerous trade-off studies were conducted, and through a series of design phases and client reviews the detailed plans and specifications were finally developed. The resulting design is for a vessel 108 feet by 35 feet with a navigational draft of 15 feet and an air draft of 45 feet.

Propulsion power is provided by a pair of CAT 3512C main engines, each developing 2,012 bhp and driving a Voith Schneider cycloidal propeller, the same drives that were chosen for the Los Angeles fireboat. An additional two CAT 3512C engines and one CAT C12 engine team up with the drive engines to power an array of seven individual firefighting pumps, ranging in size from 454 m^3/hr to 1817 m^3/hr with a total impressive aggregate capacity of 9311 m^3/hr. The vessels are equipped with ten monitors in total, the largest capable of delivering 2725 m^3/hr at a range of 183 metres (600 feet).

In late 2012 the Port of Long Beach selected Foss Maritime Company to build the vessels at Foss' Seattle shipyard, a repair and maintenance facility with no prior new ship construction experience. The construction process was anything but smooth, and there were considerable delays in the completion of the work, which was in marked contrast to the way the similar Los Angeles fireboat had been built just a few years before. No construction project is ever friction-free, but this one ended up being acrimonious and, in true US fashion, very litigious. Ultimately, however, the two vessels were built and completed their trials, albeit well behind schedule. The lead boat, *Protector*, was commissioned in June 2016, followed by *Vigilance* nearly 18 months later.

Below: The powerful Long Beach fireboat #20, *Protector*.

Outboard Profile for the Long Beach fireboats.

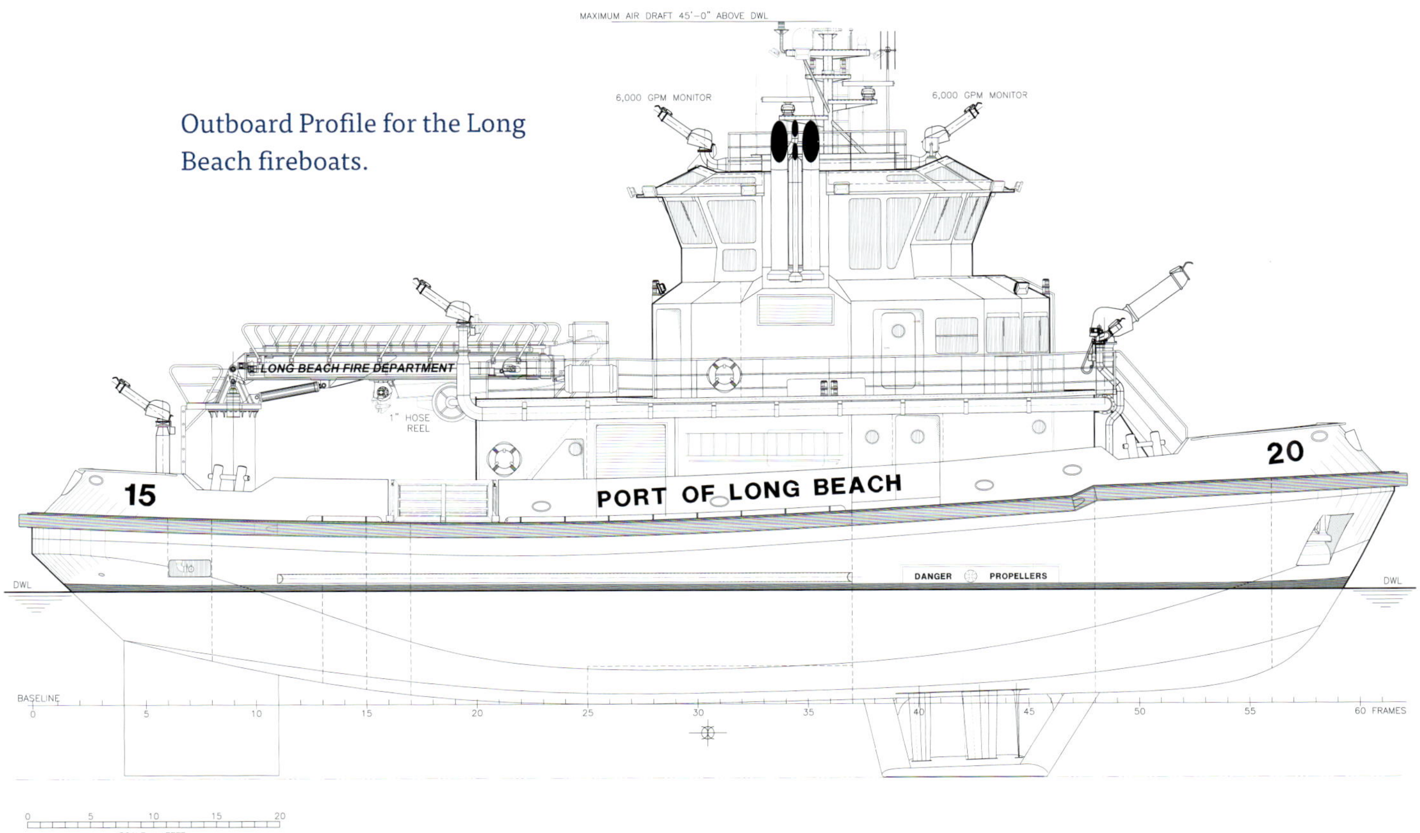

Chinese Fireboats

As Robert Allan Ltd grew its international contact network throughout the '90s, a number of interesting fireboat projects developed in Hong Kong and mainland China. These contracts reflected the different methods which various countries follow in new vessel procurement. In North America, as described in the preceding fireboat project descriptions, it is most common for owners such as fire departments to have a complete design developed by a consulting naval architect to match their individual specific needs, and then that design is put out for competitive bidding among qualified shipyards. This process is generally referred to as "design then build" and has the advantage that the owner is involved throughout the design process, reviewing the design to ensure that the specified performance and indicated arrangement address its needs. A shipyard selection can then be made based solely on shipyard qualifications and the quoted cost, knowing that the end product is clearly defined and all bidders are quoting on exactly that same end product.

In many other locations it is more common for clients to seek proposals from shipyards for a turnkey "design-build" proposal in response to either a description of the required vessel or, in some cases, a more detailed specification, but not on a design. This process has the advantage that the shipyard is contractually responsible for every aspect of the vessel from design to delivery. The primary disadvantage, however, is that the client then has to compare an array of proposals that quite likely will vary widely in concept and design details and most likely also in price. The owner must then have (or contract with) the technical resources to ensure they are comparing apples to apples in the bid evaluation process.

As China was growing rapidly throughout the '90s, and either enlarging existing ports or in some cases creating large new port cities, a demand emerged for several new fireboats. In China, the normal method of acquisition was to follow a design-build process, but Robert Allan Ltd was fortunate to have some good contacts with shipyards there that were large enough to be experienced in these specialized vessel projects and yet small enough to be price competitive.

Hong Kong *Fire Boats* 2 and 5

In 1994 Robert Allan Ltd partnered with Wang Tak Engineering and Shipbuilding Co. Ltd of Hong Kong to respond to a requirement from the Hong Kong Marine Department for a new fireboat for one of the world's busiest ports. The team was successful in this bid, and the result was the *Fire Boat 2* (yet another in that imaginatively named series). The fire department requirements in Hong Kong were interesting in that the primary objective was more related to transporting large numbers of firefighters to the scene of an incident than in providing high-capacity pumping capability.

This 30-metre twin-screw *Fire Boat 2* has a top speed of 15 knots and a total pumping capacity of just 970 m^3/hr, but it can accommodate 32 firefighters and is outfitted to act as a command centre in emergencies.

In 1995 another project, *Fire Boat 5*, was tendered for Hong Kong, and we teamed with Greenbay Marine Pte. Ltd of Singapore for the successful bid on this project. Greenbay Marine is a company with whom we made contact back in the early days of our work in Singapore, in the late '70s, and with whom we have worked frequently since. The performance objectives for this 35-metre boat were similar to *Fire Boat 2*, with a requirement for a top speed of 15 knots and facilities to accommodate a crew of 10 and (in transit) 32 firefighters.

Fire Boat 5, known as the "Tuen Mun Fireboat," was delivered to the Hong Kong Marine Department in 1997. It was built to provide fire and rescue cover to the northwestern waters, including the Hong Kong International Airport. It was powered by twin 1,078 kW diesels and had a firefighting capacity of 1800 m^3/hr from three identical pumps, one independent engine-driven and two main engine PTO-driven.

Hong Kong *Fire Boat 2*.

Bottom: Hong Kong *Fire Boat 5*.

Shenzhen Fireboats

In 1996 Robert Allan Ltd again teamed with Greenbay Marine to design and build two 36-metre fireboats for the City of Shenzhen, in southern mainland China. These 36-metre boats were intended for use as firefighting and emergency standby duties in the Shenzhen coastal region, each with a total pump capacity of 2400 m^3/hr. The two boats were named *Shen Xiao Yi Hao* and *Shen Xiao Er Hao,* which, not surprisingly, translate to *Fire Boat #1* and *Fire Boat #2* respectively.

Right and bottom: Shenzhen fireboats showing off.

Dongguan Fireboats

In 2004 we teamed again with Wang Tak Engineering and Shipbuilding Co. Ltd for a new 38.9-metre fireboat for Dongguan, China, a historic port city in the Pearl River delta, and one with the reported reputation as the "Capital of Sex" in China. The new boat was not, however, intended to dampen the ardour of its citizens, but to cope with the growing cargo shipment demands of the port itself.

The new RAnger 3900 class fireboat *Guan Xiao Yi Hao* was delivered in 2007 to the Port Authority of Dongguan. With a continual demand for similar fireboats, we had recently branded our fireboat designs as the RAnger series of emergency response vessels, a tactic we had found successful in promoting various vessel type design series.

The new fast-response fireboats were designed to specifically address the firefighting and rescue needs of the local port area, which is contained entirely within the local river estuary. Consequently, the fireboat had a water-draft limit of 2.2 metres and an air-draft constraint of 10 metres, the latter to deal with numerous local bridges. These fireboats achieved a full-load trial speed of 16 knots, with a low-wake 11-to-12-knot cruising/patrol speed. The firefighting equipment is effectively rated at a Fi-Fi I international standard, with four monitors and two 1500 m^3/hr fire pumps driven by power take-offs on the front of each main engine.

A 3D rendering of the RAnger 3900 class fireboat *Guan Xiao Yi Hao*.

At the time of this design project, we were experimenting increasingly with 3D concept development and using early design-stage renderings to verify the overall layout. Such a rendering was useful in this boat, with its distinctive front-end deckhouse treatment.

In November 2011 a second fireboat, the *Guan Xiao Er Hao,* was delivered to the Dongguan Fire Services Department, another joint project by Wang Tak Engineering and Shipbuilding Co. Ltd and Robert Allan Ltd.

Guan Xiao Er Hao, which means yet again *Dongguan Fireboat #2*, is a 46.42-metre twin-screw diesel-powered fireboat specifically designed for marine and shore-side firefighting and offshore rescue in the Pearl River estuary, the Pearl River and the Dongguan region harbour environment.

The propulsion machinery consists of a pair of CAT 3516B-HD diesel engines, each rated 1,864 kW, and each driving a fixed-pitch propeller, achieving a top speed of 16 knots.

The firefighting equipment is nominally to a Fi-Fi II-rated international standard, but with increased pump output, making this the highest capacity special-purpose fireboat designed by Robert Allan Ltd, with a total firefighting capacity of 10,800 m^3/hr. This impressive fireboat is among the most capable of its type worldwide.

Top: *Guan Xiao Yi Hao* on trials (without the designer's distinctive red stripe!); **bottom:** *Guan Xiao Er Hao* demonstrating its deluge self-protection system.

Outboard Profile of Dongguan fireboat *Guan Xiao Er Hao*.

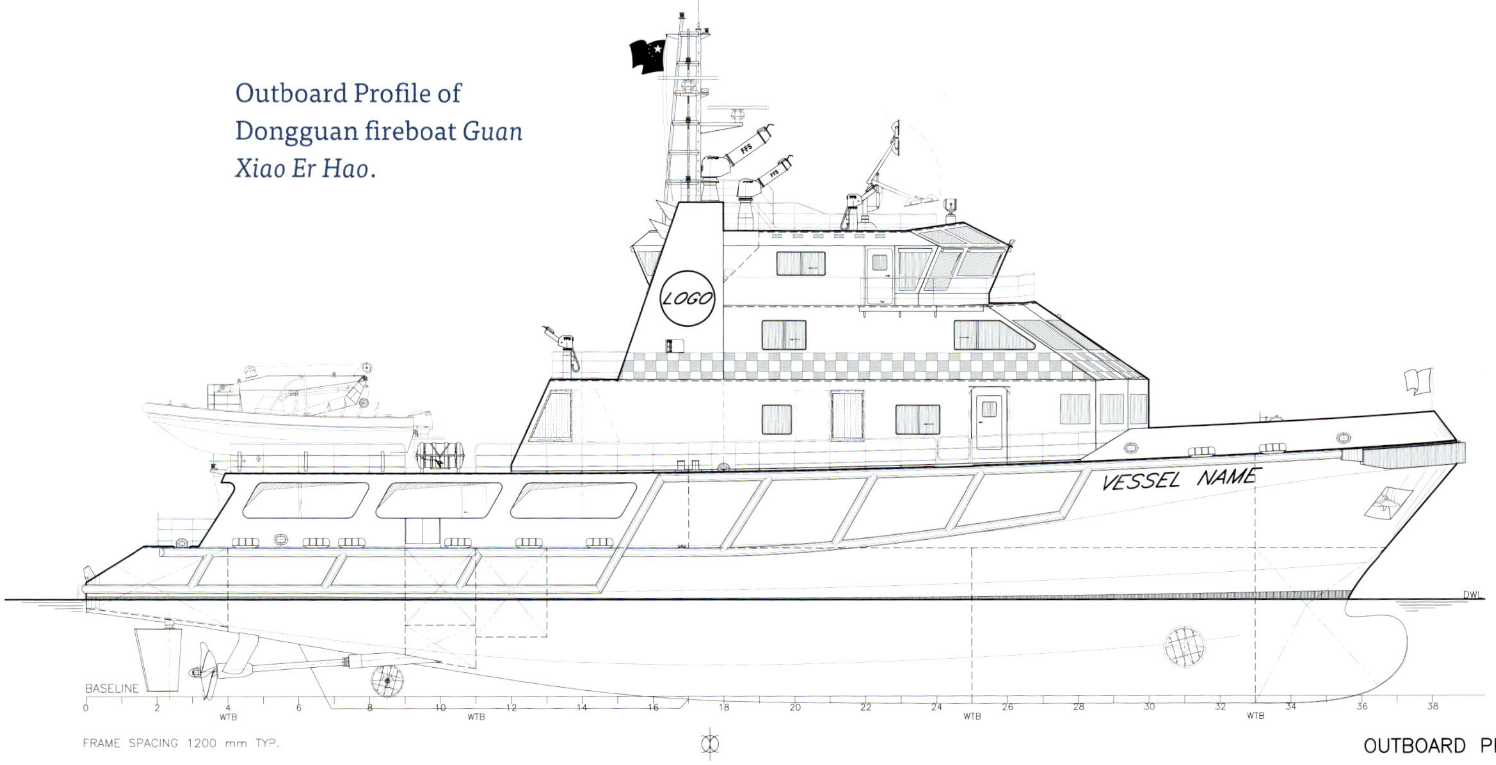

Guan Xiao Er Hao at speed.

Other Fireboat Projects

While the majority of recent fireboat projects have either been for US or Chinese cities, there have been some quite interesting projects for other more remote parts of the world.

Kuwait Firefighting Tug

When is a fireboat a tug? When is a tug a fireboat? The fitting of firefighting capacity on harbour tugs is quite common; however, when an owner requires a major terminal support tug with significant towing and ship-handling capabilities, plus the significant total fire-pump output of 8400 m^3/hr, the result is an exciting combination of the best attributes of both vessel types. With a significant record for successful designs for both tugs and fireboats, Robert Allan Ltd was an easy choice for Kuwait Fire Services to make regarding the development of a new design for a major firefighting tug for operation in the Port of Shuwaikh, in Kuwait City.

The contract for design and construction of the new vessel was awarded to the trading company Al Boom Marine of Kuwait, who retained Robert Allan Ltd to develop the design, and then awarded the construction contract to Simulation Tech Inc. (STI) of South Korea. The use of trading companies in this middleman role is common in the Middle East and in Asia. It has never been terribly clear what value is added by that process, but it is their custom and hard to change. The new 39-metre vessel, named

Mojaweb, a major firefighting tug for Kuwait.

Mojaweb, was delivered in March 2011, sailing under its own power from Korea to the Middle East.

Although primarily configured as a fireboat, the *Mojaweb* is also equipped for typical tanker support and ship-handling work, with a hawser winch on the fore deck, and a 75-tonne SWL towhook aft. The accommodation is arranged for an operating crew of up to 14 persons, with a berthed crew of nine, but it also has the capability to accommodate survivors and/or additional firefighters.

The Fi-Fi II firefighting capability is provided by a single independent pump engine, rated 2,590 kW, driving two pumps, each rated at 2400 m^3/hr, and a pair of 1800 m^3/hr main engine-driven pumps.

Firefighting Tugs

Although *Mojaweb* had a significant and clearly defined firefighting role, it is also extremely common for harbour and escort tugs in general to have a lot of Fi-Fi capability. This makes sense, as tugs are generally omnipresent in most ports, and in many cases will be first responders to a waterfront or shipboard incident. At a guess, more than 90 percent of the tugs we design today have a Fi-Fi I designation (2400 m^3/hr) and some of the larger boats have more. This represents a serious investment by tug owners, so it is assumed that the ports they serve must offer some compensation for providing that capability, but those fiscal mechanisms are seldom transparent.

A Z-Tech 30-80 class tug of Suderman & Young fighting a 2021 fire in the Houston Ship Canal.

Bangladesh Fire Float replete with camouflage paint job.

"Fire Floats" for Bangladesh

The most recent (as of 2021) fireboat project from Robert Allan Ltd was the 2019 delivery of a design for two unique "fire floats" (an anachronistic term from UK colonial days) for construction in Khulna Shipyard Ltd in Bangladesh. The boats, measuring 18.9 by 5.3 metres, are designed for river service. The all-aluminium hulls are powered by two 560 kW main engines, with a design speed of 20 knots. Two modest 120 m^3/hr fire pumps are fitted, each driven by a 55 kW diesel engine. Two 120 m^3/hr fire monitors are located on the house top within view of the operators in the pilothouse. Six hydrants for hose service are located fore and aft on the main deck. In keeping with the evolving use of 3D design and concept visualization in the office, this design was worked up to an advanced stage in 3D.

At time of writing the construction is not yet complete, more than two years after delivery of the design. The delay may be attributed to the camouflage paint job applied to the boats rather than the more traditional paint scheme envisaged by the design team!

A 3D rendering of the Bangladesh Fire Float.

Rescue Sauvetage
Coast Guard
Garde côtière

11

Emergency Response Vessels

In the marine world, the term "emergency response" can cover a multitude of operational requirements and often includes firefighting. Since that subject has been covered in detail separately in this treatise, in this instance the emergency response term applies only to vessels used for search and rescue (SAR) and pollution response.

Search-and-Rescue Lifeboats

Rigid-Hull Inflatables

At the small end of the scale of boats designed by Robert Allan Ltd for SAR service are many different types of rigid-hull inflatable boats (RHIBs). These small, fast boats are a far cry from the heavy-duty workboat designs usually associated with the firm, but the type has its origins in extensive work done in the 1970s for the C.W. Lucas company, who built and marketed a range of successful RHIBs for many years under the *Hurricane* brand name. This very successful all-Canadian company fell to the fate of many local companies and was bought by its much larger competitor, the French company Zodiac, in about 1990. Robert Allan Ltd designed numerous standard series RHIBs for Lucas and later for Zodiac Hurricane (as the new company was named and under which it still operates), from 5.5 to 8 metres in length and with various combinations of outboard or stern-drive power. These boats are widely used by the Canadian and US Coast Guards and by their navies and are now marketed internationally by the extensive Zodiac organization.

Opposite: CCGS *Gordon Reid*, see page 219.

CGR 100 Lifeboats

In the mid-1980s the Canadian Coast Guard (CCG) was considering using the RHIB concept as the basis for a much larger SAR lifeboat. Robert Allan Ltd designed the prototype vessel *CGR 100* for testing and evaluation. This 14-metre boat was powered by twin 525 hp diesels driving Parker waterjets and achieved a top speed of 32 knots. Although this specific vessel is still in service with the CCG, the concept of a "big-RHIB" was not widely accepted within the ranks of CCG at the time, and they elected instead to build replicas of the USCG 47-foot motor lifeboats. The *CGR 100*, however, was praised for its performance and handling.

CGR 100, a prototype RHIB SAR vessel.

Arun Class Lifeboats

In its seemingly endless quest to find the ideal lifeboat for the widely diverse conditions of the Canadian coastline, CCG turned to the UK Royal National Lifeboat Institute (RNLI) for guidance. The RNLI is a highly regarded organization with an excellent track record in developing craft well suited to UK coastal conditions. CCG bought the rights to the Arun class motor lifeboats, which were at the time deemed obsolete by the RNLI and were being replaced with more advanced designs. Robert Allan Ltd was awarded the contract to convert the original GRP Arun design to aluminium construction. This was no mean feat, as the fully moulded shape of the Arun class was not amenable to plating, having no developable surfaces. Further the CCG insisted (with no logical justification) on maintaining the many small watertight compartments used in the fibreglass version. In GRP construction these small pockets could be formed, sealed and filled with foam for enhanced survivability, but in welded aluminium, each pocket had to be thoroughly accessible for welding and then fitted with vents. We also had to convince Transport Canada, who in their wisdom believed that all these tiny pockets also needed bilge suctions, that the smallest compartments could exist without those without posing a safety risk to the boat. The result was a vessel much more complex than it needed to be if CCG had been more amenable to sensible design alterations to suit aluminium construction.

Ten of these 15.77-metre boats were constructed between 1990 and 1993. They were built in Canada by three different shipyards, thus ensuring in true Canadian fashion that political favours received first consideration over the practicality of allowing one competent yard to maximize the learning curve benefit of a reasonable series of the same boat from the same builder. The boats were built by Industrie Raymond Ltée of Sept-Iles, Quebec; East Isle Shipyard Ltd of Georgetown, Prince Edward Island; and Hike Metal Products Ltd of Wheatley, Ontario. All were built with aluminium hulls except for the FRP-hulled British-built prototype (CCGS *Bickerton*), which was built in the UK. They are considered high-endurance lifeboats staffed by a crew of four.

CCG version of the Arun class lifeboats, converted to suit aluminium construction by Robert Allan Ltd.

Bay Class SAR Lifeboats

CCGS *Pennant Bay*, one of a new series of modified Severn class lifeboats for CCG.

In 2012, as the aging Arun class lifeboats were nearing the end of their service life, the Canadian Coast Guard again turned to the UK's Royal National Lifeboat Institute for a "proven" lifeboat design. They negotiated for the larger Severn class lifeboat design, but thankfully recognized that to convert this boat to suit Canadian conditions, including converting to aluminium construction, the design would have to be modified extensively. The CCG's willingness to accept design input on these matters was refreshingly different this time around compared to the earlier Arun exercise. The new Canadian design is larger than the British boat, with a greater range, and was specifically designed to handle the worst weather encountered year-round in the waters off Newfoundland and Nova Scotia. CCG now refers to these as the Bay class lifeboats.

Ten of these new 19-metre-long SAR lifeboats were built. They are self-righting and capable of safe operation in extreme weather up to Beaufort Force 12 conditions and can survive in 12-metre seas. With 2400 bhp they achieve a service speed of 23.5 knots.

Fibreglass to Aluminium Conversions

One interesting project we completed recently was the design for a series of 19-metre self-righting search-and-rescue vessels for the Canadian Coast Guard in collaboration with the Royal National Lifeboat Institute (RNLI). It started as the result of a request for bids from the Canadian government. A proven hull form was required, so the successful bidder would have to contract with RNLI, who would supply the hull form. RNLI are a non-profit organization with a long and successful history, but they have been able to do extensive model testing and optimization of their hull shapes for self-righting and for speed. They have self-righting SAR vessels stationed all along the UK coast, ready to go out in the most severe weather conditions.

The CCG SAR boats are designed to be capable in very extreme weather. The design was such that if the boat was knocked down it would come back up; the boat always wants to self-right, at any degree of heel. The RNLI vessels are all-fibreglass construction and the Canadian government wanted theirs built in aluminium. We won the contract to basically convert their fibreglass hull shape into an aluminium hull.

Lawren Best was the team leader of the naval architecture team. I worked on the structural design, lightship weight estimate, and oversaw the stability work. To convert from fibreglass to aluminium, the structural philosophy had to be completely redeveloped. The bigger problem was weight. The hull dimensions were scaled up in size to accommodate the added weight of the aluminium construction.

I completed the hull structural design, and Phil Barron designed the deckhouse and superstructure. Hans Muhlert was the project director; he was very patient, sharing his extensive knowledge of small craft design and aluminium structural design with the entire project team. To date, four of what are called the Bay class vessels have been built for the West Coast.

—Andra Papuc

Search-and-Rescue Cutters

At the large end of the SAR vessel design inventory at Robert Allan Ltd are the CCG Type 500 cutters. Designed in 1987, two of these vessels were built in 1991 by Versatile Pacific Shipyards Ltd in Victoria: the CCGS *Gordon Reid* and CCGS *John Jacobson*. This was a major design commission for our company at the time, coming at the end of a lengthy dry spell without much significant work. The vessels were designed with a rather fine hull form for a good turn of speed and manoeuvrability. Key to their performance was having a relatively high GM and maintaining the design weight, which required some vigilance by the builder.

The vessels are 49.95 metres long by 11 metres beam with a load draft of 4 metres. Power is provided by four diesel engines, each rated 1184 bhp, combined in pairs through reverse-reduction gearboxes driving controllable-pitch (CP) propellers. They were designed for a service speed of at least 15 knots in Sea State 5 conditions.

These cutters were equipped with "daughter boats": small, fast 7-metre RHIBS deployed to deal with fast-response on-water rescues and similar activities. A concept was developed to use a stern configuration like a deep-water trawler to launch and recover the RHIB; however, doing this safely and repeatedly in all anticipated sea conditions was a major challenge. A manned model of the cutter (about 5 metres long) was built from which the "driver" could operate a model RHIB to test the efficacy of the proposed system. After some minor refinements the system was approved and implemented in the ships, and it works very well. About a decade after these boats went into service, the US Coast Guard was evaluating stern launch systems for its next generation cutters. They deemed the Type 500 system the best of all the options they evaluated and consulted with us extensively on their own derivative of this.

On Building in a Failing Shipyard

The two Type 500 SAR cutters were the last ships built by Versatile Pacific Shipyards. Regrettably, the economic writing was on the walls of the fading shipyard. It became apparent at an early stage of construction that neither weight control nor CG control measures were in place, and that the yard's objective was simply to finish the contract with the least possible expenditures. The previous projects in this yard had been heavy icebreakers, and clearly the yard was under internal instructions to use any piece of steel in stock before ordering new material. Thus, we noted, for example, small lube-oil tanks built from 20 mm plate rather than the 5 mm specified, and many similar issues of complete disregard for the stated design objectives. The shipyard naval architect who monitored weight and stability rejected our pleas for some caution and corrective measures, stating that the ship would still satisfy Transport Canada minimum requirements for ships of this size and that was all they (the shipyard) were obliged to do. Our design objectives for weight and stability were completely ignored. Our letters to the Department of Supply and Services (DSS), who were ostensibly responsible for overseeing the construction process, raised red flags on all these issues but also went completely ignored. The end result was a pair of vessels which just met minimum regulatory stability requirements, were overweight, and which suffered from severe rolling due to their much lower than intended GM.

This was a painful exercise for me. The boats were attractive, well laid out and a great deal of innovative work had gone into their design. To have all that effort botched by the neglect of the

Type 500 SAR cutter CCGS *Gordon Reid*.

General Arrangement for CCG Type 500 SAR cutters.

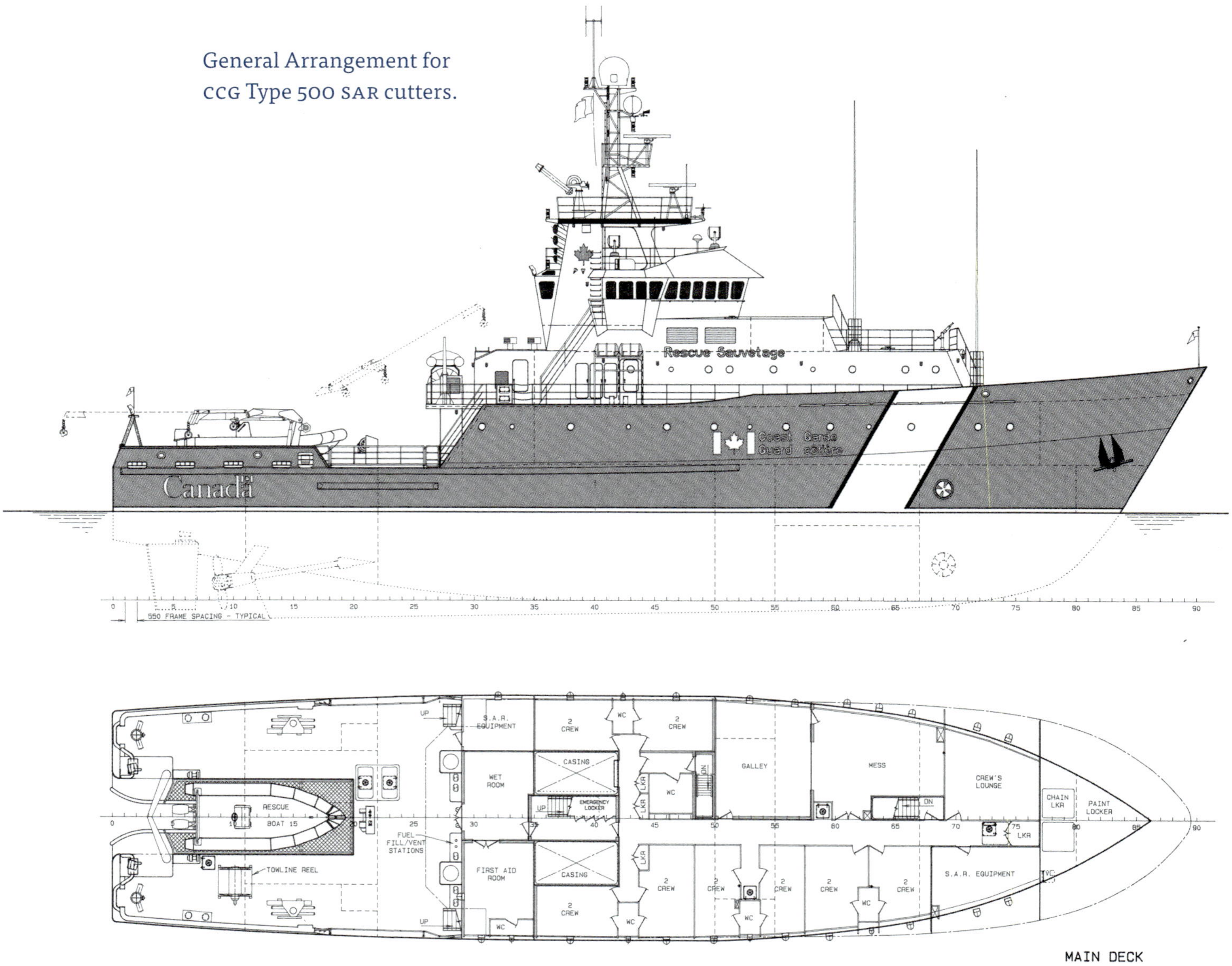

shipyard, and to a degree the parallel neglect of both the CCG program officers and DSS officials, was a huge disappointment, especially as it tended to reflect poorly on our design. After some remedial measures, the performance of the boats was improved, but they never reached their promised potential.

—R.G. Allan

Hong Kong Rescue Craft

One of the more interesting types of SAR craft we have been involved with was a pair of airport emergency response boats for the then-new Hong Kong Airport on Chek Lap Kok Island. The concept was to have two fast-response boats which could attend to a downed aircraft, as the new airport was totally on an island. For this project, Robert Allan Ltd teamed with Incat Designs of Australia, a firm with which we had collaborated on a few multi-hull projects.

The boats, designated as "command boats," were named *CB1* and *CB2*. They were built for the Hong Kong Marine Department by Greenbay Marine (Pte) Ltd of Singapore in 1997. They measure 35.3 by 12 metres and are waterjet powered, with a top speed of 28 knots. Each boat has a rescue capability of 600 persons based largely on the capacity of the large array of inflatable liferafts carried on board.

Hong Kong Airport Rescue Boat *CB1*.

Pollution-Response Boats

A much more recent type of vessel design for the company arose when both the federal and provincial governments were under pressure to increase oil-spill response capabilities on the BC coast in anticipation that projects such as the Trans-Mountain Pipeline expansion would be implemented.

PRV II and *PRV III*

Robert Allan Ltd was awarded a contract in 2009 by the Canadian Coast Guard to design two series of standard small oil-spill response craft: one (designated PRV II) at 24 feet in length with outboard power, and one (PRV III) at 40 feet in length with stern-drive power. Both classes were designed for movement by road.

In March of 2010 CCG awarded two contracts for the construction of 15 boats of each type. The 15 PRV IIs were built in Newfoundland by C&W Offshore of Mount Pearl, and by all accounts this contract was satisfactorily completed.

The 15 PRV IIIs were also awarded to a single small contractor in the Maritimes, who got into financial trouble and did not finish all the vessels. Eventually all 15 were built by a variety of alternate builders and placed in service.

Additional vessels of both these classes were also built after the initial contracts.

BRAvo 2500 Class

In 2016 Robert Allan Ltd was contracted by the Western Canada Marine Response Corporation (WCMRC) to design a new class of large spill-response boat to bolster their local capabilities. These new 25-by-10.25-metre boats are intended for use in more extended response operations with full accommodation for a crew of up to eight persons.

Designated as the BRAvo 2500 class, these vessels are equipped with a main reel, storing enough pollution-control boom to encircle a casualty the size of a Panamax ship. They also carry onboard a "Current Buster" spill-response boom and can carry additional spill-response equipment on deck and in the hold as required.

Built to avoid the various Transport Canada and SOLAS regulatory requirements that apply to vessels above 24 metres registered length, these boats pack a lot of gear into a confined space. Like all such vessels, they would have benefitted greatly by the addition of a couple more metres in length, but exceeding the regulatory hurdle represents significant added cost in construction and in ongoing operations.

Construction of these boats was tendered internationally, and ultimately three of these boats were built in Singapore/Batam, Indonesia, by ASL Shipyard.

Top left: PRV II class spill-response craft on the East Coast; **top right:** PRV III class response boat; **above:** *Coastal Sentinel*, first of the BRAvo 2500 class pollution-response boats for the Western Canada Marine Response Corporation.

HARVEY STONE
HARVEY STONE
FFS

12

Offshore Industry Vessels

The international offshore oil industry has been a dominant force in the marine industry since the 1960s, focused initially in the Gulf of Mexico and the Middle East, with a shift in focus to the North Sea in the early 1970s. The related ship-design and shipbuilding industries have been logically centred on the US Gulf Coast and in Norway. The Norwegians, and the Ulstein organization in particular, showed great acumen in developing designs for high-quality support vessels to serve their own offshore industry and were very successful in promoting those as the industry standards worldwide.

At the time, Robert Allan Ltd was too small a company and too far removed from the heart of these industries to compete realistically for design work in the mainstream of offshore supply vessels (OSVs). However, we did find a demand for more non-standard vessel designs in areas where the big players were not dominant. Accordingly, we had success with designs for specialized smaller OSVs for shallow-draft operations, and later for ice-class vessels. More recently, we had considerable success designing specialized crewboats and specialty offshore support vessels, mainly for operations in the Middle East.

Opposite: MV *Harvey Stone*, see page 238.

Many of the OSVs were very challenging designs and unique in the world at the time, thus enabling us to gain some expertise and a reputation in this important industry.

Offshore Supply Vessels

Beaufort Sea Explorer

Our entry into the offshore field began in 1970 with the design of a compact dual-draft tug/offshore supply vessel for Kaps Transport Ltd of Edmonton, to operate in the Mackenzie River delta area of the Beaufort Sea. The *Beaufort Sea Explorer* is a 2,250 bhp 118-by-36-foot twin-screw tug/OSV built by Vito Steel Boat and Barge Ltd of Delta, BC. (Note: The ship registry lists a company named Alberta Equipment Centre Ltd as the builder. That untraceable name may have been an Alberta-registered company used by Vito to separate the value of the work done in BC from that done in the Northwest Territories for taxation purposes.)

In a 1972 article in the *Vancouver Province*, marine editor Norm Hacking wrote, "The tugboat was designed for three functions: to tow barges up the Mackenzie and into the Arctic, to carry deck cargo to re-supply Arctic petroleum operations, and to shift into place and anchor huge drilling platforms for offshore work in the Beaufort Sea. Construction began at Vito's yard in Delta in mid-January. After the hull was finished, it was trucked to Hay River in 33 loads weighing an average of 22 tonnes each. Ship assembly took 17 days at Hay River..."

Robert Allan Ltd supported the shipyard by development of a complete set of construction working drawings, done in the "old" way of manual lofting and drafting, frame by frame.

In the late 1970s Justice Thomas Berger concluded, after a multi-year enquiry, that it was not the correct time to build a pipeline down the Mackenzie Valley, which effectively put the brakes on all the activity in this area. The offshore work continued into the very early 1980s, but by 1983

Beaufort Sea Explorer, an early dual-draft offshore supply vessel (OSV) for the Beaufort Sea.

virtually all the offshore drilling and exploration work in the Beaufort had ceased. Consequently, this boat, along with virtually all the other floating equipment in the North, found work in numerous other areas of Canada or internationally. The *Beaufort Sea Explorer* is still working today, but currently in Newfoundland.

John Mattson

In 1972 Kaps Transport Ltd retained Robert Allan Ltd to design another dual-draft tug/supply vessel similar to the *Beaufort Sea Explorer*. The *John Mattson*, named after Kaps' marine superintendent, who was a significant force in those days of early Beaufort work, measured 106 feet by 35 feet and was, like almost every vessel in the North in those years, powered by a pair of CAT D399 diesels, rated 1,125 bhp at 1,225 rpm. This boat was designed to both push barges and tow off a winch, hence has a rather unusual reverse-bow geometry.

"Mini" Supply Vessels for Brunei

In 1976 Robert Allan Ltd obtained a contract from Singapore Shipbuilding and Engineering Pte Ltd (SSE) to design a pair of shallow-draft supply vessels for operation in Brunei for OIL Asia (Pte) Ltd. This was one of our earliest projects in Southeast Asia and helped to establish our design reputation in that area. These boats, imaginatively named *OIL Supply 1* and *OIL Supply 2*, measured 147 feet 6 inches by 36 feet. They had to access a terminal on the Belait River in Borneo, which was only accessible across a bar at the river mouth, and thus they had a restricted operating draft of just 6 feet. They were capable of loading down to an 8-foot draft once outside that limitation. Our experience with comparable shallow-draft tunnel-stern designs in the Mackenzie River system gave SSE confidence in our ability to develop such a new design.

OIL Supply 1, one of a pair of shallow-draft "mini" supply vessels for Brunei.

Canmar Suppliers: Ice-Breaking OSVs for the Beaufort

In 1974 Dome Petroleum Ltd began their exploration work in the Beaufort Sea in earnest. Robert Allan Ltd was selected to design their first ice-breaking OSVs, which were to be capable of working in not less than 2 feet of Arctic ice. At the time Transport Canada had just introduced the new Canadian Arctic Shipping Pollution Prevention Regulations (CASPPR), and these supply boats were the first vessels ever built to these new regulations. It became a bit of an experiment at many levels.

In order to meet an operational deadline of early 1976, the project was totally fast-tracked. In addition to doing all the design work for four of these vessels, Robert Allan Ltd managed the purchase and distribution of all the heavy steel needed for their construction, so that the material would be on hand once construction contracts were awarded to shipyards. We were also involved in the selection of all the main machinery and overseeing the work in the shipyards on behalf of the owners. It was a very intense effort for our small office at the time, and the pace caused our senior shipyard superintendent, Jack Macfie, to suffer a heart attack, from which he thankfully recovered. This project also led to the hiring of Claire Johnston and Peter Woodward as shipyard overseers, two individuals who became extremely important to the long-term success of our company (Chapter 23).

As these boats were limited to a draft of just over 14 feet to access the relatively shallow entry to Tuktoyaktuk harbour, and needed high bollard pull for their anchor-handling and towing duties, the design incorporated deep tunnels at the stern for the largest possible propellers. Good success with thrust-augmenting nozzles around propellers in previous smaller boats, such as the *Beaufort Sea Explorer*, led the design team to suggest that a similar stern-gear arrangement be used, but this much larger design was in uncharted territory in terms of power and deep tunnels operating in ice. To hedge all bets, the decision was made to build two boats with open screws and two with Kort-type nozzles and assess their relative performance for consideration on future projects.

Shipyard contracts were awarded to Vito Steel Boat and Barge Ltd, Allied Shipbuilders, Burrard Dry Dock Corp. Ltd and Yarrows Ltd. The boats were named as follows:

- *Canmar Supplier*: Allied Shipbuilders (open-screw)
- *Canmar Supplier II*: Vito Steel Boat & Barge (open-screw)
- *Canmar Supplier III*: Burrard Dry Dock Corp., North Vancouver (nozzles)
- *Canmar Supplier IV*: Yarrows Ltd, Victoria (nozzles)

These unique designs, among the first ice-breaking (as opposed to just "ice-strengthened") OSVs anywhere at the time, measured 208 feet long by 45 feet beam. They had nominally 7,000 bhp each.

The open-screw boats were powered by twin Nohab Polar diesels rated 3,510 bhp each. The nozzle boats had EMD diesels rated 3,520 bhp each. In the final analysis, the nozzles proved their worth, but the thrust-generating advantage of the nozzles was considerably less than in a typical installation due to the fact that the nozzles were largely integrated into the tunnels, so much less than the full nozzle circumference was effective. The nozzles also had a tendency to choke in slush ice, but that proved easy to remedy by reversing pitch on the controllable-pitch props.

The four *Canmar Suppliers* working in the Beaufort Sea.

Below: *Canmar Supplier*, the first of four CASPPR Arctic class II OSVs for Dome Petroleum.

A Supply Chain Issue

We had an interesting project in 1974 when oil exploration started in the Arctic. We did four ice-breaking supply boats—the *Canmar Suppliers*. There was this guy from Texas who came in and talked to Mr. Allan, and he figured out that we could do it. He was a really outgoing, very loud-speaking Texan with a Texas drawl. He made sure that we could do the job on time.

At one point near the beginning there was a forecast that the price of steel was going to go up and it was going to be harder to get. That would sure throw a monkey wrench into the whole project. So he says, "Okay, you guys, tell me what steel you want for this supply boat." But we hadn't designed it yet, so we had to give him a steel list for something we hadn't designed and then we had to design using those scantlings. The North Slope Basin [Beaufort Sea] is pretty shallow, so these couldn't just be regular icebreakers, they had to be shallow draft. A 200-foot supply vessel would normally draw about 18 feet, but we designed them with 14 feet 3 inches of draft and big tunnel sterns—not too extreme, but pretty shallow, as they had to be able to get into Tuktoyaktuk harbour. This was the time when Canada was developing new regulations for vessels operating in the Arctic: what ice-strengthening you should have and other specifications depending on when you wanted to operate and where. We had to design these boats to the new regulations, so we had to get in touch with the people who wrote the regulations to get things to work out. We were the first to use those new regulations. It was a challenge, but we did it.

—Hans F. Muhlert

Vito and the Unorthodox Launch of *Canmar Supplier II*

Vito Trevisi was a thorn in the side of the conventional shipbuilders in the Lower Mainland of BC. Unorthodox in his personal style and in his management practices, he built a substantial and successful shipyard enterprise from his beginnings as a journeyman welder in small boat yards. His yard was a big building on the banks of the Fraser River, and for most of his projects a mobile crane would be brought in to simply lift and launch the smaller tugs and workboats that were his typical output. *Canmar Supplier II* was bigger and much heavier than anything he had previously built and needed a more conventional launch than he was used to. A set of steel launching ways was built in his yard, and when ready for launching, the boat was brought into position, and they awaited high tide to minimize the drop off from the ways into the river. Side launches of this type are common in US Gulf Coast yards, but most BC yards used more conventional longitudinal slipways, which are much easier on the boats being launched. When the timing was right, Vito ordered two CAT bulldozer drivers to "push." However, the ground in his yard was soft and one of the Cats started digging itself into the ground without moving the ship, while the other made a little headway but thus skewed the ship on the ways. As the tide was starting to fall, in typical fashion Vito screamed at the drivers, pulled one of them off the machine and jumped into the driver's seat himself. Through sheer bravado he managed to get the ship straightened out and pushed into the water, declaring himself the hero of the hour! As the launch was underway, I was standing in the crowd not far from Arthur McLaren of Allied Shipbuilders, who couldn't bear to watch this spectacle and simply buried his head in his hands. When the boat was finally in the water, he simply walked away shaking his head.

—R.G. Allan

Launching *Canmar Supplier II* the hard way.

Ice-Breaking Anchor-Handling Supply Vessels for Gulf Canada

Our next major offshore project came in 1981 when we won the contract to design two CASPPR Arctic class IV Anchor-Handling/Tug Supply (AHTS) vessels for Gulf Canada Resources Inc. (GCRI). At the time, the Roman numerals (e.g., IV) corresponded roughly to the ice thickness (in feet) through which an ice-breaking vessel was capable of continuously breaking at a speed of 3–4 knots. These boats were part of a four-boat support package for two major drilling platforms that Gulf had commissioned, one bottom-based caisson named the *Molliqpak*, and the *Kulluk*, a unique conical-shaped floating drilling platform. The other two support boats were larger icebreaker supply boats (ultimately named the *Terry Fox* and *Kalvik*), designed by German and Milne of Montreal, intended primarily for ice-management duties.

The ice-breaking AHTSs, christened *Ikaluk* and *Miscaroo*, were the first of their type ever built and had very little in the way of design precedents anywhere. We worked closely with Gulf Oil engineers and their naval architect, Robin Browne, in developing this novel design. Model-testing in ice was conducted at the Hamburg Ship Model Basin (HSVA) in Germany. The final design, as depicted, was for a vessel 79.25 metres long and 17.22 metres beam, with a load draft of 7.50 metres.

These ships each had four main diesel engines: Wärtsilä model 8R32s rated 3,725 bhp each, combined in pairs through reduction gearboxes and driving CP propellers in nozzles. At the time they were the most powerful AHTSs afloat and certainly had no equal in terms of ice-breaking capability, able to break 1.2 metres of ice at more than 3 knots.

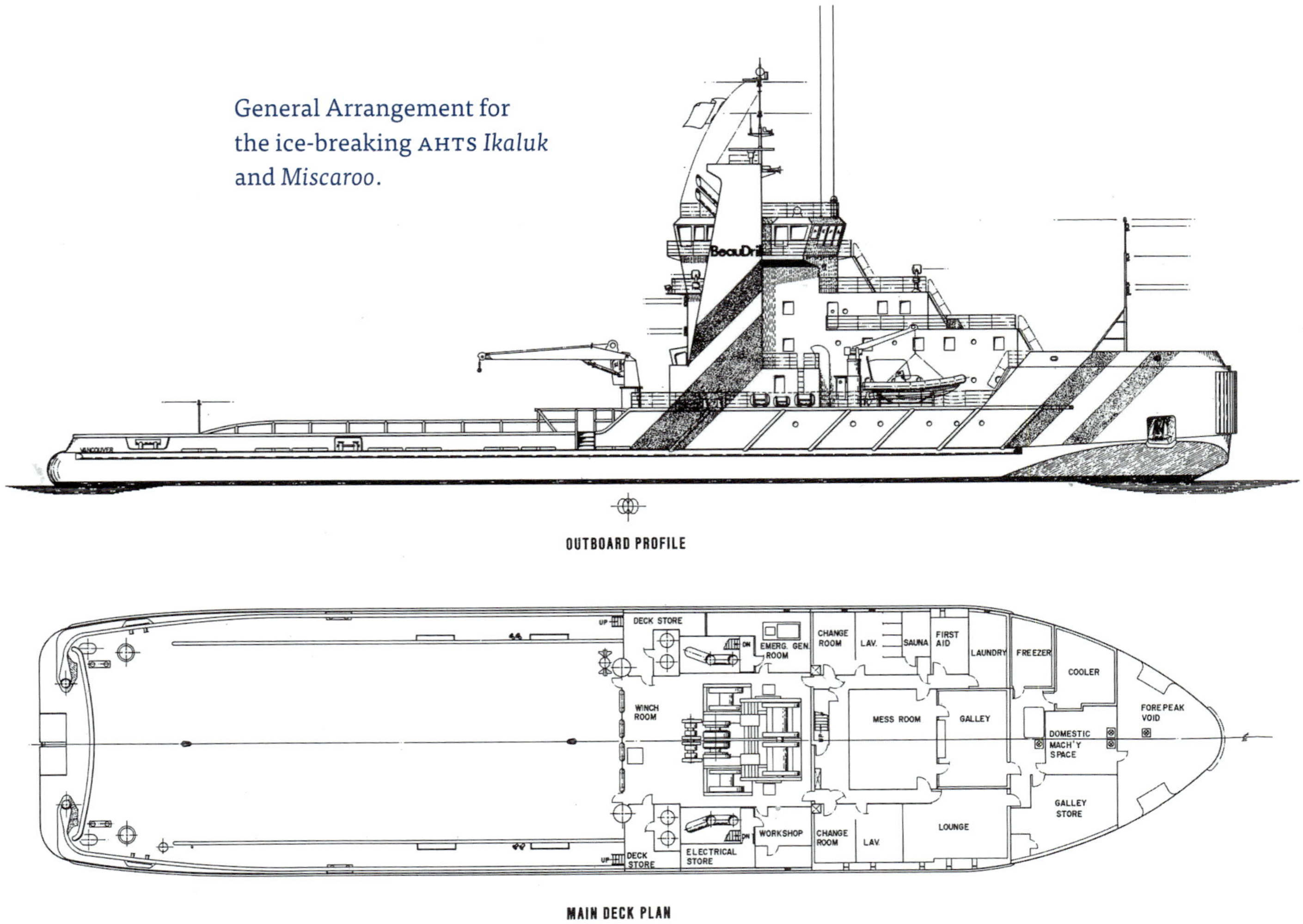

General Arrangement for the ice-breaking AHTS *Ikaluk* and *Miscaroo*.

A Lesson in Shipbuilding

The construction contracts for these boats were awarded to Vancouver Shipyards (*Miscaroo*) and to Nippon Kokan (NKK) of Japan (*Ikaluk*). This presented a most unique opportunity to compare the methods (and the results) of shipbuilders in Japan and North America. Through the course of the project I had the opportunity to visit NKK at least eight times, primarily meeting with the shipyard design teams as they worked in great detail through their production engineering process. I would go there for a week at a time with Robin Browne and with our electrical consultant, Bill Lapper. Bill would meet all week with their electrical team, and I would deal with everything else: hull design, structure, machinery, outfitting and systems. It was an intense but extremely rewarding process. NKK produced hundreds of drawings and many booklets defining the smallest of details before they cut a single piece of steel. Once everything was signed off, however, they pressed "go" on construction and the ship began to appear very quickly. The end product was excellent in every respect, and I have nothing but admiration for the way that the shipyard managed this complex project.

Unfortunately, the Canadian-built ship was a distant second in quality. The shipyard rushed to get steel into place so they could claim progress payments, and then later fell behind schedule. Their quality-control procedures were not a patch on the Japanese yard. For example, the Inerta high-build epoxy paint coating used demanded careful preparation and temperature stabilization of the steel. NKK put the boat in a drydock, sandblasted as detailed, wrapped the entire dock and the boat completely, and heated that ship for several days until the specified conditions of temperature and humidity were met and stable. VanShip picked a damp day in the spring and painted the ship like it was any other. The expensive coatings on *Miscaroo* came off in large sheets after only a few weeks in the ice; those on *Ikaluk* lasted for decades!

—R.G. Allan

Miscaroo in ice.

These vessels didn't get to work much in the Beaufort Sea as the combination of the Berger enquiry's recommendation against a Mackenzie Valley pipeline and the concurrent world oil glut of the early '80s shut down the entire offshore operation in Canada's northern ocean. These powerful AHTSs went to work in the North Sea for several years and ended up in Sakhalin (Russia) working for SMIT. *Ikaluk* continues working there, now as SMIT *Sakhalin*, and the *Miscaroo* has recently been decommissioned.

Ikaluk and *Miscaroo* tending anchors on the drill rig *Kulluk*.

A Career-Shaping Project

Some projects really stand out as milestones in my personal professional career, and the GCRI project is perhaps the most important. This was one of the largest design contracts the company had ever undertaken, and the vessels were unique in the world. At the time we were negotiating the contract, and throughout the entire design process, my father was struggling with cancer, and the responsibility for the project came to rest squarely on my shoulders. I directed all the design work, wrote the entire specification, and was heavily involved in the machinery selection process, visiting major engine and gear manufacturers across Europe. My father died just about the time we were finishing the design work, in October 1981, and I can recall being at HSVA in Germany just a few weeks before he passed away, wondering if I should be there at all. Our interactions with the Japanese shipyard that was building the *Ikaluk* were challenging but also extremely rewarding, and it was a huge boost to my confidence knowing that I was able to address all the questions from their multitudes of teams from all disciplines. To say this was a stressful time is a significant understatement. The technical aspects of the project alone were significant and very demanding, to which were added the process of coping with my father's death and then dealing with his estate, taking on the sole responsibility for the company, all while trying to do my share of raising a young family. I gained more than a few of my now many grey hairs through this time, even though I was still in my early 30s.

—R.G. Allan

Rob Allan, nearing the end of sea trials of *Ikaluk* in Japan (no time for haircuts!).

Ice-Breaking OSVs for the Caspian Sea

On numerous occasions throughout our company history, there have been projects for which an earlier Canadian design became a significant precedent for a new project in some distant place. Such a case arose in 2004 when we were approached by Keppel Singmarine in Singapore asking if we had any designs on file that might be a reasonable match for a set of operational requirements coming from the Russian company Lukoil for operation in the Caspian Sea. The requirements included operating in 0.5 metres of ice, shallow draft, a limiting air draft for bridges and a beam constraint throughout the canal system used to access the Caspian.

It just happened that the *Canmar Supplier* series of OSV designs, completed 30 years before, was a very close match, and ultimately Keppel was successful in its bid to construct four of these vessels using an updated adaptation of our 1974 design.

These boats, which we designated as the Beaufort 1000 series, measured 65 by 15 metres with an operating draft of 4.3 metres. Power was supplied from two 2,970 kW diesel engines. They had open CP screws in deep tunnels, much like two of the *Canmar Suppliers*. Like most OSVs they had cargo capacity for bulk cement, drilling water, fuel oil, fresh water, brine and recovered oil. A large anchor-handling/towing winch served the open aft deck.

The four vessels were named *Kogalym, Langepas, Svetlyy* and *Vzmorye*, which at the very least were more sonorous than *Supplier* 1 to 4!

Caspian Sea ice-breaking OSV *Kogalym*.

Multi-Purpose Offshore Support Vessels – the RAmpage Class

In the early 2000s we developed a range of Offshore Support Tugs that we designated as the RAmpage class, intended to provide a lower-cost solution to offshore operators than larger and more expensive OSVs and anchor-handlers. We still called these "tugs," and the mainstream series of those is described in Chapter 19. However, some of these vessels became considerably more complex and were definitely more AHTS than tug, and so are included in this offshore discussion.

A trio of "Infield Support Vessels" were designed in support of the Shell Oil "Prelude" Offshore LNG project in northwestern Australia. These vessels were Rotortugs, with triple Z-drive propulsion (Chapter 19).

In 2011 the first RAmpage 4500 class tug, *AOS Power*, was delivered to her owners, Atlantic Maritime Group, of Sharjah, UAE. This vessel is fully equipped for anchor-handling, ship-assist, long-distance towing operations, fire-fighting, deck cargo transfers, personnel transfer and other offshore support duties. She measures 45 metres by 15 metres and has a total power of 4,080 kW producing 70 tonnes of bollard pull.

AOS Power, first of the RAmpage 4500 class offshore tugs.

In 2014 Robert Allan Ltd was retained by Eastern Shipbuilding Group of Florida to design a quite unique vessel for Harvey Gulf International Marine, LLC, of New Orleans. Delivered in 2016, the *Harvey Stone* is a Multi-Purpose Field Support Vessel (MPFSV) that serves as a dedicated field-support vessel for the Shell "Stones" Floating Production, Storage and Offloading (FPSO) offshore terminal. Her primary duty is to operate as the dedicated pull-back tug in support of and assisting the berthing, loading and unberthing of off-take tankers of up to 46,287 TDwt. The vessel is also expected to escort the FPSO in the event of disconnection from the submerged buoy mooring, for hurricane avoidance purposes, and to tow the FPSO in case of mechanical breakdown at any time during the transit from or back to the field. Notable features include a hybrid propulsion system and a double hull that segregates all fuel, oil and sewage from the outer shell, as required by the owner's specification.

This RAmpage 6400 class vessel measures 64.8 by 18 by 7.8 metres and has a total installed power of 6,990 kW delivered by a pair of GE 12V250 MDC main engines with EPA Tier 4 emission certification. The engines drive Schottel SRP3030 CP Z-drive thrusters. The hybrid capacity is provided by a pair of 1,000 kW induction shaft motor/generators with a DC bus via bi-directional variable frequency drives.

This ship has a dynamic positioning (DP-2) designation and thus is fitted with two large bow thrusters in addition to the Z-drives. Additionally, it has a Fi-Fi 2 designation with a total pump output of 7200 m^3/hr and a monitor throw distance of 150 metres.

In 2012 the tug *Doña Clary* was delivered to her owners, Serviport S.A. of Colombia. Built by Cotecmar shipyard in Cartagena, Colombia, *Doña Clary* was the first of the RAmparts 3700 class design. This multi-tasking tug is designed with enhanced manoeuvring and sea-keeping capabilities and is fully equipped for ship-assist, long-distance towing, firefighting, anchor-handling, cargo-handling and other offshore support duties.

The *Seaways 24* is a RAmpage 5500-ZH class of Infield Support Vessel designed in 2015 and constructed at Keppel Singmarine in Singapore. The owner, Seaways International, took delivery of this versatile and adaptable vessel in August 2016. At 55 by 15 metres, this vessel incorporates almost every function imaginable for a vessel of this type and size. The powering/propulsion system offers multiple options for efficient operations, and consists of the following components:

- 2 × 3,000 kW MAK 9M25 main engines
- 2 × CAT MTA8 Z-drives each with two Power Take In (PTI) shaft connections and turning CPP propellers
- 2 × 560 kW PTI electric motors
- 2 × main engine Power Take Off (PTO) gearboxes each driving 2 × Fi-Fi pumps and two 1,200 ekW constant-speed shaft alternators
- 2 × 830-ekW and 1 × 350 ekW ship's service gensets

As illustrated, this vessel is packed with equipment and provides significant towing and gear-handling capability. The aft deck is dominated by a large offshore-rated crane with a 25-tonne SWL and an outreach of 25 metres.

In 2018 Seaways International again came to Robert Allan Ltd for a similar multi-purpose offshore support tug. The *Seaways 20* was built by Keppel Singmarine of Singapore. Like its cousin, *Seaways 24*, this RAmpage 5500 class tug was designed for

Left and below: MV *Harvey Stone*, a very sophisticated hybrid-powered Multi-Purpose Field Support Vessel; **bottom:** *Doña Clary*, a 37-metre multi-purpose ASD tug for Colombia.

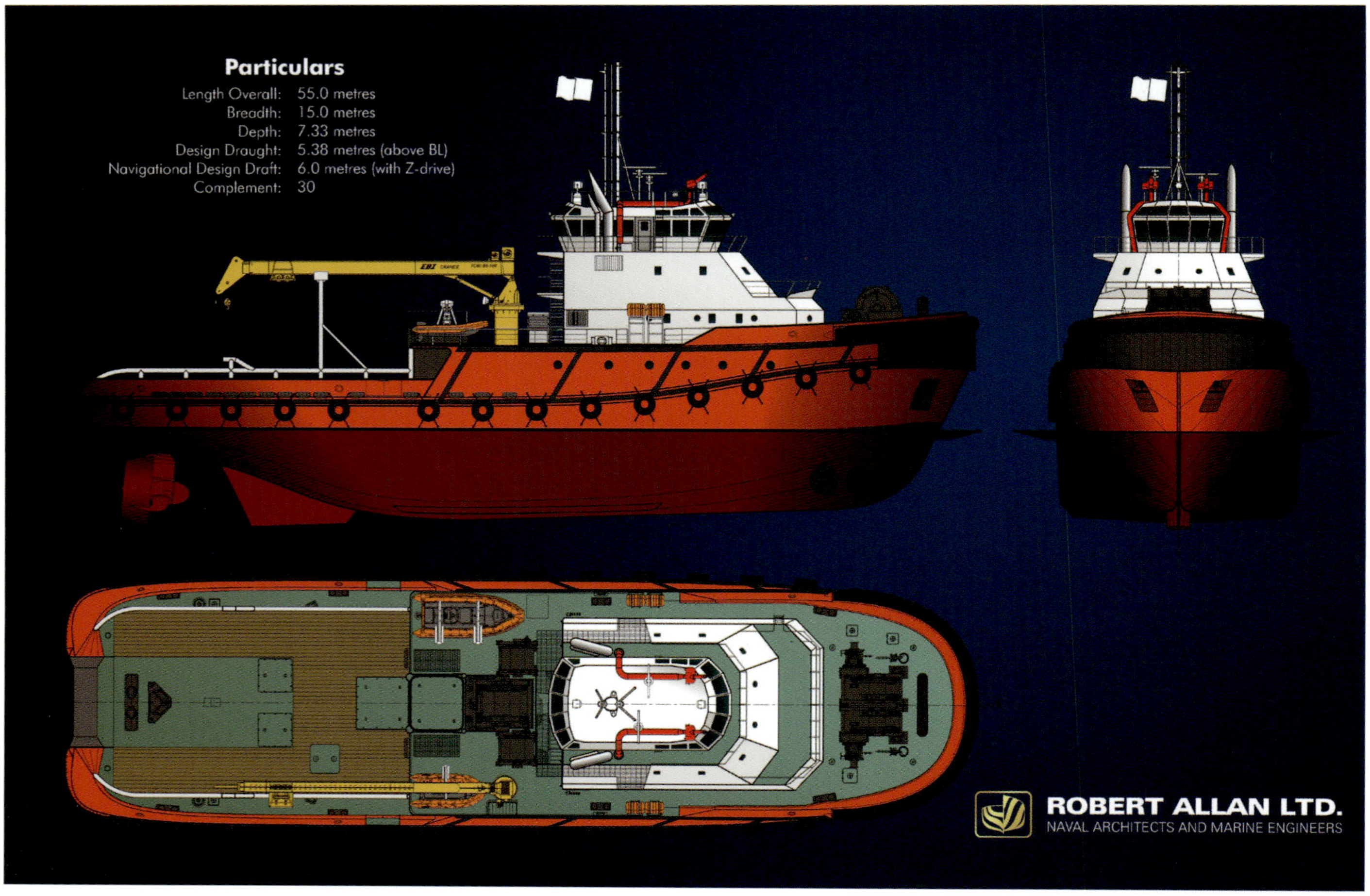

Top: Rendering of the General Arrangement of *Seaways 24*.

Bottom: *Seaways 24*, living up to the reputation of the RAmpage class tugs as the "Swiss Army knife" of the offshore.

a wide range of offshore tasks, including long-range towing, handling/positioning of tankers at FPSO or offshore mooring production buoys, anchor-handling operation, hose-handling at SBM/CALM buoys and FPSOs, etc., diving and ROV support, offshore personnel transfer, standby duties, oil-spill response, firefighting, cargo deliveries and any rescue or salvage towing operations.

In keeping with the owner's objectives of reducing their environmental footprint, the *Seaways 20* is distinguished by a Green Passport and ENVIRO+ notation by the Classification Society ABS, environmental requirements which were becoming increasingly important to owners in the 21st century. These certifications require procedures to deal with ballast water, sewage and garbage management. In addition, there is no fuel oil against the skin of the vessel, the ballast and cargo tanks are fully segregated, and the vessel's anti-fouling paint is biocide-free.

The latest edition of an offshore industry vessel from Robert Allan Ltd is the 60-metre RAmpage 6000-ZM offshore support vessel *Koc Al Zour*, built for the Kuwait Oil Company in 2021 by Uzmar Shipyards of Turkey. The vessel is equipped to undertake operations such as oil-spill response, oil-spill containment and recovery, area surveillance, offshore firefighting, back-flush capabilities, logistics supply, towing, rescue operations and related duties around Kuwait and nearby international waters.

Seaways 20 sets a high standard for environmental responsibility in offshore support vessels.

Koc Al Zour, a 60-metre multi-tasking offshore support vessel for Kuwait Oil Company, is the largest Robert Allan Ltd design built in Turkey to date.

Unique Tug for Kuwait

One of my interesting recent projects was a 60-metre oil-spill recovery vessel for the Kuwait Oil Company. It was a challenging project with very strict and demanding requirements. One of them was that the ship had to have the capacity to hold minimum 750 cubic metres of recovered oil on top of all the normal consumables, like fuel oil and fresh water, all within the vessel's allowable deadweight. The challenging thing with that was because of the large amount of recovered oil volume that was required, we had to double-hull the tanks, meaning we had to separate them from the outer shell. That meant we were struggling to achieve that volume and the separation and how to make it to meet the minimum requirements and at the same time to be buildable. If it was too narrow in some spaces, it wouldn't be possible to allow for welding inside and inspections later on. Because the owner had very strict performance requirements, the design was extensively model-tested in Vienna, something the client wanted and was willing to pay for. Another difficult aspect was that to meet their requirements, the ship was not allowed to have bow trim in any of the operational conditions. The vessel had to remain at level or aft trim only. Their statement called for no bow trim at all. That made it difficult because we have this huge amount of volume that is aft and reserved for recovered oil, but you don't recover oil all the time, just in an emergency. If the tanks are empty but everything else is full, it is hard to achieve that no bow trim requirement, especially as the whole superstructure was forward. Balancing the vessel was challenging. We did end up adding some fixed ballast to deal with that issue.

It was to be a multi-purpose vessel. Because of the nature of their industry, they need to have vessels that can respond to any oil-spill emergency response. It was also to serve as an offshore supply vessel, firefighting, and towing vessel, though the 40-tonne bollard pull towing capacity wasn't huge. The ship had to be able to do all these tasks well. The ship was built in Turkey and is currently in service in Kuwait.

—Bilyana Ivanova

Crewboats

The ubiquitous crewboat is a vessel type most associated with US offshore operations in the Gulf of Mexico. The early versions of these were typically quite simple boats but well suited to the often shallow-draft requirements of the Gulf. More recently these have become larger and faster but able to get into many more ports in that area than could more deep-draft OSVs.

Akin to the Norwegian dominance in the OSV/AHTS market, the Gulf of Mexico shipyards developed simple standard designs for these vessels and cornered their local market. Robert Allan Ltd didn't make much of an effort to challenge in that parochial US market, but we did have some more recent success in the Middle East as shipyards in the area became more familiar with our design services.

Our earliest foray into this market, however, came in Singapore in the late 1970s with a smaller shoal-draft aluminium crewboat named *Mariam*, built by our Singapore joint-venture partner Seaconstruct (Pte) Ltd for Wijsmuller Offshore, a division of the famous Dutch towing and salvage company. With a capacity for 34 passengers, *Mariam* measured 84 feet long by 30-foot beam. Driven by a trio of waterjets, each powered by a 525 hp diesel engine, she achieved a top speed of 20 knots.

Mariam, an early aluminium crewboat for Southeast Asian operations.

More recently, and as we gained a client base in the Middle East, opportunities arose for crewboat and other workboat design in those areas. In 2000 we designed the aluminium crewboat *Khulood* for Mutawa Marine of Abu Dhabi. Measuring 27.4 metres by 6.7 metres, she had capacity for 31 passengers and 20 tonnes of cargo on the open aft deck. Powered by a pair of 1,250 bhp diesels driving conventional open screws, this attractive boat achieved a top speed of 25 knots.

Not all our crewboat experience was outside of BC waters, though the lines between a crewboat and a water taxi can be rather blurry. In 2002 we designed the 21-metre aluminium crewboat *Tsimshian Storm* for North Co-Corp Ferry Services, to serve First Nations communities on the North Coast of BC. Built by ABD Enterprises of North Vancouver, she has a capacity for 45 passengers and a service speed of 20 knots.

A number of steel crewboats were designed in 2004 for major Middle East clients. Zamil Offshore commissioned the *Zamil 41* and *Zamil 42*, 33.5-by-6.8-metre boats with capacity for 30 passengers and 18 tonnes of deck cargo. Lamnalco, another important client in the area, built the *Lamnalco Oryx*, a 30-by-7.6-metre shallow-draft crewboat, at Colombo Dockyard Ltd in Sri Lanka.

Top left: *Khulood* in operation in the Arabian Sea; **top right:** *Tsimshian Storm* serves the remote communities of BC's North Coast; **bottom:** *Lamnalco Oryx*, a shallow-draft steel crewboat.

Offshore Miscellany

Among the distinct pleasures of working in the offshore industry are the many unique and demanding technical challenges with which one is presented. Although offshore businesses will seek market-available solutions whenever possible, there are many times when some creative design work is required to address a different operational need. This is the niche market in which Robert Allan Ltd frequently works, and as a consequence we have had some most interesting projects.

Ice-Breaking Emergency Evacuation Vessels for the Caspian Sea

In about 2005 we were contracted by Kazakhstan-based Agip KCO to design a unique type of craft to serve as emergency evacuation vessels for drilling platforms in the Kashagan Field in the northeast Caspian Sea. The requirements were basically to enable the safe evacuation of 340 platform workers during a blow-out or gas-leak type event, from within a toxic and potentially explosive atmosphere. The craft thus had to have a power source which could not ignite such a gas-air mixture and which could get both craft and its human cargo at least a kilometre away from the platform. It had to be able to do this in both an open-water environment and during winter ice coverage in the operational area.

Special focus was put on the development of the propulsion solution, allowing the vessels to operate safely within a toxic/hydrocarbon environment. Combustion air is provided by a specialized installation consisting of 16 high-pressure cylinders storing compressed air at 350 bar, decompressed via pressure-reducing panels and then supplied

One of the ten IBEEVs in the Caspian Sea.

to the engines. The IBEEVs have diesel-electric propulsion consisting of two 800 kW diesel-electric prime movers running at 1,500 rpm and driving twin azimuthing thrusters rated at 550 kW. The winter ice conditions demanded a hull form with the capability to operate in 0.6-metre ice. This capability was proven by extensive model tests, developed in concert with our internationally recognized ice-expert colleagues Offshore Research Ltd and AKAC Inc. The shallow waters of the Caspian Sea also required tight design control over the vessel draught: 2.0 metres in summer and 2.1 metres in winter.

The evacuees are seated in one of three hermetically sealed compartments. During evacuation, personal CO_2 scrubbers and re-breather sets are provided to each evacuee.

A series of at least ten IBEEVs were constructed by Remontowa Shipbuilding in Poland, who rather churlishly now claim design credit for this most unique design.

Dive Support Vessels for Brazil

Underwater service providers play a major role in maintaining offshore assets, especially in deep water, where they perform a variety of diving inspection operations and ROV (Remotely Operated Vehicles) surveys.

Based on anticipated demand from the Brazilian State Oil Company, the Brazilian company Belov Engenharia Ltda built two new dive-support vessels at their own newly built fabrication facility in Salvador. They engaged Robert Allan Ltd to prepare the Contract Design package for these vessels. The two vessels, *Belov Amaralina* and *Belov*

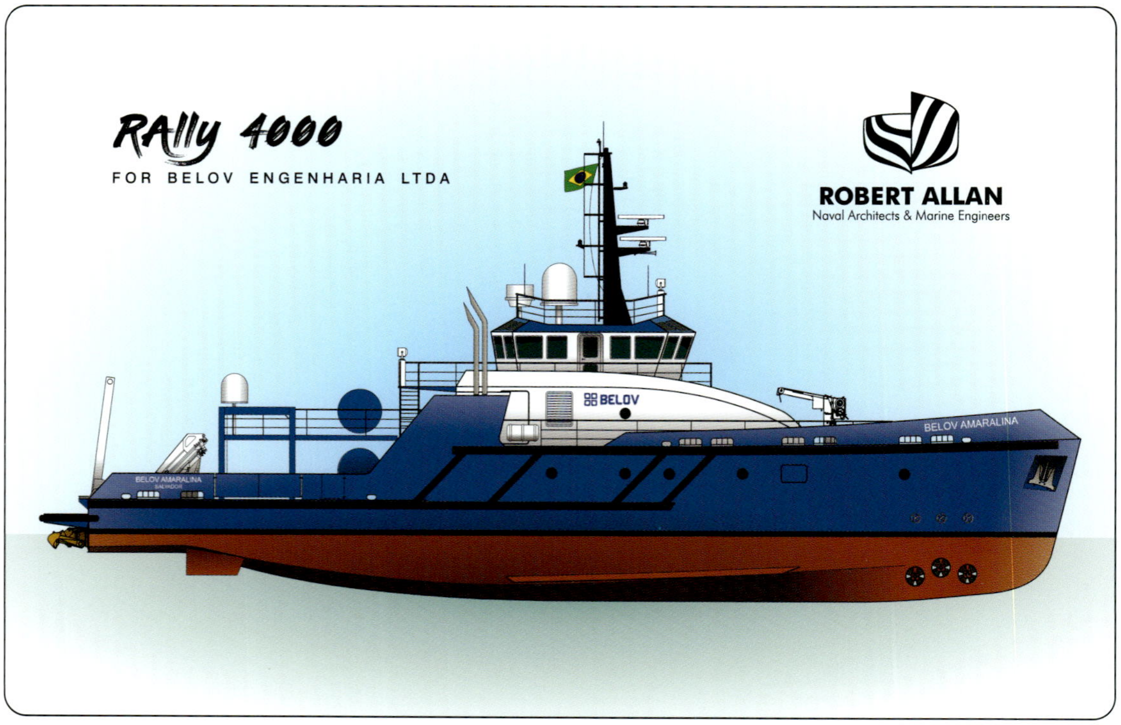

Rendering of the RAlly 4000 Dive Support Vessels for Brazil.

Humaitá, were constructed and commissioned in just over 18 months, an outstanding feat for any shipyard. The vessels are fitted with a DP-2-rated Dynamic Positioning (DP) System incorporating four diesel generators, three waterjet main propulsors and triple bow thrusters for position and heading-keeping per the DP requirements. The propulsion system is diesel-electric to improve efficiency when operating in lower-power DP modes. The vessel is certified as a Special Diving Support vessel suitable for Unrestricted Navigation. The design, designated as the RAlly 4000 SDSV, measures 40.5 metres in length by 10.98 metres beam and carries a crew of 38 persons including the dive crew.

The DSVs are fitted with an ROV Launch and Recovery System, an aft-mounted A-frame to operate the diving bell, a decompression chamber to support the divers and spacious crew quarters. Waterjets are used to ensure diver safety. Because waterjets are typically used on higher-speed vessels, extensive self-propelled CFD analysis (including waterjet tunnel flow) was performed to verify resistance and thrust at the slower speeds at which these vessels operate when on station.

Belov Humaita, a unique waterjet-powered Dive Support Vessel for Brazil.

13

A Barge for All Reasons

The late Tom Ward, a design engineer at Robert Allan Ltd in the late 1960s who went on to an illustrious career in the BC marine industry, wrote a seminal paper, "The Ubiquitous Barge: Some Trends in Barge Design and Construction in British Columbia," published in the January 1980 issue of the Society of Naval Architects and Marine Engineers' (SNAME) *Marine Technology* magazine. The paper nicely summarized the state of the art in barge design and construction at that point in time.

Not a great deal has changed since, as there has been a general dearth of new construction of significant barges for the BC industry in the interim. Barges are not considered beautiful, but they are the very backbone of the coastal towing industry in BC and are vital to the coastal economy. Without them, the coastal tug fleet would be cartless horses, and so it is important to pay homage to these humble and oft-abused load-bearing servants of the coast. The files of Robert Allan Ltd contain many hundreds of barge designs of all types, from the simplest of scows to complex self-unloading barges. They also include extensive amounts of model-testing data for ensuring reduced hull resistance and for verifying directional stability to ensure the most economic towed performance. Here's to the humble barge!

Opposite: *Island Forester*, see page 273.

In the Beginning...

It is interesting to review when in the Allan corporate history barge designs began to take place and what types those were. Our first recorded barge design was not until about 1948, a covered paper barge. Prior to that, most barges built locally were smaller wooden scows and didn't involve much "design." Larger barges were often the hulls of deep-sea ships that had outlived their ocean-going days. In 1950 F.M. Yorke & Sons commissioned a design for the first in a significant series of railcar barges, the 183-foot *Yorke #9*. In the following decade Bob Allan designed seven more railcar barges for the Yorke company, ranging from 94 feet in length to 273 feet—the *Yorke 11, 12, 14, 15, 16, 20* and *21*.

What is most notable, however, is that the third barge in the Allan design list was the first self-dumping log barge in BC, the *Powell #1*, built by Burrard Dry Dock in 1955. This is the barge type for which Robert Allan Ltd is perhaps best known, as a significant majority of these distinctive barges (and their self-loading/self-dumping derivatives) were designed in this office between 1955 and 1979, with most of those built in the mid-to-late 1960s.

In 1960 the self-unloading cement barge *St. Malo* (about which more will follow) was designed for Deeks-McBride Co. Ltd. This was a ground-breaking concept in barges on the West Coast—or anywhere else for that matter. This was followed shortly thereafter by the 10,000 TDwt barges *Island Importer* and *Island Exporter* for the transport of lime-rock. These were among the largest ocean-going barges afloat anywhere at the time.

It is noteworthy how Bob Allan, in his early years of leading the company, had launched directly into designing some of the biggest and most complex barge types around, rather than working up from basic scows and chip barges. Many dozens of those simpler types would come later, but he certainly proved his skills early on with the much tougher assignments.

Barge Design Research

Barges may look simple, but their performance in terms of overall resistance and directional stability can make or break the economics of any cargo movement. It has been a hallmark of Robert Allan Ltd from the earliest days that appropriate levels of research are conducted in order to both prove and improve the performance of all vessel types we design, and barges are certainly no exception.

Early in the days of barge design in this company, Bob Allan recognized the need to be able to assure his clients of a barge's performance and behaviour. He established a good working relationship with the University of Michigan, which was one of the leading schools of naval architecture in North America at the time, and which had a decent-sized model-testing basin. The use of students to do much of the testing made this a quite economical means of getting model tests performed. Dozens of barge designs were tested there during the 1960s, and that data still supports much of our work, though the use of computational fluid dynamics (CFD) has enabled much more detailed and specialized research today.

The Barge Book

A young graduate student at the University of Michigan named Corning Townsend III conducted many of these model tests, and he went on to a quite successful career as a consulting naval architect in the US, specializing in inland tugs and barges. I met Corning many years later, and he was always very complimentary about how good those early Allan barge designs were in comparison to many others he tested. He obviously made good use of the information in his career as well. One of Hans Muhlert's early tasks when he joined Robert Allan Ltd in 1970 was to put all this barge model test data into non-dimensional form for most efficient use for future barge designs. Hans' "The Barge Book" remains an invaluable reference in the office to this day.

—R.G. Allan

Various forms of barge bow shapes have been developed and refined by this office, depending upon the voyages and berthing/loading arrangements. Stern-rake geometry is also crucial to overall resistance, and there is a critical angle which avoids flow separation, ensuring reasonable directional stability and low resistance. It is remarkable how often one sees designs that don't seem to recognize this fact and therefore must have extremely high resistance.

Beyond basic capacity, an overriding factor in barge design is the need for directional stability. Because of their basic "box" shape, barges won't tow in a straight line without some corrective forces, generally provided by skegs at the aft outboard corners. Various shapes of these have been tried over the years, with varying degrees of success. The "skeg augment" on resistance can be 30 percent or more of the total resistance, so it is very important to provide the barge with just enough directional stability, but not so much that it tows straight at the expense of a great deal of fuel.

Barge Testing at University

I kind of knew about Robert Allan Ltd even before I went to university, but my first day there, about a week before classes, I went to the chairman of our department, Professor Richard Couch, and introduced myself and asked him for a job. I had to pay my own way because my parents didn't have any money. He took me downstairs, where there was—and still is—a huge towing tank for testing models. He introduced me to the guy running it, Mr. Moss, and I got put to work doing calculations and helping him on the carriage that pulls the models. Some of the models that were laying around there were barge models designed by Robert (Bob) Allan, and I kind of clicked onto that. I talked to Professor Couch, and he said, yes, Mr. Allan had been testing barges there for some time. So I became aware of him again, and when I got near the end of my time there, I wrote him a letter asking for a job, and he offered me one right out of school.

—Hans F. Muhlert

When under tow in the open ocean and on a long towline, it is quite typical for barges to wander a bit from a straight course over a very long period. As long as the lateral excursion off the course heading is not great, this is quite acceptable except in confined waterways, but in that situation the towline is shortened, and this phenomenon generally disappears. This long sway cycle of barge tows was a subject of significant interest to Bob Allan, and he invested considerable time and energy into ensuring his designs were the best possible in regard to directional stability, both through model tests and on ocean voyages. The following photo shows him in his element, and at his scrutinous best, on board the sailboat of our long-time "ship-science" consultant Alan Reynolds of Offshore Research Ltd, towing a barge model in Coal Harbour. Open water testing of directional stability was much more effective and indeed more practical than conventional model testing, where the constraints of limited tank length and width usually were too confining to see the full extent of barge behaviour.

With the recent advent of reasonably affordable computational fluid dynamics (CFD) in analysing vessel performance, the range and nature of performance assessment available to the consulting naval architect has changed dramatically. Robert Allan Ltd has invested heavily in its CFD capability since about year 2000 and now has a highly skilled team of specialists working solely in this discipline. We have used that capability to great effect in several barge projects where proof of performance was essential before the construction could be approved.

The best example of this was the large Río Paraguay/Río Paraná inland tug-barge transportation project completed in 2008, originally designed for Rio Tinto, then eventually taken over by Brazilian mining giant Vale. An extensive CFD analysis demonstrated that the tug-barge system designs proposed by Robert Allan Ltd could navigate the circuitous bends of the river system, particularly going downstream, where currents have a negative influence on steering capability. This demonstration was critical in the company securing one of our largest-ever projects, for eight major river pusher tugs and more than 100 barges (described later in this chapter).

A 50,000 TDwt tug-barge convoy on the Río Paraguay/Río Paraná system.

Left: Bob Allan in his element, studying barge behaviour in an open-water test; **below:** A CFD image of a triple Z-drive towboat with a barge train. The towboat has to reverse thrust when manoeuvring the barges downstream around a river bend, hence the high reverse flow rates shown in red.

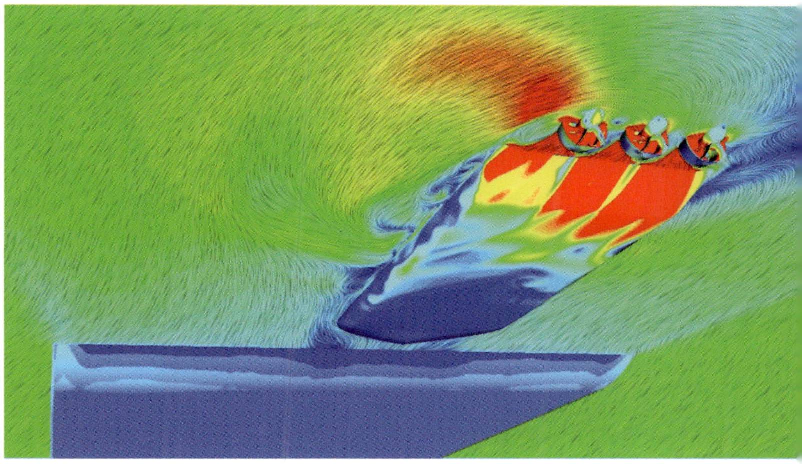

Chip Barges

No barge type is more emblematic of the BC coast than the humble chip barge. Often towed in pairs or triplets, these barges carry wood chips, sawdust and hog fuel to pulp-and-paper mills in coastal British Columbia and Puget Sound.

Early variants of these barges were all wood, but in the 1960s a series of steel-hulled barges were designed by Bob Allan for Island Tug & Barge Ltd (ITB). These barges had wooden cargo boxes supported by steel stanchions, but later versions evolved to have all-steel boxes. Unloading methods evolved as well, from crane-operated grabs to the use of large front-end loaders. This change involved fitting large removable doors at the aft end of the barges for loader access and significantly increasing barge deck strength to cope with the high wheel loading. Specialized floating unloading platforms are utilized, into which the chips are dumped and then conveyed ashore.

The evolution of the BC chip scow is well traced through the Robert Allan Ltd archives. In 1962 a series of fourteen 800-unit chip barges were designed for ITB. (A unit of wood chips is 200 cubic feet.) Measuring 168 feet by 46 feet by 10 feet 10 inches depth, these barges (*ITB #80* through *ITB #93* inclusive), were built primarily at Victoria Machinery Depot, with a pair built at Yarrows and a trio at Bel-Aire Shipyard, all from 1963 through 1966.

In 1964 another batch was designed, also for ITB, but this time they were slightly larger, with a capacity of approximately 1,000 units. Eighteen of these barges were built, each measuring 175 feet by 48 feet by 11 feet 3 inches depth. *ITB 110* to *ITB 115* were built in 1964 and 1965, with *ITB 122* and *123* built in 1967. These barge fleets were significant capital investments for ITB at the time.

From 1979 through 1982 there was another boom in chip scow building as the older generation was wearing out in this demanding service, and the need for larger capacities and faster loading led to a newer generation of all-steel barges. Larger 1,200-unit barges were designed for both Maple Leaf Shipping and Crown Zellerbach Ltd. These barges measured 208 feet 7 inches by 50 feet by 12 feet depth and had much stronger decks than the earlier barges. Fifteen of these barges were built by Burrard Yarrows Corporation in a three-year period: *Maple Leaf 1201, 1202* and *1203*, and *Crown Zellerbach #11* through *#22* inclusive.

It is noteworthy that until about the late 1980s, local BC shipyards were very price competitive with straightforward steel construction, and it was economical for owners to build in Canada, given the various import duties and construction incentives that were in effect at the time. By the time the next wave of construction came around, in the late 1990s, the tide had turned. Many BC shipyards had closed, and China was emerging as a shipbuilding powerhouse, particularly for basic steel construction.

In 1997/98 RivTow was among the first to build in Asia, with a series of 1,200-unit Chinese-built barges delivered to Vancouver on a heavy lift ship. Since those days sadly almost all BC-owned barges have been built in China.

A triple tow of Seaspan chip barges exiting Vancouver harbour.

Below: *ITB 92*, one of an early series of fourteen 800-unit chip barges for Island Tug & Barge Ltd.

Railcar Barges

As noted in the introduction, some of the earliest barge designs from this office were the railcar barges designed for F.M. Yorke and Sons. Railcar barges are an interesting design challenge, since the point-loading from railcar wheels is extremely high and the rails on deck must be supported entirely by heavy girders underneath. The rails also have to converge at the load/unload end(s), so the supporting deck structure becomes quite complex.

The Yorke railcar barge designs developed by Bob Allan were the following:

Rail Car barge designs for F.M. Yorke & Sons from Robert Allan Ltd

Built	RAL Ref.	Barge name	Loa	Beam	Depth	Builder
1950	8-50	*Yorke No. 9*	182.00	43.00	10.00	Yarrows Ltd
1957	28-56	*Yorke No. 11*	272.00	43.10	11.00	Yarrows Ltd
1958	17-58	*Yorke No. 12*	94.00	34.00	9.30	Allied Shipbuilders Ltd
1958	17-58	*Yorke No. 14*	94.00	34.00	9.30	Allied Shipbuilders Ltd
1958	17-58	*Yorke No. 15*	94.00	34.00	9.30	Allied Shipbuilders Ltd
1958	n/a	*Yorke No. 16*	100.32	42.00	9.30	Allied Shipbuilders Ltd
1959	19-59	*Yorke No. 20*	100.25	42.00	9.25	Allied Shipbuilders Ltd
1963	6-62	*Yorke No. 21*	273.00	43.10	12.00	Allied Shipbuilders Ltd

It is noted in *Ships of Steel: A British Columbia Shipbuilder's Story* by T.A. McLaren and Vickie Jensen, 2000, that the credit for the design of all the Yorke barges built at Allied is given to McLaren & Sons; however, the presence of all the above design drawings in the Robert Allan Ltd archives rather refutes that claim.

In the early 1960s the major towing companies began entering the Lower Mainland to Vancouver Island railcar barge market to compete with the Yorke organization. Island Tug & Barge built the 264-foot *ITB 106* in 1964 and the *ITB 210* in 1966, both to Allan designs. In 1971 Seaspan had Robert Allan Ltd design the largest local railcar barge to date, the *Seaspan 930*, with a capacity for 30 railcars on a platform 339 feet long with a 64-foot beam and a depth of 16 feet 6 inches. This vessel was built in 1971 at Mackenzie Barge & Derrick Co. Ltd. In 1978 an updated sister barge, the *Seaspan 931*, was built by Vancouver Shipyards to the same dimensions.

Seaspan 930 under tow.

Deck Barges

Perhaps less obvious than chip barges in our coastal waters are the many hundreds of simple deck barges (often deprecatingly referred to as "scows"); less obvious only because they lack the large billboard-sized sidewalls that make their chipper cousins more noticeable. These barges are typically designed for carrying sand, gravel or rock to construction sites or to cement factories and have sidewalls that typically range from 4 to 8 feet in height. They also are used for carrying just about anything that needs moving on coastal waters.

The barge designs of this type in the Robert Allan Ltd files are too numerous to mention, ranging from as small as the 36-foot *Coast Marine #20*, designed for John Manly Ltd in 1960, to the 16,000 TDwt ocean-going barge *Seaspan 252*, built in 2015. The table below identifies some of the more typical coastal deck barges from the Allan files, generally less than 4,000 TDwt.

Simple Deck Barge Designs from Robert Allan Ltd

Year Built	Barge Name	Description	Builder	Owner	Loa	Beam	Depth	DWT
1960	*Coast Marine #2*	Steel Work Scow	John Manly Ltd	John Manly Ltd	36.00	20.00	5.00	50
1961	*GofG 230* to *GofG 233* incl.	Steel Scow	Burrard Dry Dock Co. Ltd	Gulf of Georgia Towing Ltd	120.00	44.10	12.10	1000
1962	*EVCO 80* to *EVCO 85* incl.	Gravel Scow	Yarrows Ltd	Evans Coleman & Evans	126.00	38.00	11.00	900
1963	*V.T. 200* to *V.T. 202* incl.	Deck Cargo Barge	Yarrows Ltd	Vancouver Tugboat Co.	290.00	68.00	22.00	7500
1963	*Gulf Stav-Ray*	Deck Cargo Barge	Yarrows Ltd	Gulf of Georgia Towing Ltd	236.00	60.00	17.50	5000
1963	*GofG 237 / 238*	Steel Scow	Allied Shipbuilders Ltd	Gulf of Georgia Towing Ltd	120.00	44.00	12.50	1000
1968	*ITB No. 2 & No. 4*	Steel Scow	Yarrows Ltd	Island Tug & Barge Ltd	130.00	46.00	11.25	1100
1969	*EVCO 90/91/92*	Deck Cargo Barge	Bel-Aire Shipyard Ltd	Ocean Cement Ltd	216.00	56.00	16.00	3500
1981	unknown	Gravel barge w. 20′ ramp	unknown	Ritchie Transportation Co. Inc.	120.00	40.00	7.00	500
1988	unknown	Deck Cargo Barge	unknown	Ritchie Transportation Co. Inc.	150.00	50.00	10.00	1200

Gulf Stav-Ray, a 5,000 TDwt combination deck/oil barge built for Gulf of Georgia Towing Ltd.

There are some quite notable designs within this class that deserve more than passing mention, and the most significant of these are the specialized lime-rock-carrying barges. Lime-rock is a key constituent in the manufacture of cement, and it is quarried at various places on the BC coast, but most notably on Texada Island, about 70 miles northwest of Vancouver in the Strait of Georgia. The crushed lime-rock is transported to various cement plants in the BC Lower Mainland, in Puget Sound, and in Portland, Oregon.

The development of large ocean-going lime-rock barges began in 1961 with the design of two 10,000 TDwt barges for Island Tug & Barge Ltd. Built in 1962 at Yarrows Ltd, the *Island Importer* and *Island Exporter* were among the very first ocean-going bulk-cargo barges anywhere, and Robert Allan Ltd worked closely with the classification society American Bureau of Shipping to establish the criteria for structural scantlings which should apply for this service. Previous barge rules were all based on inland sheltered waters voyages. These 360-by-75-foot barges each moved 10,000 tons of lime-rock every week from Texada Island to Portland, Oregon.

The distinguishing feature of these and successive barges in this service was the use of front-end loaders to discharge the cargo, using a small hopper on deck feeding into a stub conveyor, which then fed a shore-based discharge conveyor system. The loaders drive on ramps onto the top of the cargo to start unloading, work their way down to an empty deck, then exit through the large gates located in the sidewalls that are provided for that purpose. It's a very slick and economical system which provides a significantly faster discharge rate than any previous grab-based system.

By 1976 the *Importer* and *Exporter* (now named *Seaspan 240* and *241*) were well used, and Seaspan ordered two replacement barges, *Seaspan 250* and *251*, from Mitsubishi Heavy Industries of Japan. Slightly larger than their predecessors, these barges measure 360 feet by 80 feet by 25 feet depth, with a cargo capacity of 11,350 long tons. Numerous "lessons learned" were built into this design, but the basic concept was unchanged. A notable change was the transition to a spoon-shaped bow rather than the earlier more ship-shaped bow. These heavily built barges continue in that same service to this day.

Front-end loader beginning work discharging a load of lime-rock from *Seaspan 250*.

General Arrangement of the 10,000 TDwt 1961-era *Island Importer* and *Island Exporter* lime-rock barges.

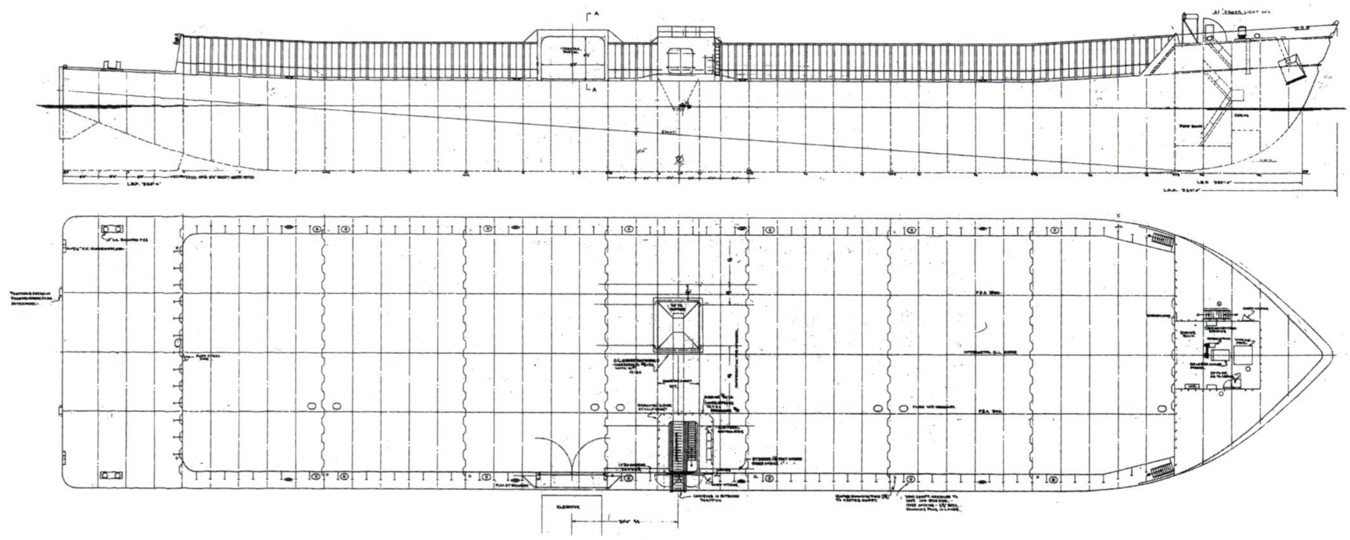

General Arrangement for *Seaspan 250* and *Seaspan 251*, 11,350 TDwt deck cargo barges, ca. 1976.

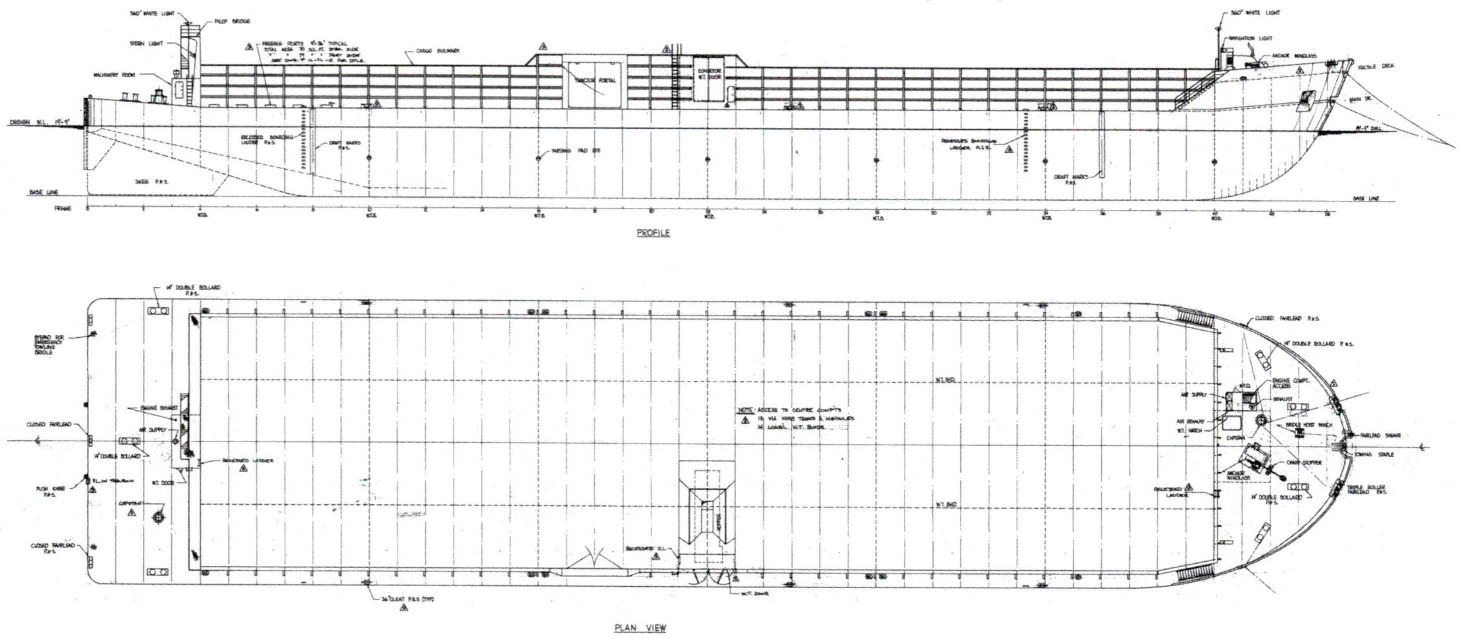

Barges in Court

Over the years I have been called upon several times to act as an expert witness in legal cases involving marine incidents. One of the more memorable of these cases involved the *Seaspan 240*, formerly the original *Island Importer*. In 1986 this barge capsized alongside the unloading dock in the Fraser River, causing considerable damage to the barge and to the facilities, not to mention the loss of 10,000 tons of lime-rock. I was retained by Mr. Nils Daugulis of Bull, Housser & Tupper, counsel acting for the defendant barge owners, and given the task of trying to determine what could have caused the capsize. The plaintiff's lawyer was Mr. David McEwen. Both these gentlemen were well-known members of the local maritime law fraternity.

Critical to my theory was the fact that the barge had sat unattended alongside the facility for several days, during which time it had rained heavily and almost continuously. Obviously, a very wet cargo weighs more than a dry one, and the barge had settled deeper in the water, to the point that when unloading started and the loaders both migrated to the hopper side, the deck edge of the barge became submerged and over it went. It was a perfectly plausible and logical explanation.

In order to calculate the added weight of water in the cargo, I obtained a sample of about a cubic foot of the lime-rock cargo, weighed it when dry and then wet it thoroughly and weighed it again, a relatively simple and straightforward exercise. The weight difference was appreciable, as one might expect.

I was on the witness stand in court for three days dealing with the minutiae of the theoretical capsizing process. After my presentation, under cross-examination McEwen challenged my qualifications to weigh wet and dry rocks! "Have you taken any university courses in Geology?" he asked. "Yes," I replied, throwing him slightly off-stride, as I had taken Geology 100 in my first year at UBC. "Do you have a degree in Geology?" he asked. "No," I replied. McEwen claimed that because I had no degree in "rock science" I was not qualified to opine on the difference in weight between wet and dry rocks, a matter of simple physics. To my utter amazement, the foolish judge agreed with him, and the essence of my argument was stricken from evidence. Ultimately it didn't matter too much, as the insurance companies haggled out a settlement, but I felt quite wounded by this decision and learned to cover every possible base when dealing with subsequent similar cases.

—R.G. Allan

The latest edition of this type of barge is the 16,000 TDwt *Seaspan 252*, built in 2015 by Jiangsu Yangzijiang Shipbuilding Co. Ltd in China. The same fundamental design and load/unload system still applies, as does the lime-rock transportation business, gradually nibbling away at the core of Texada Island, a process in its 60th year as of 2022!

Another interesting variant on the large-deck cargo barge is the *NT 12000*, a 12,000 TDwt barge designed primarily to carry containers on deck, serving the northern Arctic coast of Canada. This barge was built in Indonesia in 2009 for Northern Transportation Co. Ltd but was subsequently sold to the BC-based Amix Marine Services Ltd.

Above: *Seaspan 252*, the latest variant of West Coast lime-rock barges at 16,000 TDwt.

Left: The launching of *NT 12000* in Indonesia, 2009.

Cement Industry Barges

In the early 1960s Robert Allan Ltd began a long-lasting relationship with the cement-construction industry operating in BC waters. Barge transportation is an ideal method for transporting low-value, high-density cargoes such as gravel, lime-rock, clinker and bulk cement. (According to the European Cement Association, "Clinker is the backbone of cement production. It is essentially a mix of limestone and minerals that have been heated in a kiln and have been transformed by this heat...The clinker is finely ground and mixed with gypsum and often with alternative raw materials to make cement.")

The introduction in the early '60s of self-unloading barge systems for the transport of aggregate made a remarkable impact on the industry, cutting transportation costs by a factor of three. Barges such as the *St. Malo* are particularly efficient where a large volume of graded material must be handled over a relatively short voyage.

Transportation of dry bulk cement is typically handled by barges fitted with gravity air-slides that "fluidize" the cement powder, feeding into pneumatic offloading systems, or in some instances by screw conveyors. Other products are handled by systems not unlike their land-based counterparts, such as drag-scrapers, but all these innovations meant significant improvements in marine bulk handling operations on BC's west coast.

The various barges built for this industry in the 1960s, and as described below, are nearly all still in service and working hard today—and look to be in very good condition—a testament to their design and the excellent maintenance performed by their owners.

St. Malo, built by Yarrows Ltd for Deeks-McBride Ltd in 1961, was the very first of this barge type. With a capacity of 2,000 long tons, it has multiple cargo "bins" that can be selected remotely by the operator for discharge via a transverse belt conveyor discharging to shore silos.

The self-unloading barge *St. Malo* running light.

General Arrangement for *St. Malo*.

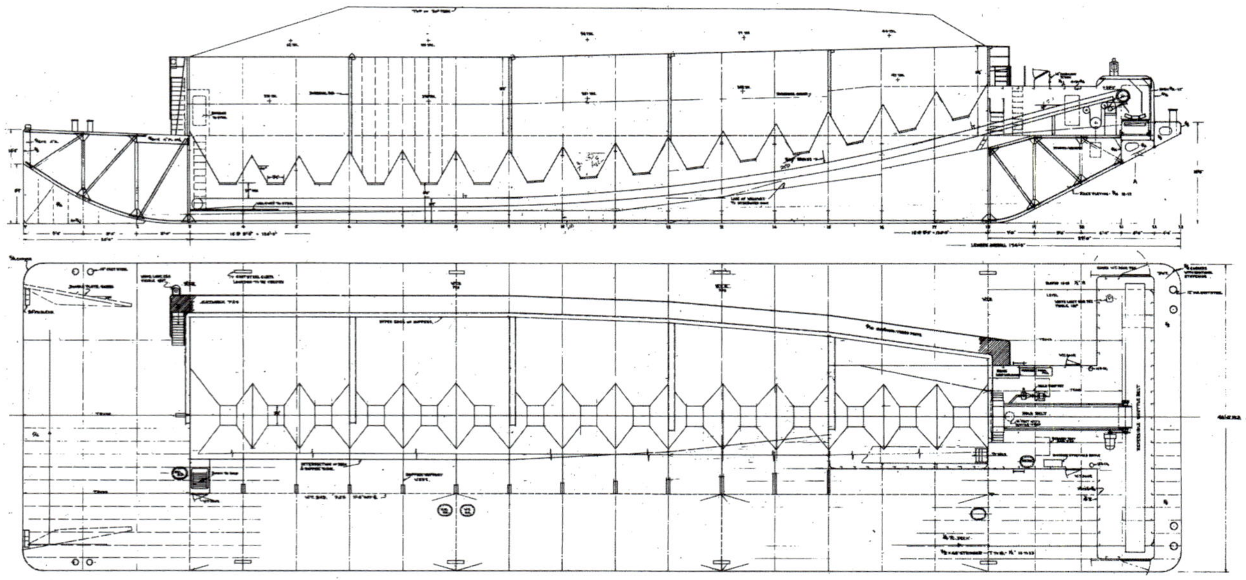

General Arrangement for *EVCO 14* and *15*.

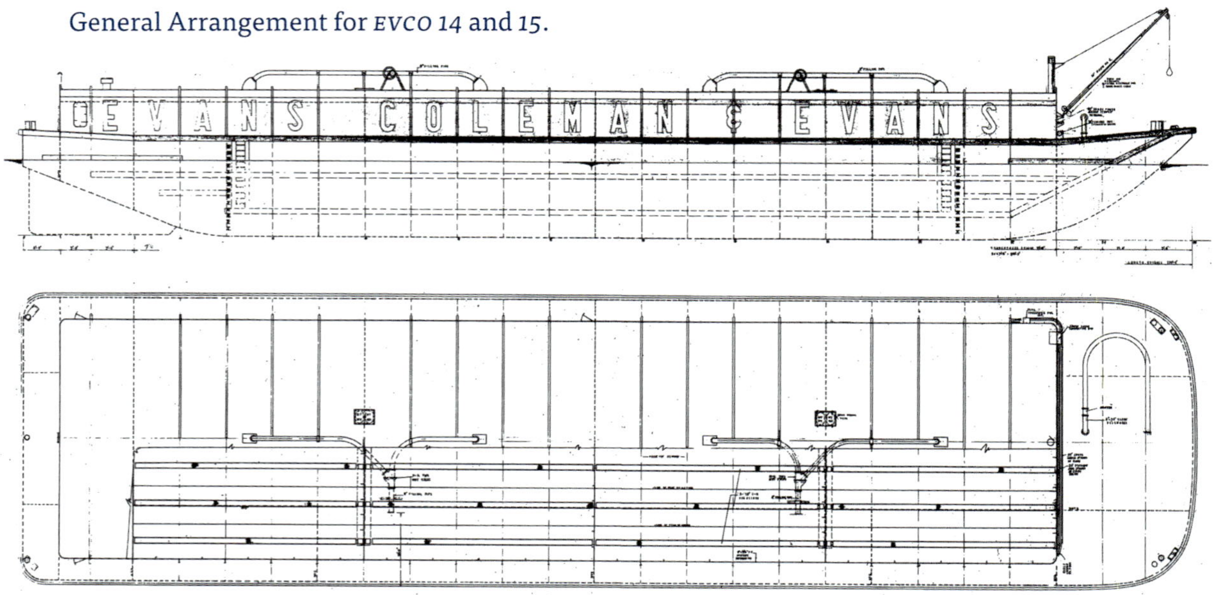

General Arrangement for *Plymouth*, showing its simple drag-scraper unloading system.

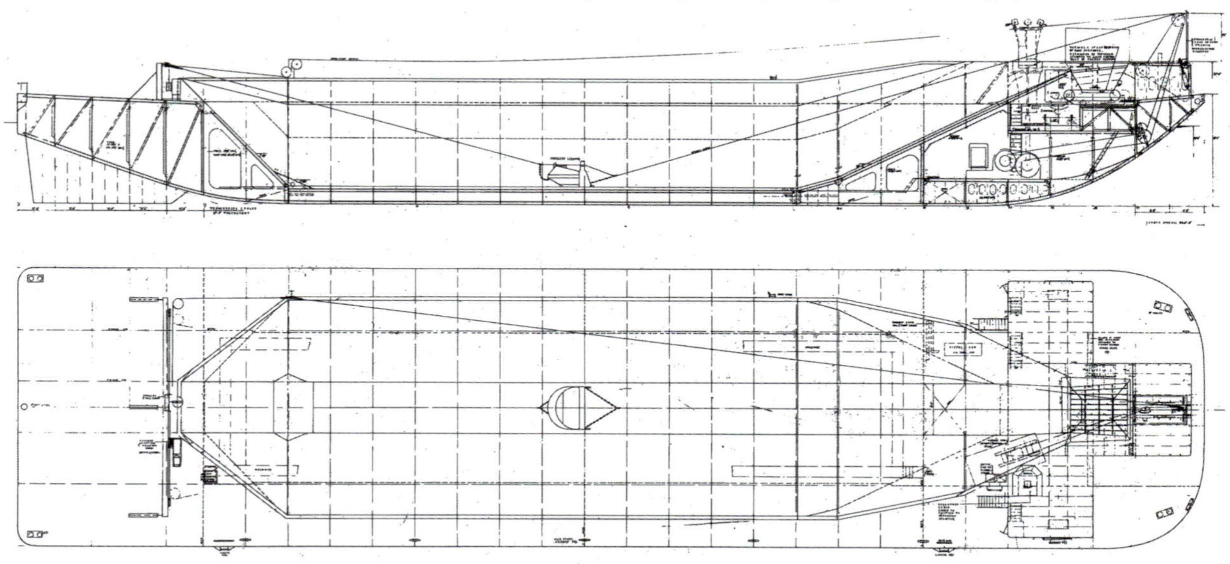

EVCO 14 and EVCO 15 were built in 1964 (Yarrows Ltd) and 1968 (Bel-Aire Shipyard Ltd), respectively, for Ocean Cement Ltd. These are dry bulk cement carriers with a capacity of about 12 million barrels (approx. 49 million cu. ft.) in four major compartments. The barge's beam and air draft are restricted to suit bridge spans in the operating area. The unloading system consists of a Fuller air-slide and a pneumatic conveyor system, powered from shore, with a discharge rate of 125 tons per hour.

The *Plymouth* was built by Victoria Machinery Depot in 1964 for Canada Cement Lafarge Ltd. This unique barge is designed as a single-product carrier of 4,000 long tons capacity, with an open hopper and a drag-scraper system for unloading.

EVCO 2 and EVCO 3 were built in 1967 for Ocean Cement Ltd by Victoria Machinery Depot and Bel-Aire Shipyard, respectively. These self-unloading barges were designed for hauling aggregate in an open-hopper system. They have a capacity of 2,000 short tons each and an unloading rate of 900 tons per hour.

L'Etoile was built for Canada Cement Lafarge Ltd by Mackenzie Barge & Marine Ways Ltd in 1968. This barge was designed for the transport of surplus clinker from the Richmond, BC, Lafarge plant to a cement plant in Seattle. This 263-foot-long barge has a capacity of 5,000 tons, handling clinker loaded at 160 degrees F. It is unloaded by a belt conveyor system with a capacity of 1,200 tons per hour. To control dust, the conveyor is fully enclosed and positive air pressure is applied on the external side in conjunction with a powerful dust-collection system.

The *L'Etoile* was a major project for Robert Allan Ltd. In addition to the basic design drawings, the company was retained by the builders to develop the complete set of construction working drawings. This was quite complex, as it involved not only a quite convoluted structural design, but also all the mechanical systems involved to define the material-handling equipment.

The latest (and largest) evolution of this type of barge is the *Pacific Sounder*, designed by Robert Allan Ltd for Lehigh Cement Company (the successor company to Evans, Coleman

EVCO 2 being towed by *Evco Breeze* into Vancouver's False Creek.

L'Etoile, a 5,000 TDwt self-unloading barge for Canada Cement Lafarge Ltd.

Below: General Arrangement of the self-unloading clinker barge *L'Etoile*.

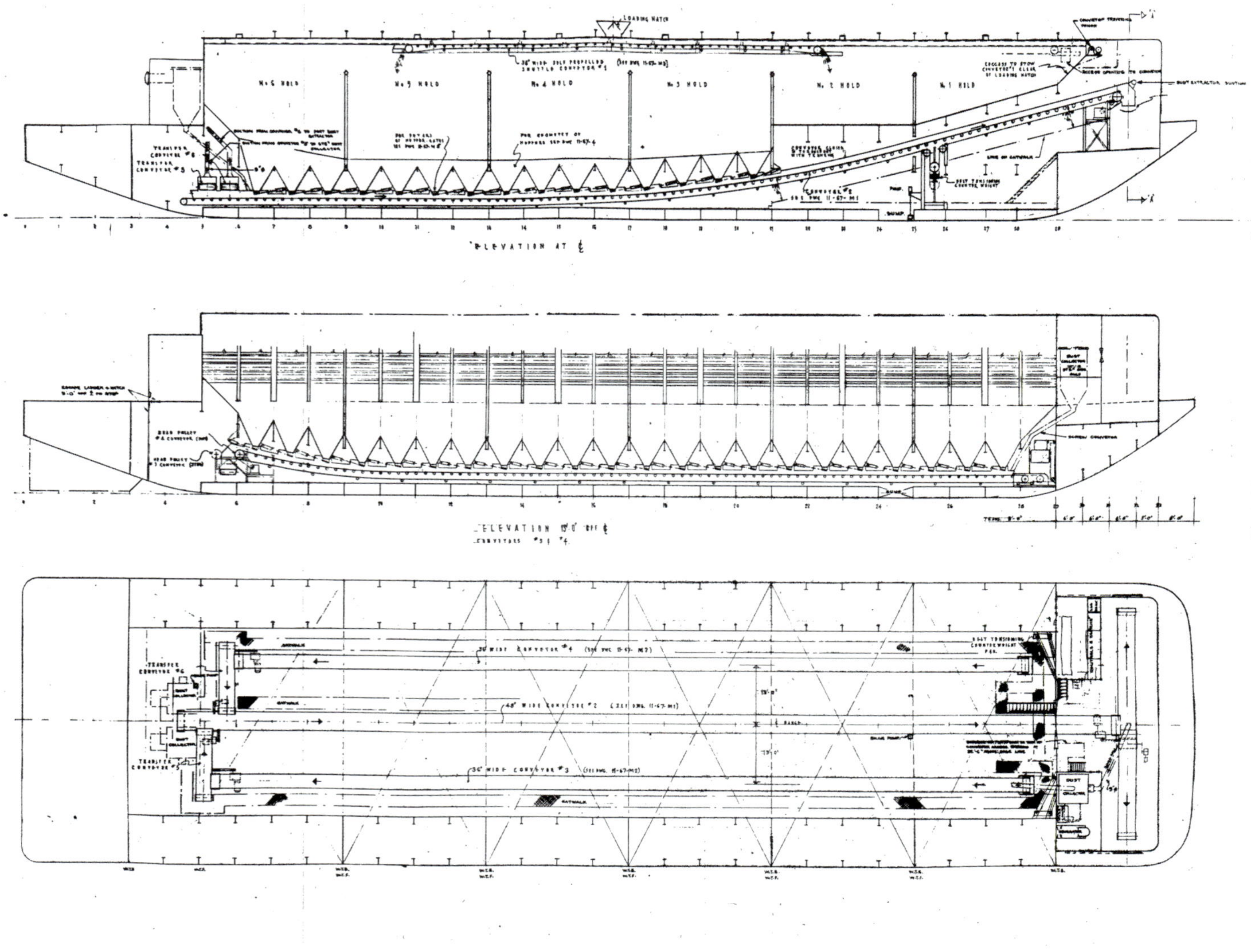

and Evans and Ocean Cement) in 2002 and built by Jinling Shipyard, China, in 2005. Designed to carry a variety of cement products, this 85-metre barge has four distinct holds, each served by an air-slide-type bulk-handling system. Loading takes place through connections on the deckhouse top. The discharge involves pumping very low-pressure air into the bottom of the load-carrying hoppers, which "fluidizes" the dry cement, causing it to migrate to a vertical and then a fore-and-aft screw conveyor, and thence into a discharge hopper, where it is pressurized and blown into a discharge pipe connection on the foredeck at a discharge rate of 225 tonnes per hour.

Below: General Arrangement for the bulk cement barge *Pacific Sounder*.

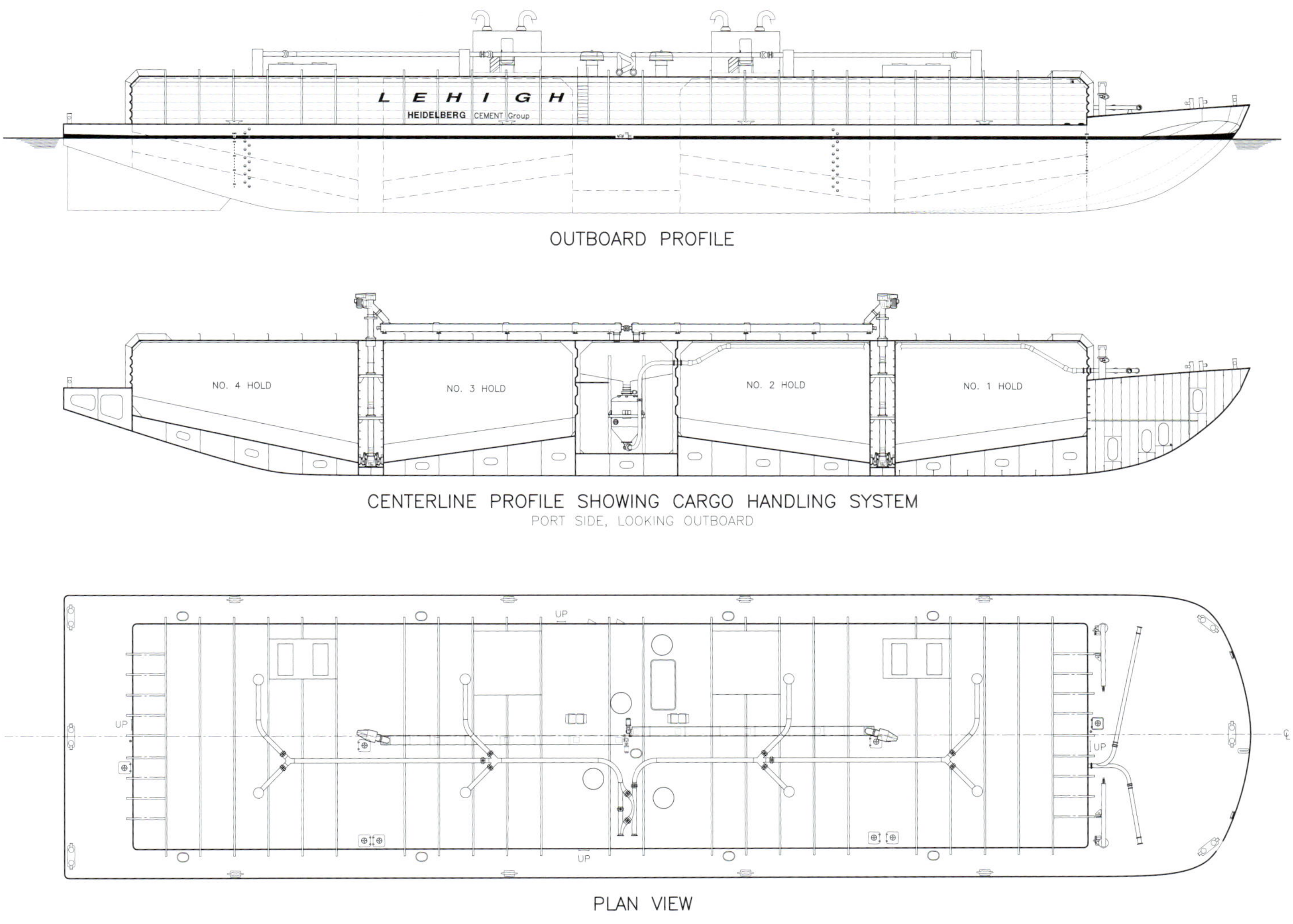

Right: Lehigh's most recent cement barge, *Pacific Sounder*.

Newsprint Carriers

Somewhat unique to the Pacific Northwest are large, covered deck cargo barges specially designed to carry newsprint and similar paper products. Obviously, these cargoes, packaged in large rolls, need to be kept fully dry and undamaged in transit, so the deckhouse geometry is tailored to tight tolerances to match the cargo roll dimensions with minimal dunnage required. The barges are loaded and unloaded by forklifts, so multiple large doors and ramp systems are fitted to facilitate the fastest possible load/unload rates of the palletized cargo.

As mentioned previously, the very first barge in the Allan corporate design record was such a paper-carrying barge, the 144-foot *Harmac 50*, designed in 1948 for Kingcome Navigation Co. Ltd (the marine transportation arm of the MacMillan Bloedel Company). Kingcome became a steadfast client for many years and commissioned several major and innovative projects. The most significant of these was the development in 1964 of one of the world's first ocean-going tug-barge transportation systems, where a single tug operated with three identical barges moving newsprint from Port Alberni, BC, to Los Angeles. This is often referred to as a "swap-and-drop" system, where simultaneously one barge is in transit, one loading and one unloading, and the tug almost never stops working. This operation was well described in R.F. Allan's paper "Barge Transportation Systems & Barge Towing Tugs," presented to the first International Tug Conference in London in 1969. The deckhouses on the barges were designed with short gaps in the steel structure at intervals along the length, bridged by flexible sections of heavy neoprene to avoid the transmission of hull-bending stresses to the house, and thus the upper structure could be built quite lightly.

The barges in this transportation system were the 7,000 TDwt *Nahmint Carrier*, *Nanoose Carrier* and *Nootka Carrier*, all measuring 356 feet long by 82 feet beam on deck (78 feet at the waterline). The *Nahmint Carrier* was slightly different from her sisters, having the capacity to back-haul oil cargoes to the mill in Port Alberni, thus earning revenue on both legs of the voyage and eliminating the need for ballast in a light condition.

This system was highly successful and operated for decades. Soon after the West Coast system was operating, a comparable system was commissioned on the East Coast by the Irving organization to move newsprint from their company paper mills in New Brunswick to a variety of major US East Coast ports. The barges *Rothesay Carrier* and *St. John Carrier* were configured essentially identically to their West Coast cousins but were slightly larger at 363 feet by 82 feet with a capacity of 7,250 long tons.

Smaller and simpler variants of these paper barges operate on more sheltered routes within the Strait of Georgia, including the 202-foot *Island Tug 150* and *Island Tug 151*.

Left: Tug *Haida Brave* outbound from Vancouver with the newsprint barge *Nanoose Carrier* in tow; **below:** Smaller paper barges such as the *Island Tug 150* are used on more sheltered routes on the BC coast.

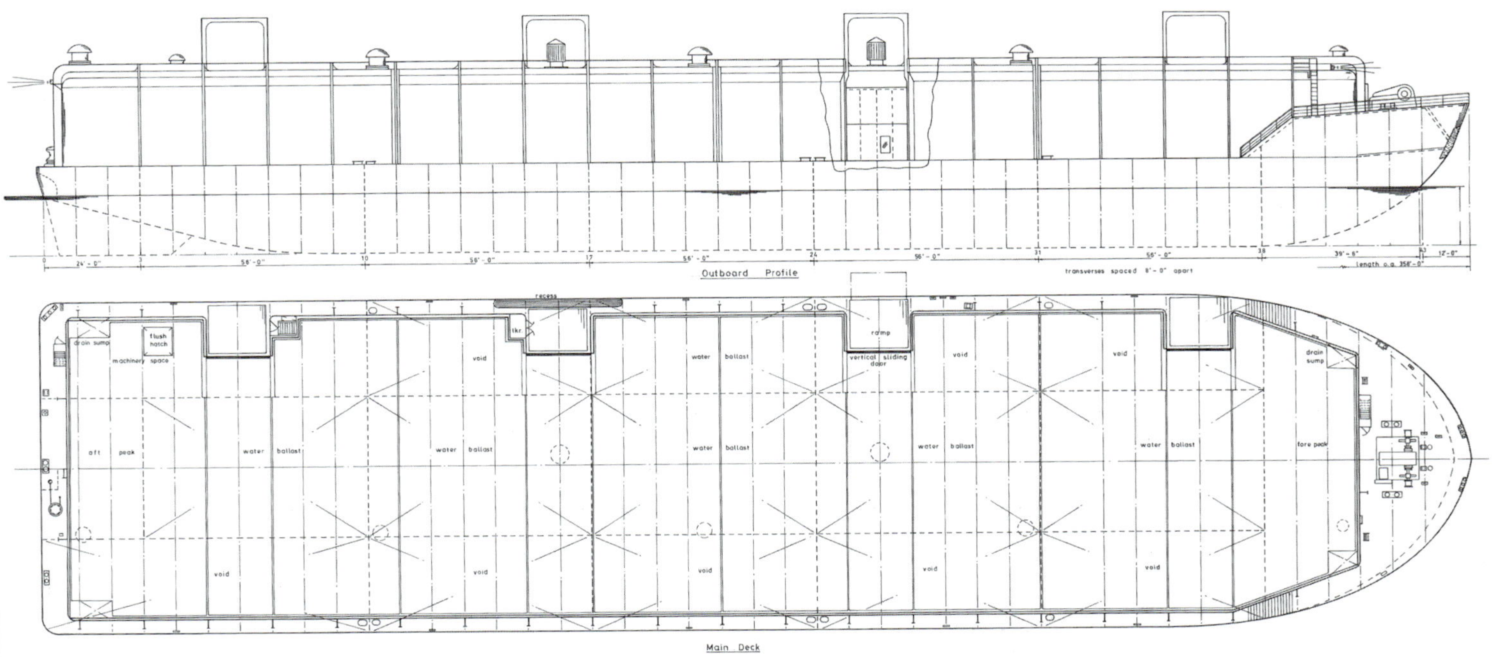

Above: General Arrangement for the 7,000 TDwt covered newsprint barges operating on the Pacific Coast.

Specialty Carriers

Among the many dozens of more typical bulk and deck barges are a few very specialized barge designs for product-specific movements. One commodity moved by barge in large quantities on the BC coast is caustic soda, a major constituent in pulp-and-paper manufacturing. This hazardous product is typically carried in hull tanks of barges which may carry other deck cargoes such as railcars, but Hooker Chemical Corp. commissioned Robert Allan Ltd to design a specialized barge to carry bulk caustic in hull tanks as well as large volumes of chlorine, another major chemical used in paper production, in three large tanks on deck. The 270-foot barge *Metlakatla* was built by Allied Shipbuilders in 1967 and served the BC coast in this role for many decades.

Another very specialized barge was the multi-product tank barge *Chevron Haida*, designed for Chevron Canada Ltd and built by Mackenzie Barge and Marine Ways Ltd in 1982. This was an interesting design, effectively a small product tanker carrying diesel, gasoline, aviation fuel and other volatile petroleum products to various communities on the BC coast. It utilizes a system of stainless-steel hydraulically driven deep-well pumps for cargo discharge. The barge was designed to be pushed as an articulated tug-barge (ATB) system, but the owners did not want to use a dedicated tug for working with this barge, as their towing services were chartered. This meant a generic stern notch shape had to be used to accommodate a possible variety of tugs, but the barge also had to have skegs for directional stability if the tug was unable to continue pushing in the notch due to poor weather and thus revert to using a towline. This became an ongoing problem, as the presence of the notch created turbulence, which disrupted the flow over the skegs, rendering them ineffective in some load and trim conditions. A lot of effort went into trying to devise a solution for this, which really should have been the use of one of the various patented pin systems available from suppliers in Japan or Europe to connect tug and barge, thus effectively eliminating the need to tow.

Robert Allan Ltd has designed numerous other tank barges, including many for the Mackenzie River system, where oil movements during the brief summer season are critical to the energy needs of all the northern communities. Many of these oil-carrying barges are also designed to carry deck cargoes such as containers or palletized cargo to maximize their utility in this remote area. The most recent of these were the *Deh Cho* series of barges for ITB Marine Group Ltd (ITB here not to be confused with the Island Tug & Barge Ltd "ITB" of the 1960s).

In September 2012, ITB North Ltd, a division of ITB, contracted with Robert Allan Ltd to design a new class of shallow-draft double-hull combination bulk-oil and deck-cargo barges for operation on the Mackenzie River, Great Slave Lake and the Beaufort Sea. An extremely tight design-and-build schedule was required in order to get the barges delivered to the Canadian Arctic in time for the 2013 summer operating season. Steel-cutting commenced just three months after the award of the design contract. In all, four barges were built to this design, named the *ITB Deh Cho 1, 2, 3 and 4;* "Deh Cho" being the Dene First Nation's words for "Big River," their name for the Mackenzie River. The barges were constructed at Jiangsu Yangzijiang Shipbuilding Company in China and were put into service in the North less than 11 months after the original design contract was placed. These barges represented a significant milestone as the first double-hull oil barges on the Mackenzie River.

Metlakatla being towed through the Second Narrows railway bridge, 1979. (A portion of the bridge had collapsed following a cargo ship collision.)

Above: *Chevron Haida*, a 37,000-barrel-capacity ATB multi-product carrier for Chevron Canada Ltd.

Left: ITB *Deh Cho 2*, one of a quartet of the first double-hulled oil barges on the Mackenzie River.

Log Barges

Next to tugboats, the vessel type perhaps most closely associated with Robert Allan Ltd is the self-dumping log barge. This is a vessel almost unique to British Columbia, reflecting the importance of the coastal forest industry to the provincial economy and the innovations that have evolved here to improve the safety and economy of the forest sector transportation systems.

There are a total of 16 purpose-built log barges in BC credited to Robert Allan Ltd, out of a total fleet ever built of 25, and the design concept used in all of them is generally credited to Bob Allan. The largest of these barges is the *Island Forester* (later *Seaspan Forester*) at 18,000 long tons Dwt. The detailed story of all these barges and more is well documented in *The Log Barges*, a most informative little (5" × 7") book by Syd Heal (Cordillera Books, 2007).

Log movements by barge, rather than by rafts or booms, began with the use of surplus cargo ships as deck-cargo barges. Logs were loaded and unloaded by crane, a slow and laborious process, and the typical ship hull lacked sufficient stability to carry any significant deck-cargo load. In 1953 that all changed with Bob Allan's design of the first self-dumping log barges, *Powell No. 1* and *Powell No. 2*, for Kingcome Navigation. These barges each had a capacity of 7,000 tons and featured a tipping tank that flooded naturally when sea valves were opened in the fully laden state, heeling the barge to the angle at which the logs would slide off into the water. The tipping tank would then, with the barge floating at a much lighter draft, self-drain to empty under gravity alone. This elegantly simple system, illustrated in the accompanying figure, was used for many years. It was not, however, foolproof, as the centre of gravity of the log load could vary considerably. Also, if there was considerable side overhang of the load, then those logs would pick up buoyancy before the full dump angle was achieved and the dump could stall. In later barges, this problem was overcome by introducing pumped water ballast systems, and the dumping process was thus fully controlled and much less dynamic than the many dumps where the barge literally jumped completely from the water. A controlled dump meant much less damage to both the log cargo and to the barge.

With the Robert Allan Ltd design of the *Forest Prince* for BC Forest Products in 1960, a new generation of log barges began. These were fitted with large cranes for self-loading, in addition to self-dumping capability. This enabled the barges to work in many more areas of the coast without the need for land-based log-loading infrastructure. As log sizes began to decrease, these cranes were fitted with bundle grabs with a capacity of about 50 tons at an operating radius of 100 feet.

Typical dumping sequence of a self-dumping log barge.

Designing Log Barges

Rob Allan came into the company just as I was leaving. We sort of passed each other, but Rob came in at a time when Robert Allan Ltd was growing again, particularly with Crown Zellerbach. I think one of Rob's early projects was being project manager on tug construction for Crown Zellerbach. At the time, log barges were generally 10,000 tons or thereabouts, but getting bigger. The *Seaspan Forester* was Bob Allan's really big one, built for Island Tug & Barge, and that was 18,000 tons. The log barge business was changing then, in part because a couple of cranes had fallen off log barges. Log barges had traditionally loaded the logs by using a grapple, which picked up 6 or 8 tons of loose logs at a time from the water and put them on the deck, but because of environmental concerns, the industry was forced to find a better way to handle the logs, and Robert Allan Ltd was right at the front of that. I was working on changing the existing log barges with 12-ton grapple cranes to 50-ton cranes. At the same time the industry started building 50-ton steel cribs, which they'd bring down to the water and the log barge grapple would then pick up 50-ton bundles from the crib. Add to that the water that was trapped in the logs, which was about 10 tons, and the grapple itself, which I think was about 12 tons, and we actually ended up with 70-ton cranes—and they're still around. This new system was more efficient and profitable simply because the end scale at the mill would be 8 percent heavier than the loose logs, because with loose logs, eight percent of the load would sink or drift off. The tugs used to tow the barges also went from 1,800 hp to 4,000 hp.

—Tom Ward

An iconic Murray McLellan photo of the 18,000 TDwt *Island Forester* dumping logs at Gambier Island.

BC Log Barge Designs by Robert Allan Ltd

Year Built	Name	Owner	Type	Loa (feet)	Beam (feet)	App. Dwt. (L.Tons)	Builder
1954	*Powell No. 1 / Powell Carrier*	Kingcome Navigation Co. Ltd	S/D	342.7	63.0	6000	Burrard Dry Dock Co. Ltd.
1954	*Powell No. 2 / Alberni Carrier*	Kingcome Navigation Co. Ltd	S/D	342.7	63.0	6000	Burrard Dry Dock Co. Ltd
1957	*Straits Cold Decker*	Straits Towing Ltd	S/D	285.3	50.0	3500	Burrard Dry Dock Co. Ltd.
1957	*Straits Water Skidder*	Straits Towing Ltd	S/D	285.3	50.0	3500	Burrard Dry Dock. Co. Ltd
1957	*Crown Zellerbach No. 5*	Crown Zellerbach Canada Ltd	S/D	285.3	60.0	4500	Burrard Dry Dock. Co. Ltd
1957	*Crown Zellerbach No. 6*	Crown Zellerbach Canada Ltd	S/D	285.3	60.0	4500	Burrard Dry Dock. Co. Ltd
1960	*Forest Prince*	BC Forest Products Ltd	S/L - S/D	305.0	60.0	5000	Yarrows Ltd
1961	*Haida Carrier*	Kingcome Navigation Co. Ltd	S/L - S/D	344.0	64.0	6250	Yarrows Ltd
1963	*Crown Zellerbach No. 4*	Crown Zellerbach Canada Ltd	S/D	345.0	64.0	6250	Yarrows Ltd
1964	*Straits Logger*	Straits Towing Ltd	S/L - S/D	369.0	76.0	9000	Yarrows Ltd
1965	*Island Yarder*	Island Tug & Barge Ltd	S/L - S/D	364.0	80.0	10000	Yarrows Ltd
1965	*Rivtow Carrier*	RivTow Straits Ltd	S/L - S/D	369.0	76.0	10000	Yarrows Ltd
1970	*Swiftsure Prince*	BC Forest Products Ltd	S/L - S/D	345.9	70.0	7500	Yarrows Ltd
1970	*Crown Zellerbach No. 1*	Crown Zellerbach Canada Ltd	S/L - S/D	363.0	80.0	10000	Allied Shipbuilders Ltd
1970	*Island Forester*	Island Tug & Barge Ltd	S/L - S/D	453.0	96.0	18000	Yarrows Ltd
1980	*Seaspan Rigger*	Seaspan International Ltd	S/L - S/D	396.0	88.0	12700	Vancouver Shipyards Ltd

The BC log barges attracted international interest. Back in 1966 Robert Allan Ltd was contracted by the French company Société Navale Delmas-Vieljeux to design a self-propelled log ship for carrying hardwood logs to Europe from West Africa. An extensive design was completed and model-tested. However, the owners got cold feet and unfortunately elected not to proceed with construction, missing the chance to claim the world's first self-propelled, self-loading, self-dumping log barge.

In 1968 the Swedish company Neptune Towing and Salvage approached Bob Allan with the idea to adapt BC log barge technology to their log-handling operations in the northern Baltic. Their logs were much smaller, of generally uniform diameter, and were cleaned and cut to standard lengths before any shipping took place. This necessitated a much different and very large grab design, foreshadowing the type of grabs that would later be used in BC. Two 9,000 TDwt barges, *Hera* and *Juno*, were built in Germany to this design. This was a milestone project for Bob Allan, one of his first major projects for an overseas client cementing his reputation as an innovative designer.

Exporting Canadian barge design technology: Swedish log barge *Hera* arriving in Goteborg, 1970.

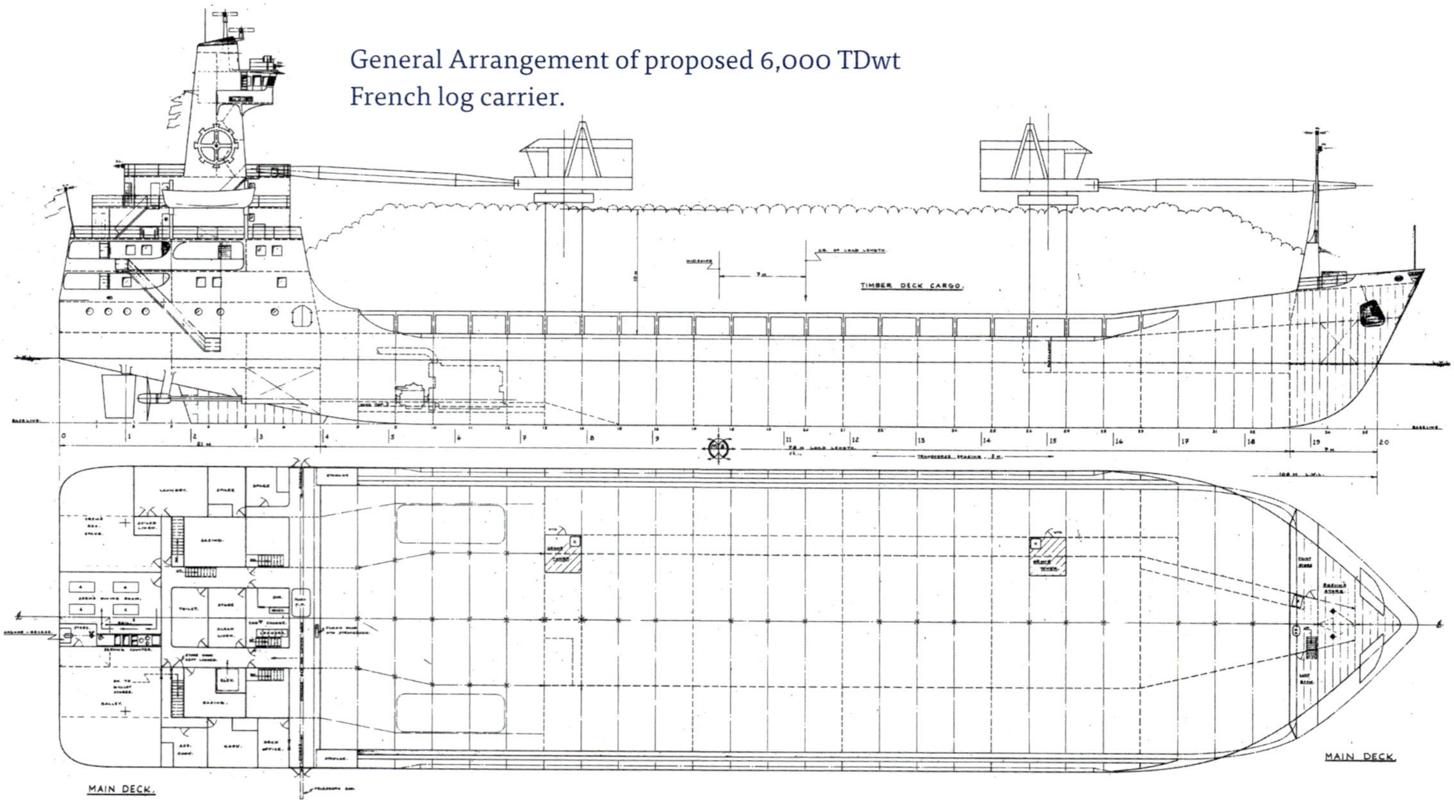

General Arrangement of proposed 6,000 TDwt French log carrier.

Backyard Barge Modelling

The design program for the Swedish barges was interesting, comprehensive and included model-testing in Germany, a process well described by Capt. Jochim E. Brix of the Hamburg Ship Model Basin (HSVA) in a paper to the 2nd International Tug Conference in 1971. As the design work on this barge was starting, I was working in the office for the summer of 1968 before heading off to Glasgow to begin my studies there. A concern about the viability of this specific design was whether the barge would dump with the geometry and density of these logs, which were much smaller in size than the typical BC cargo. We constructed a small plywood pool in my father's backyard and built a section of the barge to test the dumping process. My task was to model the logs in size and density. Allowing for scale effects, that meant that half-inch-diameter maple dowelling was the best choice. I spent days buying up every foot of maple dowel that could be found in the city, and then cutting it into six-inch lengths. That amounted to several thousand dowels when all was said and done. We then loaded the barge section and tested dump angles versus load geometry. Thankfully it all worked; however, the owners ultimately elected not to dump, as the straps securing the log bundles would not withstand the dumping impacts, and they unloaded more slowly but safely using the three cranes. The 6-inch-long maple dowels, however, persist to this day and have formed the basis of the many tool boards and garage tool racks I have built ever since!

—R.G Allan

Right: Backyard modelling of the dumping of the Swedish log barges.

Left: *Seaspan Rigger* (renamed *Seaspan Phoenix*), the last BC log barge built.

HB FILHOTE

14

Working the Rivers

Although the majority of our design work has been for harbour, coastal and ocean-going vessels, a not insignificant number of designs have been developed for working in the world's big (and some quite small) rivers. Due to the challenges imposed by these environments, and especially in very shallow rivers, this portion of our portfolio includes some of our most interesting and innovative custom designs. The beginnings of this work can be traced to early designs for small BC tugs and workboats operating on the lower reaches of the Fraser River, but the Fraser is a rather deep river, and those local boats did not need to be especially shoal draft.

Our most challenging river projects began with designs for the northern rivers of Canada, and in particular for the Mackenzie River, which wends its way north for more than 1,000 miles from Great Slave Lake to the Arctic Ocean. Navigable for its entire length, the Mackenzie is the longest river in Canada and second in length in North America only to the Mississippi. In many places it is quite deep, but there are a few very difficult shallow areas, often associated with rapids, which limit maximum operating draft to only about 1 metre.

Opposite: The HB *Filhote*, see page 300.

Virtually all transportation on this remote river and its tributaries is performed by shallow-draft tugs pushing trains of multiple barges. There are virtually no services for vessel repairs or docking anywhere along the river, so the designs must be extremely simple and reliable and have the means incorporated for anticipated tasks such as propeller and rudder repair or replacement. Consequently, all boats on this river carry spare rudders and propellers, as grounding on the constantly shifting sandbars in the river is almost routine.

The Mackenzie River runs through northern Canada to the Arctic Ocean for more than 1,600 kilometres.

Sandy Jane

The earliest Robert Allan design for this remote river was created in 1945 for the *Sandy Jane,* a pusher tug for Yellowknife Transportation Co., previously mentioned in Chapter 2. This 75-by-23-foot shallow-draft tug drew just 30 inches of water.

Built by Yarrows Ltd in Victoria, this boat was then shipped north in sections by rail and road. An article in *Harbour and Shipping* from April 1946 describes that activity nicely: "It was a notable undertaking as some sections of the tug weigh as much as 17 tons and are 50 feet in length, and apart from railway transportation and moving these prefabricated parts by truck over the Alaska Highway, they have to be taken 35 miles over a newly built road to a yard hewed out of the forest on the shore of the Fort Nelson River." This method of building was common for almost every vessel built for the North before the mid-1970s, when it became more feasible (largely due to moderating ice conditions) to move vessels around Point Barrow and into the Beaufort Sea on their own bottoms.

Sandy Jane pushing a barge in 1954, the first of many ultra-shallow-draft vessels designed by Robert Allan Ltd for Canada's northern rivers.

Below: General Arrangement for *Sandy Jane*.

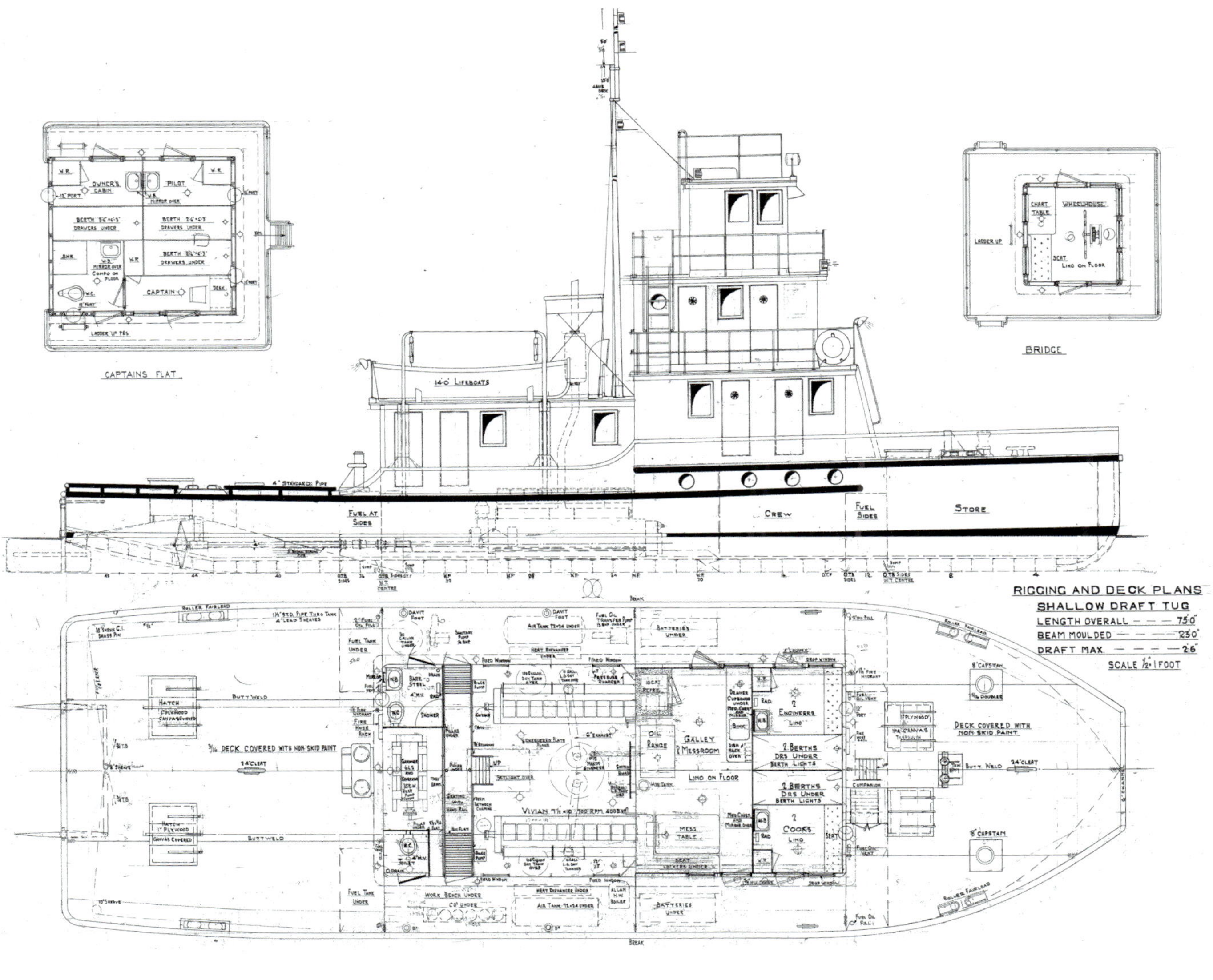

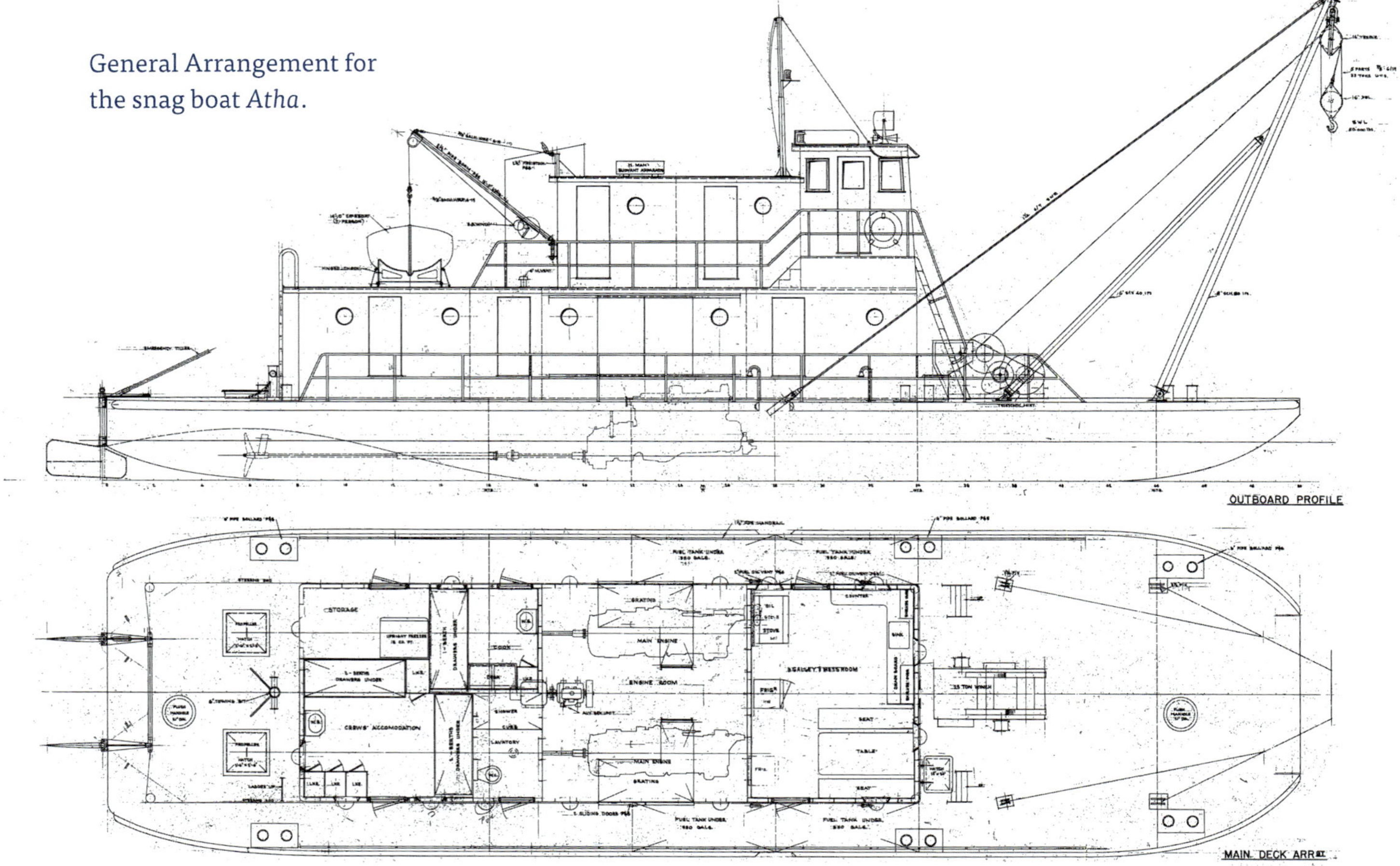

General Arrangement for the snag boat *Atha*.

Atha

In 1957 Bob Allan was commissioned by the Canadian Department of Public Works to design a "snag boat" for the Athabasca River, which flows from the Rocky Mountains near Jasper northeast to Lake Athabasca in northern Alberta, the lower reaches of which are a major transportation corridor. Today this river flows through the heart of the Alberta Oil Sands region. The twin-screw *Atha*, illustrated in the General Arrangement drawing, was built by John Manly Ltd in New Westminster, BC, and shipped north in sections for assembly on the shores of Lake Athabasca. She is 75 feet long by 20 feet wide and draws about 30 inches of water.

Sans Sault

The next shallow-draft vessel design for the North was the *Sans Sault*, developed in 1963 for the Canadian Department of Public Works (DPW). This was a twin-screw 700 bhp dredge-tending tug to support DPW work on the Mackenzie. She is named for one of the most difficult stretches of rapids on the river. At 77 feet 6 inches by 24 feet, the *Sans Sault* was light and fast, with a draft of only 45 inches. Built by John Manly Ltd in New Westminster, BC, she was segmented, shipped by rail and reconstructed in Hay River to enter service in the summer of 1965.

Anticipating the Mackenzie Valley Pipeline

The period between 1968 and 1973 represented one of the busiest periods of work ever for Robert Allan Ltd. There was a lot of coastal tug and barge work in BC, but the prospect of construction of the Mackenzie Valley Pipeline generated an unprecedented level of new vessel construction for the North. The major players in this area at the time were Kaps Transport Ltd, Arctic Transportation Co. Ltd (ATL), Northern Construction Co. Ltd (NCCL) and the Crown Corporation Northern Transportation Co. Ltd (NTCL). These companies were to one degree or another direct competitors for work on the Mackenzie, with the exception of NTCL, whose mandate was simply to supply the various small communities along the Mackenzie and the Arctic coast. All these operators were Robert Allan's clients at various times. Virtually every project was done in a great rush to get the vessel built for the next season of work. The boats were generally simple and unglamorous, but they were all highly effective, designed to meet the challenges of working in this very remote area, devoid of support services. Some of the largest of these vessels are described in Chapter 12 (Offshore Support Vessels), as those were among the earliest designs of that type built in Canada. The table on the following pages summarizes particulars of some of the many smaller boats designed for this demanding remote service (excluding the extensive fleet of NTCL tugs), followed by illustrations of a few of them. (Good photos of boats built on a remote riverbank are hard to come by!)

Launch of the *Arctic Circle No. 1*, a quadruple-screw tug with just 36 inches draft.

Shallow-Draft Vessels Designed for the Mackenzie River System

Year	Name	Vessel Type	Client	Loa (feet)	Beam (feet)
1946	*Sandy Jane*	Pusher Tug	Yellowknife Transportation Ltd	75.0	23.0
1957	*Atha*	Tugboat/Snag Puller	Dept. of Public Works, Canada	75.0	20.0
1960	*Hugh A. Young*	River Survey Vessel	Dept. of Public Works, Canada	87.0	22.0
1963	*Sans Sault*	Pusher Tug	Dept. of Public Works, Canada	77.5	24.0
1968	*Arctic Circle #1*	Pusher Tug	Kap's Transport Ltd	117.3	26.0
1969	*W.H. Horton*	Pusher Tug	Kap's Transport Ltd	69.0	25.0
1970	*Takla Transporter*	Self-Propelled Barge	Walter Dagenais	70.0	22.1
1971	*Mary-Lu Spencer*	Pusher Tug	Northern Construction Co. Ltd	68.0	24.0
1971	*Kap IV*	Self-Propelled Barge	Kap's Transport Ltd	125.0	35.0
1971	*Kap V*	Self-Propelled Barge	Kap's Transport Ltd	125.0	35.0
1972	*Kakisa*	Yarding Tug	Northern Construction Co. Ltd	84.0	36.0
1972	*Cecelia Hall*	Pusher Tug	Northern Construction Co. Ltd	66.0	25.0
1972	*Kap I*	Pusher Tug	Kap's Transport Ltd	45.8	12.0
1972	*Kap II*	Pusher Tug	Kap's Transport Ltd	45.8	12.0
1972	*Kap III*	Pusher Tug	Kap's Transport Ltd	48.5	15.5
1973	*Beverly Lambert*	Pusher Tug	North-Western Dredging Ltd	70.0	25.0
1973	*Carlyn McMurren*	Pusher Tug	North-Western Dredging Ltd	70.0	25.0
1973	*Dorothy Robinson*	Pusher Tug	North-Western Dredging Ltd	70.0	25.0
1973	*Thruster 3*	Thruster Barge	Northern Transportation Co. Ltd	105.0	35.0
1973	*Thruster 4*	Thruster Barge	Northern Transportation Co. Ltd	105.0	35.0
1974	*Betty Coulter*	Pusher Tug	North-Western Dredging Ltd	68.0	25.5
1974	*Arctic Sun*	Pusher Tug	Arctic Navigation & Transportation Ltd	74.0	25.0

Draft (feet)	Power (BHP)	Builder	Notes
2.5	1200	Yarrows Ltd	
2.5	230	John Manly Ltd	
3.5	592	Allied Shipbuilders Ltd	
3.0	700	John Manly Ltd	
3.0	2000	Bel-Aire Shipyard Ltd	quadruple-screw S/D tug
3.0	984	Vito Steel Boat & Barge Ltd	quadruple-screw S/D tug
3.0	280	John Manly Ltd	
3.5	960	Allied Shipbuilders Ltd	designed to split down CL for remote transport
~3.0	680	Alberta Equipment Centre Ltd	"White Gill" steerable waterjet propulsion
~3.0	680	Alberta Equipment Centre Ltd	"White Gill" steerable waterjet propulsion
3.5	2900	Bel-Aire Shipyard Ltd	"White Gill" steerable waterjet propulsion
4.2	960	Vito Steel Boat & Barge Ltd	twin-screw river tug
1.4	710	Yarrows Ltd	simple twin-screw yarding tug
1.4	710	Yarrows Ltd	simple twin-screw yarding tug
1.4	1065	Erickson Boatworks	simple twin-screw yarding tug
4.5	1406	Allied Shipbuilders Ltd	shallow-draft Arctic coast tug
4.5	1406	Allied Shipbuilders Ltd	shallow-draft Arctic coast tug
4.5	1406	Allied Shipbuilders Ltd	shallow-draft Arctic coast tug
~3.0	720	Allied Shipbuilders Ltd	barge train remote steering barge
~3.0	720	Allied Shipbuilders Ltd	barge train remote steering barge
4.5	2800	Zenith Steel Fabricators	shallow-draft Arctic Coast tug
4.2	1450	BC Marine Shipbuilders Ltd	shallow-draft Arctic Coast tug

As noted in the chapter on offshore vessels, with the collapse of oil exploration activity in the Arctic, the vast majority of these boats became surplus to local northern needs. Many were brought down to Vancouver and put into service on the BC coast, though they were not designed for, nor terribly well suited to, coastal towing duties. Some found employment in the Fraser River, a more germane environment. A few of them even made the return trip north to work for some of the smaller companies operating on the Mackenzie system.

Mary-Lu Spencer, a simple twin-screw yarding tug designed in parts for bolted assembly.

Below: *Miller Delta*, built in 1974 for NCCL as the *Beverly Lambert*.

The Northern Transportation Co. Ltd Fleet

In the aforementioned period, between 1968 and 1973, NTCL became one of our major clients, and in fact remained so for many years afterward. Robert Allan Ltd designed numerous tugs and barges for NTCL's operations on the Mackenzie, the Arctic coast and in Hudson Bay, as well as advising on innumerable refits and modifications, and some challenging loading options.

This relationship began in 1969 with the design of a new mainline tug for NTCL's Mackenzie River service. The *Vic Ingraham* was a prototype quadruple-screw tug built by Allied Shipbuilders in 1970. At 149 feet 6 inches in length by 50-foot beam, she was designed to operate at a draft of 3 feet 3 inches.

The development of this quite complex design and its successors is described in detail in a paper by Bob Allan for the 2nd International Tug Conference.

Business end of the quadruple-screw mainline pusher tug *Vic Ingraham*.

Below: *Vic Ingraham* at work on the Mackenzie River.

Northern Transportation Tugs

Northern Transportation was the primary operator on the Mackenzie River, and they'd been there during WWII. It was a Crown corporation set up for the transportation—from Great Slave Lake and down 1,000 miles to the Arctic Ocean—of the uranium used in atomic bombs. In some places on the Mackenzie, such as the Sans Sault Rapids, the water is only a bit more than 3 feet deep, and it runs very fast. In 1969, shortly after I joined Robert Allan Ltd, Northern Transportation started a program to modernize their fleet.

A fellow called Lionel Montpetit was the executive vice-president of Northern Transportation. He actually ran the Crown company from Ottawa. The company had initially hired Allied Shipbuilders to design three new tugs, but they ended up drawing too much water for the Mackenzie, and they dragged their propellers in the shallow rapids section and damaged things. Because of this problem, Lionel came out to have a meeting with Bob Allan, and in the course of the meeting he asked him to take over the rest of the program to rebuild the fleet. So Bob Allan became the architect. In 1969, when the tenders went out to build the prototype tug, the *Vic Ingraham*, the low bid was from Allied. When the tugs were built, I went north to help put them together.

We had designed the tugs with a draft of 3 feet 6 inches, and they ended up at 3 feet 3 inches. They were the first to have four propellers instead of two. They were designed with very light steel to keep the weight down, and that involved designing a structure that was able to take the stresses and impacts. It was the very first boat up there to have nozzles and a tunnel. With four engines and four props, the props could be smaller in diameter, and this made the tugs able to operate with shallower drafts. They ended up being really good pushers. The first one, the *Vic Ingraham*, was built in 1970, and on its first tow, I think it pushed 50 barges down the river. It was a great success. In 1972/73, Northern Transportation ended up building a further four of them, all designed as pusher tugs and slightly different than the prototype vessel. Bob Allan was the design agent for Kort nozzles, and we did a lot of designs with them. The nozzles allowed the propellers to be loaded heavier before they started to cavitate and get air on the face—the bubble zone. It was a tremendous leap forward for Bob Allan. Bob Allan was the expert, and of course that information got passed on to his son Rob when he came to work for him.

NTCL mainline river tugs chilling out for the winter in Hay River.

The same boats are still up there today, because the Arctic is just like a desert: it freezes in the winter and humidity is zero, so nothing rusts.

—Tom Ward

The next major project for NTCL began in late 1972 with the design of a quartet of mainline pusher tugs. These 149-foot 6-inch by 52-foot tugs had 4,500 hp in four screws, with an operating draft of 3 feet 3 inches. Building on the direct experiences and lessons learned with the prototype tug *Vic Ingraham*, they were quite unique for their type and packed a lot of power into that shallow draft by using complex tunnel shapes to accommodate the propellers and nozzles. This was another of those multi-vessel projects where the merits of thrust-augmenting nozzles were evaluated by comparing the performance of tugs with open screws to that of boats built with nozzles. Two tugs of each configuration were built. The four tugs were *Matt Berry*, built by Vito Steel Boat & Barge Ltd; *Jock McNiven*, built by Vito Steel Boat & Barge Ltd; *Henry Christoffersen*, built by Yarrows Ltd; and *Johnny Hope*, built by Burrard Dry Dock Corp. Ltd.

These four tugs were part of a major flotilla that was built in record-fast time in 1973 and delivered to the North in convoy fashion via the sea route around Point Barrow. The convoy, shown departing Vancouver harbour in July 1973, consisted of the four mainline tugs, each towing a variety of barge loads; the 84-foot Gill-Jet-propelled yarding tug *Kakisa*; about a dozen 1500 series barges (some of which were stacked three deep); two powered thruster barges, all of which were designed by Robert Allan Ltd; plus a number of other smaller bits of floating equipment.

Matt Berry, one of a quartet of powerful quadruple-screw shallow-draft mainline tugs for the Mackenzie River and Arctic coast.

Below: The northbound convoy of NTCL tugs and barges departing Vancouver harbour in 1973.

The thruster barges, shown clearly as the deck cargo in the foreground of the convoy photo, were very interesting, remote-operated, unmanned barges that were connected to the head of a barge train to provide better steering control when transiting the more extreme bends on the river course. These pre-dated accurate radio remote controls and used a direct electrical connection to control the thrusters. This presented some difficulties, as the cables were subject to mechanical damage, and so the concept, though quite valid, did not live up to its promise. Some decades later, when control technology had evolved to a higher degree of reliability, the same steering-barge concept was introduced on the Mississippi River to some fanfare.

The next major project for the North was in 1973, when we designed a shallow-draft coastal tug and two multi-purpose barges to support NTCL's operations on the western shore of Hudson Bay, based in Churchill, Manitoba. The triple-screw tug *Keewatin*, 125 by 36 feet with a 6-foot draft and 3,375 hp, was built by Yarrows Ltd in Victoria. This vessel was christened by Bob Allan's wife, Nancy, an honour she much cherished. The barges were designated as the 1800 series (approx. 1,800 TDwt) and featured a partial house for cargoes needing protection from the elements, an open deck for general cargo and tanks below deck for a variety of oil cargoes.

Kakisa, NTCL's Gill-Jet-propelled yarding tug, built by Bel-Aire Shipyard.

Below: Triple-screw tug *Keewatin* for NTCL's Hudson Bay service.

The Border Train Shuttle

Although the majority of our river work was for northern Canada, a most interesting pair of pusher tugs was built in 1971 to operate out of Windsor, Ontario. Their mission was to move barges laden with railcars across the swift-flowing St. Clair River to Detroit, Michigan. This was a year-round operation and had to cope with strong currents as well as significant ice floe conditions in the winter.

This operation was run for CN Rail by the F.M. Yorke Company out of Vancouver, a firm with whom Bob Allan had a long and productive association. The tugs, built by Hike Metal Products Ltd of Wheatley, Ontario, were named the *Margaret Yorke* and *Phyllis Yorke* and had identical hulls measuring 100 by 35 feet. Their distinguishing feature was a tripod-supported wheelhouse that towered over the hull, giving the operators excellent visibility in a crowded and difficult waterway, and creating a futuristic look for a workboat of this type.

The other feature of these boats was the fact that they were powered by deck-mounted diesel-driven Z-drive units, creating an extremely simple engine room and enabling quick replacement of a propulsion unit if damaged in ice. The fully azimuthing drives also gave the tugs the extreme manoeuvrability and control needed to work on that challenging route. The *Margaret* had two drive units rated at 850 bhp each, and the *Phyllis* had three of the same units.

Margaret Yorke (left) and *Phyllis Yorke* (right), unique pusher tugs with deck-mounted Z-drives for railcar barge operations on the St. Clair River.

Spin-Off Projects

The expertise developed in working in these remote and challenging areas of northern Canada gave Robert Allan Ltd a distinct advantage in offering our services to clients who required comparable vessels in other parts of the world. As many clients look for a proven design, we were sometimes able to utilize an existing design with little modification, but mostly we would use that experience and expertise to show that a "proven designer" was much more important than a "proven design" and that we were up to the challenge of creating customized, efficient designs for almost any application. Some examples of spin-off shallow-draft designs from Canadian roots follow.

The Exmibal Project

In 1974 we were retained by the well-known Vancouver consulting firm Swan Wooster Engineering Ltd to develop designs for a small tug and barge system to support a remote-site nickel mine/smelter they were developing in Guatemala. The system consisted of a single small tug and two oil/deck cargo barges used to move crude oil into the smelter as fuel for its operations and to carry refined product out to the coast in containers. The pusher tug *Rio Sauce*, measuring 21.5 by 7.93 metres with an operating draft of 1.53 metres, was built in a small shipyard on England's south coast then shipped to Guatemala on a heavy-lift cargo ship and deposited at the mouth of the river where it was intended to operate.

The two barges, named *Setal* and *Boqueron*, were built by Vancouver Shipyards in Vancouver, BC. They were 63.8 metres long by 15.88 metres beam and had a total cargo capacity of 1,070 tonnes each: 1,000 tonnes of bunker C for refinery fuel and 70 tonnes of diesel oil. The deck was outfitted to carry 30 standard 20-foot containers on the outbound voyage. These two barges were the very first vessels in Guatemalan registry! Unfortunately, the world nickel market slumped badly within a very few years of this smelter going into service, and it was shut down and decommissioned. The subsequent fate of the tug and barges is unknown, but they are undoubtedly still operating in Central or South America.

The *Rio Sauce* being launched into Guatemalan waters.

South American River Transportation Systems

Perhaps the best example of Intellectual Property (IP) transfer from the Canadian North to other regions of the world is the work that Robert Allan Ltd has been doing in South America for the past 15 years or so. This began in 2006, when we were retained by Sandwell Engineering Inc. (the successor to Swan-Wooster Engineering) to consult on marine transportation options for a major iron ore development in southwestern Brazil, at the time known as the Corumbá Project of Rio Tinto.

The basic requirements were to move 50,000 TDwt barge trains of ore downstream to tidewater over approximately 2,500 miles of navigable river of the Río Paraguay–Río Paraná system, flowing through Brazil, Paraguay and Argentina, and terminating in the Río de la Plata at Montevideo, Uruguay. From there, the ore would be trans-shipped into deep-sea bulk carriers. This river system is the second-longest waterway in South America after the Amazon and carries traffic from the eastern seaboard deep into the heart of the continent.

After initial studies and a variety of concept designs, with related analyses of transportation costs, preliminary designs were developed for large pusher tugs and

The Río Paraguay–Río Paraná system serves the Corumbá mining project in Brazil, very near the Bolivian border.

cargo barges. As the project was being readied for international bidding in 2009, Rio Tinto sold the mine property to the Brazilian mining giant Vale. The building of the mine and its associated transportation elements went into a prolonged holding pattern while the new owners re-evaluated every aspect of the project.

In about 2012 the project was rekindled, though with a quite different approach to the transportation plan. Vale did not want to own the transportation system, as Rio Tinto had planned, instead putting that into the hands of independent contractors. Eventually Robert Allan Ltd connected with Hidrovias do Brasil S.A., who were the successful bidders for this project. Modified designs were then developed for eight major pusher tugs and a fleet of 144 open-hopper barges, each with a deadweight capacity of approximately 3,100 tonnes at a 3.96-metre draft.

The design constraints on the tugs were not dissimilar to those of the Mackenzie River, but the South American river was generally deeper, allowing more draft. The barge trains planned were also significantly larger than those used in the Canadian North, and the ability of a tug to control that heavily loaded barge train on the downstream voyage was a major concern. This naturally led to the conclusion that Z-drive propulsion was much preferred for such an operation. This would provide enhanced steering but, most importantly, more than double the astern thrust capability, essential in stemming the effects of the river flow from astern. To prove the ability of our design to accomplish this, an extensive CFD analysis was undertaken by Robert Allan Ltd, which clearly demonstrated that the concept of triple Z-drives would provide the propulsion and steering capacity needed to safely navigate the river system.

Rio Tinto in South America

The project that was most defining for me after starting at Robert Allan was for Rio Tinto Brazil. They were developing a new mine in Corumbá, Brazil—said to be the highest-grade iron ore mine in the world—and they wanted to move 10 million tons of iron ore annually over almost 4,000 kilometres down the Río Paraguay and Río Paraná, where the water depth at times is only about 2 metres. Between 2005 and 2007 that was pretty well the only thing I was working on. I was flying down to Brazil once a month and flying to Australia, where Rio Tinto had its headquarters. We got to the point where we had signed shipyard contracts and steel was being cut for the first of the barges when the whole project was cancelled because of the economic collapse of 2007, so those boats didn't get built at the time.

Ten years later, we ended up redesigning those boats for Vale S.A., who bought the mine from Rio Tinto. They hired another company, Hidrovias do Brasil, to operate the vessels. Hidrovias do Brasil has been one of our best clients for more than 10 years now. The company also operates riverboats and barges throughout Brazil and South America. Those tugs had to push 50,000-ton convoys of sixteen 60-metre-long barges arranged four wide and four long. The river has constantly shifting shoals and, despite currents of usually less than 3 knots, it is a very difficult navigational environment, but also a pretty important thoroughfare. Certainly, those boats that we designed have performed very well and set all kinds of speed records. They were the first Z-drive tugs to successfully operate on that river. They were far more manoeuvrable and far safer than more conventional types because they could stop that much quicker.

—Mike Fitzpatrick

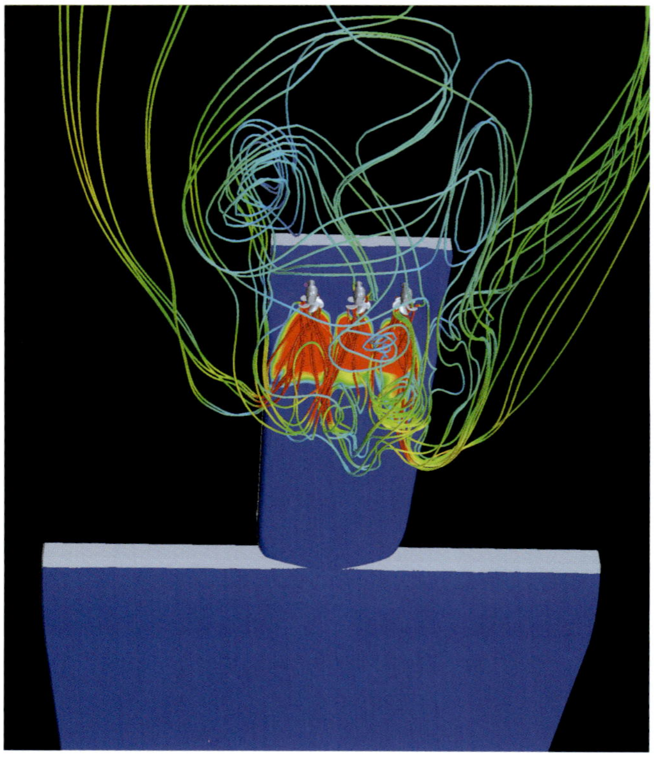

Left: CFD image of the triple Z-drive towboat manoeuvring downstream with a barge flotilla, with thrust directed in reverse "braking" the tow.

Right: Pusher tugs delivered to Argentina from Turkey by RollDock.

Below: The first shipment of 32 hopper barges from China arriving in Argentina.

The final tug design, designated the RApide 4600-Z3 series, was for a vessel 45.6 by 16.5 metres, with an operating draft of 2.1 to 2.5 metres. The propulsion system is diesel-electric—quite unique for a workboat—consisting of three diesel gen-sets, each of 1,710 ekW, running on heavy fuel oil (HFO), a most unusual configuration for any inland or coastal marine operation. The reason to use heavy fuel was quite pragmatic: diesel fuel was in such demand in the transited area that the crew had a history on other vessels of "dealing" with the locals for that product. HFO wasn't such a marketable commodity. The choice of HFO essentially dictated the use of a diesel-electric system, as engines running on HFO are not amenable to varying load demand. Each engine drives a fixed-pitch Z-drive propeller, Schottel model SRP 1215 FP, driven by an electric motor. The pusher tugs were all built in Turkey by Uzmar Shipbuilding and delivered by semi-submersible RollDock ship to Buenos Aires in 2014.

Of the 144 barges, each 61 by 15 metres, 128 were built in China by Shanghai Zhenhua Heavy Industries Co., Ltd (ZPMC) and 16 units were built in Asunción-Paraguay by Consorcio de Ingeniería Electromecánica S.A. (CIE), who built the unit blocks, and Astillero Chaco, who was responsible for the final assembly and launch. The Chinese-built barges were delivered in batches by semi-submersible lift ship to the Río de la Plata. The barges were designed for convoys four units long by four wide, so the lead barges have shaped bow rakes, but the aft set of 12 barges are a simple box shape. A fully laden convoy thus carries in excess of 51,200 tonnes of ore.

Moving 50,000 tonnes of iron ore southbound through the Río Paraguay–Río Paraná waterway.

HB *Hydra*, one of eight RApide 4600-Z3 class triple Z-drive pusher tugs built by Uzmar in Turkey for Hidrovias do Brasil.

With the successful completion of this large and rather complex project, Robert Allan Ltd established a significant and trusted working relationship with Hidrovias do Brasil, which led to more projects with this valued client and attracted the attention of other owners in Brazil and elsewhere in South America.

The following photos illustrate other river tug projects completed between 2014 and 2018 and operating on the Amazon and other South American rivers for companies such as Cargill, Louis Dreyfus Company (LDC) and Hermasa Navegação da Amazônia (Hermasa), now AMAGGI Logística e Operações (ALO), part of the Amaggi Group. It is notable that all these tugs are using Z-drive propulsion, which was not at all common on the rivers of either North or South America before this series of designs was introduced. American river pushboat owners are only just now (about 2020) starting to "test the waters" with Z-drives on the Mississippi system, having been using open propellers and flanking rudders for many decades, even resisting the many advantages of using Kort nozzles. Before the Corumbá project, the vast majority of pushboats in South America were second-hand boats from the American waterways, so they too were locked into 1950s technology. The innovative pusher tug designs from Robert Allan Ltd have revolutionized the thinking among vessel owners in South America and elsewhere.

Three RApide 4000-Z3 pusher tugs, *LDC Tapajós* (shown here), *LDC Amazonas* and *LDC Para*, were designed in 2014 and built by INACE in Fortaleza, Brazil, for the Louis Dreyfus Company to push 16 barge convoys. They each measure 39.6 by 18 metres with an operating draught of 3 metres. They are powered by three 1,600 kW diesel engines driving FPP Z-drives.

The *Josué Fabris*, a RApide 3400-Z2 river tug designed for Hermasa in 2014. It is 33.8 by 13.5 metres with an operating draft of 2.7 metres. This is another example of an arrangement with deck-mounted main engines to facilitate Z-drive propulsion in shallow-draft vessels. It was built by ERIN (Black River Shipyard) in Manaus, Brazil.

Above: The HB *Filhote*, a RApide 2400-Z2 pusher tug for Hidrovias do Brasil. It measures 23.6 by 11.0 metres with an operating draft of 2.1 metres and is powered by two 559 kW diesel engines driving FPP Z-drives. It was designed in 2013 and built by Estaleiro Rio Maguari.

In 2013 two RApide 4000-Z3 mainline pusher tugs, HB *Pirarara* (shown above) and HB *Pirarucu*, were designed for Hidrovias do Brasil. (*Pirarara* and *pirarucu* are two large fish found in the Amazon River system. Hidrovias have adopted the use of fish names for all their vessels.) These tugs are 39.6 by 13.8 metres with an operating draft of 2.8 metres, powered by three 1,600 kW diesel engines driving fixed-pitch propeller Z-drives. They were built by Estaleiro Rio Maguari in Belém, Brazil.

Designed in 2014, the *Cargill Jaú* is a RApide 2000-Z2 yarding tug for Cargill Brasil, measuring 19.5 by 9 metres with a 2.1-metre operating draft. Power comprises two 447 kW diesel engines driving FPP Z-drives. The builder was Estaleiro Rio Maguari.

Above: *Cargill Tucunaré* (shown) and *Cargill Cachará*, two RApide 2800-Z2 pusher tugs, were designed in 2014 and built by INACE for Cargill Brasil. They measure 28 by 10.5 metres with an operating draft of 2.5 metres. The propulsion configuration is interesting, with the two 820 kW main engines, mounted at main deck level, driving into the top of the Z-drive units.

The LDC *Tocantins*, a RApide 2600-z3 pusher tug for Louis Dreyfus. It measures 25.6 by 10.5 metres with a draft of 2.1 metres. Powered by three 447 kW diesel engines driving FPP Z-drives, it was designed in 2015 and built by Estaleiro Rio Maguari.

Above: In 2014 three RApide 2000-z2 yarding/pusher tugs, LDC *Guama* (shown here), LDC *Capim* and LDC *Estreitos*, were designed for Louis Dreyfus and built by Estaleiro Rio Maguari. They measure 19.5 by 9 metres with a draft of 2.2 metres and are powered by twin 447 kW diesel engines driving FPP Z-drives.

NO SMOKING
SAFETY FIRST
PROTECT THE ENVIRONMENT
SIGNET MAGIC
SIGNET POLARIS
SIGNET ARCTURUS
ANNE

15

A World of Tugboats

Although it hasn't always been the case, no other vessel type defines the work of Robert Allan Ltd as does the humble (but increasingly sophisticated) tugboat. The following chapters trace the evolution of our tug designs from the beginning of the BC tug fleet to our position today (2022) as the dominant force in international tug design. It is important to provide some basic background to the world of tugboats and the diverse range of work that they do, each of which demands differences in hull shape, power and deck gear if they are to operate safely and efficiently. There is really no such thing as a "typical" tug, though the thousands of tugs working today to our many designs may bely that fact. We use the following as the basic classifications of tugs, by function:

Opposite: Multiple tugs from Signet Maritime Corp. of Texas berthing a large tanker: three pushing and two holding back on lines to control the contact speed.

Coastal Towing

Coastal towing involves moving barges or towing log booms and similar floating gear from place to place in more-or-less sheltered waters. A subset of this might be the typical small tugs used to support dredging and floating crane operations.

Ocean Towing

These are larger and usually much more powerful tugs used for towing barges, floating platforms or similar structures over long distances in open-sea conditions. These tasks are also frequently performed by large anchor-handling tug/supply (AHTS) vessels.

Typical BC coastal tow of a trio of chip scows.

An ocean-going AHTS towing an offshore platform. (Not a RAL design.)

Ship-Assist Tugs

Ship-assist (or "ship-handling" or often just "harbour") tugs represent the quintessential image of a tugboat, used to berth and unberth ships in ports, nearly always at slow speed, and generally in sheltered waters.

Terminal Support Tugs

These are a relatively new variant of the ship-assist tug, developed for use mostly at oil and LNG terminals built in more exposed locations, and thus typically are much larger and more powerful than tugs for harbour operations.

Terminal support tugs moving a large LNG carrier onto a berth.

Escort Tugs

This newest class of tugs, developed in the mid-1990s and refined through extensive research over the subsequent years, represents probably the most capable type of tug afloat today, and they certainly have the most challenging duty. Escort tugs are purpose-designed to enable them to apply very high steering and braking forces to a ship at speeds in the 8–10-knot range in the event of a loss of power or steering on the attended ship. They accomplish this by working in what is known as the "indirect" mode. Escort tugs work with ships carrying hazardous or other critical cargoes where any off-track incident would severely impair the port's function, and thus by definition in near-coastal waters on approaches to ports or terminals.

The Norwegian escort tug *Ajax* applying indirect steering forces to a tanker in stormy conditions.

Salvage Tugs

These are a type of large purpose-designed tug seldom seen anymore but common in the last half of the 20th century. Only a few dedicated salvage tugs are still in service today, in strategic areas such as the Cape of Good Hope. These tugs are large and seaworthy and equipped for rescuing and towing large ships in distress on the open sea, often in extreme conditions.

Branding the New Tug Designs

Robert Allan Ltd's involvement with tugs began in the 1930s with the design of a few small wooden tugs in BC. That expanded with the rebirth of the BC coastal towing fleet in the 1960s and 1970s, and then, after a slow decade, really proliferated and became internationally recognized in the 1990s.

MV *Singapore*, now under the Svitzer flag. Originally built as SMIT *Singapore*, one of a quartet of large, powerful salvage tugs built by SMIT for worldwide service. (Not a RAL design.)

In the course of the latter work, we began to "brand" our tug types to help both ourselves and our clients understand our interpretations of the differences in tug roles. The first of these brands was the RAmparts series, which was intended to describe tugs built for standard series production. The name was an acronym for Robert Allan Modular *Parts*. The objective of this series as a modular design never fully materialized, but it did start a trend whereby we branded the majority of our various design series with words that began with or which incorporated "RA." The tugs series used today, and which have become globally recognizable, are the following:

ART: the designation given to the series of Advanced Rotor Tugs designed under our exclusive design agreement with Rotortug (KST) B.V. of the Netherlands for these unique triple Z-drive tugboats.

RAmparts: a series of well-proven azimuthing stern-drive (ASD) tug designs which have coalesced into a set of relatively standardized configurations, representing one of the most versatile multi-purpose type of tugs in use today. Used for harbour operations and coastal towing.

RApport: compact, flush-decked and cost-effective ASD tugs developed specifically for ship-handling in smaller harbour and port operations.

RAscal: high-performance but relatively simple line-handling tugs less than 24 metres in length.

RAmbler: small, simple line-handling/mooring boats.

TRAder: a broad-based series of coastal towing or "line-haul" tugs, typically twin-screw, equipped with heavy-duty towing gear aft and designed to maximize thrust at a towing speed typically in the 6–10-knot range. (The majority of our older generation tug designs now fall under this classification.)

TundRA: tugs for operation in the full range of ice conditions, uniquely designed for each specific ice class and set of operating conditions.

RApide: very specialized shallow-draft pusher tugs propelled by dual, triple or quadruple conventional screws, Z-drives, or in some cases by waterjets.

RAmpage: larger tugs developed to address the demand for high-performance towing and anchor-handling tugs for critical offshore terminal and oilfield support duties without reverting to full-sized OSV/AHTS vessels.

RANGLer: escort and ship-handling tugs powered by LNG or dual fuel, conceived to take maximum advantage of the very limited space available for LNG systems in tugs.

ElectRA: a series of electric-propulsion tugs designed to maximize the potential of electric propulsion in all aspects of tug operations.

RAstar: a distinctive class of very high-performance ASD escort/offshore terminal tugs, designed with a unique sponsoned hull form, which has been proven in both model and full-scale to provide significantly enhanced escort towing performance and sea-keeping compared to more conventional hull forms.

RAZer: another development in the evolution of high-performance Z-drive ship-assist/escort tugs. These tugs are optimized for every aspect of ship-handling and tanker escort service, but due to their "aft-biased" layout are not generally recommended as a configuration suited for towing astern.

TRAktor: combines Voith Schneider propeller (TRAktor-V) and Z-drive (TRAktor-Z) tractor tugs (tugs with the drive units positioned in the forward part of the vessel, typically about one-third of the length from the stem) into a distinctive series.

RAsalvor: tugs designed to provide long-range towing, anchor-handling, rescue, and salvage capabilities, and typically also equipped with significant firefighting and spill-response capacity.

RAVE: an acronym for the unique Robert Allan Voith Escort tug, a new concept introduced in 2013 for a highly manoeuvrable high-performance escort and ship-handling tug, jointly developed by Robert Allan Ltd and Voith Turbo Marine. The RAVE series features the longitudinal alignment of the two VSP drives, in contrast to the more conventional side-by-side tractor-drive configuration.

Z-Tech: The most notable exception to the "RA" branding scheme, this Red Dot award-winning series of tugs was initially developed specifically for PSA Marine of Singapore. The design incorporates the best handling and operational characteristics of both Z-drive tractor tugs and azimuthing stern-drive tugs. Now often shamelessly imitated, but never equalled, Z-Tech tugs are intended primarily for operation in major ports and marine terminals. They have a relatively low forward sheer and a wide, heavily fendered bow. The wheelhouse is set well aft and inboard, offering excellent visibility over the entire working deck and to the sides. With the low bow, the tug is designed to make transit or open-water voyages stern-first in tractor mode.

The following chapters describe the evolution of these various series and more of our tug designs for the world.

Custom Designs

All our projects are to some extent custom designs. We work specifically with shipyards and owners, and they all have their own different requirements. Even on a "standard design," if, for example, a different engine is chosen or more fuel capacity is required, these trigger knock-on effects in the remainder of the entire design that need to be carefully engineered. The vessel's area of operation and operational profile also play significant roles. There are standards among the various classification societies around the world, both international and for every flag state, such as Transport Canada. We must deal with a lot of different requirements that are not standard, as each classification society has some things that they want done differently.

—Andra Papuc

HAIDA BRAVE

16

The BC Coastal Tugboat Fleet: 1960s and 1970s

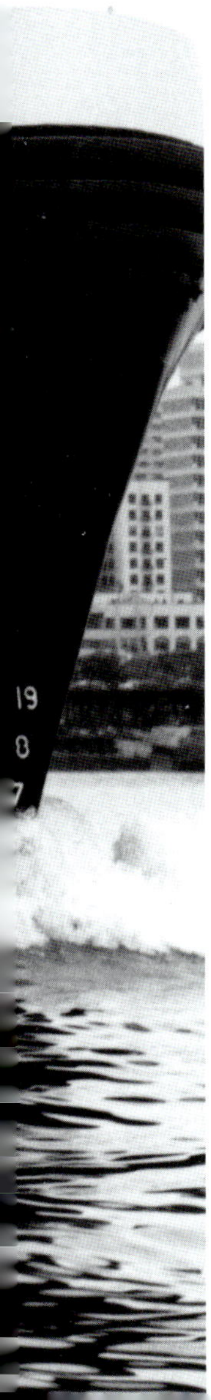

The coast of British Columbia is extensive, rugged, mostly remote and has no comprehensive means of access to its numerous small coastal communities except by air or, primarily, by water. The coast is for the most part well protected from Pacific storms by the barrier of islands that extends for hundreds of miles from the mainland US border in the south to the Alaskan border in the north, making this an ideal marine transportation corridor.

Industries along the coast have traditionally been largely resource-based, primarily forestry, fisheries, and, to a lesser extent, mining. Eco-tourism and sport-fishing are becoming increasingly important as low-impact, high-revenue small businesses. The movement of goods along the BC coast has evolved over the last century from a mixture of coastal freighters, smaller passenger vessels and tug-barge tows to consist now almost exclusively of tug and barge transportation. Barges typically carry logs, lumber, paper, pulp, wood chips, gravel, lime-rock, bulk cement, clinker, oil,

Opposite: The ocean-going tug *Haida Brave* introduced a new era in ocean tug-barge transportation.

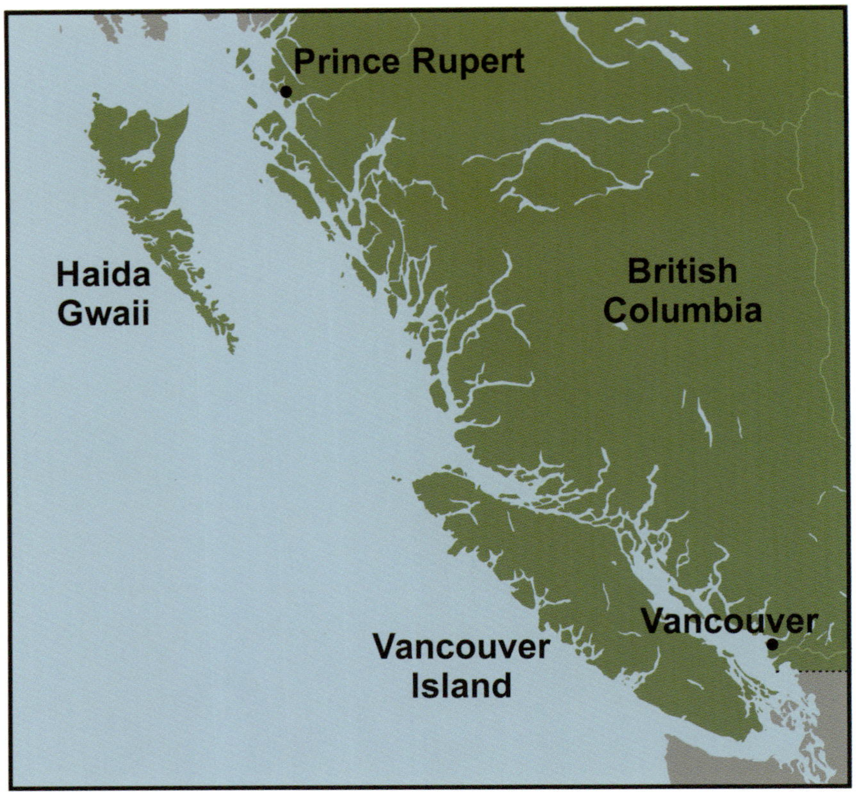

The British Columbia coastline offers more than 500 miles of protected waterway between the mainland US border and Alaska.

general cargo and, in limited quantities, containers. Tugs tow one, two or three barges at a time, depending on the cargo and the exposure of the voyage. The products moved by barge are mostly of relatively low unit value, and volumes are relatively small, hence ideal candidates for this form of economical low-speed transport. Cargoes can be stored on barges for extended periods, avoiding warehousing costs. The major developments in the BC towing industry in the 1960s are well described in Bob Allan's paper to the first International Tugboat Conference in 1969 and in the Inaugural Paper presented by the BC Towboat Owners Association to the 3rd International/1st North American Tug Conference held in Vancouver in 1973.

The majority of tugs operating in the BC coastal trade today were built between 1963 and 1975, and the majority of those over 15 metres in length came from the drawing boards of Robert Allan Ltd. The tugs are, with a few notable exceptions, less than 28 metres in length and less than 1,500 kW in power. Barges are generally less than 5,000 TDwt, except for a few large bulk cargo and log barges (which are up to 18,000 TDwt). Many BC-built barges from the 1960s and '70s are still in service, though their replacements over the past 10–15 years are largely being sourced in China. The ravages of time and a demanding service are taking their ultimate toll on the older vessels, and freight rates have not kept pace with the cost of transportation, making capital reinvestment in modern, more fuel-efficient vessels a significant challenge to the industry. It is, however, a testament to the quality of the tugs (and many of the barges) built during the 1960–1980 period that many of them are still operating today.

Tugs and barges are the mainstay of the BC coastal economy.

Early Tugs

The earliest tug design in our archives was only identified a short time ago as we dug back in our history in 2017 to try to identify "Tug #1" on the way to identifying the 1,000th tug built to a Robert Allan design. As noted in Chapter 3, among a drawer of old tracings was a very small but also very pretty ink-on-linen drawing, shown here, only about 12 inches wide by 7 inches high, showing the outboard profile of a simple wooden tug/workboat. This little drawing lacks any identifiers other than "Robert Allan, NA, Vancouver" and

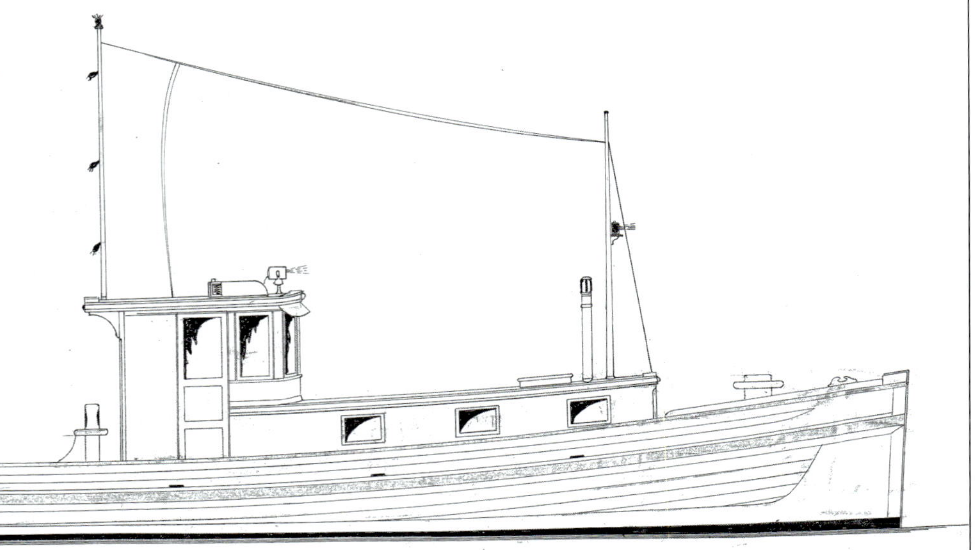

Outboard Profile of *Weaver Lake*, the very first tugboat design in the Robert Allan archives.

Below: Inboard Profile and Deck Plan of the *Weaver Lake*.

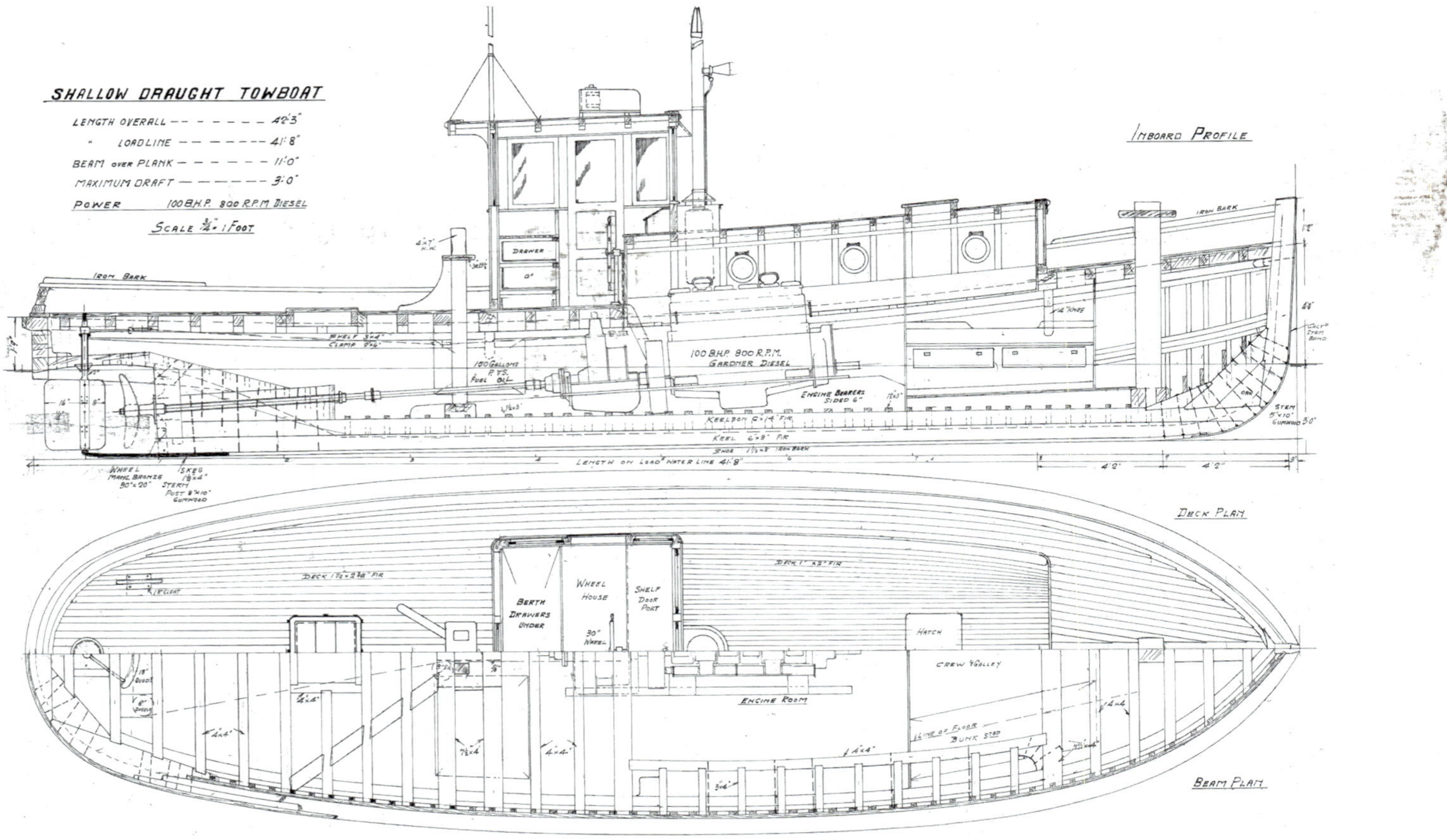

has no date. However, some further digging uncovered a couple of other drawings for the same boat, and some library research identified this as the river tug *Weaver Lake*.

Robert Allan's Tug Design #1, *Weaver Lake*, ca. 1934.

Although this design predates the formal incorporation of the company as Robert Allan Ltd, it represents the first tugboat designed by "Robert Allan, B.Sc." in his venture into private practice as a consulting naval architect. The *Weaver Lake* was designed in 1934 and built that same year by Benson Brothers Shipbuilding Co. Ltd. Of local wood construction, the tug was a simple 42-by-11-foot shallow-draft river towboat, named for Weaver Lake, west of Harrison Lake. Power was supplied by a single 102 bhp Gardner diesel engine.

Historical registry records indicate that in 1934 she was owned by Sparky New's Coastal Towing Co. Ltd of Vancouver. In 1937 she was owned by Ed Merchant of Mission, BC. From 1949 to 1959 she was owned by the Mission Towing Co. Ltd. In 1959 she sadly was sunk at Andy's Bay on Gambier Island, crushed between a log boom and a Davis raft.

Only a few more tug designs appear in the company record between 1934 and 1959, including the following:

1937	*Hyak*	52' × 15'	wooden tug for Coastal Towing Ltd
1944	*Green Light*	32' × 9' 6"	(unknown)
1946	*Sandy Jane*	75' × 23'	shallow-draft steel river tug (see Ch. 14)
1950	*Black Bear*	45' × 15'	steel tug for Black Ball Towing
1954	*Larbro*	44' × 14'	wooden tug for Bert Larsen
1955	*Tugger Yorke*	38' 9" × 13'	wooden tug for F.M. Yorke & Sons
1955	(unknown)	42' × 14'	steel tug for Dolmage Towing
1956	*Iris G*	48' × 15'	steel S/D tug for Kootenay Water Transport
1956	*Lake Monarch*	39' × 12' 6"	steel tug for Pioneer Towing
1957	*Atha*	75' × 20'	shallow-draft steel tug/snagboat for DPW, Canada (see Ch. 14)
1957	*Melinda Jane*	56' × 17'	steel shallow-draft tug for Kootenay Water Transport

Of the list above, one of the most interesting is the 45-foot *Black Bear*, built in 1950 for Gulf of Georgia Towing Ltd by Victoria Machinery Depot (VMD). No actual photos of this tug can be found showing her original configuration with a hydraulically elevating wheelhouse, but she is well shown in an advertisement from VMD, heralding the unique fact that she was built in steel!

Tugboat *Hyak*, ca. 1938.

Left: Tug *Green Light* working a log boom; **below:** *Tugger Yorke* in Vancouver harbour, ca. 1955. (R.F. "Bob" Allan at the aft window.)

Bottom right: The *Melinda Jane*, built in 1957 for Ivan Horie for service on Kootenay Lake, was brought out to the coast in 1979 and entered coastal service as *Storm Spray*; **bottom left:** A 1950 advertisement from *Harbour and Shipping* magazine showing the new barge-pushing tug *Black Bear*.

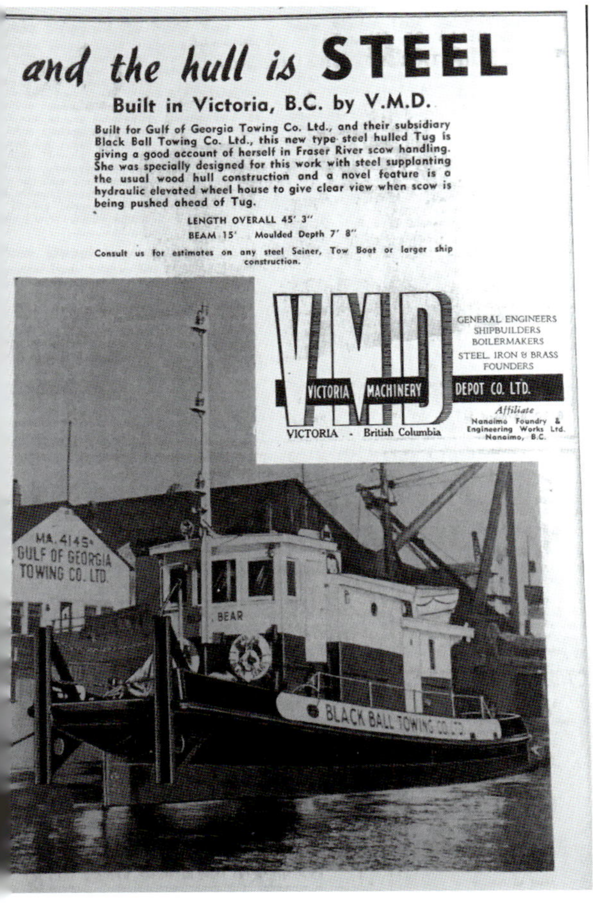

Rebuilding the Coastal Tug Fleet

With a few exceptions, prior to 1959 tugs in BC were built of wood and in modern terms were quite modestly powered, with engines of 300–400 hp being typical. A fairly extensive fleet of wooden steam-powered tugs was built before 1930, and they served the coast well for decades. Today the sole survivor of that era that has not been converted to a private yacht or charter vessel is the steam tug *Master,* designed and built by the famed Arthur Moscrop in his False Creek shipyard in 1922. A dedicated team of volunteers labour tirelessly to care for this lovely old boat and hope to have it designated as a BC/Canadian heritage vessel. She is currently the *only* wooden-hulled steam-powered tug afloat in the world and deserves preservation.

By 1959 the tide in the Robert Allan office had definitely turned from the fishboats prevalent in the '40s and early '50s towards tugboat designs. That began with the ocean-going tug *Ocean Master*, designed for Great-West Towing and Salvage Ltd of Vancouver. This 126-by-31-foot 1,850 bhp steel tug was a very large tug in BC for its time, and the owners elected to have experienced Dutch shipbuilders N.V. Scheepswerf Gebr. van der Werf construct it. That shipyard is still in business and is now a well-regarded luxury yacht builder. That project also was the door opener for Bob Allan's international exposure and gave him a taste of the world of opportunity beyond BC's borders, from which he never looked back.

The elegant SS *Master*, sole survivor of BC's steam tug era.

Ocean Master, the first Allan-designed tug to be built overseas.

The design package for the *Ocean Master* consisted of a comprehensive set of drawings and specifications. Notably, it is the first instance when a set of detailed stability calculations appear formally in the design documentation, presumably for formal approval by class or Canada Steamship Inspection (as Transport Canada's marine branch was known at the time), as opposed to more informally in design notebooks, which was the norm previously.

In 1960 the 92-foot 6-inch steel tug *Lorne Yorke* was designed for F.M. Yorke & Sons. She was billed in the local marine press as "the first modern twin-screw steel tug in BC" when delivered from Star Shipyard (Mercer's) Ltd the following year. Twin-screw vessels of any type were rare at this time, but as powers increased and the demand for better manoeuvrability arose, twin-screw tugs gradually entered the market.

In a study done in our office in 2009 to establish the best hull form and stability characteristics of coastal tugs (based on operating crew feedback), and as reported in a paper on this topic by Allan and Papuc to the 2010 International Tug Conference, this vintage vessel (then operating as the *Seaspan Lorne*) ranked extremely high among BC coastal tugs for its sea-keeping and comfort, even after 50 years of hard service.

Above: *The Lorne Yorke*, cited in the local press as "the first modern twin-screw steel tug in BC."

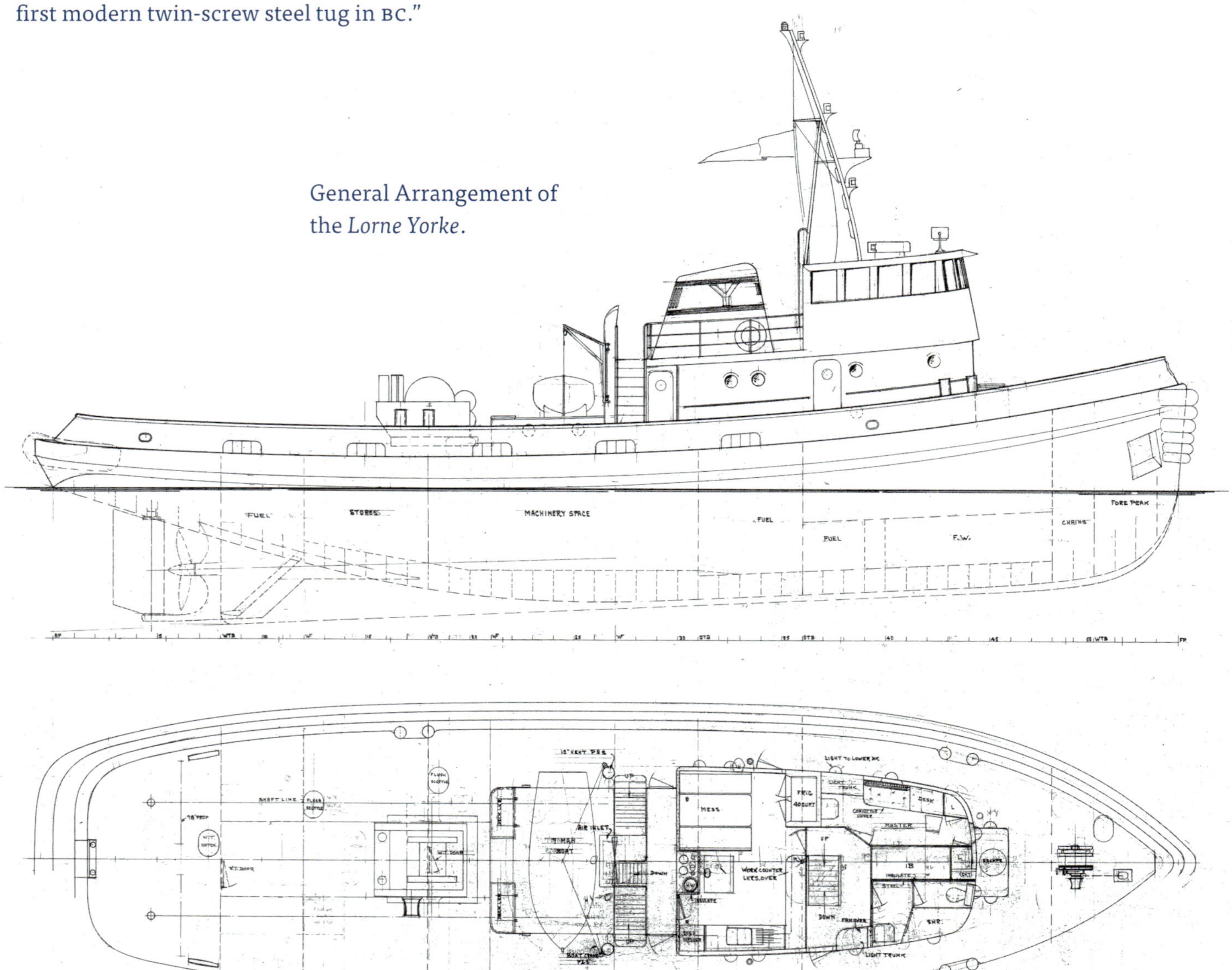

General Arrangement of the *Lorne Yorke*.

The "Hot-Rods"

From 1961 through 1964 a series of similar, nominally 65-foot single-screw steel tugs were built to Allan designs with 600 to 800 hp. This power-to-size ratio was a huge change from the earlier generation of larger wooden tugs and really represented a sea change in tug design and construction on the coast. Crews who had been used to working on bigger (and quieter!) wooden tugs with modest power did not take kindly to these new boats, which became known on the coast as the "Hot-Rods." It has proven difficult to determine exactly which boats earned this epithet, but the following certainly fit the size and power range. The series began with the *Jacques Cartier*, built for Deeks McBride in 1961.

1961: *Gulf Dianne* and *Gulf Muriel*, 64 feet 2 inches by 19 feet, 765 bhp, for Gulf of Georgia Towing.

1962: *Harmac Spruce* and *Harmac Pine*, 64 feet 7 inches by 19 feet 2 inches, 750 bhp open-screw tugs for Kingcome Navigation. These were the first of four tugs which revolutionized tug propulsion design on the BC coast (about which more follows).

1963: *Rosario Straits, Malaspina Straits, Haro Straits* and *Georgia Straits*, all 65 feet 3 inches by 19 feet 7 inches, but with a range of powers between 600 and 800 bhp, built for Straits Towing Ltd.

1963: *Harmac Cedar and Harmac Fir*, the second pair of virtually identical 750 bhp tugs for Kingcome Navigation, but these two were fitted with Kort nozzles in their original configuration.

These boats were built before a lot was known about the resilient mounting of main engines and other noise-mitigating measures common today and, as a consequence, were indeed quite noisy to live aboard. The industry was changing rapidly, and crew unions were finding their voice. These boats, welcomed by owners for their significant efficiency towing logs and barges in comparison to their predecessors, nevertheless became a target for much abuse from tug crews.

The *Harmac Pine* was one of a series of four tugs which revolutionized tug propulsion design on this coast.

Built in 1961, the *Jacques Cartier* was the first of the new generation of steel "Hot-Rods."

General Arrangement for the 600 bhp *Jacques Cartier*, 65 feet by 18 feet 6 inches, designed for Deeks-McBride Ltd to tow the clinker barge *St. Malo*.

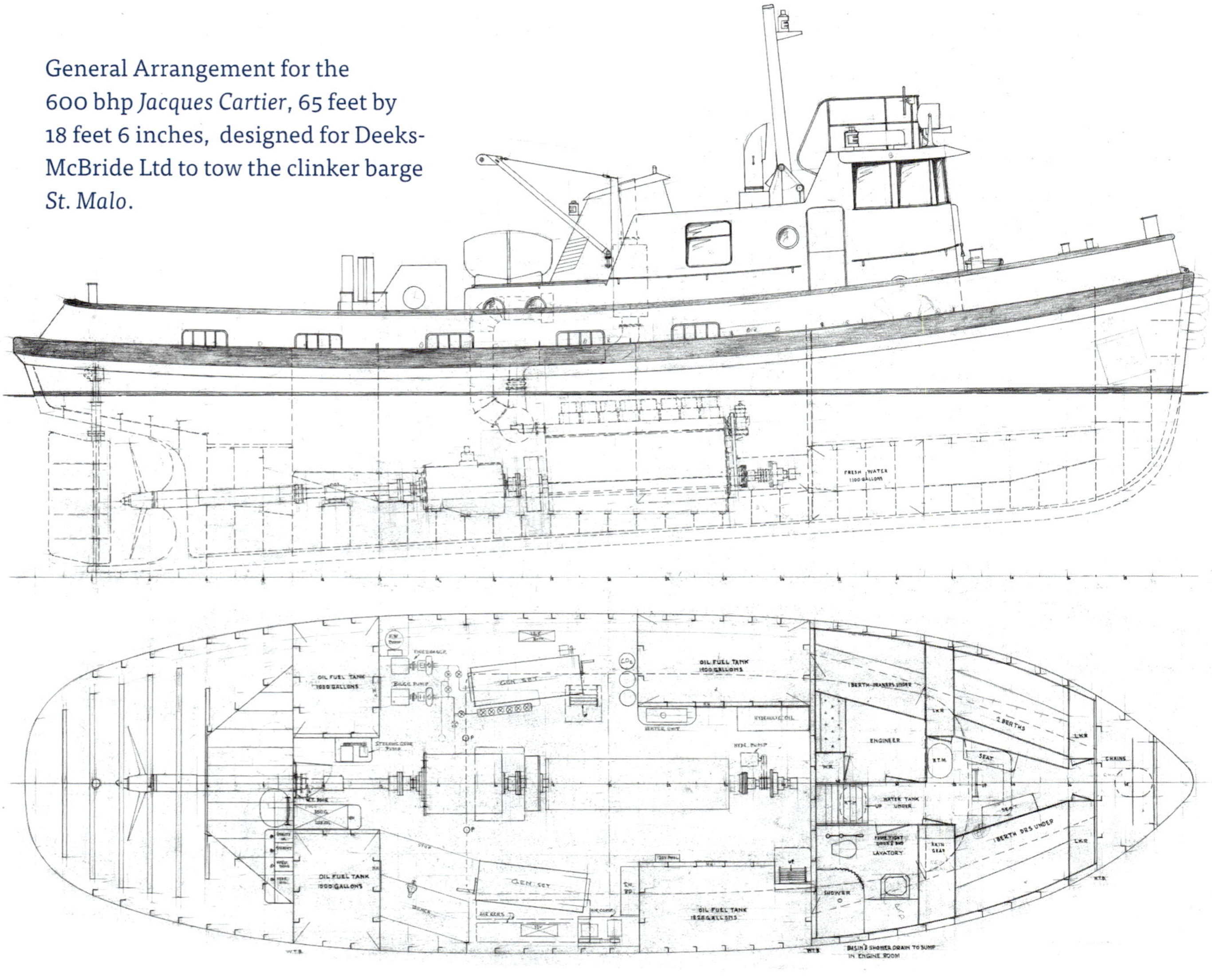

Steel vs. Wood

In the summer of 1967, I worked as a "day-man" (considered subhuman, less than a deckhand) aboard the venerable old wooden tug *Sea Lion*. I recall clearly the crew talking about their experiences recently viewing a steel tug under construction, where the steel plates were being pulled into place for welding using chain blocks, a very common technique. The crew were convinced that if that tug hit a log the welds would crack and the steel plate would spring back into its original flat plate shape! No arguing about material properties would convince them otherwise, and so stories about "death traps" circulated, a lot like the misinformation common on the internet today when people often don't understand and don't want to listen.

—R.G. Allan

Above: *Malaspina Straits*, one of four sister tugs for Straits Towing Ltd.

Following pages: *Harmac Fir* (shown), and her sister *Harmac Cedar* were the first tugs built in BC to use Kort nozzles.

HARMAC FIR

K

Kort Nozzles in BC Waters

No device has had a greater positive impact on the towing industry in BC than the Kort nozzle and its locally designed derivatives. Invented and patented by German engineer Ludwig Kort in 1934, this device consists of a foil-shaped ring surrounding the propeller, with limited tip clearance. The nozzle can be fixed or steering. The nozzle accelerates and concentrates the water flow over the propeller, increasing overall efficiency, particularly at lower speeds. Typically, at bollard condition (zero speed) a nozzle system will deliver 20–25 percent more thrust than an open propeller of the same diameter, with the same input power.

In about 1961 Bob Allan negotiated rights with the Kort Propulsion Co. to become the only licensed designer of Kort nozzles in North America. At the same time, John Manly Ltd had the license to be the sole builder of Kort nozzles, certainly within BC and maybe for a broader geographic base. Over the next decade or so, Robert Allan Ltd designed (and Manly built) more than 100 nozzle installations for both new tugs and for refits. Ultimately, the license agreement with Kort was allowed to lapse, perhaps in timing with the nozzle patent expiry, but also because Bob Allan had ideas for improving nozzle performance which the parent company would not entertain, and they agreed to terminate the agreement. Robert Allan Ltd has continued to successfully design its own nozzles ever since.

The aforementioned Harmac tugboats were the first vessels built in BC to use Kort nozzles, all built by John Manly Ltd between December 1963 and February 1965. The first two tugs (*Harmac Pine* and *Harmac Spruce*) were built as conventional open-screw single-screw tugs. The second pair (*Harmac Fir* and *Harmac Cedar*) were built with steerable nozzles. The difference in towing performance was immediately noticeable, and not many months after the nozzle tugs went into service, the open-screw boats were converted to nozzles. The whole project and details of the performance results are described in a 1966 paper by Bob Allan to the Society of Naval Architects and Marine Engineers (SNAME) Pacific Northwest Section, titled "Considerations on the Use of Kort Nozzles in BC Tugs."

It is notable that since this successful experiment, virtually every tug built in BC has had nozzles. Steering nozzles, initially quite popular, were eventually replaced by fixed nozzles with multiple high-aspect ratio rudders, as the latter arrangement is much lighter and less expensive than the steerable device. This also reflected the parallel, gradual trend to twin-screw tugs rather than single-screw.

Left: An early Kort steering nozzle. The early propellers used with nozzles were not dissimilar to conventional open props, but with some squaring off at the tips. Later propellers, known as Kaplan style, were developed with much more "fan-shaped" blades to maximize area at the outer radius for improved performance.

Above: A later twin-screw nozzle installation (likely one of the Cates 60-foot tugs) showing the typical twin-rudder/fixed-nozzle configuration that became the standard on the BC coast. Note the much wider tips on these propellers.

General Arrangement for the initial open-screw Kingcome tugs *Harmac Spruce* and *Harmac Pine*.

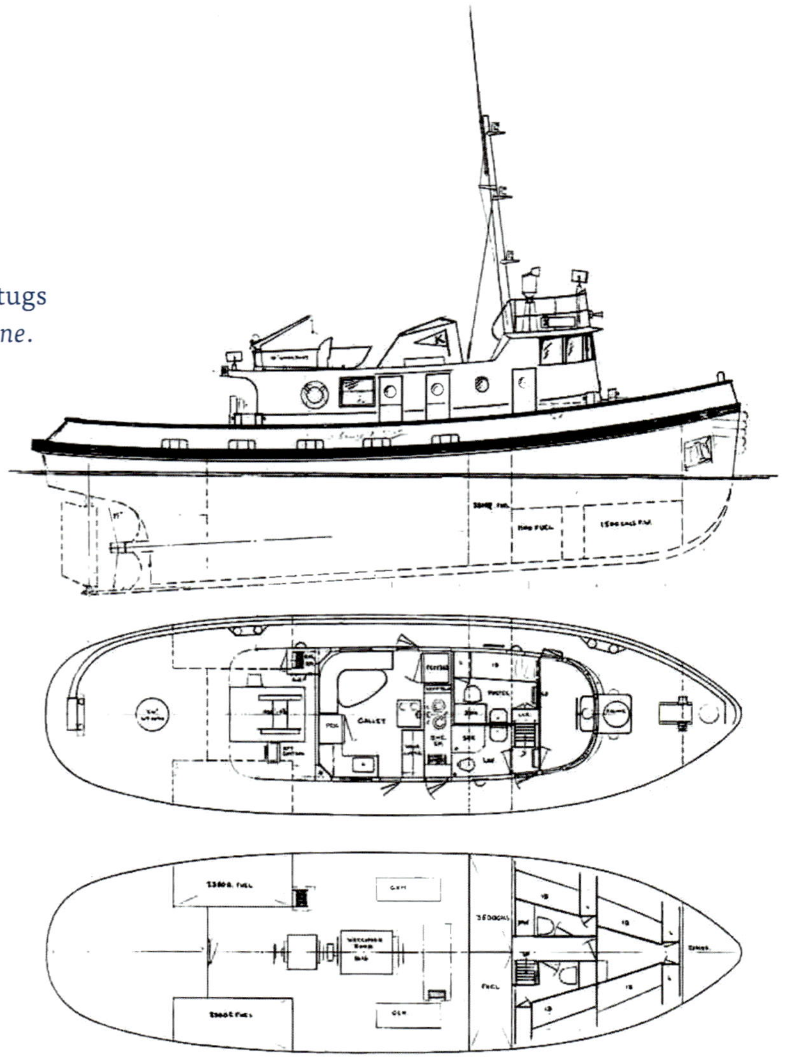

Below: General Arrangement of the nozzle versions of Kingcome's 65-foot tugs *Harmac Fir* and *Harmac Cedar*.

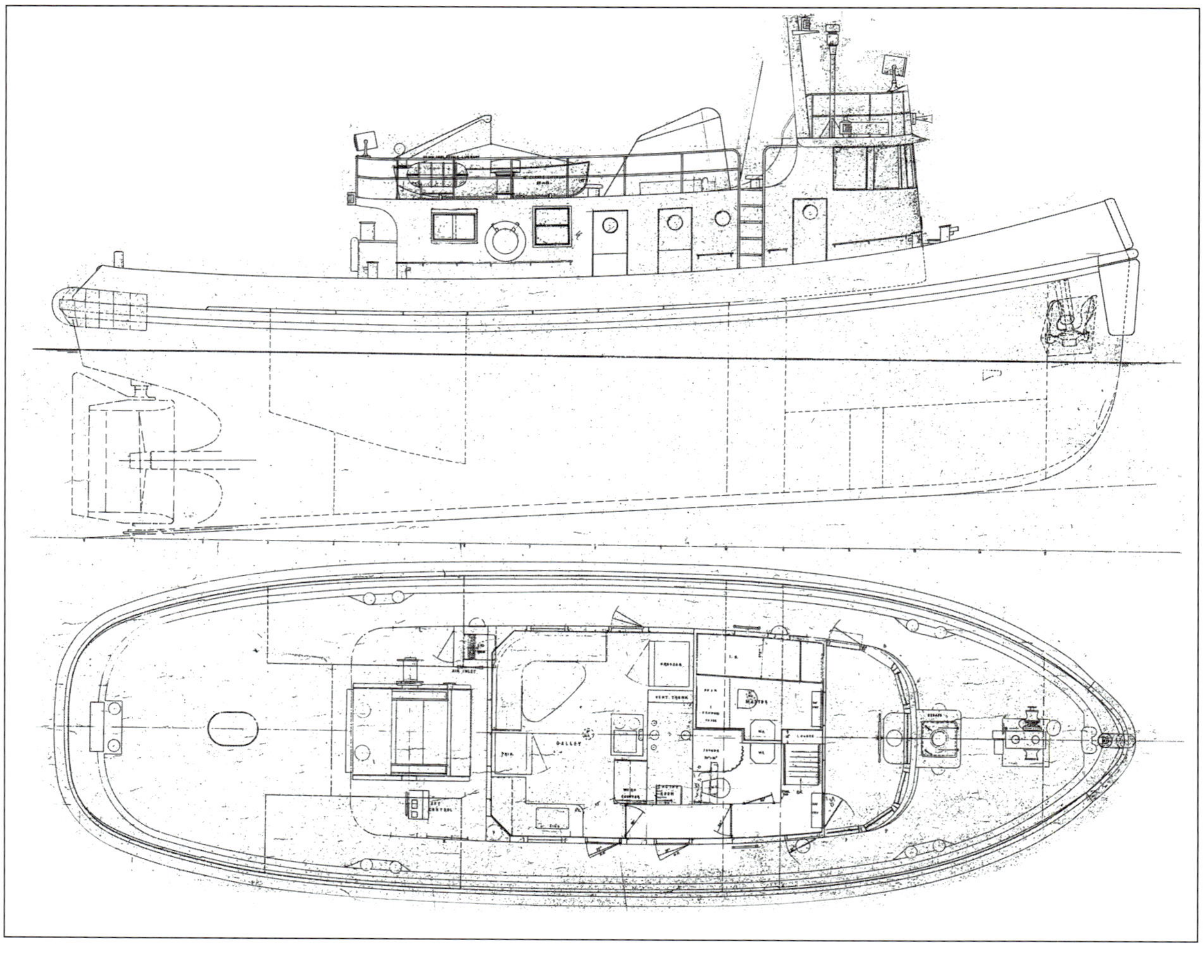

Construction Industry Tugs

Much of the growth in the coastal industry in the 1960s was driven by significant expansion in the construction industry in BC, spurred on by fleet developments among major cement industry players such as Evans Coleman and Evans (Evco), LaFarge Cement and Deeks-McBride, a company later absorbed into the LaFarge group of companies. The innovations in barge design for these companies during this era are described in Chapter 13; however, a new fleet of steel tugs was built in the early 1960s to tow them, replacing older and lower-powered wooden tonnage.

Tugs of this period for LaFarge include the previously mentioned *Jacques Cartier*, the 73-foot 950 hp *Captain Cook* (1964) and the 87-foot 1,500 hp *Jose Narvaez* (1968).

Captain Cook, a coastal workhorse.

Below: The handsome and purposeful *Jose Narvaez* photographed in 2021, impressively belying her 53-year age!

Evans Coleman and Evans was the major player in this business at the time, and certainly the largest investor in new equipment. In addition to dozens of new barges of various types, they built a string of modern tugs, which included a trio of 68-foot 760 hp sisters, all built at Star Shipyard (Mercer's) Ltd: *Evco Wave* (1962), *Evco Breeze* (1963) and the *Evco Spray* (1965). These three tugs are still in service today, but under different owners.

In 1968 Evco built another pair of slightly smaller tugs, the 52-foot 800 hp *Evco Crest*, used primarily for work in the river, and the 60-foot 800 hp *Evco Foam*. The entire fleet of Evco tugs and barges was designed by Bob Allan.

The *Evco Crest* was modelled on a number of similar local "yarding" tugs. That term describes a class of tugs intended for short-range work, typically moving barges and often log booms into position for longer tows by bigger tugs, and sometimes doing simple ship-assist work. These boats had their origins with the slightly larger *La Fille* (56' × 18'), built for Vancouver Tug in 1962 (and her sister *Le Prince* in 1964), and included the *Bering Straits* (53' 3" × 18' 4") for Straits Towing in 1964; the twin-screw 670 bhp *Seymour Crown* (53' × 18' 8"), built in 1966 for Crown Zellerbach; and later the twin-screw 940 bhp *C.T. Titan* (53' × 19' 3"), built in 1979 for Chemainus Towing Ltd.

Evco Breeze, one of a trio of long-serving 68-foot 760 bhp barge-towing workhorses, shown in the Fraser River.

Evco Crest, a 52-foot yarding tug.

Below: *Evco Foam*, a 60-foot utility tug.

La Fille, built for Vancouver Tugboat Co. in 1962 by John Manly Ltd. This 567 bhp single-screw tug became the prototype for a series of similar-sized yarding tugs used up and down the BC coast.

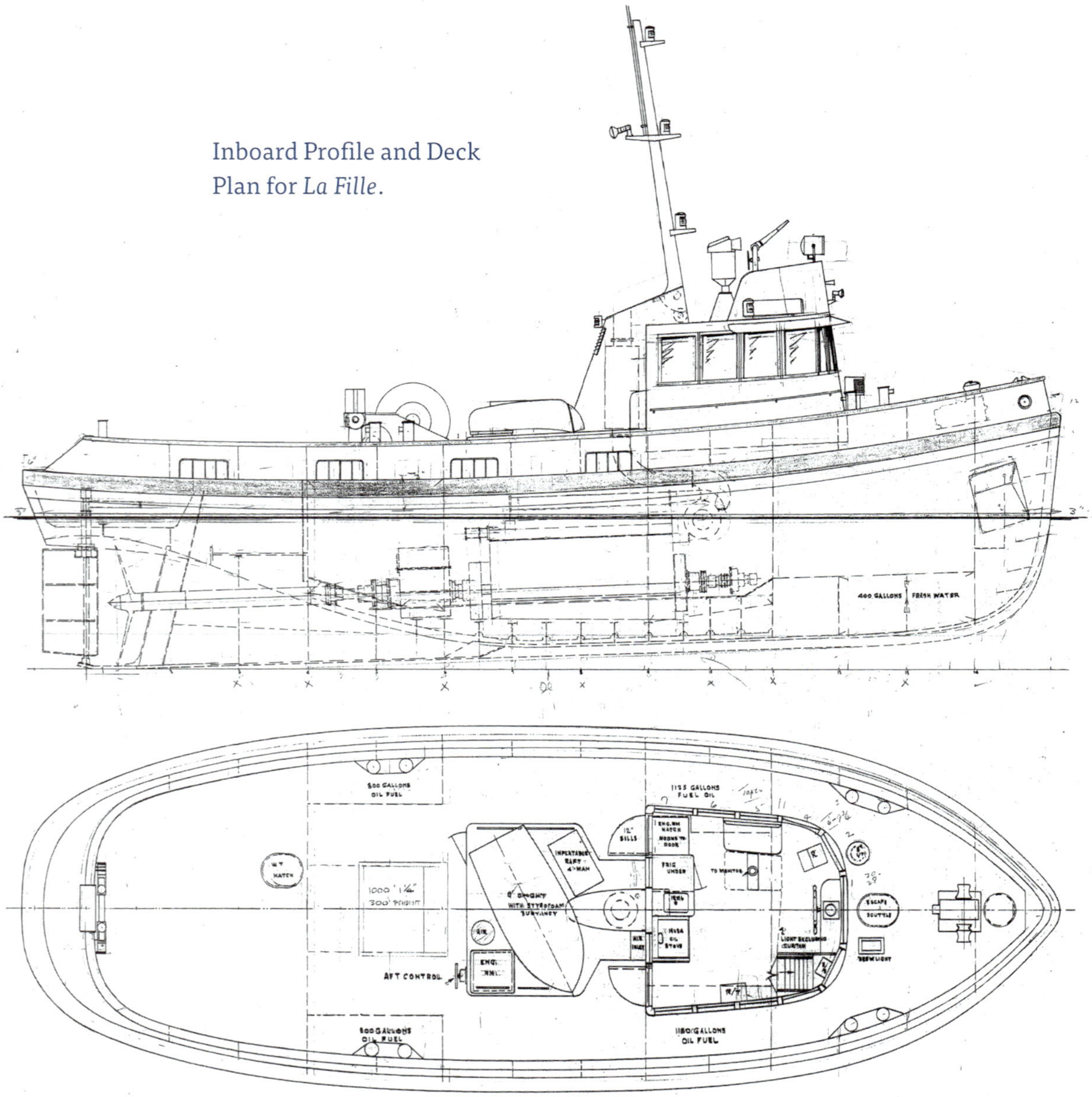

Inboard Profile and Deck Plan for *La Fille*.

Below: The 590 bhp single-screw *Bering Straits*, built by BC Marine Shipbuilders in 1964 for Straits Towing; **right:** Launch of the *Seymour Crown* at John Manly Ltd in 1966. The exceptional visibility from the wheelhouse of this twin-screw tug was a notable design feature.

Bottom: *C.T. Titan*, a 940 bhp twin-screw yarding tug for Chemainus Towing, built in 1979 at Allied Shipbuilders to a Rob Allan design.

Towing Bigger and Better

The longer and larger barge tows on the coast required tugs with better sea-keeping, more power and more fuel capacity than the emerging fleet of smaller tugs in the 65–70-foot range. Robert Allan Ltd was the consultant of choice of all the major tug companies at the time, and that is reflected in the diverse group of tugs which emerged in the mid-late '60s for these more heavy-duty tows. The following images show many of this group of hard-working tugs, a good number of which are still in service on the BC coast.

Several were equipped with controllable-pitch propellers (CPPs), quite unusual at the time, which gave them a good turn of speed when running light and enabled more efficient operation with a range of barge loads. Notably, all were still single-screw boats, for maximum towing efficiency, with owners favouring a single big, deeply immersed propeller. Twin-screw tugs became more common in the 1970s.

Neva Straits, designed in 1961 for Straits Towing and built the next year by Allied Shipbuilders.

Below: *La Reine* (later *Seaspan Queen*), a 95-foot 1,300 hp single-screw coastal tug designed in 1963 for Vancouver Tugboat and built by Star Shipyard.

Island Chief and *Island Master* on trials. Identical twins, they were designed in 1965, 87 feet long by 23-foot 6-inch beam, with a single 1,200 hp diesel driving a CP propeller in a steering nozzle. They were built in 1966 by McKay-Cormack for Island Tug & Barge Ltd.

Designed in 1965, *Le Mars* (later *Seaspan Pacer*) is a 95-foot 1,800 hp coastal tug built by Star Shipyard for Vancouver Tugboat.

Below: *Island Champion* (later *Seaspan Champion*) in her original Island Tug & Barge colours. Designed in 1969, she was single-screw, 87 feet long, with 1,450 hp and a controllable-pitch propeller. She was built by A.C. Benson Shipyard for ITB and was a virtual twin to the *Jose Narvaez*.

The Big Guns

The year 1964 was a banner one for Robert Allan Ltd, during which three of the largest and most powerful tugs on the BC coast were designed, among numerous other large projects. These tugs were all designed for long-distance barge towing, for either the emerging fleet of large log barges or, in one case, for a major ocean-going tug-barge transportation system.

The *Haida Brave* was designed for Kingcome Navigation, the marine branch of the MacMillan-Bloedel Co. Built by Halifax Shipyards on the East Coast of Canada, this handsome twin-screw tug, 140 feet in length by 34 feet beam, was powered by a pair of Werkspoor TAMBS 398 diesel engines, each rated 1,735 bhp at 250 rpm, driving 9-foot-diameter open propellers. This tug was the motive power for the novel ocean-going bulk transportation system with three 7,000 TDwt newsprint barges, described in Chapter 13. The *Haida Brave* worked continuously on the open Pacific, transiting back and forth between Port Alberni and Los Angeles/Long Beach, California, for more than 20 years, logging in excess of 8,000 hours annually—more than 90 percent utilization—stopping only for annual dockings and fuelling. This "swap-and-drop" system was a perfect marine transportation model: one barge loading, one unloading and one in transit, ensuring that a barge was always ready to go when the tug arrived. In 1978 the contract was given to an outside contractor instead of maintaining this in-house operation. The *Haida Brave* was then sold to Rivtow Straits and renamed *Commodore Straits*.

In that same year Straits Towing commissioned a design for another big twin-screw tug, launched from Yarrows shipyard as the *Gibraltar Straits*. Not dissimilar in configuration to the *Haida Brave*, she measured

The handsome 3,700 hp *Haida Brave*, a trendsetting design for ocean-going tug-and-barge transportation systems.

140 feet by 34 feet, with 3,840 bhp. She continued to serve Straits Towing and subsequently the Rivtow Straits organization until retired from service in BC waters in 1990.

The third big twin-screw BC tug from the office of Robert Allan in 1964 was the *Harold A. Jones*. She was designed for Vancouver Tugboat Co. and built by Star Shipyard (Mercer's) Ltd. Slightly smaller than the previous big boats, she measured 131 feet 3 inches long with a 32-foot beam and was fitted with two 1,750 hp Stork-Werkspoor diesels. With the merger of Vancouver Tug, Island Tug & Barge and various other local towing companies into the Seaspan organization in the early '70s, she became the *Seaspan Monarch*. Today she operates under the (new) Island Tug & Barge flag as the *Island Monarch*, modified to a pusher tug for their fuel-barge operations.

Soon afterward, two more big tugs were designed and built locally for barge towing on the BC coast. In 1966 the *Island King* was designed for Island Tug & Barge Ltd. This 3,600 hp single-screw tug measures 131 feet by 32 feet and features a single large steering nozzle. She was built by Star Shipyard (Mercer's) Ltd in 1968. Joining the Seaspan fleet in 1971, she became the *Seaspan King*.

Robert Allan Ltd was retained by Star Shipyard to develop the complete set of construction working drawings for the *Island King*, an exceedingly detailed and comprehensive package for pre-CAD days, which survives in our archives. The existence of that detailed production design package undoubtedly was an advantage for Star Shipyard in 1969, when it won the contract to build a virtual sister tug for Crown Zellerbach (Canada) Ltd, the *Hecate Crown*.

Like most of these big tugs, their principal duty was towing log barges from the North Coast down to mills in the Lower Mainland area of BC.

The *Hecate Crown* served Crown Zellerbach in this role until 1985, when that company ceased to operate on the BC coast, after which she was sold to Seaspan and became the *Seaspan Challenger*.

Harold A. Jones in her original Vancouver Tugboat Co. markings.

Gibraltar Straits, another big tug for the BC coast from a 1964 Allan design.

Below: *Island Monarch* (ex. *Harold A. Jones*) is currently set up for pushing various oil barges on the BC coast.

Island King (later *Seaspan King*).

Top: *Hecate Crown*, a 3,600 hp barge-towing tug.

Right: The *Hecate Crown* under construction in the rustic Star Shipyard, showing her fine moulded lines and that big steering nozzle.

Beating the Tonnage Rules

A Canadian steamship inspector once said in our presence that "our aim is to have the best marine regulations in the world." He didn't much care if anyone could build a boat within that regulatory framework, or if the end result would actually be a safe or sensible boat to operate; his concern was only that they had "good" regulations that wouldn't expose the individual inspectors to any potential liability for poor decisions. That mindset has unfortunately dominated the Canadian marine regulation/inspection world throughout the better part of my 50-year career and, most regrettably, is currently getting worse, to the utter exasperation of most of our Canadian tug-owning clients.

The regulators set seemingly arbitrary limits on the application of certain regulations that govern inspections, manning and equipment requirements. While many of these limits or boundaries coincide with international standards, there are often Canadian "enhancements" to them, reflecting the mindset of individuals like the one quoted above.

In the Canadian tugboat world, there exist regulatory boundaries that define the general size of tug, as defined by its Gross Registered Tonnage (GRT or GT), which is subject to various levels of inspection or approvals, and the associated manning qualifications. (GRT is a measure of the vessel's internal volume and is not in any way a measure of its mass or displacement.) The principal regulatory hurdles for tugboats are as follows:

5 GRT: Below which a tugboat is not subject to any inspection or approvals.

15 GRT: No mandatory Transport Canada plan approval for tugs of less than 15 GRT. Vessels greater than 15 GRT are subject to plan approval and, if between 15 GRT and 150 GRT, are subject to quadrennial inspections.

150 GRT: Above which a vessel is subject to plan approval and annual inspections.

500 GRT: Above which vessels are subject additionally to various international regulations, including SOLAS (if making international voyages), and a higher standard of manning qualifications.

Such regulations clearly provide huge incentives for owners to reduce their initial and operating costs by avoiding the nuisance, downtime and costs associated with crossing any of the GRT boundaries, thereby initiating regular inspections by government officials. Accordingly, in the 1960s and throughout the '70s, there was a big demand for small tugs which were designed to be the maximum physical size possible and yet still measure under 15 GRT. At the time, the tonnage admeasurement rules permitted deductions for various sizes of frames and floors, and some spaces were exempt from measurement, so there was considerable room for manipulation of the design to create the largest boat possible under this GRT limit.

The trend to building tugs to just squeak under the 15 GRT limit (typically about 14.9 GRT) actually began in the late 1940s with the construction of several small tugs at John Manly Ltd. These boats were typically about 36 feet by 12 feet and were modestly powered at around 150–240 bhp.

The first tug in our archives built to challenge the tonnage rules was the *Syringa*. This early boat, built by John Manly Ltd in 1962 for Columbia Cellulose (Celgar), was smaller than her later rivals in this category, at 38 feet by 12 feet 6 inches, but also notably larger than the majority of her predecessors.

Outboard Profile of the *Syringa* for Celgar, the earliest sub-15 GRT tug in the Robert Allan archives and an early Kort steering-nozzle application.

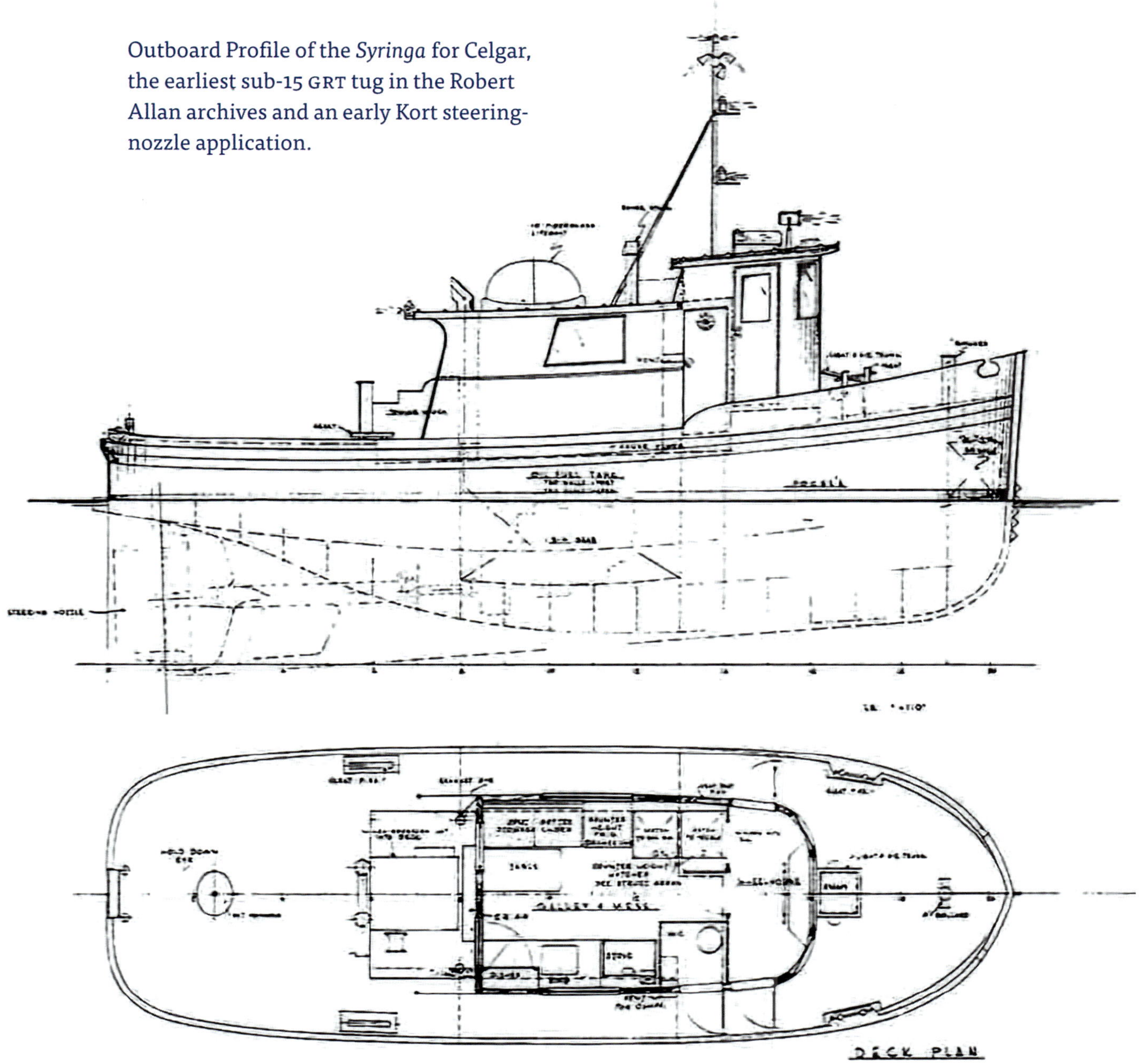

The first of the series of 41-foot tugs built as serious rule-beaters was the *Papco III*, designed in 1963 for the Pacific Pine Co. She was built by John Manly Ltd and measured 41 feet length overall by 14-foot beam, with a GRT of 14.91. She was single-screw with a steering nozzle and powered with 325 bhp.

The creation of this design stimulated a lot of interest from owners and builders alike. John Manly Ltd built many, and the ensuing series of tonnage-limit tugs also gave rise to Vito Trevisi's emerging shipbuilding enterprise, and a lengthy cooperation with our design office. Those two shipyards accounted for the majority of a fleet of at least 42 of these "41-footers." At least 22 of these were designed by Robert Allan Ltd, and almost certainly there were a few more.

The under-tonnage tugs built by John Manly Ltd were numerous. Manly was the early leader in the field of small tugs, including probably hundreds of boom-boats, especially throughout the 1960s. Allan-designed sub-15 GT tugs that were built there include *Syringa, Papco III, Nanaimo Yarder, Westminster Brave, H & R, Naskeena 7, Westminster Warrior, Renata II, Sea Cap III, Naskeena 4* and the *Westminster Chief.*

Sea Cap III of Valley Towing, still hard at work after 55 years.

Below: *Westminster Warrior*, a 41-footer built by John Manly Ltd for Westminster Tug Boat Ltd.

Papco III, first of the 41-foot sub-15 GRT tugs for the BC coast.

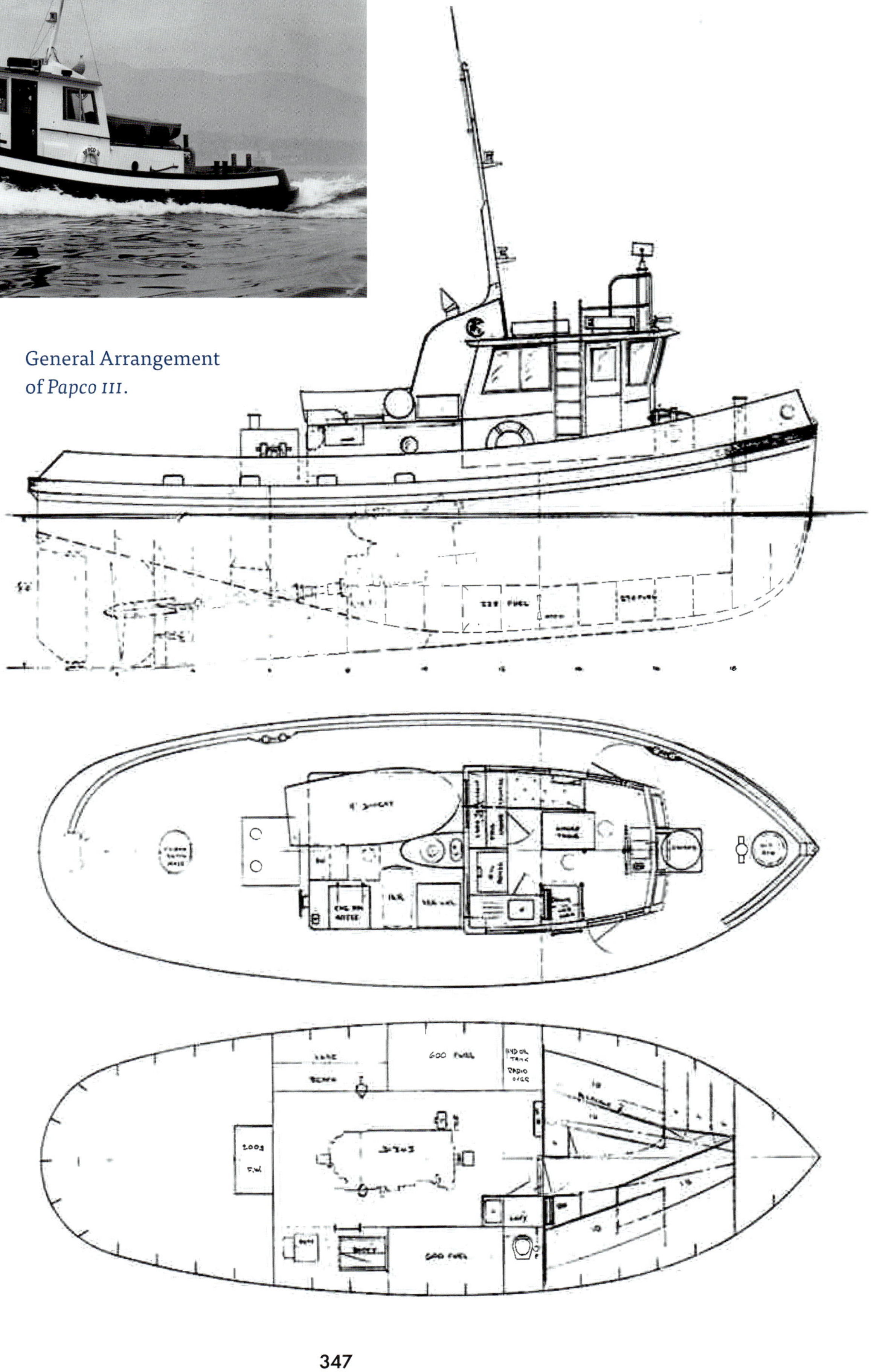

General Arrangement of *Papco III*.

Helga T., built by Vito for W.E. Thuveson.

Vito Steel Boat and Barge Ltd also built a whole series of these under-tonnage tugs between 1966 and 1970, taking a significant market share from Manly. These were quite similar, with varying powers; typically they measured 41 feet overall by 14-foot beam, and were single-screw with Kort steering nozzles. Allan-designed tugs recorded to Vito's credit include the *Debra K*, *Nanoose Yarder*, *Pont Grey Endeavour*, *Helga T.*, *Squamish Chief*, *Rivtow Rambler*, *Victory III*, *Regent*, *Squamish Warrior*, *Sea Tow* and the *Point Grey Providour*.

In addition to those two most prolific builders of this niche market of tugs, there were a few quite interesting self-builds to the Robert Allan sub-15 GRT tug design by small owner-operators. One of these, Dirk Kirkwood, built the tug *Ocean Iris* for his own purposes, a project described in the November 1963 issue of *Harbour and Shipping*: "The spanking new 41-foot steel tug *Ocean Iris*, now on charter to Texada Towing Co. Ltd of Vancouver is a rare example of initiative and enterprise. She was built by her owner, Dirk Kirkwood, on a vacant lot in North Surrey, with assistance from Carl Bruce, Cloverdale, on engineering; Fritz Molman, Whonnock, electrical work; Jack Hyland of le Clerc and Collins Shipyard, Coquitlam, woodwork; and Roy Stevens, Cloverdale, finishing...The tug, built to a design by naval architect Robert Allan, took about a year to complete, and is of the same basic design as the tugs *Papco III*, *Nanaimo Yarder* and *Westminster Brave*."

After this extended series of 41-foot tugs, the overall size of the sub-15 GRT tugs began to grow in the late 1970s. Another local practitioner exploited a loophole in the Tonnage Measurement Regulations by incorporating "false floors" into their designs, which dramatically reduced the measured tonnage depth of the tugs, thus allowing much larger lengths and beams in the boats while still measuring under 15 GRT. Transport Canada ought never to have allowed this subterfuge. The sheet-metal floors were all removed after the boats were in service and thus simply cheated the rules. This was a practice in which Robert Allan Ltd refused to engage, and hence we soon lost our share of this local small-tug market. It was a time, however, when our focus had shifted to the North and to larger vessels in any case, and so our business did not suffer.

Right: *Nanoose Yarder*, built by Vito in 1966 for Chemainus Towing Ltd; **below:** The *Ocean Iris*, still at work in 2021; **bottom:** *Squamish Chief*, another under-tonnage 41-footer built by Vito.

Harbour Tugs for the Navy

Robert Allan Ltd has never had much involvement with naval vessels. Warships are not in our area of interest, but we have designed a few naval auxiliary vessels, such as the aforementioned fireboats (Chapter 10). Our expertise in small tug design, however, won us a contract in 1971 for a series of small tugs for both of Canada's major naval seaports, in Esquimalt and Halifax.

These "Ville" class tugs were not dissimilar to the under-tonnage tugs but, as the Navy doesn't have to comply with Transport Canada regulations, they opted for slightly larger boats. These tugs measured 45 feet 4 inches long by 14 feet 6 inches beam, were single-screw, 335 bhp, and were fitted with a single Kort steering nozzle. Five of these boats were built. Two, the *Lawrenceville* and *Parksville*, were built by Vito Steel Boat & Barge Ltd in Delta, BC, with the hulls' steel fabrication sub-contracted to Hike Metal Products of Wheatley Ontario. The remaining three, *Listerville, Merrickville* and *Marysville*, were built at Georgetown Shipyard Ltd on Prince Edward Island. These tidy little tugs compile very few operating hours and thus continue in service to this day.

Fifty years later, we were successful once again in designing a new series of tugs for the Navy, after a long and rather painstaking procurement process. This was dubbed the Naval Large Tug Project. These new vessels, under construction in Quebec at time of writing, are based on the highly successful RAmparts 2400 series design. Delivery of the four new 60-tonne bollard pull tugs is scheduled to begin in the fall of 2022.

A New Generation of Tugs

As a consequence of a series of sinkings and related fatalities in the BC towing industry in the late 1960s (none of which involved Robert Allan Ltd-designed tugs), Transport Canada invoked a new set of regulations in 1970 governing the design and construction of tugs. The major changes included:

- Canada Shipping Act/Crew Accommodation Regulations cite that "In every new ship 22.9 m in length or over (a) sleeping rooms and galleys shall be wholly situated above the deepest waterline..." *but* the Crew Accommodation Standards inherent in the Union Agreements state that in vessels between 50 feet (15.2 m) and 90 feet (27.4 m) in length "no sleeping rooms shall be below the main deck," and "all accommodation space to be situated above the deepest waterline." The Agreement therefore effectively overrides the Regulation.
- Down-flooding of compartments aft of the engine room should not immerse a margin line 75 mm below the main deck. (This requirement arose because it was perceived that tugs could be swamped by seas from astern *if* a manhole or similar were left open.)
- Watertight doors to be remotely actuated from above deck.

As virtually every tug in BC at the time was "flush-decked" and had at least some crew accommodation in the forecastle, meeting these new requirements demanded a totally new approach to tug layout, and a new generation of raised fo'c'sle tugs was spawned.

The first tug ever built to these new regulations was the *G.D. Hemmingsen*, designed by Robert Allan Ltd in 1971 for Chemainus Towing Ltd. Named for the founder of

YTS 590 Lawrenceville, built by Vito for the Canadian Navy.

Profile of the new Canadian Naval Large Tug design, for delivery in 2022/23.

Chemainus Towing and built by Vito Steel Boat and Barge Ltd, she was a compact but powerful tug designed for barge and log towing and a limited amount of ship-assist, as per her owner's business profile. The *Hemmingsen* measured 68 feet 9 inches by 22 feet 6 inches and was powered by two 490 hp CAT 346 diesels. She had two fixed nozzles with twin high-aspect-ratio rudders behind each, a configuration which soon became the norm in coastal tugs and which persists to this day.

A few years later, in 1977, a virtual sister to the *Hemmingsen* was built and christened as the *Hamilton Baillie*, recognizing the other founder of Chemainus Towing. The *Baillie*, also built by Vito, measured 69 feet by 22 feet 6 inches and had two 490 hp CAT diesels. The design reflected only a few minor improvements over the prototype design.

The new rules severely influenced a new generation of tugboats whose designs were evolving to take over the small to medium barge towing duties on the coast from the older generation of wooden tugs. Flush-deck tugs built just a few years before, such as the *Evco Wave* series and larger boats like *Jose Narvaez* and *Island Champion*, could

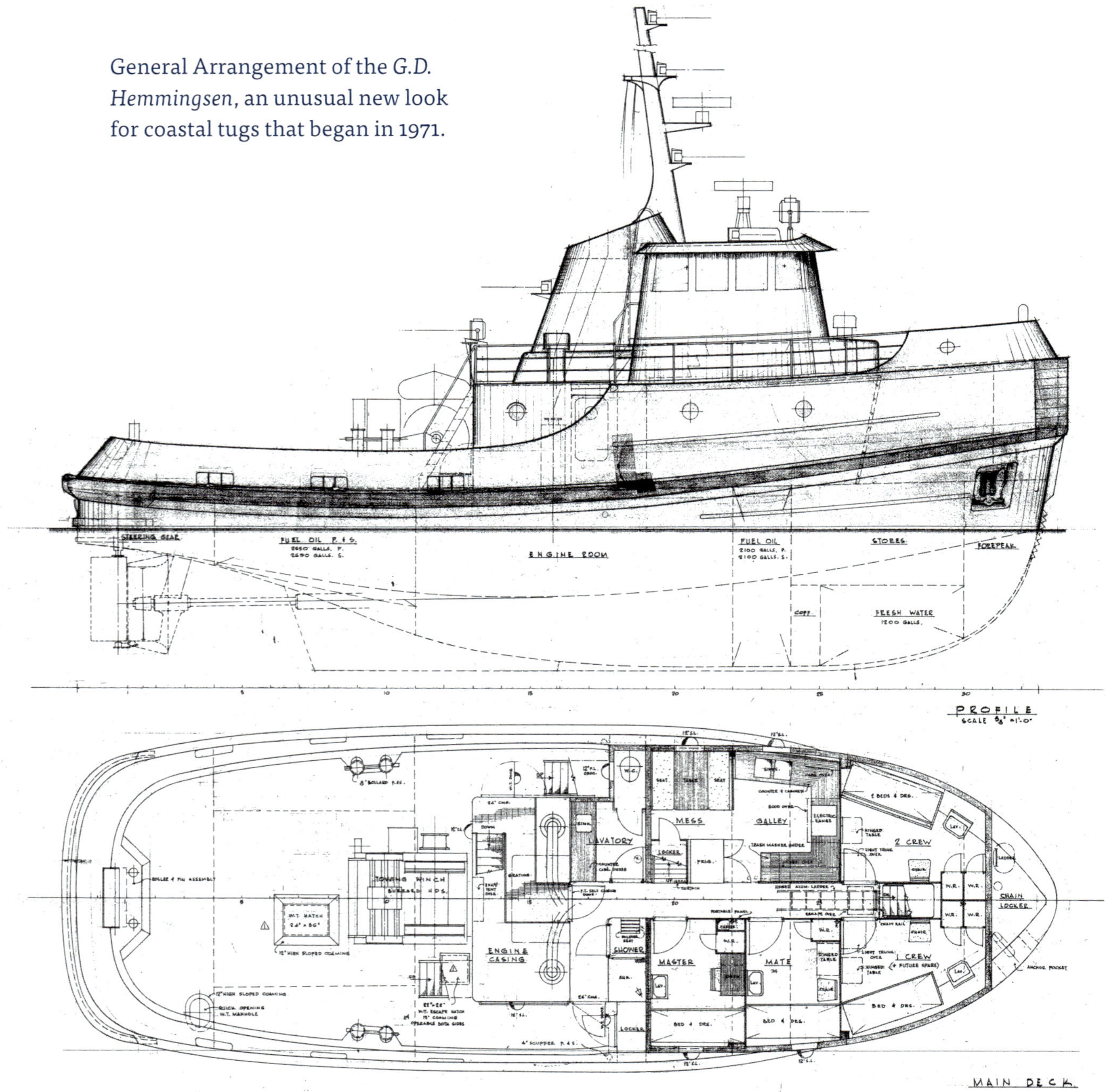

General Arrangement of the *G.D. Hemmingsen*, an unusual new look for coastal tugs that began in 1971.

Above: *G.D. Hemmingsen*, the first of a new class of BC tugboats to address a new set of regulations.

Left: *Hamilton Baillie*, built in 1977, a virtual sister to the *G.D. Hemmingsen*.

no longer be built under the new regulations, in spite of their excellent stability and performance. Owners also recognized that there were considerable cost savings by keeping new boats under 150 GRT, above which vessels were subject to annual rather than quadrennial inspections. Again, much lower costs for downtime and inspection-related repairs or modifications could be achieved by staying below a somewhat arbitrary GRT threshold.

In 1973 Gulf of Georgia Towing Ltd retained Robert Allan Ltd to design a series of three identical sub-150 GRT tugs. The resulting vessels, *Gulf Julia*, *Gulf Irene* and *Gulf Sally*, measured 77 feet 9 inches length overall by 24-foot beam and were powered by a pair of GM 12V-149 diesels totalling 1,450 bhp. These boats also had twin fixed nozzles with multiple rudders. Not long after entering service with Gulf of Georgia Towing, these boats became part of the new Seaspan fleet and were renamed the *Seaspan Defender*, *Seaspan Guardian* and *Seaspan Protector* respectively.

At the same time, Chemainus Towing ordered yet another new tug to virtually the same design as the Gulf of Georgia Towing tugs, named the *C.T. Champion*. Crown Zellerbach ordered the slightly shorter *Comox Crown*, also in 1973. Very similar in style and configuration, she was only 73 feet long and had 1,350 bhp. The last and most powerful of this series was the *Cindy Mozel*, built that same year for Capt. John Mozel. The *Cindy* was also 77 feet by 24 feet but had a pair of Detroit Diesel 16V-149 diesels producing a total of 1,800 bhp. All the tugs listed here were built by Vito Steel Boat & Barge Ltd in Delta, BC, and marked the closest thing to series production of identical commercial boats seen in BC to that time. The "Hot-Rods" were probably greater in numbers, but from a variety of builders. Shown are some of the fleet of sub-150-GRT tugs built by Vito Steel Boat & Barge Ltd in 1974 and 1975 to Robert Allan Ltd designs.

C.T. Champion for Chemainus Towing.

Seaspan Protector (ex. *Gulf Sally*).

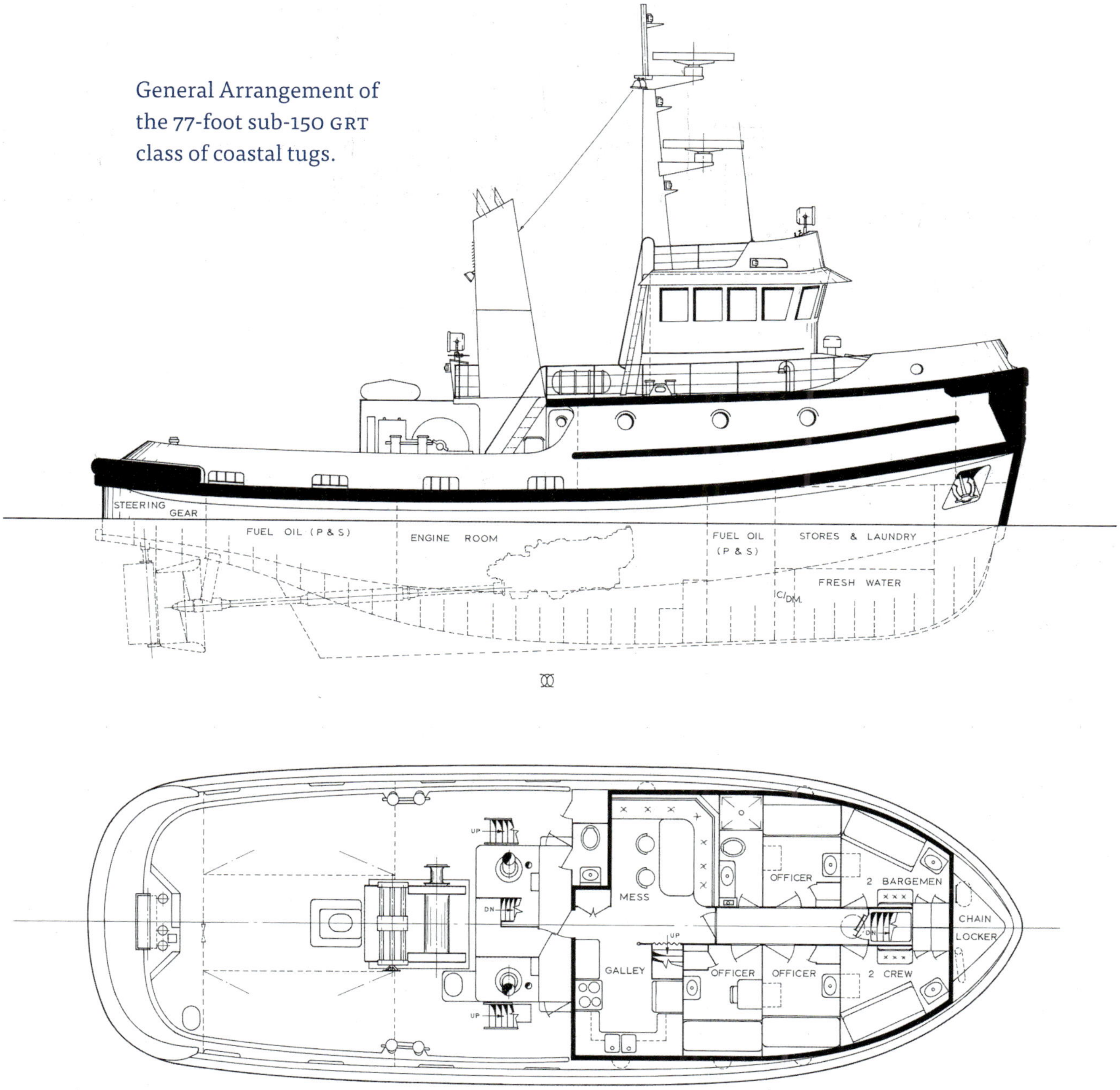

General Arrangement of the 77-foot sub-150 GRT class of coastal tugs.

Comox Crown, a 73-foot-long variant in this series.

Below: *Cindy Mozel*, the most powerful tug in this series at 1,800 bhp.

Jervis Crown

After somewhat of a hiatus in local tug-building, Crown Zellerbach commissioned Robert Allan Ltd to design a new coastal tug for service with log barges up to 10,000 TDwt. The result was the *Jervis Crown*. She had twin Nohab Polar F26R diesels with a total of 2,640 bhp driving fixed-pitch props in large, fixed nozzles. This tug was built at John Manly Ltd and was delivered to considerable fanfare in 1977. Also memorable was her purchase price of $2.6 million, a rather high number in the day, but consistent with the rule-of-thumb estimate of $1,000 per hp used at the time.

The *Jervis Crown* was a unique size of tug on the BC coast, falling in size and power between the series of 85-to-95-footers and the 130-footers built in the late '60s and the larger 125-to-140-foot ocean-going tugs of that same era. She measured 114 by 30 feet with a load draught of 14 feet. She was the last BC coastal tug designed by our office for several decades, as the industry began to shrink throughout the 1980s and '90s and no new equipment was built locally.

Pulling Together

The *Jervis Crown* project was another milestone in my personal career and to this day remains one of my all-time favourite projects. In the first instance, I was given almost total responsibility for the design by my father (though his watchful eye was ever-present). This was also one of those projects where all the players seemed to be on the same side and working toward the same end result of a top-quality project. Joe Upex was the general manager at Manly's and nearing the end of his long and illustrious career as a yard manager. He was a pleasure to work with and was very accommodating of our wishes to make this boat special. The owners were represented by their local manager, John Stenstrum, who was a relatively young person in the towing business, but smart and personable. John and I were not very different in age, and we got along extremely well. The tradesmen at Manly's also sensed that this was a project somewhat out of the ordinary, and it became obvious that all were doing their very best work. It showed in the end product, which was a handsome and extremely well-finished tug. I remember clearly at the commissioning ceremonies, standing in the wheelhouse behind one of the finishing carpenters who had worked on the boat. He was showing his young son one of the teak window frames he had worked on and showing where he had so carefully matched the colour and grain in two abutting pieces of wood, such that one couldn't see that there was a joint at all. That sort of pride showed everywhere in her construction.

—R.G. Allan

Jervis Crown leaving English Bay on a blustery day.

General Arrangement of the *Jervis Crown*.

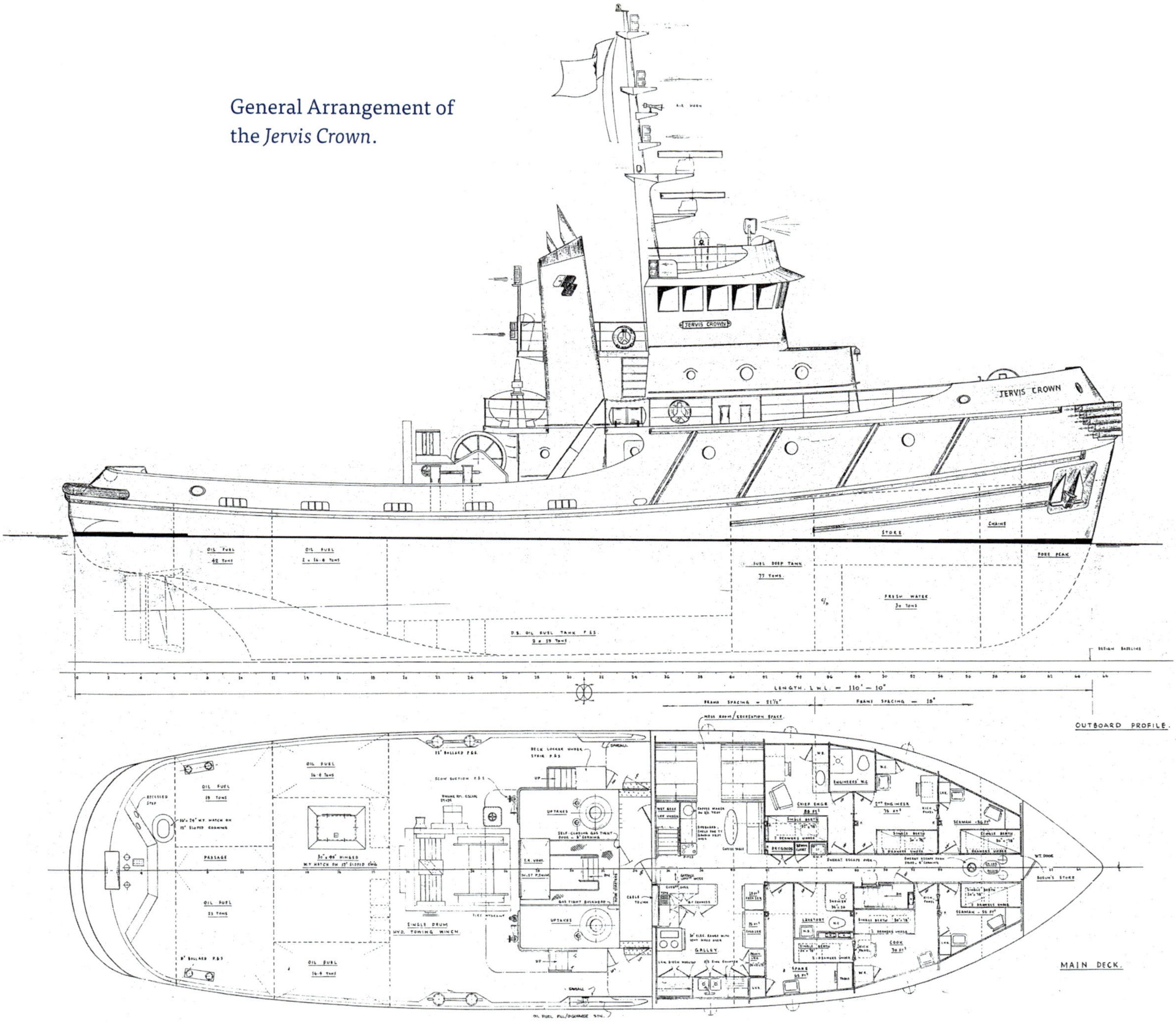

Tugs Built to Robert Allan Ltd Designs 1960–1970

Build year	RAL Proj. No.	Vessel Name	Vessel Reg. Number	Builder	Builder Hull No.
1960	11-59	*Ocean Master*	313730	N.V. Scheepswerf Gebr. van der Werf	
1960	3-60	*Syringa*	313658	John Manly Ltd	200
1961	6-60	*Lorne Yorke*	313751	Star Shipyard (Mercer's) Ltd	352
1961	1-61	*Jacques Cartier*	314837	Victoria Machinery Depot	
1962	4-61	*La Fille*	314838	John Manly Ltd	224
1962	15-61	*Neva Straits*	319346	Allied Shipbuilders Ltd	124
1962	24-61(a)	*Gulf Dianne*	319436	Allied Shipbuilders Ltd	127
1962	24-61(b)	*Gulf Muriel*	319446	Star Shipyard (Mercer's) Ltd	
1962	16-62	*Rosario Straits*	319429	John Manly Ltd	250
1963	19-62	*Evco Wave*	319165	Star Shipyard (Mercer's) Ltd	
1963	31-62(a)	*Harmac Pine*	320283	John Manly Ltd	260
1964	31-62(b)	*Harmac Spruce*	322349	John Manly Ltd	261
1963	3-63(a)	*Georgia Straits*	320253	BC Marine Shipbuilders Ltd	1
1963	3-63(b)	*Haro Straits*	320295	BC Marine Shipbuilders Ltd	2
1964	3-63(c)	*Malaspina Straits*	322437	BC Marine Shipbuilders Ltd	3
1963	7-63	*Papco III*	320318	John Manly Ltd	279
1964	16-63	*La Reine*	323220	Star Shipyard (Mercer's) Ltd	
1965	32-63	*Evco Breeze*	320352	Star Shipyard (Mercer's) Ltd	376
1965	40-63(a)	*Harmac Cedar*	323250	John Manly Ltd	294
1964	40-63(b)	*Harmac Fir*	323241	John Manly Ltd	295
1965	1-64	*Bering Straits*	323263	BC Marine Shipbuilders Ltd	7
1964	10-64	*Le Prince*	322492	John Manly Ltd	297
1964	17-64(a)	*La Dene*	323201	Yarrows Ltd	274
1964	17-64(b)	*La Belle*	323230	Yarrows Ltd	275
1965	24-64	*Gibraltar Straits*	325685	Yarrows Ltd	280
1964	27-64	*Nanaimo Yarder*	322084	John Manly Ltd	300
1965	28-64	*Ocean Iris*	320358	Built by Owner	
1966	42-64	*Harold A. Jones*	326548	Star Shipyard (Mercer's) Ltd	
1966	45-64	*Haida Brave*	326535	Halifax Shipyards Ltd	
1966	1-65	*Irving Maple*	326466	Saint John Shipbuilding Co. Ltd	1063
1967	2-65	*Irving Birch*	326472	Saint John Shipbuilding Co. Ltd	1066
1967	10-65(a)	*Island Master*	327100	McKay-Cormack Ltd	68
1967	10-65(b)	*Island Chief*	327101	McKay-Cormack Ltd	69
1965	19-65	*Westminster Brave*	323580	John Manly Ltd	320
1967	39-65	*Le Mars*	328843	Star Shipyard (Mercer's) Ltd	378

Owner/Client	Particulars				Power (BHP)
	Loa	L reg	Beam	Depth	
Great-West Towing & Salvage Ltd	126.00	119.70	31.00	16.58	1850
Columbia Cellulose Ltd	38.00	35.50	12.50	7.00	335
F.M. Yorke & Son Ltd	92.50	82.70	24.00	12.00	800
Deeks-McBride Co. Ltd	65.25	60.40	18.50	10.50	600
Vancouver Tugboat Co.	56.25	52.50	18.00	8.00	567
Straits Towing Ltd	92.54	86.60	24.75	12.83	900
Gulf of Georgia Towing Ltd	64.17	59.67	19.00	10.50	765
Gulf of Georgia Towing Ltd	64.17	59.67	19.00	10.50	765
Straits Towing Ltd	65.25	61.00	18.50	10.00	750
Evans Coleman & Evans	68.00	63.00	20.33	10.00	760
Kingcome Navigation Co. Ltd	64.58	59.90	19.17	11.42	750
Kingcome Navigation Co. Ltd	64.58	59.90	19.17	11.42	750
Straits Towing Ltd	65.25	61.00	19.67	10.00	600
Straits Towing Ltd	65.25	61.00	19.67	10.00	800
Straits Towing Ltd	65.25	61.00	19.00	10.00	765
Pacific Pine Co. Ltd.	41.00	n/a	14.00	7.75	325
Vancouver Tugboat Co.	95.25	88.60	26.00	12.00	1300
Evans Coleman & Evans	68.00	63.00	20.33	10.00	760
Kingcome Navigation Co. Ltd	64.58	59.90	19.17	11.42	750
Kingcome Navigation Co. Ltd	64.58	59.90	19.17	11.42	750
Straits Towing Ltd	53.75	52.00	18.33	10.10	600
Vancouver Tugboat Co.	56.25	52.50	18.00	9.25	650
Vancouver Tugboat Co.	65.08	60.00	19.17	11.42	765
Vancouver Tugboat Co.	65.08	60.00	19.17	11.42	765
Straits Towing Ltd	140.00	131.20	34.10	15.00	3840
Chemainus Towing Ltd	41.00	38.40	14.00	8.00	450
Dirk Kirkwood	41.00	38.40	14.00	8.00	380
Vancouver Tugboat Co.	131.25	125.80	32.00	17.50	3500
Kingcome Navigation Co. Ltd	140.00	129.10	34.10	18.88	3700
J.D. Irving Ltd	124.50	n/a	31.00	17.50	3200
J.D. Irving Ltd	160.00	150.00	36.50	19.25	3750
Island Tug & Barge Ltd	87.00	80.20	23.50	13.25	1200
Island Tug & Barge Ltd	87.00	80.20	23.50	13.25	1200
Westminster Tugboat Co. Ltd	41.00	38.50	14.00	9.75	350
Vancouver Tugboat Co.	95.25	88.20	25.00	13.92	1800

Tugs Built to Robert Allan Ltd Designs 1960–1970 (continued)

Build year	RAL Proj. No.	Vessel Name	Vessel Reg. Number	Builder	Builder Hull No.
1967	40-65	*Evco Spray*	323624	Star Shipyard (Mercer's) Ltd	
1966	48-65(a)	*Nanoose Yarder*	328427	Vito Steel Boat & Barge Ltd	7
1966	48-65(b)	*Debra K*	327199	Vito Steel Boat & Barge Ltd	6
1966	5-66	*H & R*	323602	John Manly Ltd	332
	10-66	??			
1966	13-66	*Seymour Crown*	327261	John Manly Ltd	342
1967	25-66	*Island Rustler*	327104	BC Marine Shipbuilders Ltd	
1968	28-66	*Naskeena 4*	330252	John Manly Ltd	387
1967	39-66	*Point Grey Endeavour*	328849	Vito Steel Boat & Barge Ltd	8
1968	47-66	*Island King*	330317	Star Shipyard (Mercer's) Ltd	392
1967	8-67	*Helga T.*	328417	Vito Steel Boat & Barge Ltd	9
1967	19-67	*Renata II*	328855	John Manly Ltd	364
1968	27-67	*Gulf Jean*	329465	BC Marine Shipbuilders Ltd	15
1967	35-67	*Sea Cap III*	346357	John Manly Ltd	375
1969	54-67	*Irving Beech*	330504	Saint John Shipbuilding Co. Ltd	1093
1968	3-68(a)	*Squamish Chief*	329527	Vito Steel Boat & Barge Ltd	10
1968	3-68(b)	*Gulf Coaster*	331112	Vito Steel Boat & Barge Ltd	12
1969	3-68(c)	*Regent*	331923	Vito Steel Boat & Barge Ltd	14
1969	4-68(a)	*Island Valiant*	331847	Bel-Aire Shipyard Ltd	
1970	4-68(b)	*Island Victor*	331229	Bel-Aire Shipyard Ltd	
1969	15-68	*Jose Narvaez*	331836	Benson Brothers Shipbuilding Co Ltd	
1969	20-68	*Charles H. Cates XX*	330470	John Manly Ltd	411
1969	52-68	*Arctic Circle No. 1*	330852	Bel-Aire Shipyard Ltd	
1970	19-69	*Cedar King*	529534 (US)		
1969	21-69	*Victory III*	331832	Vito Steel Boat & Barge Ltd	13
1970	22-69	*Swiftsure 3*	344613	BC Marine Shipbuilders Ltd	
1970	24-69	*Island Champion*	344714	A.C. Benson Shipyard Ltd.	
1970	30-69	*Hecate Crown*	344602	Star Shipyard (Mercer's) Ltd	
1970	54-69	*W.H. Horton*	344611	Vito Steel Boat & Barge Ltd	18
1970	60-69	*Vic Ingraham*	344735	Allied Shipbuilders Ltd	173

Owner/Client	Particulars				Power (BHP)
	Loa	L reg	Beam	Depth	
Evans Coleman & Evans	68.00	63.00	20.33	10.00	760
Chemainus Towing Ltd	41.00	37.50	14.00	9.50	463
Gordon Kleaman	41.00	38.40	14.00	9.50	325
H & R Towing	41.00	38.40	14.00	9.75	325
Gisborne, Larry	50.00	n/a	18.00	9.50	n/a
Crown Zellerbach Co. Ltd	53.00	49.50	18.67	9.25	670
Island Tug & Barge Ltd	47.50	43.10	16.50	9.67	465
Armour Salvage	38.75	36.25	14.00	6.00	480
Point Grey Towing Ltd	40.00	38.30	14.00	9.50	463
Island Tug & Barge Ltd	131.00	122.00	32.00	17.70	3600
Thuveson, W.E.	41.00	38.40	14.00	8.00	600
Columbia Cellulose Ltd	38.00	36.00	12.50	5.50	365
Valdez Marine Ltd	78.00	72.00	24.00	18.75	1125
Valley Towing Ltd	40.00	37.00	14.00	9.00	420
Atlantic Towing Ltd	101.00	97.50	28.00	14.42	2250
Quadrant Investments	41.00	38.00	14.00	8.00	425
Jarl Towing/Harvie Forest Industries Ltd	41.00	38.40	14.00	8.00	550
Regent Towing	41.00	38.40	14.00	8.00	490
Island Tug & Barge Ltd	70.00	67.00	22.50	10.25	765
Island Tug & Barge Ltd	70.00	67.00	22.50	10.25	765
LaFarge Cement Ltd	87.08	81.00	24.00	14.00	1450
C.H. Cates & Sons Ltd	45.00	41.00	14.50	6.00	700
Kap's Transport Ltd	117.25	108.00	26.00	8.00	4 × 500
Dunlap Towing Ltd (US)	37.58	n/a	13.75	5.58	350
Stone Brothers Ltd	41.00	38.40	14.00	8.00	455
BC Forest Products Ltd	54.67	49.00	19.00	10.16	850
Island Tug & Barge Ltd	87.00	81.00	24.00	14.00	1450
Crown Zellerbach Co. Ltd	131.00	122.00	32.00	17.75	3600
Kap's Transport Ltd	69.00	n/a	25.00	6.00	4 × 250
Northern Transportation Co. Ltd	154.58	149.50	50.00	9.50	4500

Tugs Built to Robert Allan Ltd Designs 1970–1979

Build year	Proj. No. / Dwg. No.	Vessel Name	Vessel Reg. No.	Builder	Builder Hull No.
1970	18-70	*Margaret Yorke*	345163	Hike Metal Products	
1970	18-70(b)	*Phyllis Yorke*	345141	Hike Metal Products	
1970	19-70	*Anne Carlander*	835681	Marine Const'n & Design Co. (MARCO)	
1971	19-70(b)	*Edith Lovejoy*	821125	Marine Const'n & Design Co. (MARCO)	
1972	41-70	*Charles H. Cates XVIII*	346375	John Manly Ltd	
1971	43-70	*Beaufort Sea Explorer*	344932	Vito Steel Boat & Barge Ltd	122
1971	1-71	*Mary Lu Spencer*	345584	Allied Shipbuilders Ltd	176
1971	9-71	*G.D. Hemmingsen*	346293	Vito Steel Boat & Barge Ltd	
1974	33-71(a)	*Lawrenceville*	YTL 590	Vito Steel Boat & Barge Ltd	134
1971	33-71(b)	*Parksville*	YTL 591	Vito Steel Boat & Barge Ltd	135
1974	33-71(c)	*Listerville*	YTL 592	Georgetown Shipyard Ltd (PEI)	
1974	33-71(d)	*Merrickville*	YTL 593	Georgetown Shipyard Ltd (PEI)	
1971	33-71(e)	*Marysville*	YTL 594	Georgetown Shipyard Ltd (PEI)	
1972	1-72	*Cecelia Hall*	346574	Vito Steel Boat & Barge Ltd	132
1973	24-72	*Kap I*	347951	Yarrows Ltd	364
1973	24-72(b)	*Kap II*	347952	Yarrows Ltd	365
1969	24-72(c)	*Kap III*	331152	Erickson Boat Building	
1973	43-72	*J. Mattson*	347543	Dominion Bridge Company	
1973	50-72(a)	*Matt Berry*	347535	Vito Steel Boat & Barge Ltd	
1973	50-72(b)	*Henry Christoffersen*	347544	Yarrows Ltd	
1973	50-72(c)	*Johnny Hope*	347537	Burrard Dry Dock Co. Ltd	
1973	50-72(d)	*Jock McNiven*	347538	Vito Steel Boat & Barge Ltd	
1973	56-72	*Kakisa*	347536	Bel-Aire Shipyard	
1974	18-73	*Comox Crown*	348790	Vito Steel Boat & Barge Ltd	138
1974	26-73	*Cindy Mozel*	369170	Vito Steel Boat & Barge Ltd	144
1974	34-73	*Charles H. Cates IV*	369221	John Manly Ltd	
1973	38-73	*Kaipokok*	348104	Selkirk Machine Works Ltd	
1974	48-73	*C.T. Champion*	348791	Vito Steel Boat & Barge Ltd	140
1974	49-73(c)	*Gulf Sally*	368710	Vito Steel Boat & Barge Ltd	141
1974	49-73(b)	*Gulf Julia*	368711	Vito Steel Boat & Barge Ltd	142
1974	49-73(a)	*Gulf Irene*	369175	Vito Steel Boat & Barge Ltd	143
1974	59-73	*Keewatin*	348028	Yarrows Ltd	
1974	67-73	*Beverly Lambert*	348855	Allied Shipbuilders Ltd	191
1974	67-73(b)	*Carlyn McMurren*	348856	Allied Shipbuilders Ltd	189

Owner/Client	Particulars Loa	L reg	Beam	Depth	Draft	Power (BHP)
F.M. Yorke & Sons Ltd	99.75	99.75	35.00	10.50	5.75	2 × 850
F.M. Yorke & Sons Ltd	99.75	99.75	35.00	10.50	5.75	3 × 850
Puget Sound Freight Lines (US)	74.42	67.95	23.00	12.90	12.25	1125
Puget Sound Freight Lines (US)	74.42	67.95	23.00	12.90	12.25	1125
C.H. Cates & Sons Ltd	57.00	53.00	18.00	7.33	7.00	2 × 487
Kap's Transport Ltd	118.00	110.00	36.00	10.50	5.5 / 7.0	2250
Northern Construction Co. Ltd	58.00	55.00	24.00	7.00	3.50	960
Chemainus Towing Ltd	68.75	65.00	22.50	11.83	10.50	980
Dep't. of National Defence, Canada	45.00	n/a	15.00	8.25	6.25	365
Dep't. of National Defence, Canada	45.00	n/a	15.00	8.25	6.25	365
Dep't. of National Defence, Canada	45.00	n/a	15.00	8.25	6.25	365
Dep't. of National Defence, Canada	45.00	n/a	15.00	8.25	6.25	365
Dep't. of National Defence, Canada	45.00	n/a	15.00	8.25	6.25	365
Northern Construction Co. Ltd	66.00	60.00	25.00	8.00	4.17	960
Kap's Transport Ltd	45.75	n/a	12.00	3.17	1.42	2 × 355
Kap's Transport Ltd	45.75	n/a	12.00	3.17	1.42	2 × 355
Kap's Transport Ltd	48.50	n/a	15.50	4.00	2.00	3 × 355
Kap's Transport Ltd	106.00	n/a	35.00	9.00	4.5 / 6.5	2250
Northern Transportation Co. Ltd	153.25	149.50	52.00	9.50	3.25	4500
Northern Transportation Co. Ltd	153.25	149.50	52.00	9.50	3.25	4500
Northern Transportation Co. Ltd	153.25	149.50	52.00	9.50	3.25	4500
Northern Transportation Co. Ltd	153.25	149.50	52.00	9.50	3.25	4500
Northern Transportation Co. Ltd	84.00	73.33	36.00	9.50	3.50	1840
Crown Zellerbach Co. Ltd	72.33	69.00	24.00	12.25	11.50	2 × 675
Mozel, Capt. John	77.75	73.00	24.00	12.75	11.50	2 × 900
C.H. Cates & Sons Ltd	50.50	47.00	16.50	7.29	6.58	2 × 340
Dept. of Public Works, Canada	37.50	35.00	12.00	5.38	3.75	2 × 190
Chemainus Towing Ltd	76.50	73.00	24.00	12.50	11.58	2 × 670
Gulf of Georgia Towing Ltd	77.75	73.00	24.00	12.75	11.50	2 × 725
Gulf of Georgia Towing Ltd	77.75	73.00	24.00	12.75	11.50	2 × 725
Gulf of Georgia Towing Ltd	77.75	73.00	24.00	12.75	11.50	2 × 725
Northern Transportation Co. Ltd	126.50	n/a	38.00	12.00	6.50	3 × 1125
North Western Dredging Ltd	70.00	65.00	25.00	9.25	4.50	2 × 703
North Western Dredging Ltd	70.00	65.00	25.00	9.25	4.50	2 × 703

Tugs Built to Robert Allan Ltd Designs 1970–1979 (continued)

Build year	Proj. No. / Dwg. No.	Vessel Name	Vessel Reg. No.	Builder	Builder Hull No.
1974	67-73(c)	*Dorothy Robinson*	348857	Allied Shipbuilders Ltd	190
1974	4-74	*Betty Coulter*	348884	Zenith Steel Fabricators	
1974	49-74	*Arrowhead*	370606	Alberni Engineering	
1977	55-75	*Charles H. Cates VII*	383382	Bel-Aire Shipyard	
1977	34-76	*Jervis Crown*	383384	John Manly Ltd	
1977	36-77	*Hamilton Baillie*	383412	Bel-Aire Shipyard	
1980	58-78	*Annie M. Dean*	370097	Built by Owner	
1979	59-78	*Charles H. Cates V*	391885	Bel-Aire Shipyard	
1979	70-78	*C.T. Titan*	391718	Allied Shipbuilders Ltd	213
1980	89-78	*Irving Elm*	369826	Saint John Shipbuilding Co. Ltd	
1980	71-79	*Charles H. Cates VI*	395353	Allied Shipbuilders Ltd	225

Owner/Client	Particulars Loa	L reg	Beam	Depth	Draft	Power (BHP)
North Western Dredging Ltd	70.00	65.00	25.00	9.25	4.50	2 × 703
North Western Dredging Ltd	68.00	63.32	25.47	7.32	n/a	2 × 703
Canadian Cellulose Ltd	31.25	28.00	12.00	7.52	6.00	230
C.H. Cates & Sons Ltd	60.00	56.92	22.00	9.42	9.00	2 × 930
Crown Zellerbach Co. Ltd	113.67	106.00	30.00	15.72	14.00	2 × 1300
Chemainus Towing Ltd	69.00	65.52	22.50	8.83	10.67	2 × 480
Dean Construction Ltd	50.50	n/a	19.00	8.00	5.00	2 × 267
C.H. Cates & Sons Ltd	60.00	56.92	22.00	9.42	9.00	2 × 930
Chemainus Towing Ltd	53.00	50.00	19.00	9.25	8.50	2 × 470
J.D. Irving Ltd	113.67	106.81	30.00	15.72	14.00	2 × 1750
C.H. Cates & Sons Ltd	60.00	56.92	22.00	9.42	9.00	2 × 930

Opposite: This 2020 image nicely captures multiple generations of tugs and barges within the Seaspan Marine fleet, all designed by Robert Allan Ltd.

丸 町
HARAMACHI M

17

Ship-Handling Tugboats

The majority of tug designs created by Robert Allan Ltd prior to 1980 were primarily log-towing or barge-towing tugs for the BC coast. A few of these occasionally got used in ship-docking duties, but in general they were not well-suited to that task. Their bow geometry was not intended for pushing, and they lacked suitable fendering all around, but especially forward.

The distinct separation of ship-handling and barge-towing services on this coast arose primarily due to a gentleman's agreement between the historic company of C.H. Cates & Sons Ltd and the other main tug operators. They agreed that Cates would handle all the ship-docking duties in Vancouver harbour, and the others would do what they did best, serving the coastal resource industries with barge and log movements, and neither would interfere with the other. This arrangement worked extremely well for many decades, until the owners of Cates decided to sell their business in 1992. After that, first Seaspan and then RivTow decided to compete in that market, and the very profitable harbour ship-handling business, which had been Cates' sole domain for over a century, got sliced up by competition and undoubtedly became much less profitable.

Opposite: *Cates I*, see page 383.

This chapter begins with the story of our long and most fortuitous relationship with the Cates company, and how those very successful designs evolved into a worldwide phenomenon of an entire genre of ship-handling/harbour tugs. The evolution of the Cates fleet and its derivatives is thoroughly described in a paper entitled "Compact Tugs: Their Evolution, and Recent Developments" by R.G. Allan, presented to the International Tug Conference in 2002. The term "compact tugs" was coined with this paper and is now in common usage, referring generally to tugs which fall under the 24-metre registered length limit for compliance with SOLAS and other international "big ship" regulations.

The Cates Twin-Screw Tugs

Cates had for many years operated with a fleet of small, simple but handy wooden tugs.

Terry Waghorn was a son-in-law of Capt. John Cates, and he was the president and GM of the tugboat company for many years. Terry had a vision of what the future of their company should look like, and he realized early on that as cargo ships were growing in size and complexity, the future of his business didn't include small, low-powered wooden tugs. In 1968 he contacted Bob Allan to discuss his ideas for a new generation of steel tugs for the future. That concept was for what Terry always referred to as a "slippery dish," a tug that must be simple, highly stable, very manoeuvrable, relatively powerful and able to move laterally at a good speed when tending a moving ship. This concept was first realized in the design for the *Charles H. Cates XX*, a 45 foot 600 hp twin-screw tug built by John Manly Ltd in 1969.

A Chance Meeting

The first boat we built with Robert F. (Bob) Allan as the designer came about sort of by chance. In the late 1960s, I'd gone with my father-in-law, John Henry Cates, who was the president of Cates, to a BC tugboat convention. The speaker was Bob Allan. I was pretty green, but this guy just fascinated me because he talked so good. I have always admired people who can talk properly, which I can't, but Bob was really something. He talked about the towing industry and stuff. Afterwards, I said to myself, Geez, if I was ever going to build a tug, that would be the guy I'd get to design it. When I got into a management position in the late 1960s, we approached Robert Allan to design a modern ship-berthing tug. Later I was given full rein and title by John Henry Cates to take the company in this new direction.

I went to Bob Allan and said I was interested in building some new tugs. I said, I'd like you to come down and go out on the tugs to observe two or three jobs. I said, I'll go with you and point out the strengths and weaknesses and show him what I wanted in the new tugs.

After that experience, Bob understood, and he was amazed at the different demands of a ship-berthing tug as opposed to a regular tug, and that they were two different animals. We wanted a tug designed and powered for berthing of ships as opposed to towing barges or logs. I was happy working with Bob. He was really excited about a new opportunity to revolutionize ship-berthing tug design. The old designs were just simple—hook on and pull from A to B, no one was thinking about a tug sliding sideways.

I decided I wanted to go with twin screws. This was really new. I think there were maybe one or two other twin-screw tugs around at the time. We went with twin screws because you wanted to give your customer the biggest bang for his buck [meaning the most manoeuvrable

The wooden-tug fleet of C.H. Cates & Sons at the foot of Lonsdale in North Vancouver in 1973, with the *C.H. Cates XVIII*, one of their new steel tug fleet, at the pier end.

and powerful tug available]. I wanted this tug to be as shallow draft as it could be, and free of a lot of cumbersome stuff, like too much hull. Kort nozzles were new at the time. These are a fixed or steerable ring, like a barrel or tube, around the propeller. Twin rudders were attached to the aft end of the nozzle. These Kort nozzles were thought to give you 20 or 25 percent more thrust, and you had twin rudders, which were foil rudders (like an airplane wing)—another Robert Allan innovation. We went ahead with fixed nozzles, as I didn't like the steerable nozzles because I thought they'd block off too much of the water so you wouldn't get that clean thrust. I wanted it to be our first steel hull, and all the superstructure was to be aluminium, as I didn't want to be fighting rust. So Bob went ahead and designed the *Cates XX* following the same basic lines and size as the wooden 45-foot tugs with jog steering that we already had. I put the same power in it as we had in the wooden tugs: two 300 hp GM 8V-71s.

I was really pleased with the "20." We put it to work and it was really good. However, it still didn't slide as anticipated. We were up on the drydock, and I was talking to Bob, and we were looking at its big skeg. Thinking about it, if you're sliding sideways, a big skeg would block the flow and the tug wouldn't slide, so we decided to take the skeg right off. I wanted all the thrust—as much of the thrust as you could get—applied to pushing, not keeping the tug in position. That was the trouble with a conventional tug: they'd spend all the time keeping themselves in position against the ship, not doing the job they were hired for. The *20* was a really big breakthrough. The Kort nozzles were marvelous; they were way more responsive and when you wanted power, you got it. The pilots were really happy with this tug. They always said, "Send the *20*. Send the *20*."

—Terry Waghorn

Charles H. Cates XX—a new look for harbour towing in Vancouver.

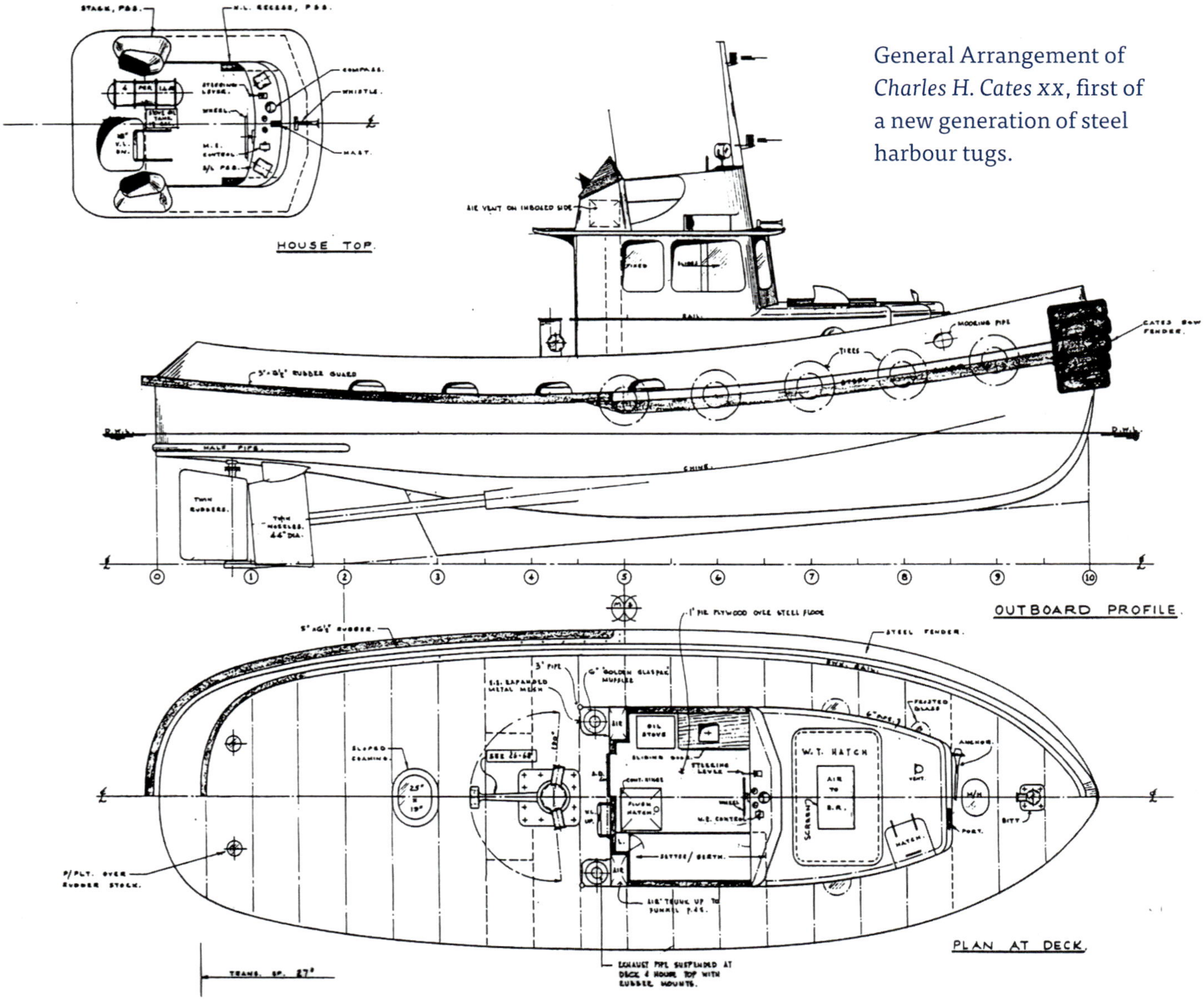

General Arrangement of *Charles H. Cates* XX, first of a new generation of steel harbour tugs.

Over the next 12 years, Cates built six more new tugs of varying sizes and powers in this same general style, each new vessel embodying the successful characteristics of its predecessors and incorporating refinements considered to enhance performance, efficiency and reliability. All of the Cates tugs were twin-screw, designed to be extremely manoeuvrable, responsive and very cost-effective platforms, and all were characterized by extremely low projected lateral areas below the waterline. All were fitted with Kort nozzles for improved thrust efficiency, and the steering systems evolved from single to twin rudders behind each nozzle. The hull form evolved to a double chine, and skegs were eliminated altogether.

The mechanics of these boats were very simple and very cost-effective: two main diesel engines; 24-volt batteries for electric starting and domestic services (lighting, etc.); an oil-fired stove in the wheelhouse for heat (and a coffee pot); and all the auxiliary systems, such as hydraulics for steering gear, etc., were run off cam-shaft-driven PTOs on the main engines. The engine-room ventilation was induced through large intake cowls on the casing top and a Venturi system, by which the engine exhaust flow created a suction for exhausting the engine room air. After the *Cates XX*, the progression of the new steel tug fleet was as follows:

The *C.H. Cates IV* designed in 1973 and also built by John Manly Ltd. She measured 50 feet 6 inches by 16 feet 6 inches by 7 feet 3 inches draft, with two 400 bhp diesels.

Following pages: Built by John Manly Ltd in 1970, the *Charles H. Cates XVIII* measured 55 by 18 feet with a 7-foot 4-inch draft and was fitted with two 487 bhp GM diesels.

18
CATES

CHARLES H CATES

The last four of the twin-screw tugs, *Cates VII* (1975, Bel-Aire Shipyard Ltd), *Cates V* (1978, Bel-Aire Shipyard Ltd), *Cates VI* (1979, Allied Shipbuilders Ltd) and *Cates VIII* (1980, Allied Shipbuilders Ltd), were virtually identical 60-foot 1,800 hp tugs, as illustrated below. The bollard pull output of these four tugs ranged between 24.9 and 25.5 tonnes, or more than 30.6 lbs/bhp. This is considered very good thrust performance in such relatively shallow-draft hulls.

Interestingly, this hull design subsequently became the basis for a whole new series of harbour tugs operating throughout Turkey and subsequently giving rise to the rather incredible scope of business developed by Robert Allan Ltd in that country (Chapter 19). The size and power of the Cates twin-screw tugs, however, could not keep pace with the growth in ship sizes calling at the Port of Vancouver, and larger, more powerful tugs were necessary. More manoeuvrability and control in the propulsion systems were also required, giving rise to the earliest use of Z-drive propulsion in tugs in Canada.

The 1,800 hp class Cates tugs were 60 feet in length overall by 22-foot beam, with a load draft of 6 feet 6 inches. They were fitted with twin 900 bhp GM 16V-149 diesels.

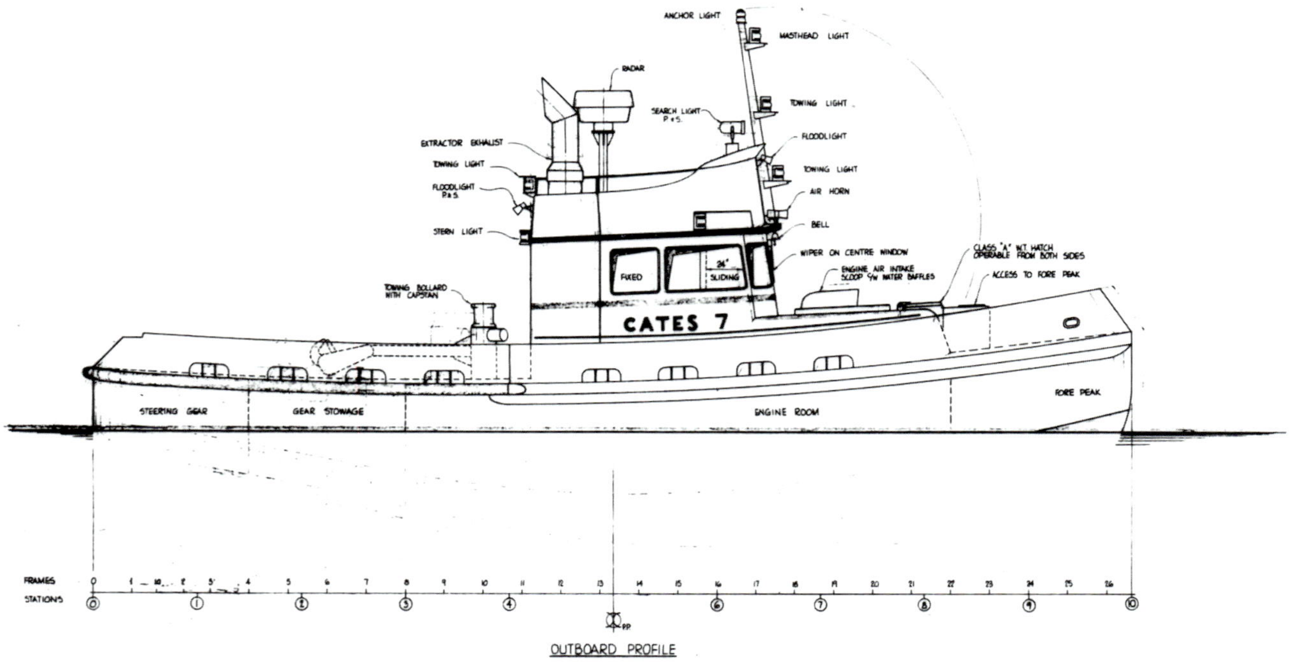

OUTBOARD PROFILE

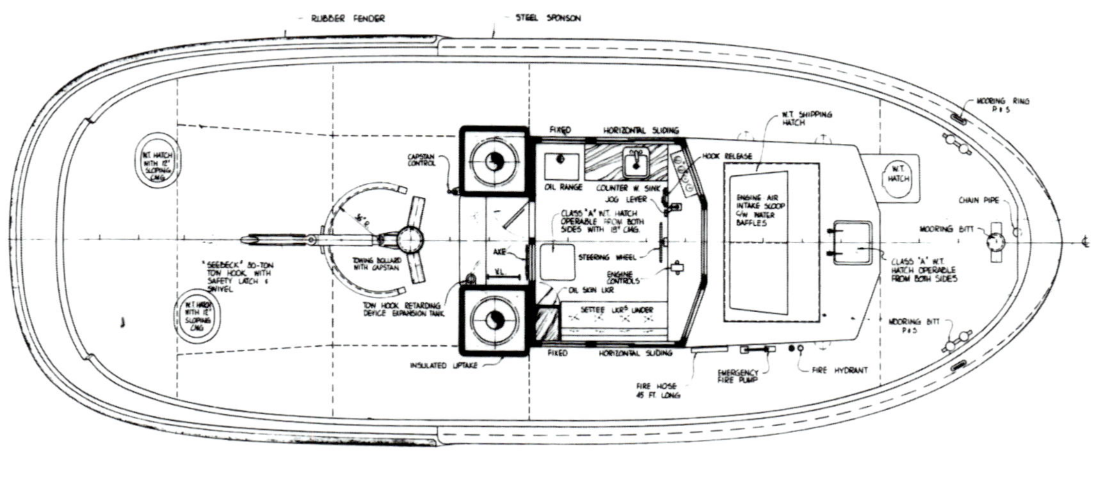

DECK PLAN

Cates v and *Cates* vii, two of the high-performance 1,800 hp twin-screw tugs in the Cates fleet.

The Cates Z-Drive Tugs

Z-drives, or steerable rudder-propellers, were invented by Joseph Becker in Germany in 1950, and the first commercial units came into service on European inland waterways vessels in the early 1960s. Becker was the founder of Schottel GmbH, still today one of the leading manufacturers of this type of propulsion unit. These drives, in very basic terms, consist of an upper right-angle gearbox to which the drive shaft is connected, then transmitted through a lower right-angle gear to a short propeller shaft. The entire system is designed to azimuth through 360 degrees, thus enabling full thrust in almost all directions (except of course for hull interference in the astern direction).

The first harbour tug with Z-drives was the *Janus*, a small pusher tug designed and built in Germany in 1967 to shuttle LASH barges (a floating container technology that was short-lived and no longer used) from larger ships to shallow ports in the Netherlands. This simple but distinctive boat was the "Adam" for the many thousands of Z-drive-powered tugs that followed.

In 1981 Cates commissioned Robert Allan Ltd to design the first Z-drive tug in Canada (the second in North America), the 2,400 hp *Charles H. Cates II*, built by Allied Shipbuilders Ltd in 1983. The particulars of this highly innovative tug design were a length overall of 22.44 metres, a beam of 8.53 metres and a load draft of 3.66 metres.

The hull form of the *Cates II*, though obviously larger than its twin-screw cousins, was designed to incorporate all of those features that had distinguished the earlier conventional screw tugs, namely a high standard of stability, low lateral resistance and a centroid of lateral area forward of midships to enhance manoeuvrability.

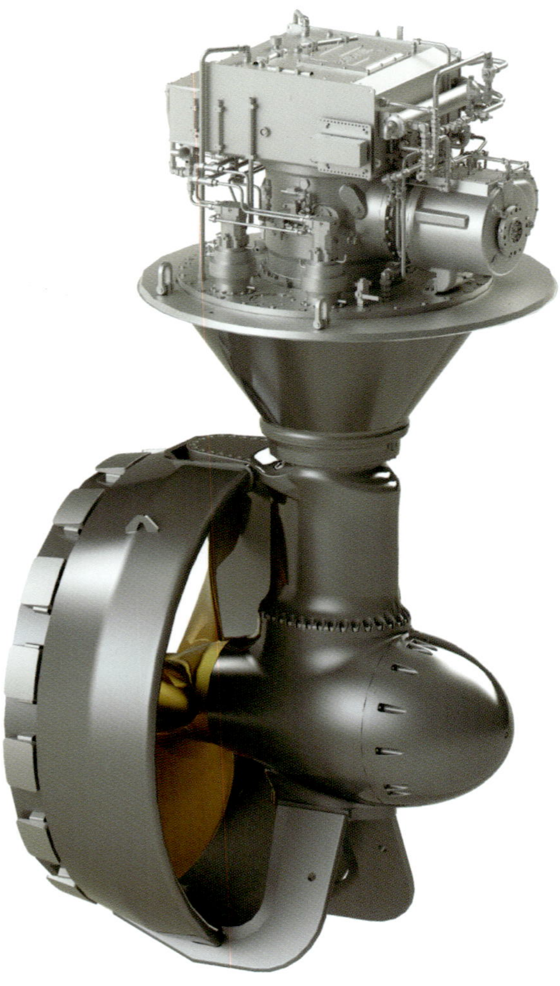

Typical modern Z-drive steerable propulsion unit, courtesy of Schottel GmbH.

Above: *Charles H. Cates II*, the first Z-drive tug in Canada.

Left: Commissioning the *Cates II* in 1983 (left to right): Jane Bradwell-Wilson, the sponsor (oldest granddaughter of Capt. John Cates), Anne and Terry Waghorn, Claire Johnston and Rob Allan.

The Evolution of Z-Drives

Ken Harford

The original technology behind the Z-drive is somewhat lost. It's like the diesel engine: some people say it was invented in the UK, some people say it was invented by Mr. Diesel. Wikipedia claims that the Z-drive was invented in 1950 by Joseph Becker, the founder of Schottel. However, Murray & Tregurtha Inc. filed a US patent in 1948. There were Z-drives for boats being built in Europe and the US, but in the 1970s it seemed that it was the Japanese who were using them most in ship-berthing tugs (though their hull designs were terrible). Z-drives are capable of directing the full propeller thrust through 360 degrees for steering (rather than just deflecting thrust plus or minus 45 degrees from straight ahead, as a conventional propeller and rudder does).

With conventional shaft, prop and rudder tugs, there are a lot of losses in efficiency. Particularly if you are trying to go sideways or reverse if you're trying to back off something. As an example, the BC coast pilots set the requirements for places such as British Columbia's Roberts Bank superport. In the 1980s it was common for the pilots to rate the capacity of tugs that you needed for berthing a ship of a certain size to the gross tonnage of the ship they were handling. They came up with a rule that said what you needed in horsepower was 10 percent of the gross tonnage of the ship, but if you had a Z-drive tug, you only needed 5 percent, because they were so much more effective. The greater efficiency is achieved because you're turning the whole propellers rather than deflecting the thrust with a rudder. That's why Seaspan built the *Seaspan Discovery*. They could build a smaller tug, with half the horsepower, and thus win the contract. You can't get from pushing to pulling quickly in a conventional tug without turning the propellers to backwards. The efficiency of a propeller turning in reverse is very low. It would be possible to turn the tug around to pull on the towline, but that takes too much time. On a Z-drive tug, you just turn the drives around and that will take about 10 seconds.

After I graduated from the University of British Columbia with a mechanical engineering degree in 1970, I formed a company with Tom Siddon (one of my professors) called Siddon Harford and Associates (SHA). Tom Siddon was a professional engineer and went on to be a Member of Parliament and cabinet minister. SHA specialized in acoustical engineering services, including architectural acoustics and industrial noise and vibration control. As explained in Chapter 20, that work included pioneering resilient mounting systems for the main propulsion engines for workboats and particularly for the BC tugboat fleet.

In 1972 one of my classmates, Doug Kennedy, joined the firm and eventually bought out Tom Siddon's interest, and the firm became Harford Kennedy Ltd (HKL) in 1973. In 1975 HKL formed a separate company, Mechtronics Research Corporation (now Mechtronics Technology Inc.), to pursue opportunities involving the integration of mechanical and electrical engineering technologies. In the marine sector, Mechtronics Research designed and oversaw installations of electro-mechanical systems to control steerable thrusters (commonly called Z-drives). The initial Z-drive controls were developed to work with the Z-drive units manufactured by Maritime Industries Ltd in Burnaby, BC.

The first Mechtronics steering controls were pneumatic, built by Kobelt, and installed on the Department of National Defence (DND) fireboat called the *Firebrand* (a Robert Allan Ltd. design, see Chapter 10). But they were having a lot of problems, and they were unacceptable to DND and holding up the delivery of these fireboats, which were being built at Vancouver Shipyards. Bob Haggarty was the design engineer at Maritime Industries. He was a classmate of mine and also knew Steve Hui from our time together at UBC, when Steve was Tom Siddon's research technician. Steve is an electronics genius with strong mechanical skills. Bob contacted Steve to come up with a design to solve the steering problems. He did, and the new Mechtronics steering system solved the problem on the fireboats and allowed Vancouver Shipyards to finally deliver those boats. Subsequently, Mechtronics steering controls were installed on the Vancouver Seabus Project (*Burrard Beaver* and *Burrard Otter*, 1976), followed by all the Z-drive-powered ferries in the BC Ferries and Ministry of Highways fleets in BC.

In the late 1960s the Japanese government funded a big program to develop the concept of the Z-drive as potential replacement for the Voith Schneider propulsion unit. Voith Schneider was very aggressive in protecting their patents and, instead of being reliant on one vendor, the Japanese funded a program for several companies to research, design, develop and build Z-drives. The Niigata Z-Peller was one of these designs that survives today. The Z-drive system took off because, in part, the Japanese would only allow Japanese-built tugs with Japanese Z-drives in their harbours. The drives were mounted in the stern and the tow point was at the bow. Niigata was the company that went out to develop foreign markets for their Z-drives. The others did a little bit of that, but Niigata was the most aggressive, in part because they were selling the whole package, including their engines.

At one point, the management of Maritime Industries felt that the company should have more than one supplier of steering systems for their drives, so they went to Wagner Engineering, who built one set for them. We felt that if Maritime Industries wanted more than one supplier, then we wanted more than one customer, as we were building our steering controls solely for them.

Around this time, John Chard of Seaspan approached us, as they had contracted for a couple of tugs and were just getting into the Z-drive era. It came up in conversation that they were thinking they didn't want to buy the Maritime Industries drives because of a history of problems with their drives. But they wanted to build Z-drive boats and were looking at buying these Niigata drives from Japan. They were concerned, however, that Niigata didn't have a distributor here. So John asked if Mechtronics would be interested in becoming the reps for Niigata, and if we did he'd give us an order. It was kind of a gift and, as a result, in 1982 the Industrial Marine Power Engineering Group (IMPEG) was formed within Mechtronics to become the OEM distributor for the Niigata Z-Pellers and medium-speed diesel engines manufactured in Japan. Cates was close behind in developing its concept for Z-drive tugs, and in 1983 C.H. Cates and Sons Ltd was the first to install Niigata Z-Pellers from IMPEG, followed in 1984 by Seaspan for the *Seaspan Discovery*.

We struck a deal with Niigata that we would put our steering systems on their drives. They had a steering system, but their steering system wasn't very good. Mechanically, though, their drives were excellent—bulletproof. We brought in the upper and lower gear cases from Niigata, and Osborne built the propellers and, of course, Mechtronics did the steering system.

We were also involved in the installation of all the Z-Pellers we sold. The mounting and machining of the Z-Peller beds were part of my mechanical engineering role. Our first installation was for the *Cates II*, being built at Allied Shipbuilders. We explained to Jim McLaren, son of owner Arthur McLaren, that it was best to build the whole 8-foot-diameter mounting bed in the hull and then machine it so it was perfectly flat. That way, when you bolt the Z-drive in place it doesn't twist anything, and you get a good watertight seal. Sealing the drive to the hull involved a combination of about a million bolts and Permatex sealant.

Allied figured they could do it much easier by prefabricating this unit, sending it out to EBCO Industries, who would machine the top, and then Allied would very carefully weld it into place—because they knew how to do all this. However, when they welded it into place, the top surface of the mounting ring was canted outboard. I had a conversation with Jim McLaren, who was saying it is perfectly good, it's all acceptable and so on. However, Hiroshi Yamada, the acceptance engineer from the Niigata factory, was standing by looking very stern. He wasn't happy at all. I said to Jim, "Fine with me. If you just want to sign an undertaking to take on all the warranty responsibility, we can accept your bed." He didn't want that responsibility, so he kept his guys going around and welding again in little places, and they managed to pull it flat enough that Hiroshi was happy.

For a subsequent installation, Niigata brought their big machining rig over from Japan—it was like a big fly-cutter that could machine out to a diameter of about 12 feet. They didn't want to ship it back, and we said we'd buy it from them. So I got into a lot of that aspect of the machinery installations when I was with IMPEG. After that, Maritime Industries didn't sell another set of drives, except to the BC government for ferries, and the business was ultimately taken over by Ulstein.

Cates I at work in Vancouver harbour. The *Cates I, II & III* were the prototypes for a new generation of compact Z-drive tugs.

A First for Cates

Ships started to get big—really big! They started coming into Vancouver because it had a lot to offer as a port. In the past, most cargo ships had a wheelhouse pretty well midships or a bit forward. With the new bulk carriers and such, the wheelhouse was aft. They came to Vancouver for coal, potash, lumber and wheat. Then there were container ships, and the big car carriers that don't even look like ships. When I started seeing these coming into the harbour, I realized that we had to accelerate in building bigger, more powerful tugs.

With this in mind, I knew I was going to need more help. One of Bob Allan's staff, Claire Johnston, had been involved in the engineering and overseeing for the *Cates VII*, one of our 60-foot steel tugs, at Bel-Aire Shipyards. He was a pretty good engineer and we got along well together, so I asked him if he wanted to come work for me. I then got hold of Bob Allan and told him I was romancing Claire and what did he think if I offered him a job. Bob said that would be a great move for Claire, so he came to work for me. I made him manager and he did a good job. He had fresh ideas about things, he was a very smart guy, and we got along really well.

It is interesting how I got the idea for a different propulsion system for tugs. We used to have a family camp up on Woolridge Island, in Howe Sound, and we had a little 3-horse Johnson outboard. It turned 360 degrees and, looking at that, I thought, this is what I need on the tug, this kind of propulsion. Then the Z-Peller came out and I said, I'm going to build a ship-berthing tug with that propulsion system.

In 1981 I went to Rob Allan (who had then succeeded his father) and said, I want to build a Z-Peller tug with more power and maneuverability, so we basically used the same proven hull configuration as the *18*, only modernized and with this new propulsion system called Z-Peller. We increased the size from 60 to 75.5 feet, and we put in twin 1,200 hp GM 16V-149 engines. Rob found a way to modify these Z-Peller drives so they would work on our tugs. We even made a mock-up wheelhouse out of plywood and got the guys to be in it to get a feel of the visibility and operating controls in a 360-degree environment. The *Cates II* was the first Z-drive tug in Canada. It was revolutionary and only the second to be built in North America.

—Terry Waghorn

The success and cost-effectiveness of this tug, and the obvious benefits of the higher power and manoeuvrability offered by the Z-drive configuration, led to orders for two subsequent tugs, *Charles H. Cates I* in 1985 (Rivtow Industries Ltd: West Coast Manly Shipyard) and *Charles H. Cates III* in 1989 (Allied Shipbuilders Ltd). These two latter tugs, though outwardly essentially identical to the *Charles H. Cates II*, incorporated all-electric steering motors and winch drive, utilizing frequency-modulating motor controllers to achieve fully variable speed control from standard AC motors. This innovative design change led to a much cleaner, simpler machinery installation than in the tugs with hydraulic steering, and with greatly reduced maintenance requirements. This was a fairly ground-breaking innovation in tug design in the 1980s.

In the midst of the series of 2,400 hp Z-drive tugs, Cates believed that there was a niche in the Vancouver market for a smaller and lower-powered Z-drive tug for working smaller ships at a more attractive rate. They commissioned Robert Allan Ltd to design the 55-foot 1,300 hp *Cates X* in 1987. Although a handy little tug, it was not a highly successful experiment, as the pilots had become used to the power and performance of the bigger 2,400 hp tugs and weren't particularly interested in giving up that level of support to their operations.

Charles H. Cates X, a 55-foot ASD tug for the Port of Vancouver.

Below: Compact one-man wheelhouse arrangement for 2,600 hp Cates Z-drive tugs; **below inset:** The compact but highly efficient wheelhouse on a Cates Z-drive tug.

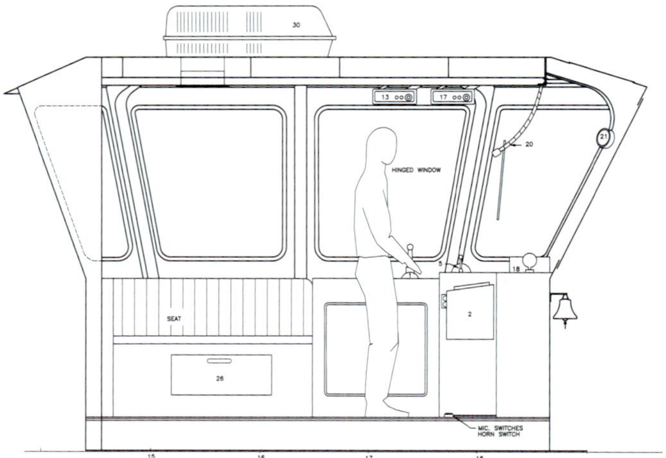

CENTERLINE ELEVATION

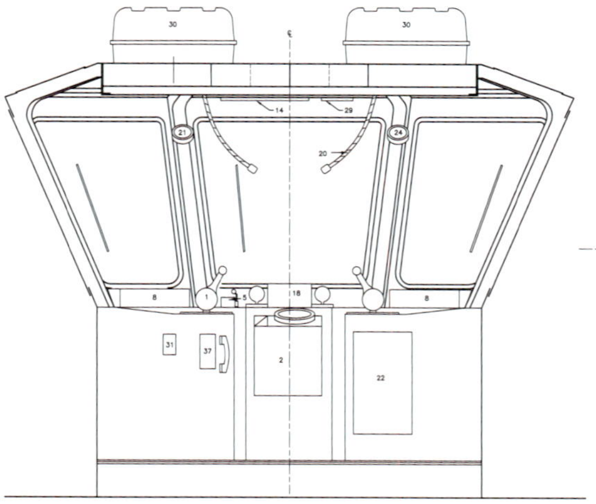

SECTION THROUGH PILOTHOUSE
LOOKING FORWARD

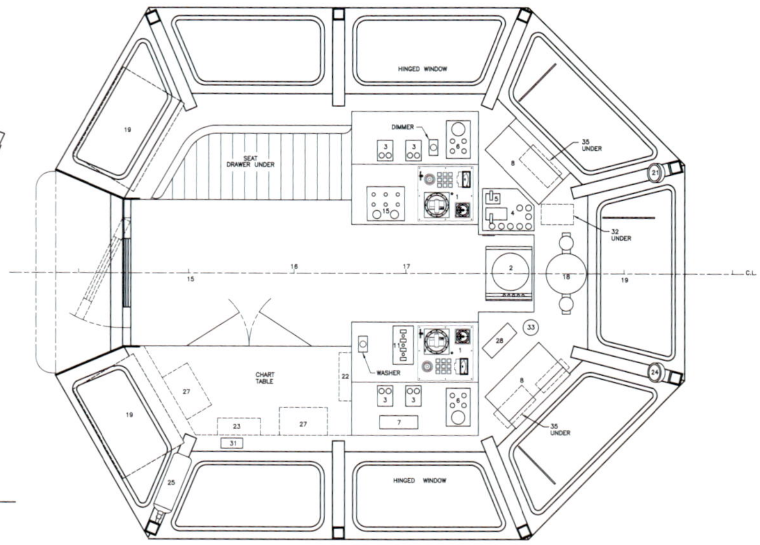

PLAN VIEW

The Cates tugs can rightly lay claim to being the first anywhere worldwide with a compact one-man wheelhouse. They also laid the groundwork for two-man tug operations in Canada based on their strong shore-based support operations and the fact there were always multiple tugs engaged in their ship-handling operations.

The vessels are equipped strictly as shift boats, with a minimal day room and a compact wheelhouse layout, where all controls are within fingertip reach of the tugmaster. The resulting visibility to all areas of the working deck is excellent.

Cates developed this "phone booth" wheelhouse concept through extensive discussion with their crews. They built a full-size wooden mock-up of the entire wheelhouse in their own workshops and worked with their tugmasters to establish a consensus on exactly what they wanted and where all the components were to be located. That mock-up was then shipped to the shipyard, who were told to "build this!" It was a superb example of collaboration between operators, owners and designers, also saving the shipyard many hours of deliberation and planning.

One of the most distinctive visible features of the early versions of this class of tug was the installation of a unique hydraulic line-handling crane on the foredeck. This device, controllable from either the wheelhouse or the winch station, was used to take the weight of the heavy polyester hawser and pass it conveniently and quickly to the deck crew on an attended ship, typically some 40 to 50 feet above the tug. As the majority of foreign-going ships have deck crews often small in both numbers and stature, lifting 50 to 60 feet of wet, heavy synthetic line was hard and difficult work. The crane made connecting a towline to these ships much faster and safer. However, the later advent of lighter high-strength lines made these cranes redundant, and they were removed. It is notable, however, that in 2022, as we explore the merits of autonomous tugboats (Chapter 26), this very line-deployment system may again become a viable system on un-manned harbour tugs.

A line-handling crane used on first-generation Cates Z-drive tugs, a possible model for autonomous tugs of the future.

Tugs for the USA

The initial success of modern cost-effective Cates Z-drive tugs soon attracted significant attention outside of British Columbia, and several variants on the design were subsequently built in the '90s for US clients, particularly Brusco Tug & Barge Inc. of Port Hueneme, California, who had a series of essentially identical tugs built by Diversified Marine Inc. in Portland between 2013 and 2017. The US fleet of these compact tugs includes the following vessels, demonstrating a clear "power creep" over time.

Eleu	1989	Hawaiian Tug & Barge Corp.	2,600 bhp
Vancouver	1992	Shaver Transportation Co.	3,000 bhp
Mamo	1996	Hawaiian Tug & Barge Corp.	3,200 bhp
Wynema Spirit	2001	Brusco Tug & Barge, Inc	3,600 bhp
Lulapin	2005	Brusco Tug & Barge, Inc	3,000 bhp
Simone Brusco	2013	Brusco Tug & Barge, Inc	4,024 bhp
Peter J Brix	2014	Brusco Tug & Barge, Inc	4,024 bhp
Bo Brusco	2014	Brusco Tug & Barge, Inc	4,732 bhp
Sarah	2015	Brusco Tug & Barge, Inc	4,732 bhp
Teresa Brusco	2017	Brusco Tug & Barge, Inc	4,732 bhp

Some of these Cates-style tugs built for US clients are illustrated below and on the following pages.

To accommodate the increased power, the later tugs were increased in beam, and the length was also increased slightly to provide larger day-room accommodation. However, the essential features of the design were unaltered, and these tugs continue to attract a lot of attention for their power and agility. The US variants of this highly successful design really marked the first serious excursions for Robert Allan Ltd into tugboat designs for an international client base. They became the launch pad for a generation of unanticipated growth for our business.

The 3,200 bhp *Mamo* for Hawaiian Tug & Barge.

Top: The 3,000 bhp tug *Vancouver*, built for Shaver Transportation of Portland; **above:** Brusco's 3,600 bhp *Wynema Spirit*.

Simone Brusco, another of the seven-tug Brusco fleet.

Compact Tugs

The next stage of the development of compact ASD tugs came with the entry of Seaspan into the Vancouver harbour towage business in 1992 with the *Seaspan Hawk* and *Seaspan Falcon*. At 24.4 metres length overall, these tugs are sufficiently large to be able to work in the more exposed conditions of the Roberts Bank terminal and yet still small enough to work efficiently in the confines of Vancouver harbour. These 3,000 bhp tugs, built at Seaspan's own Vancouver Shipyards, are outfitted only as day-boats and operate with a crew of two.

Seaspan Falcon and *Seaspan Hawk*, 3,000 bhp 40-tonne bollard pull Z-drive tugs for Vancouver harbour.

The 24-Metre Class Tugs

The success of the *Hawk* and *Falcon* led to numerous enquiries for a comparable size and power of tug, but generally clients were looking for more crew accommodation and more power. This led in 1998 to the development of a similar class of tugs for Groupe Ocean of Quebec City that were initially designated as the ASD 24/50 class (24 metres/50 tonnes bollard pull). Numerous tugs of this class were built, both for the company's own operations and for sale to other operators. Their dimensions were essentially the same as the Seaspan tugs, but the power was increased to 4,000 bhp, delivering 50 tonnes bollard pull. Although principally operated as two-man harbour tugs, these vessels could also be equipped with full accommodation for a crew of five persons.

This size and power of tug also attracted a lot of international attention, and we soon received orders throughout the '90s for similar tugs from clients in the US, South America and New Zealand.

Ocean Intrepide, one of the first ASD 24/50 class of tugs built for Groupe Ocean.

Top: *Point Valiant*, built for Eastern Towing Company of Halifax; **bottom:** A trio of 24-metre ASD tugs for Sulnorte, operating in Chile.

Blackadder, Capt. Harry and Derivatives

Two of the most interesting compact tug designs for international clients were the *Capt. Harry* for Hick Rowland of Wilmington Tug in Wilmington, Delaware, and the *Blackadder* for the Lyttelton Port Corporation of Lyttelton (Christchurch), New Zealand.

Blackadder was a challenging but interesting design, as it had to be under 24 metres length to avoid SOLAS rules, have a relatively shallow draft and deliver 55 tonnes bollard pull. It also presented some unique challenges to comply with New Zealand's unusual domestic regulations, but ultimately it evolved into a successful, innovative design. The final dimensions were 23.8 by 11 metres beam, with a maximum draft of 4.6 metres. This tug features a fully resiliently mounted deckhouse (mounted directly above the main engine casing) to limit noise in the wheelhouse to a maximum of 65 dBA. The isolation of the wheelhouse in this manner also enabled a single winch to serve both fore and aft directions of working, with the towline leading from the forward winch under the wheelhouse for aft towing, a feature clearly visible in the photo at right.

This successful design spawned a series of derivatives that is in use throughout the world, some of which are shown below.

The 60-tonne bollard pull *Bulani*, built by Sanmar in 2013 for P&O Maritime for service in smaller South African ports.

The 42-tonne bollard pull *Kaori*, built in 2004 by President Marine of Singapore for service in French Polynesia.

Blackadder, showing clearly the aft fairlead for a towline mounted on the forward winch.

General Arrangement for the tug *Blackadder* for Lyttelton Port Corporation, New Zealand.

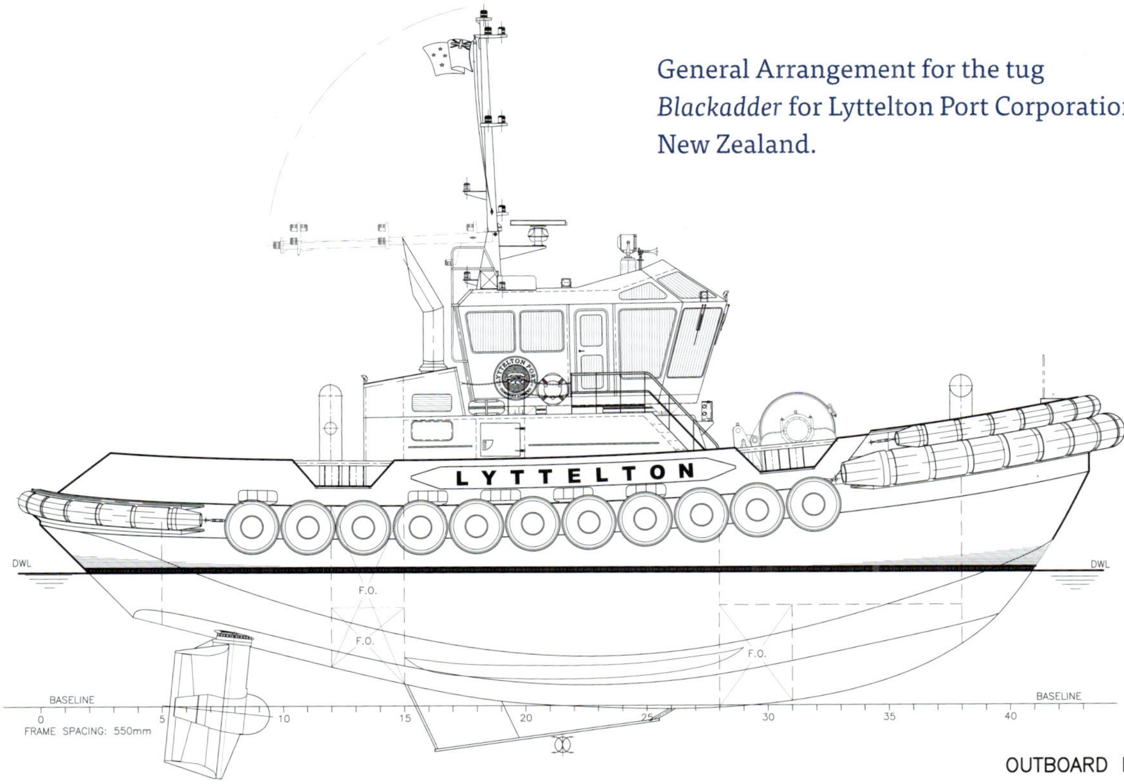

OUTBOARD PROFILE

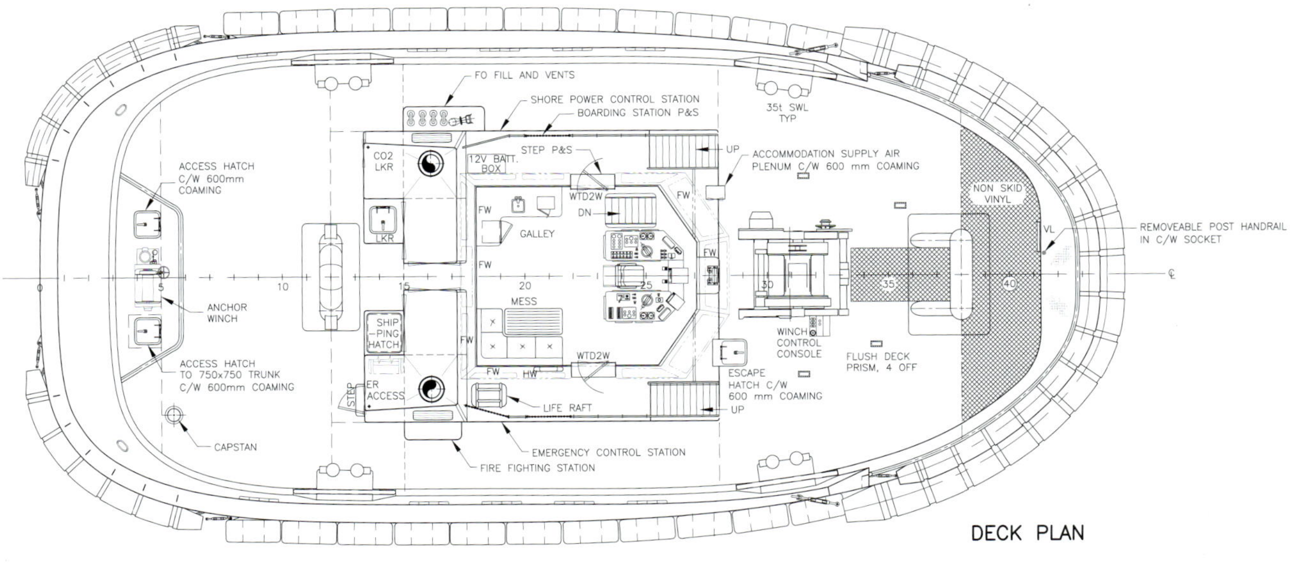

DECK PLAN

The *Capt. Harry* was the first of several compact tugs designed for Wilmington Tug of Delaware. Hick Rowland, who sadly was killed in a tragic accident some years later, was a gentleman and an insightful owner, always a distinct pleasure to work with. Serving large tankers and container ships on the Delaware River, this tug had to be capable of handling large (and empty) barges in the area, hence a minimum height of eye of 7.5 metres above water was specified, as well as a pair of barge connector winches aft. Although primarily operating in a river environment, it also had to be able to work in more exposed conditions. A semi-raised fo'c'sle design was developed to provide both a spacious accommodation area and freeboard and bow height suitable for operating comfortably in 2-to-3-metre seas. *Capt. Harry* was built by Washburn & Doughty of East Boothbay, Maine, in 2001. She was followed some years later by virtual sister tugs *Sonie* and *Madeline*.

This is another example of a design type that had broad appeal, and several similar tugs followed for clients on the US Pacific coast. The biggest uptake however was in Brazil, where SMIT and several other operators were expanding their towing operations. The Detroit Brasil Shipyard in Itajai, in southeast Brazil, started series construction of these tugs, now designated as the RAmparts 2400 class, giving Robert Allan Ltd a most useful connection into this important South American market.

Sonie, a sister to *Capt. Harry*.

Capt. Harry, named for Hick Rowland's father, became the prototype for another successful series of 24-metre tugs.

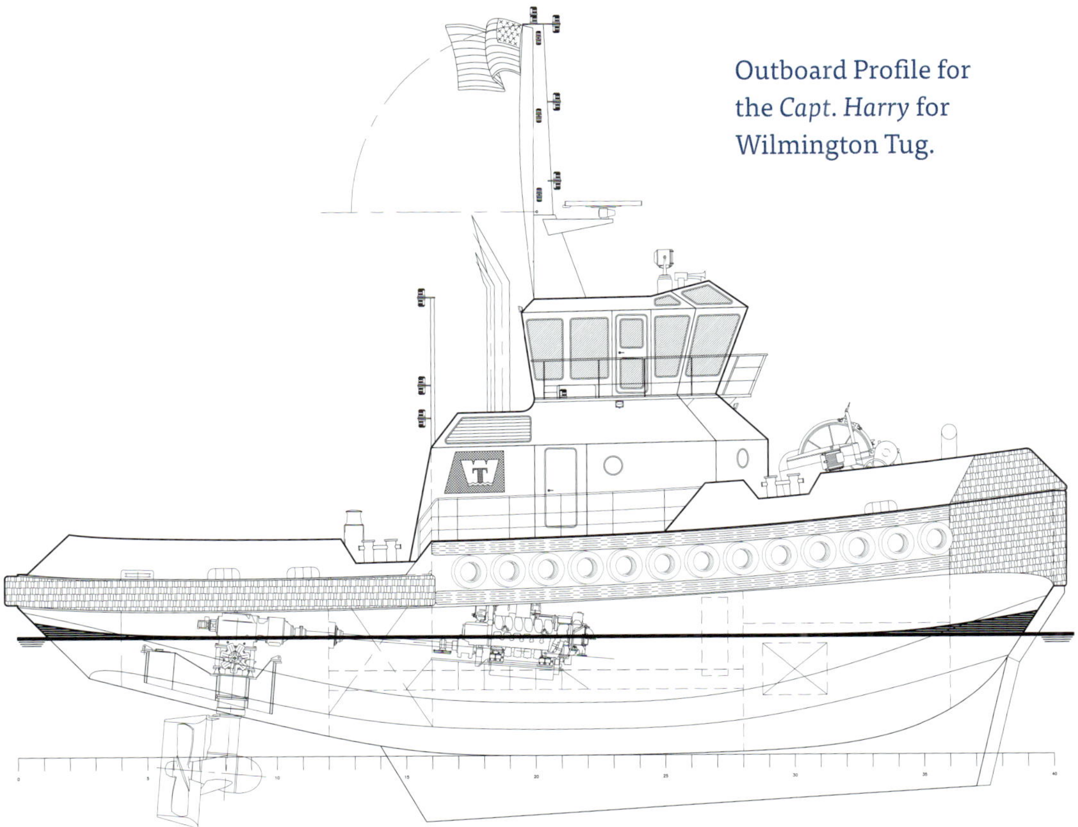

Outboard Profile for the *Capt. Harry* for Wilmington Tug.

Top: Series production of RAmparts 2400 class tugs for SMIT at the Detroit Brasil Shipyard in Itajai; **above left and right:** *Tim Quigg* and *Lela Franco*, US West Coast variants of the RAmparts 2400 design.

Dolphin Class Tugs for Foss

Another important design in the compact tug genre was for a series of ten Dolphin class tugs designed in 2003 for Foss Maritime of Seattle. These tugs measured 78 feet by 34 feet and were powered by a pair of CAT 3512 diesels rated at 2,335 bhp each, producing 65 tonnes bollard pull. That is a lot of performance from a compact tug. These were elegantly simple boats, reminiscent of the earlier Cates tugs. They had space for crew accommodation in the fo'c'sle if required, but the basic design was for a two-man day-boat. These tugs were built in Foss' own small shipyard in Astoria, Oregon, and Robert Allan Ltd provided complete production-level construction drawings.

General Arrangement for the Foss Dolphin class tugs.

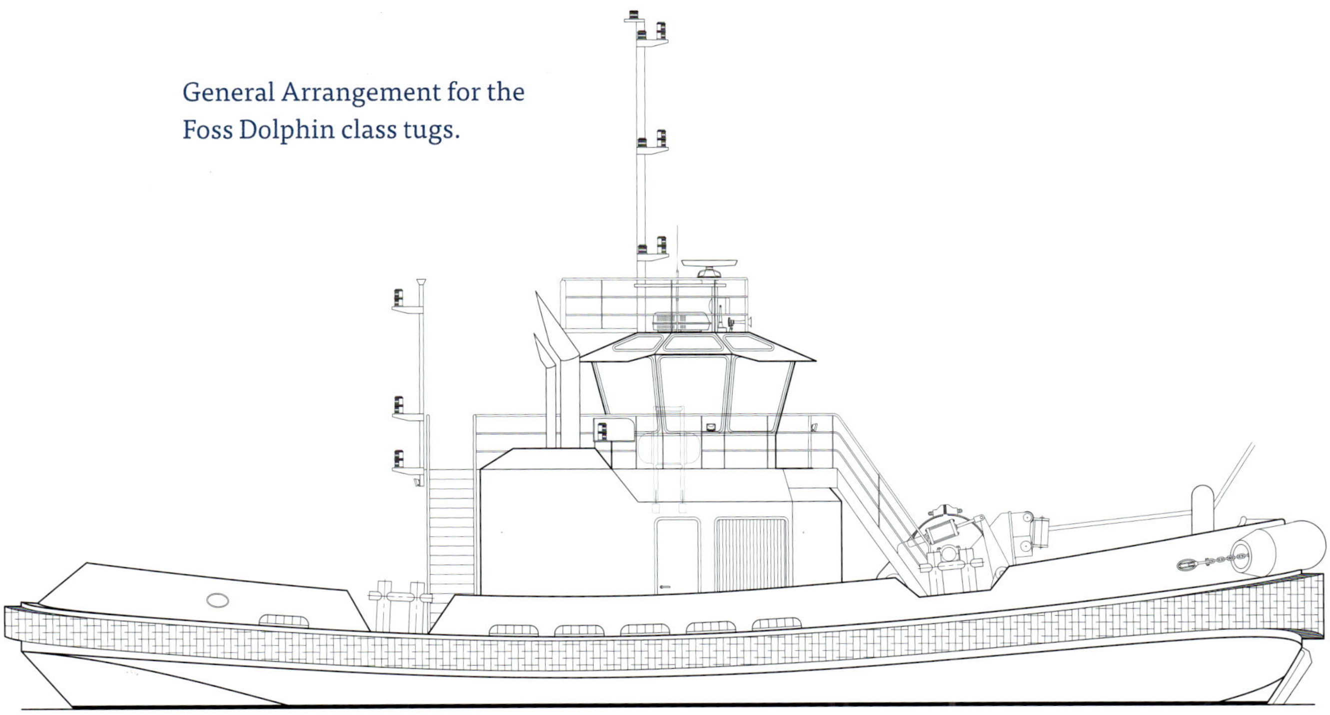

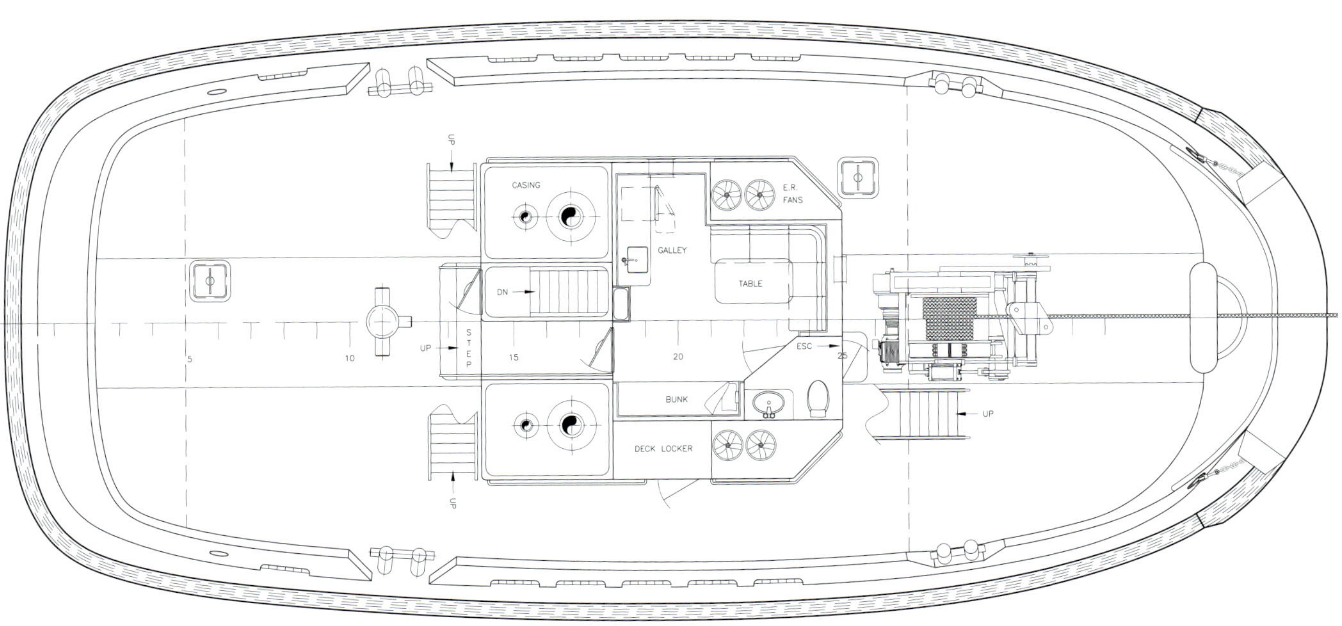

Purposeful Dolphin class tug series built for Foss Maritime, showing their clean, uncluttered layout.

World's First Hybrid Tug

In 2007 one of the Dolphin series tugs, the *Carolyn Dorothy,* was selected to become the world's first hybrid-powered tug ever built. With substantial financial support from the Ports of Los Angeles and Long Beach and the State of California, Foss embarked on an important experiment in hybrid technology in the marine environment. This tug's diesel power was complemented by a large battery bank.

A harbour tug is an ideal candidate for hybrid propulsion. Typically ship-handling tugs spend 95 percent of their time at low power, or idle, waiting to deliver 95 percent of their power for 5 percent of the time, when called upon to push or pull, so the low power demand time can be effectively met by batteries or alternatively by a motor running off one of the generator sets.

The Canadian firm of Aspin Kemp & Associates developed the unique hybrid power management system, and the batteries were intended to be from the nascent firm of Corvus Energy, another Canadian company and an emerging leader in battery design for the maritime world. Ultimately, the lithium-ion batteries intended for this project could not be delivered in the necessary quantity in time, and lead-acid batteries were used instead.

Foss has subsequently converted other tugs of the Dolphin series to hybrid propulsion.

Carolyn Dorothy, the world's first hybrid-powered tugboat.

First Z-Drives on the Mississippi

Gladys B, the first Z-drive ship-handling tug on the Mississippi River, for E.N. Bisso & Son.

In 2019 and 2020 Brusco Tug & Barge added further to its fleet with a new generation of slightly larger and more powerful tugs, most of which have been chartered to other West Coast US operators. These four tugs are 25 metres long with an excess of 6,000 bhp. Similar tugs have been designed for US operator E.N. Bisso & Son, performing ship-docking on the Mississippi River. For several years their 5,400 bhp tug *Gladys B* was the only Z-drive tug working on the "Big Muddy." This class of tug, 80 feet by 38 feet, with approximately 5,000 bhp, performs ship-assist and escort duties over 230 miles of the Mississippi River in Louisiana, and ocean towing primarily in the Gulf of Mexico and the Atlantic coast. In 2020 the *Gladys B* was joined in this arena by her near-sisters *C.D. White* and *Thomas Higgins*.

A Global Compact

The descriptions above outline the "origins of the species" but show only a fraction of the several hundreds of tugs of this compact type that have been constructed to Robert Allan Ltd designs in the past 30 years. Our shipyard clients have been building this type of tug in series and on a speculative basis, as well as to order. Sanmar of Turkey has been extremely successful with our RAmparts 2400-SX design, marketed as 24-metre Boğaçay class tugs, with its 50th edition of that design scheduled for delivery early in 2022.

Clearly there is a significant global market for these smaller-sized yet powerful tugs, reflecting a smaller capital investment for owners. The term "compact tug" has become common parlance, but we claim title to originating the term and to being the major designer of this type of tug. Compact tugs can be found around the globe today, an amazing growth since the early days of the Cates tugs in Vancouver.

Left: *Norvegia*, the first of the RAmparts 2400-SX class tugs built by Sanmar for Rimorchiatori Riuniti of Italy.

Right: The 24-metre tug *Ponga*, built in Spain for Remolques Gijonenses, S.A.

IRVING MAPLE

18

Canadian East Coast Tugboats

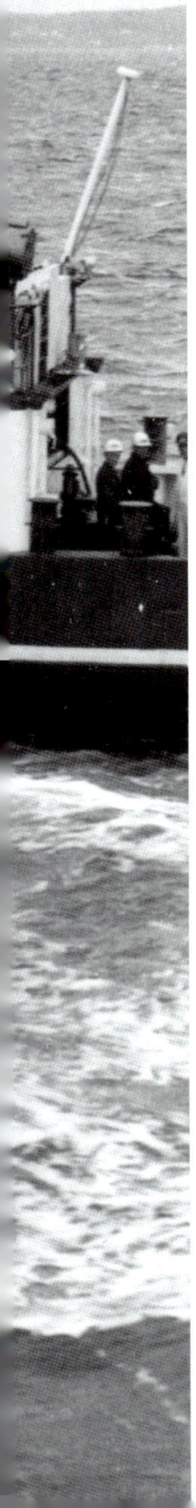

The majority of the early design work from Robert Allan Ltd was destined for the harbours and coastal waters of BC, rich in tug and barge transportation systems. That work, however, ultimately attracted the attention of tug owners elsewhere, one of the earliest of which was the Irving organization of New Brunswick, in Atlantic Canada. Bob Allan was contacted in 1965 by J.K. (Jim) Irving, who at the time was in charge of the marine side of the Irving empire. A very positive relationship was formed between these two men and a significant amount of work was done for Irving's Atlantic Towing operations over the ensuing decade.

Two Big Tugs for Irving

The first two projects for Irving were significant tug designs and started 1965 in great style for our office with Project #1-65 for a 125-foot coastal tug and project #2-65 for a 160-foot ocean-going, ice-breaking tug.

The *Irving Maple* (later *Atlantic Maple*) was a 3,200 hp ice-strengthened coastal barge-towing tug, 124 feet 6 inches by 31 feet, single-screw with a steering nozzle. The main engine was a Nohab Polar type M68T—an 8-cylinder direct-reversing diesel—driving a controllable-pitch propeller. According to Mac Mackay's *Tugfax*: "*Irving Maple* was delivered in 1966 and set to work towing oil barges for Irving Oil. Some very large barges, such as *Irving Whale* and *Irving Sealion*, were the usual consorts of this tug, which was also available for salvage and other towing work. Her fine lines and high bow made her an excellent sea boat, and she proved herself many times over in bad weather and awkward tows." After a long and distinguished career, she was scrapped in 2013.

Opposite: *Irving Maple*.

The *Irving Maple*, first of a series of powerful East Coast tugs designed for the Irving organization.

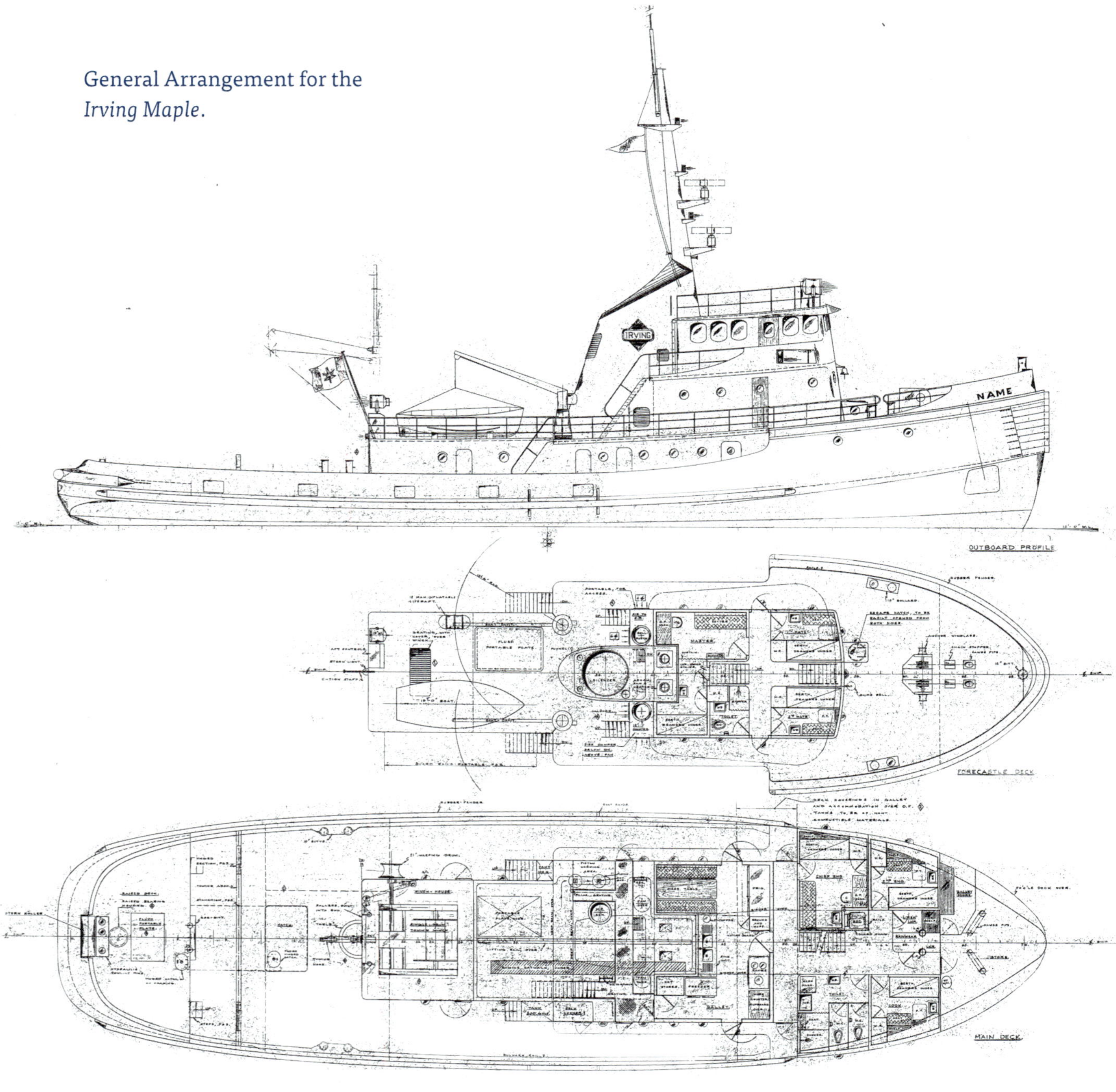

General Arrangement for the *Irving Maple*.

The *Irving Birch* (later *Atlantic Birch*) was the most powerful tug in Canadian registry for many years. Though quite modestly powered by current standards, she was a big, powerful ocean-going tug, 160 feet long by 36 feet 6 inches and rated ice class 1A Super.

She was powered by a large slow-speed, direct-reversing Nohab M69TS 9-cylinder diesel, rated 3,750 bhp at 200 rpm. The inboard profile drawing indicates the scale of this massive machine relative to the overall size of the tug. She was fitted with an open, controllable-pitch propeller for working in ice.

The *Irving Birch* was built by the Irving-owned Saint John Shipbuilding & Drydock Co. Ltd in 1967. Her typical duties included barge tows, long-distance ocean tows, and several salvage jobs. She was part of the escort to the tanker SS *Manhattan* on the eastern portion of its epic voyage through the Northwest Passage in August 1969.

General Arrangement for the ice-breaking tug *Irving Birch*.

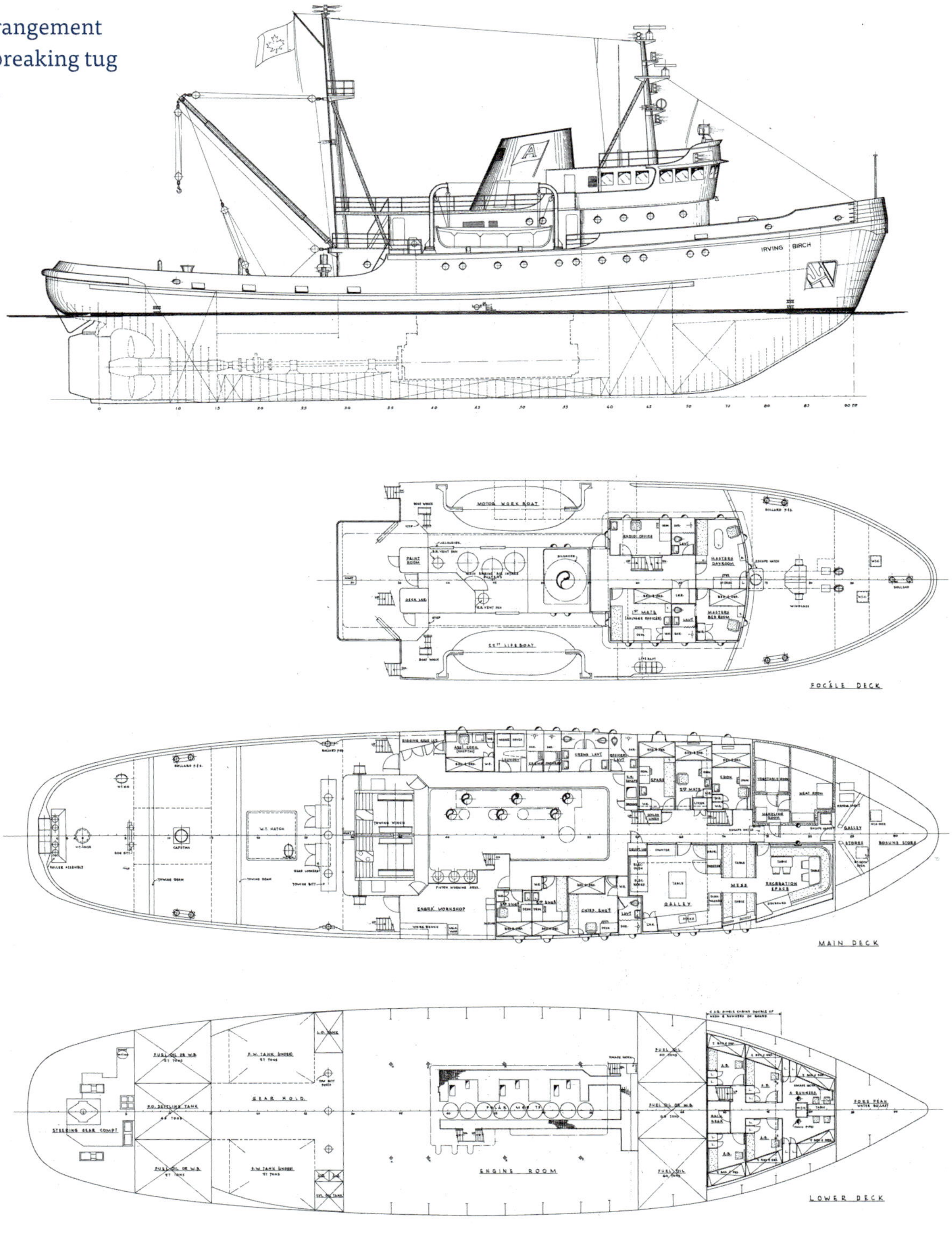

Irving Birch under way.

Inboard Profile of the ice-breaking tug *Irving Birch*.

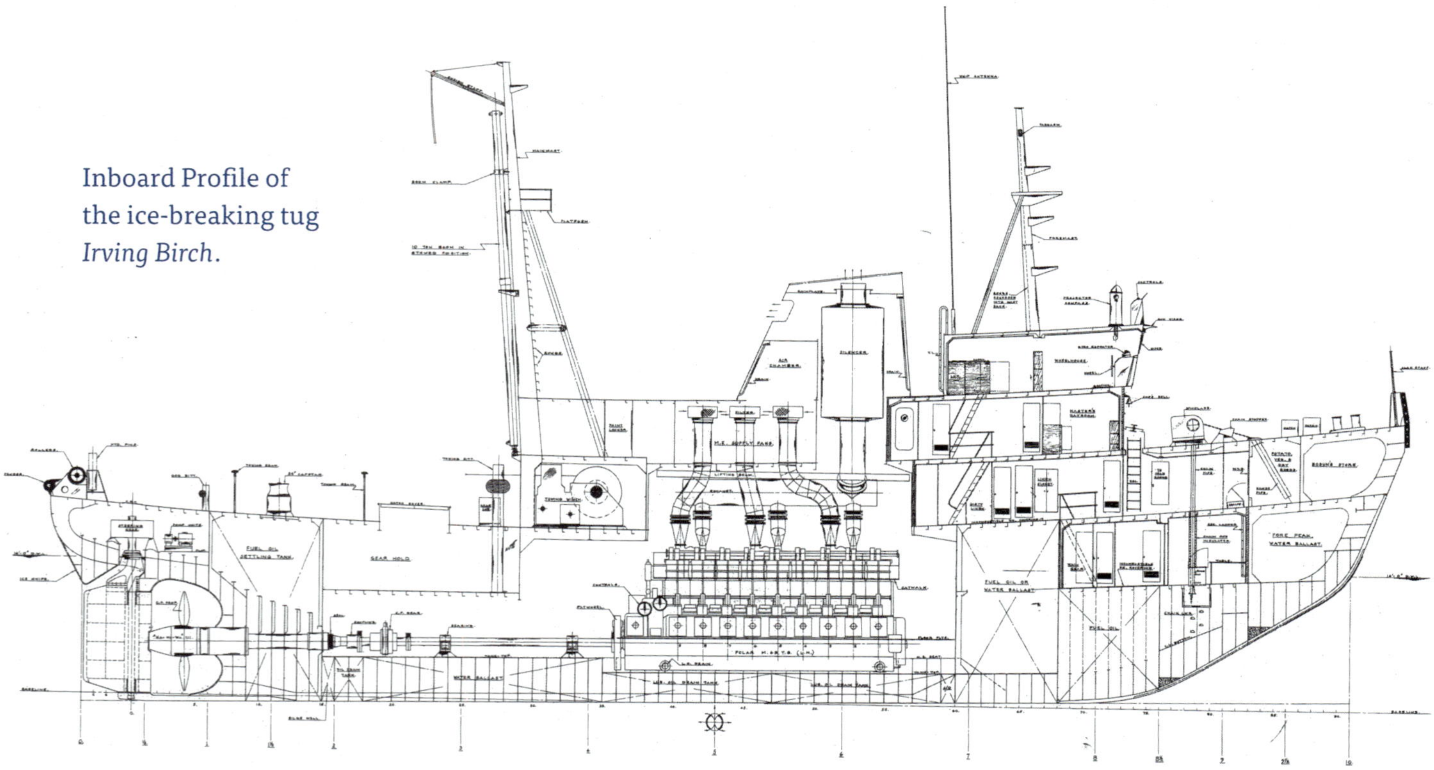

Irving Beech, a Seakindly Reputation

In 1967 Irving commissioned Bob Allan to design a near coastal/harbour tug, the *Irving Beech*. This was an interesting tug, as it had superb sea-keeping characteristics for a relatively small boat operating in often quite stormy waters. The profile and the VCG of this boat were very deliberately kept as low as possible, the result being a boat with extremely comfortable motions. These characteristics were used to good effect on numerous subsequent designs. The *Irving Beech* measured 101 feet by 28 feet and was powered by a pair of CAT D399 diesels, each rated 1,125 bhp. She was fitted with CP props in nozzles, as shown in the General Arrangement.

Irving Beech, a coastal tug with a great sea-keeping reputation.

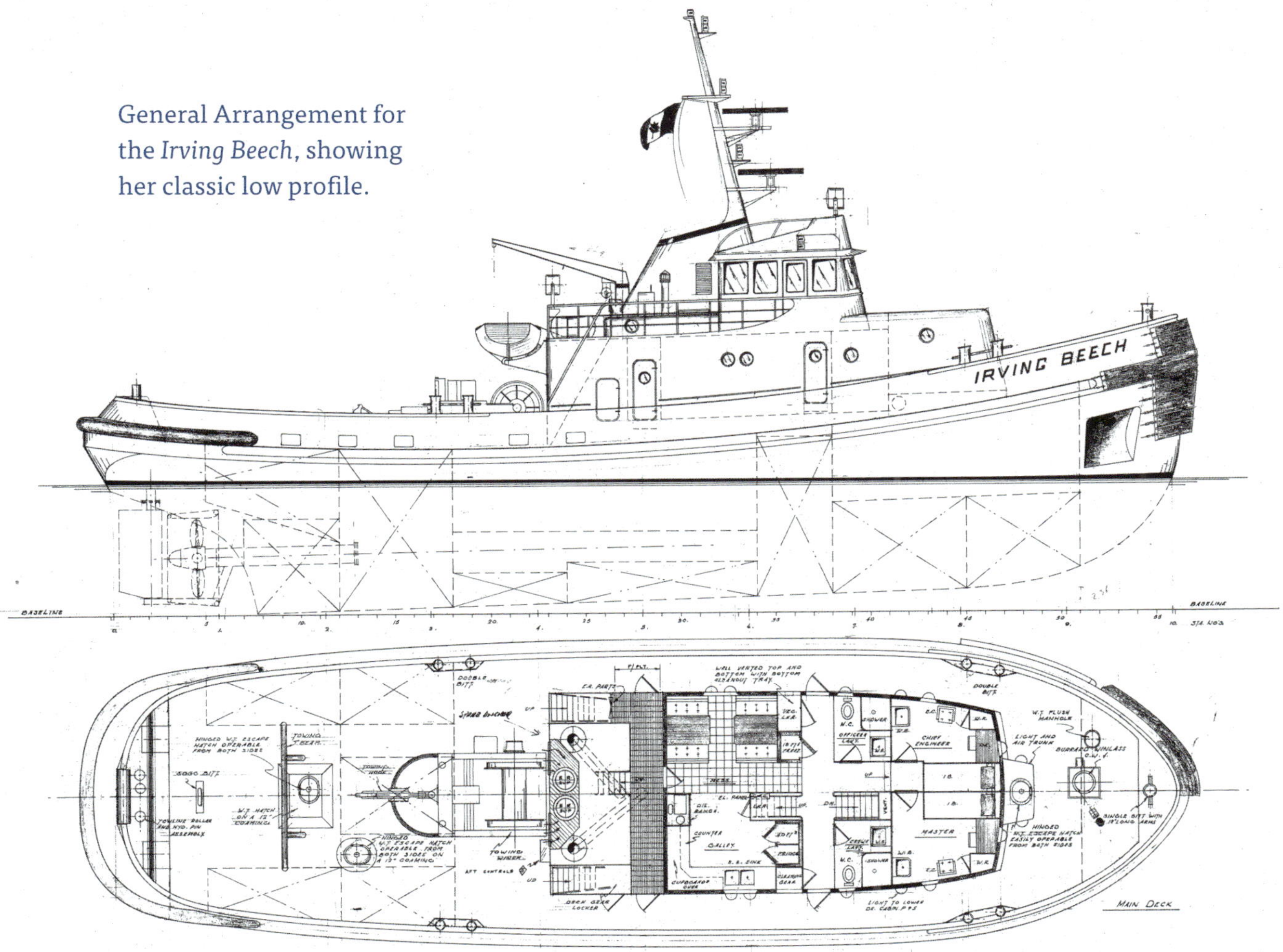

General Arrangement for the *Irving Beech*, showing her classic low profile.

Barges for the Irving Organization

Throughout the 1960s Robert Allan Ltd also designed a number of large barges for the Irving organization. This included two ocean-going newsprint carriers (Chapter 13) and a number of oil and oil/cargo combination barges, such as the *Irving Dolphin* (1965) and *Irving Whale* (1966). The latter had a rather infamous career. It sank, loaded with 4,200 tons of bunker C heavy fuel oil, during a storm on September 7, 1970, en route from Halifax, Nova Scotia, to Bathurst, New Brunswick. Oil leaked into the Atlantic Ocean from the wreck, albeit slowly after the initial sinking, but over the next 26 years as Canadian government officials deliberated on a salvage plan. (One really wonders why this could have taken so long.) Eventually a plan was devised, and the barge was recovered from the depths of the Gulf of St. Lawrence in July 1996. The *Irving Whale* was then refurbished and re-entered service as a simple deck-cargo barge, renamed the *ATL 2701*. According to the *Globe and Mail*, December 11, 2007, "The federal government sued J. D. Irving Ltd in an effort to recover the $42.4 million it spent on the salvage effort. In 2000, Ottawa settled out of court for a $5-million contribution from Irving, which had argued that the recovery mission was reckless."

Irving Elm, Modelled on the *Jervis Crown*

There was little new design work from the Irving organization throughout the balance of the 1970s, until 1978, when they contracted with Robert Allan Ltd for a new tug design, intended to be based on the *Jervis Crown* (Chapter 16). However, the owners wanted a substantial increase in power to 3,500 bhp, from a pair of MaK diesels, and the hull had to be ice-strengthened. A few other modifications were necessary to make the tug more suitable for the less temperate East Coast climate. The resulting design was the *Irving Elm* (later *Atlantic Elm*) and, like all the Irving fleet, built by Saint John Shipbuilding and Drydock Co. Ltd, in 1980. The execution on this project was not nearly a match for the workmanship in her West Coast cousin.

Atlantic Elm (ex. *Irving Elm*), a cousin to the West Coast's *Jervis Crown*.

A Production Run of Z-Drive Tugs

After another long hiatus, the Irving organization again contacted Robert Allan Ltd, in 1993, with the idea of constructing a series of Z-drive tugboats at their newly acquired East Isle Shipyard on Prince Edward Island, formerly Georgetown Shipyard. Their plan was to modernize the shipyard and construct tugs and similar workboats for the international market. Having recently completed building a series of patrol frigates for the Canadian Navy at their Saint John shipyard, they felt they could apply the same production design and block pre-outfitting techniques used on that large project to a much smaller vessel type. Robert Allan Ltd was contracted to produce a new design for a 4,000 bhp 50-tonne bollard pull tug, which at the time was considered a powerful tug in the ship-handling world. The resulting design, as illustrated, was for a tug 30.8 metres long by 11.1 metres beam, with a load draft of 4.78 metres. Propulsion was provided by a pair of CAT 3516 diesel engines rated at 2,002 bhp each, driving Aquamaster fixed-pitch Z-drive units.

The design of these tugs was somewhat of a milestone for our company. It was the first time we had been contracted directly by a shipyard for a series production run of tugs, but it was also the biggest Z-drive tug we had designed until that time. The shipyard also asked us to quote for the development of a full set of production engineering drawings. We were not unused to that task, at

General Arrangement for the Irving 30.8-metre 50-tonne bollard pull ASD tugs.

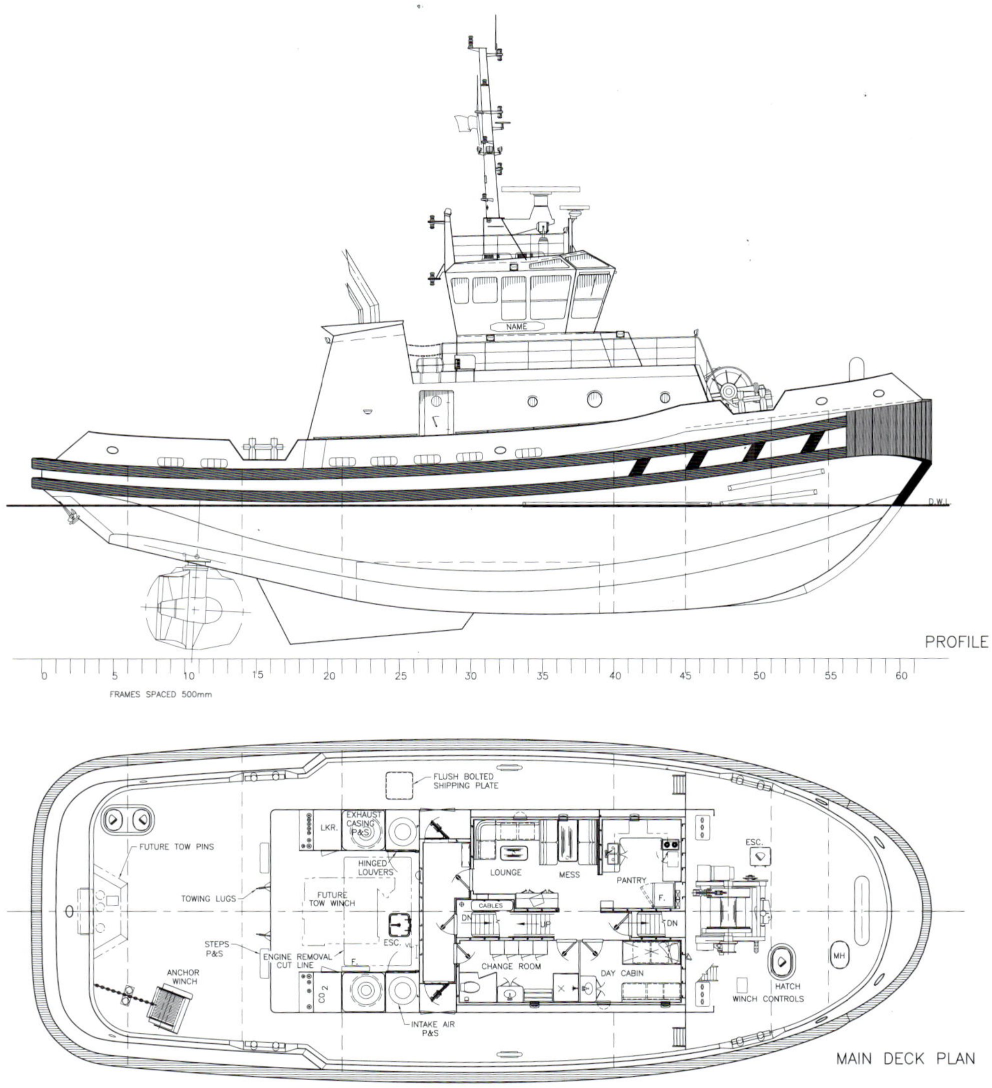

least for doing steel parts design (for CNC cutting), but these tugs were to be built using advanced zone pre-outfitting and product-by-stage construction techniques.

In order to accomplish this, it was necessary to pre-define and design absolutely every single component in the ship. This applied to all major and minor steelwork, outfitting, machinery, piping and electrical work. All this design work was performed by Robert Allan Ltd using advanced CAD techniques, with all structural parts lofted and fully developed in our offices in Vancouver, then transmitted electronically to the shipyard more than 6,000 kilometres distant. This in-depth involvement with the production design of a tug was a valuable, if somewhat painstaking, process and has occasioned our design staff to consider the construction process more thoroughly in the course of our normal design work. We had, however, grossly underestimated the level of effort involved in such a task and had to go "back to the well" a couple of times throughout the process in order to avoid completely collapsing our company. This was an unpleasant and painful exercise, but we ultimately completed all the work and the shipyard began production.

Much of the work done in this process was the subject of yet another paper presented by Rob Allan to an International Tug Conference, this time in 1994, entitled "The Design of Ship Assist Tugs: Towards More Cost-Effective Construction." The paper highlighted many of the innovative features we introduced with this design to facilitate pre-outfitting and ideally reduce overall construction costs.

Many shipboard systems, such as bilge, fire and ballast water, fuel-oil treatment, compressed air and similar systems, lend themselves readily to modularization, where all elements of machinery and fittings required to perform a particular function are preassembled on a common base, complete with motors, starters, manifolds, etc. The geometry of the base must of course be defined to fit sensibly into the machinery space. The performance of these systems could then be tested and accepted in the shop. Shipboard installation then consists of simply bolting a module to a foundation and connecting electrical and piping terminals.

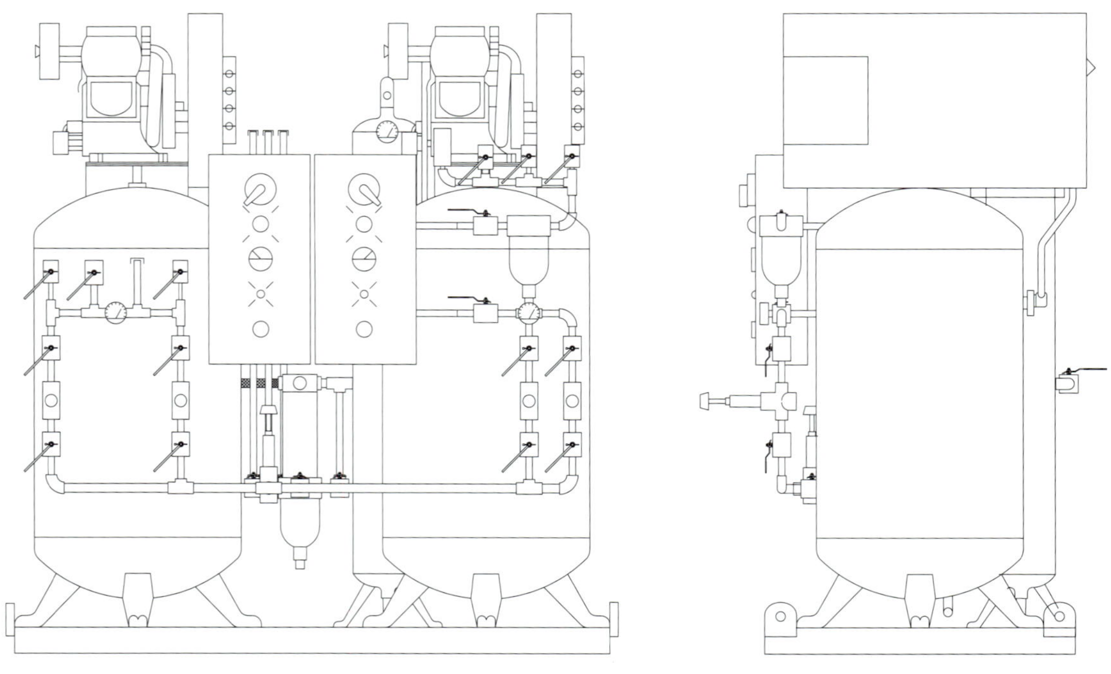

Example of a machinery system module designed for installation in this series of 4,000 hp ASD tugs for East Isle Shipyard.

The figure shown here illustrates a module of this type created for the project (without a lot of the drawing detail). The supplier was charged with creating these modules to meet a predefined set of criteria that included performance specifications and space and weight limits. After some initial resistance to this concept from suppliers, it worked exceptionally well, avoiding a lot of time-consuming fitting and rework on board the tug during construction.

East Isle shipyard built about 30 of these tugs before the yard closed permanently in 2011, which included some export success selling to clients in Europe, the Caribbean and to the Panama Canal Authority; ten of them remain today in Atlantic Canada as the backbone of Irving's Atlantic Towing Ltd (ATL) fleet.

Three of the earliest tugs of the ATL series, *Atlantic Fir*, *Atlantic Spruce* and *Atlantic Hemlock*, in St. John's, Newfoundland.

On Shipyards Taking Liberties

After all the effort involved in developing this design into a highly effective and good-looking tugboat, the shipyard took some liberties with this design with the objective of reducing their construction costs, without considering the functional impact of those changes. A good example of this was their removal of the chamfer at the top of the deckhouse beside and forward of the wheelhouse to save a few man-hours of work. The chamfer was a feature necessary for good visibility from the wheelhouse to the crew on the working deck and to the side fenders. When I was subsequently dealing with the Panama Canal Authority on other projects, this "design error" was rather emphatically brought to my attention by Capt. Max Newman, at that time the head of the Panama Canal Authority (ACP) tug division. I was able to quickly explain to Max, who subsequently became a good friend, that this was entirely out of our control and not reflective of our original design.

—R.G. Allan

This was a rather enigmatic project. The experience with a demanding client was often unpleasant, but in the final analysis the tugs were extremely effective and got great performance reviews (in spite of the shipyard modifications). But the real silver lining from this project was the fact that the Irving organization sold two of the first tugs, *Atlantic Fir* and *Atlantic Spruce*, to a tug owner in Norway, Johannes Østensjø, who became one of our most favoured clients and with whom we did some really ground-breaking design work (Chapter 19). The *Atlantic Fir* and *Atlantic Spruce* went on to work in Ireland and in Norway as the *Alex* and *Felix*, respectively (all Østensjø's tugs had names ending in "X").

These tugs travelled to Europe on their own bottoms through the North Atlantic in winter. In spite of encountering some large seas and storm conditions, the skipper of one tug described his charge as "a really fast little bas*ard" which "soared like a bird" in the heavy seas. He was a fan!

Top: *Atlantic Spruce* as *Felix*, chilling out in her charming home port of Haugesund, Norway; **bottom:** *Spitfire III*, one of the last of the 50-tonne bollard pull tug series built by East Isle Shipyard, showing some of the shipyard-initiated design alterations, such as boxy superstructure and too vertical bulwarks.

Atlantic Fir, now *Alex*, in service in UK waters for Johannes Østensjø.

Tug Series for Groupe Ocean

The Irving organization was not our only East Coast client, but it was certainly the largest. We had also established a relationship with Groupe Ocean of Quebec in about 1994, when they constructed a series of 24-metre tugs at their own small shipyard on L'Isle-aux-Coudres, in the St. Lawrence River. These tugs were extremely well built under the guidance of Serge Genois, who was the GM of the yard at that time. Serge had some clever ideas about efficient construction of such vessels, and he was a pleasure to work with. The workmanship from this small yard, in an area noted for its history of boatbuilding craftsmen, was excellent, probably the best we have seen from a Canadian shipyard.

We subsequently did further work with Groupe Ocean, who also purchased a few of the 50-tonne bollard pull East Isle-built tugs from Irving. The latest venture was in 2011, for a pair of 36-metre 6,000 kW high-ice-class tugs for Groupe Ocean's own fleet, the *Ocean Tundra* and *Ocean Taiga*, also built at the Isles aux Coudres shipyard. These tugs had the conflicting requirements of heavy icebreaking and performing tanker escort, which require rather different hull form characteristics. However, as with many projects, a successful compromise was achieved and the end result is an extremely powerful and capable tug design that became the prototype for several subsequent heavy-duty icebreaking tugs (Chapter 19).

Ocean Intrepide, one of numerous 25 metre ASD tugs built by Groupe Ocean for their own service and for others.

RT BEAGLE BAY
NASSAU
RESCUE

19

Taking a Global Perspective

The Outlook in the '90s

In the early 1990s the overall outlook for our small business was certainly not rosy. We were managing to keep moderately busy (but marginally surviving in economic terms), with a wide range of bits and pieces of consulting work, some refit designs and a few studies for the likes of Transport Canada, mostly related to the towing industry and vessel safety. There were very few new designs completed, except for a couple of small tugs. One notable exception was the design of the 60-foot fast commuter yacht *Malaspina Express* for local businessman Dick Barr. Yacht design wasn't really in our wheelhouse, but thanks to Hans Muhlert's expertise in fast-craft design, this turned out to be a successful if modestly outfitted boat, in accordance with the owner's wishes.

Opposite: RT *Beagle Bay*, see page 464.

At the end of '92 and in early '93, however, things slowly began to improve. Late in '92 Seaspan commissioned the design of the 80-foot Seaspan Hawk class ASD tugs, and early in '93 the Irving organization commissioned the design of the 30.8-metre 50-tonne class of Z-drive tugs (Chapter 18). Both tug projects serendipitously brought significant amounts of follow-up work from offshore. Gradually, an increasing number of enquiries for new tug designs was fielded. Significant work was still quite sporadic, but prospects were definitely looking brighter, and more and more enquiries were coming from overseas.

Top: *Malaspina Express*, a 40-knot commuter yacht for the Vancouver–Pender Harbour run; **bottom:** *Seaspan Hawk* and *Seaspan Falcon*, a duo of 80-foot 40-tonne bollard pull tugs for Vancouver that struck a chord with many owners around the world, becoming the prototype for an extensive series of derivative designs.

Business from South of the Border

The most encouraging signs of expanding horizons came initially from the US, with clients in Hawaii and the US Pacific Northwest ordering smaller tug designs based largely on the Cates 78-foot Z-drive designs (Chapter 17). The first of these clients was Hawaiian Tug & Barge Ltd (HTB) with their order for the 2,600 bhp *Eleu*, almost a duplicate of the *Cates I*. After the tug was delivered on a barge to Hawaii, Cates offered the services of Capt. Hugh Wallace to provide extensive training for the HTB tug skippers in the operation of these Z-drive boats, something never seen in those waters before but a pretty nice gig for Hugh, as he was able to pass on some of his considerable ship-handling expertise to the Hawaiian tugmasters.

Eleu was followed by the *Vancouver*, for Shaver Transportation of Portland, the *Mamo* for HTB, and later a series of similar tugs for Brusco Tug & Barge of Port Hueneme, California. The acceptance of a Canadian designer by these American owners opened many doors for us.

Eleu, modelled on the Vancouver Cates tugs, was the start of our expanding export tug design business.

Chapter 19

Tanker Escort Tug Developments

The old saw "every cloud has a silver lining" was never truer than in the case of the *Exxon Valdez* grounding and the associated 1989 oil spill in Prince William Sound, Alaska. As catastrophic an incident as it was, the resulting legislation, known as the Oil Pollution Act of 1990 (OPA '90), effectively brought the genre of tanker escort tugs into existence. Similar legislation was enacted shortly afterward in Europe after closely spaced oil-spill incidents with the tankers *Aegean Sea* and *Braer*, in late 1992 and early 1993, respectively.

In essence, OPA '90 and other similar legislation required that all tankers in near-coastal waters must be accompanied by a tug of sufficient size and power to provide emergency steering and braking forces to control a tanker (and prevent a grounding) in the event of a failure of the tanker's main propulsion or steering system. The challenge was that conventional forms of tugs were not well suited to this task, which involves applying large heeling forces to the tug far in excess of anything contemplated in normal towing operations. Many in the towing business predicted dire results for tugs in this service, but a few companies, Robert Allan Ltd prominent among them, rose to the challenge of identifying what sort of hull form and propulsion configuration would be best and safest in this new escort role.

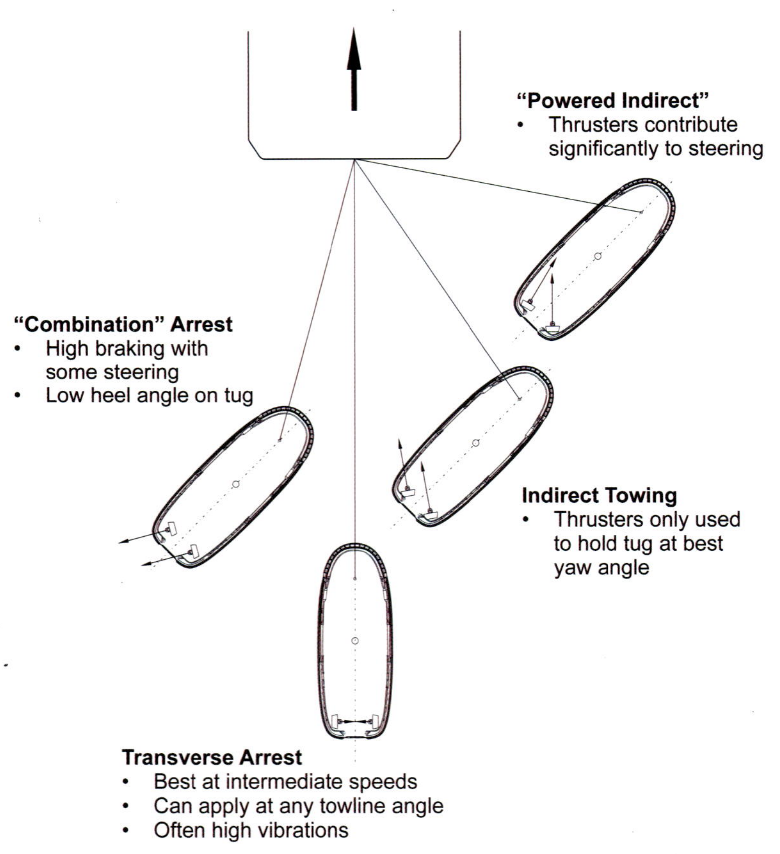

Diagram illustrating the different modes of towing for escort operations.

The basic challenge to designers was to develop a hull form and the associated propulsion that could safely generate high lateral and braking forces on the attended ship at fairly high speeds in the range of 8–10 knots. High bollard pull, which is by definition measured at zero speed, simply isn't enough, as there is not much excess thrust available when the tug is moving fast itself. The solution was to design tugs which could work in the "indirect" mode, using the hull to generate large hydrodynamic forces that can be developed when the tug is held at oblique angles to the attended ship by its own thruster forces, as illustrated showing an ASD tug. In this attitude the tug generates Steering Forces at right angles to the ship and Braking Forces aligned with the longitudinal axis of the ship.

These were nascent days in the field of designing specialized tugs for tanker escort duty. Tugs with cycloidal propulsion, by default the Voith Schneider Propeller (VSP), had an inherent advantage in this work, as they needed a large skeg for directional stability with the VSP drives, and the quick response and controllable pitch of the Voith drive system were necessary attributes for effective performance in this totally new mode of tug operation. Obviously, these transverse forces imposed high heeling loads on the tug and so a high standard of stability was also essential for safe indirect operations. The indirect mode of tug operation was inherent in the way that a VSP-propelled tugboat operated, using its large skeg to develop side forces. Voith worked hard and very effectively to convince oil companies and project influencers around the world that their propulsion system was the *only* solution for escort tug propulsion, and the very earliest big escort tugs indeed had large VSP drives, considered by many (and rightly so) to be the Best Available Technology at the time. But the times were a'changing!

To address the requirements of OPA '90, the State of Alaska required that a new generation of escort tugs be built in order to prevent another *Exxon Valdez*-type incident from occurring again. Robert Allan Ltd was contacted for design proposals by several American tug owners vying for this large contract, but almost unanimously they wanted to explore how Z-drives could be used to achieve the specified escort performance. Our clients wanted tugs that could be more versatile, should the escort contracts not continue, and Z-drives were much preferred over VSP drives for that versatility, in addition to being substantially less expensive.

Unfortunately, the earliest Z-drive tugs designed and built in Europe to address escort requirements were not well conceived for the task. Built as large deep-sea tugs, they all had towing points on high forecastles and low freeboard on their aft decks. They had no skegs to help develop indirect forces and simply could not compete with similar sizes and powers of VSP-propelled tugs. Though these were powerful deep-sea towing or rescue tugs, they simply weren't suited to indirect escort operations. The use of Z-drives for escort-tug service was set back at least a decade due to this false start.

To prove that tugs with Z-drives could be developed effectively for this important role, Robert Allan Ltd embarked on an extensive in-house research effort. Using an existing physical model of a conventional ASD tug, we worked with Vancouver-based Offshore Research Ltd at the modest tank test facilities then-available at the University of British Columbia to examine the merits of a wide range of hull modifications and appendages in increasing the indirect-force-generating capability of a tug hull. This was quite rudimentary research, but it enabled us to see the path forward to what later became our RAstar hull form, a sponsoned hull shape with a large forward-biased skeg, which has redefined escort tug geometry and performance since the late '90s.

Ultimately, the contract for these escort services in Alaska went to Crowley Maritime for a VSP-powered tug not to our design, but the ASD hull form concepts developed through our research attracted the attention of several major tug operators. The associated analysis also clearly demonstrated that the entire escort scenario demanded not only high indirect forces at the higher speeds but also, as ship speed comes down, higher bollard pull, critically important to bring about the entire ship-stopping operation in less time and less distance.

Chapter 19

Breaking into Europe: The Østensjø Connection

At about this same time, the Atlantic Towing 4,000 hp tug series was in production in Atlantic Canada and attracting attention, but the most significant aspect of that was the sale of two of those tugs to Østensjø Rederi AS of Norway. Rob Allan had met Johannes Østensjø a few years earlier at an international tug conference, and Johannes was very keen on the Cates-style tugs but didn't believe he could sell the concept of such a small boat to his European crews (or his clients). The Atlantic Towing tugs were much closer to the norm of European tugs in terms of size and power in those days, and so were more readily accepted, and Johannes bought several of those tugs for service in Norway and in the UK. That established one of our most fortuitous connections into the European market, an arena which we had previously considered impossible to breach, with numerous well-regarded designers operating close to most of the world's major tug owners. But the Østensjø connection was made and an opportunity to work directly for them arose in 1997 when they were bidding to provide a major new tanker-escort tug for the Sture oil terminal on Norway's west coast. The technical director at Østensjø, Carl-Johan Amundsen, was another third-generation naval architect with whom Rob established an immediate rapport. Østensjø won the bid for tug services at Sture, but they were told in no uncertain terms that the tug they were to provide must be VSP propelled, regardless of our proposals showing equivalent performance from other options.

The Pilgrimage to Heidenheim

Having established that this big tug was certainly going to be VSP propelled, Carl-Johan Amundsen and I travelled to Heidenheim, Germany, to meet with Voith and discuss the technical aspects of the project. Since the earliest days of my involvement in this industry, Voith's marine division had been governed by gentlemen who might be characterized as zealots about their product. In the 1960s that person was Dipl. Ing. W. Baer, who made quite a name for himself at the early tug conferences by decrying the attempts by Schottel to intrude into "his" ship-handling tug market using the then newly developed Z-drive units and daring to call these "tractor tugs," a term originally coined by Voith. Baer was succeeded by Ulrich Sturmhofel, who was cast from the same mould.

Carl-Johan and I had both been advocating publicly that properly designed Z-drive tugs could also do this escort work, but needed new and innovative hull forms, and we entered Sturmhofel's offices with that reputation. For the first three-quarters of an hour, he ranted at the pair of us for being so ignorant as to believe we could design anything equal to a Voith-propelled tug, and he as much as called us both fools. I was prepared (begrudgingly) to accept this as a mere consultant, but Carl-Johan was there as a client with the authority to sign a multi-million-Euro contract. Strange sales technique! We accepted our rebuke in silence and eventually went on to a more amicable discussion of the technicalities involved.

The later stages of this contract were handled for Voith by Dr. Jens Bartels, with whom I ended up doing a considerable amount of work on later projects. Jens was a much more pragmatic person and a pleasure to deal with always.

—R.G. Allan

The Escort Tug *Ajax*

The design that evolved after the discussions with Voith was for the *Ajax*, a big, powerful and thoroughly researched escort tug that significantly raised the bar for tanker-escort tug designs worldwide and is still rated as one of the most powerful escort tugs afloat today. The design was extensively model-tested at the National Research Council of Canada ship-model test-tank facility in St. John's, Newfoundland, to verify that the performance in all escort modes could be achieved. The novel hull design exceeded all expectations, and the test program and results were the subject of a major paper, "The Development of a New Generation of High-Performance Escort Tug," by Allan, Bartels & Molyneux, presented to ITS 2000 in Jersey.

Ajax measures 41.6 metres length overall and has a beam at deck of 15.9 metres. Her waterline beam is only 14.2 metres however, reflecting the presence of the large sponsons that start just below the waterline amidships. The hull depth is 6.25 metres, and the overall draft to the underside of the VSP drives is 6.8 metres. She is powered by a pair of CAT 3612 diesel engines rated 3,460 kW each and driving Voith model 36GII/270 cycloidal propellers through Voith turbo couplings. Each engine also drives, through a front-end PTO, a firefighting pump with an output of 1500 m^3/hr, satisfying a class Fi-Fi I rating. *Ajax* develops an impressive 93 tonnes of bollard pull and generates 160 tonnes of steering force at 10 knots in the indirect mode.

Ajax performing indirect towing at 10 knots in storm conditions. The towline is visible off the port stern, as this type of VSP tug escorts running stern-first.

General Arrangement drawings for *Ajax* showing the very effective "split-level" layout to keep the overall profile as low as possible.

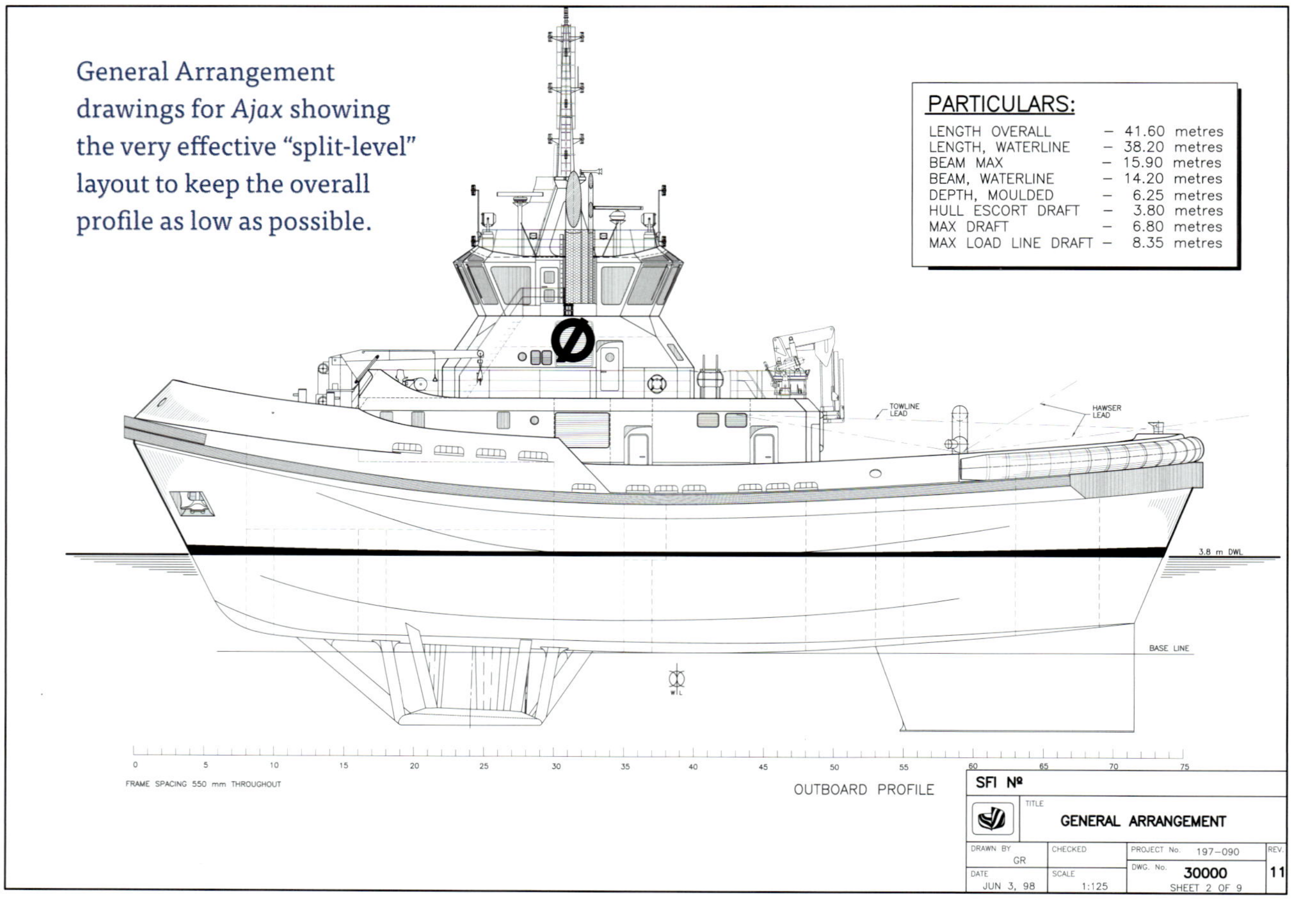

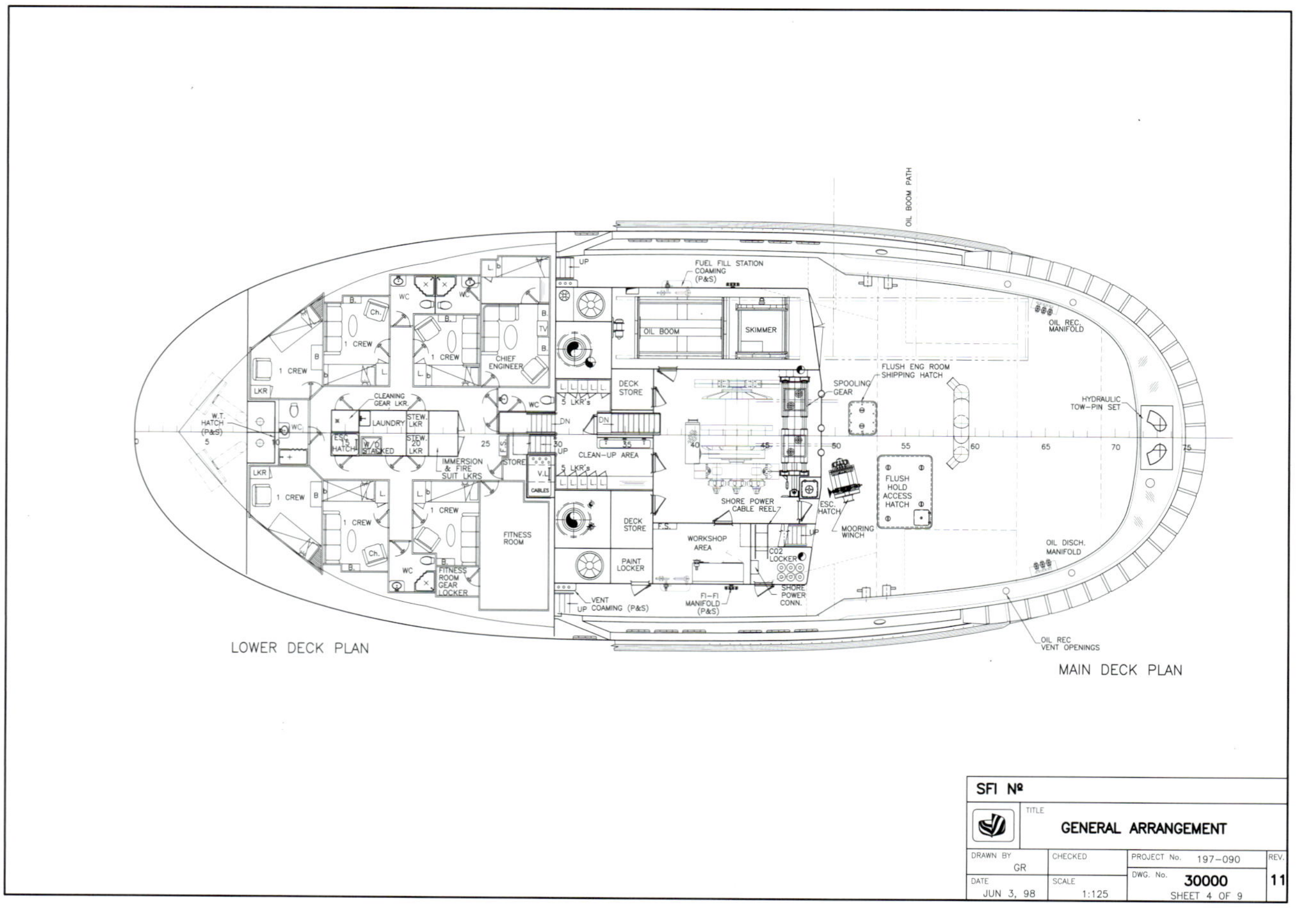

Velox Class Escort Tugs

The success of *Ajax* led to further collaborative projects with Østensjø Rederi. In 2001 it won a contract for tanker-escort services in Southampton, UK, for which a tug smaller than *Ajax* was required, but with similar performance. The result was the 36-metre Velox class of tugs, which had a refined hull form, a very large skeg with a rotating cylinder in the leading edge for enhanced indirect performance, as illustrated in the Outboard Profile drawing. Four of these novel tugs, *Velox*, *Tenax*, *Phenix* and *Apex*, were built by the Spanish shipyard Astilleros Gondan from 2005 through 2008. These high-performance escort tugs, 36 metres length overall by 14 metres beam, were, like *Ajax*, also VSP propelled, using Voith model 32 GII/225 drive units driven by Bergen C25:33L8 diesels rated 2,500 kW each.

These tugs achieved a more modest 65 tonnes of bollard pull but developed 131 tonnes of indirect steering force at 10 knots. Another innovation on these tugs was the split towing staple that served the divided towing drum, one part of which has a steel towline cable and the other a HDPE synthetic hawser. The two towline materials are incompatible on a common bearing surface, so the staple, as illustrated, had an upper eye for the wire and a wide slot just above deck level for the escort hawser. This slot keeps the hawser low and enables it to migrate outboard under load to reduce the heeling arm, thus increasing the allowable escort steering force.

Østensjø has proven to be a model client and ship-owner. The company shops foremost for quality in the builders it uses and deals only with those it trusts the most. It also outfits its boats to the very highest standards, believing that if the boats are well-built and look like a quality "home," the crews will treat the boats accordingly, which certainly seems to be the case. Johannes personally chooses original art to be mounted in all his boats, enhancing that personal effect.

The wide-based dual-port escort-towing staple on the Velox class escort tugs.

Velox in dock, showing the large Hi-Lift skeg and the Voith "Turbo-Fin" rotating cylinder in the leading edge. The cylinder produces lift locally using the Magnus effect, improving the overall efficiency of the skeg.

Outboard Profile for Velox class escort tugs showing the large escort skeg. The steering flap on the trailing edge of the skeg indicated on the drawing was not kept in the final design but could have been, as the big skeg was a challenge to the directional stability of these tugs.

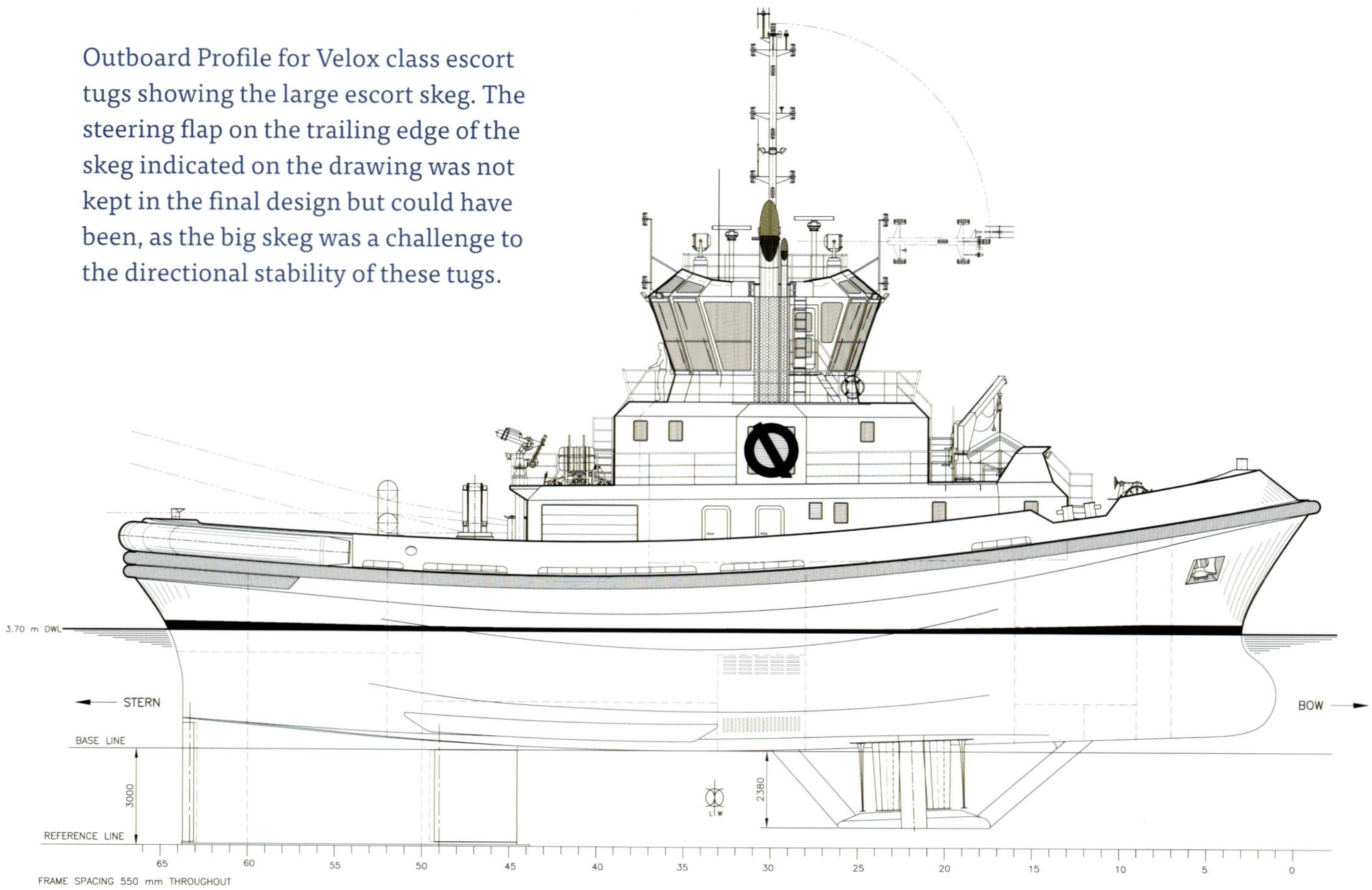

Traktor 3600v class tug *Velox* in both directions, showing the more traditional sea-going bow (top) and its "business end" running stern-first. The side sponson shape is quite evident in these photos.

The quality outfitting aboard *Tenax*, indicative of all the Østensjø tugs.

Below: Trio of Østensjø's Velox class escort tugboats.

First-Class Owner Meets First-Class Builder

One of the great pleasures in this business occurs when a good ship-owner client begins to work with one of our favoured shipbuilding clients. The results are almost always positive and can result in strong long-term relationships. Such a case began in 2006, when Østensjø Rederi commissioned the 32-metre ASD tug *Vivax* from the Turkish shipyard Sanmar Denizcilik Ltd.

Delivered in 2008, *Vivax* is one of Robert Allan Ltd's RAmparts class of ship-handling/multi-purpose tugs developed for series production in shipyards.

This 32-by-12-metre 4,800 kW tug has a non-sponsoned hull form designed for general harbour and ocean towage, as well as for limited escort operations. It is equipped with oil-recovery and firefighting equipment, and fully equipped for open-water salvage.

In 2012 Østensjø returned again to Sanmar, this time contracting for a somewhat smaller but versatile RAstar 2800 class tug, the 28-by-12.6-metre *Lomax*. Its 4,700 kW produces 80 tonnes of bollard pull.

Sanmar-built RAmparts 3200 class tug *Vivax* for Østensjø Rederi.

Below: *Lomax*, another Østensjø-Sanmar-Robert Allan collaboration.

A Multi-Functional VSP Tug

In the continuing tradition of building vessels only to the highest standards, Solent Towage Ltd of Southampton, UK, a division of Østensjø Rederi, took delivery of the AVT class tug *Vortex* from Spanish builder Astilleros Gondan SA in 2010. The VSP-powered *Vortex*, 37.5 metres long by 14 metres beam, was designed to perform a wide range of tasks, including ship-handling, tanker escort, ocean-towing, salvage, and anchor-handling duties. Accordingly, the design is somewhat unique for a true tractor, where the "stern" is heavily fendered for ship-handling operations but then is also fitted with a stern roller and Karm-Forks, etc., for anchor-handling operations. The aft-end geometry is thus also quite different to other tugs of this AVT series, with a much broader stern. Propulsion consists of a pair of Wärtsilä 8L26 diesel engines, each rated 2,650 kW and each driving a Voith model 32R5/265-2 propeller unit. With this propulsion system, the vessel recorded a bollard pull of 73 tonnes and a free-running speed of 14.5 knots. The indirect steering and braking force capabilities are rated at 130 tonnes and 170 tonnes, respectively, at 10 knots. Because of its multiple roles, this tug has a unique hydraulically retractable escort-towing staple, a technical challenge under such high loads.

The 37.5-metre multi-purpose VSP tug *Vortex*, another high-performance vessel for Østensjø Rederi.

Dual-Fuel Escorts for the Far North

In 2014 Østensjø called upon Robert Allan Ltd to develop another ground-breaking design, for a set of three terminal-support escort tugs for the LNG terminal in Hammerfest, at the very far north of Norway. The defining requirement was for tugs which could operate on LNG for at least the major part of their ship-handling and escort duties. Since these tugs also had to be large enough and capable enough to do deep-sea rescue towage, and because it is virtually impossible to fit sufficient LNG storage into typical tug-sized packages, these boats were designed to be "dual fuel," capable of operating on both LNG and regular diesel fuel. The environment in the area also posed some unusual design challenges, as snow, ice and freezing rain are common at these latitudes. Vital deck machinery components, such as the escort hawser winch and the aft towing winch, are therefore totally enclosed, leading to a very tidy-looking arrangement, but necessitating some innovative designing. The resulting 40-by-16-metre escort tug also features the RAstar-type sponsoned hull form described previously and as illustrated in the General Arrangement drawing.

Projects of this complexity, especially with a completely new design, demand extensive analysis in the design stage to assure the owner and their clients that the required performance will be delivered. At this time Robert Allan Ltd was working extensively with computational fluid dynamics (CFD), but it was considered prudent to benchmark that computer-based analysis with some physical model-testing, which was performed at Vienna Model Basin with our long-time consulting associate Alan Reynolds from Offshore Research Ltd.

Scale-model testing of the RAstar 4000 DF (dual-fuel) tug design at the Vienna Model Basin.

The test program verified that all escort performance objectives would be met or exceeded, with the impressive results of 163.5 tonnes steering force and 220 tonnes of braking force at 10 knots.

Powered by a pair of Wärtsilä dual-fuel 6L34DF diesels, each rated 3,000 kW and driving Schottel 3030 CPP Z-drives, these tugs produce a bollard pull of 105 tonnes and have a free-running speed in excess of 15 knots. For these custom-designed vessels, Østensjø returned to Gondan Shipbuilders in Spain, who did their usual high-quality work producing the *Dux*, *Pax* and *Audax*, three of the most powerful and capable escort tugs in the world, and at the time of their launch, certainly the largest and most powerful among a small handful of dual-fuel tugs afloat.

One of the many tugs delivered in 2017 would become the 1000th tug built to a Robert Allan design, and the *Dux* emerged with that title, a more than fitting and distinguished recipient of this honour.

Following pages: *Pax* and *Dux*, Østensjø's distinctive RAstar 4000 DF class escort tugs at work in Hammerfest, at the northern tip of Norway.

LNG CLEAN POWER
PAX

HAMMERFEST
PAX

General Arrangement for the RAstar 4000 class dual-fuel escort tugs for Østensjø Rederi, highlighting the fully enclosed winches fore and aft.

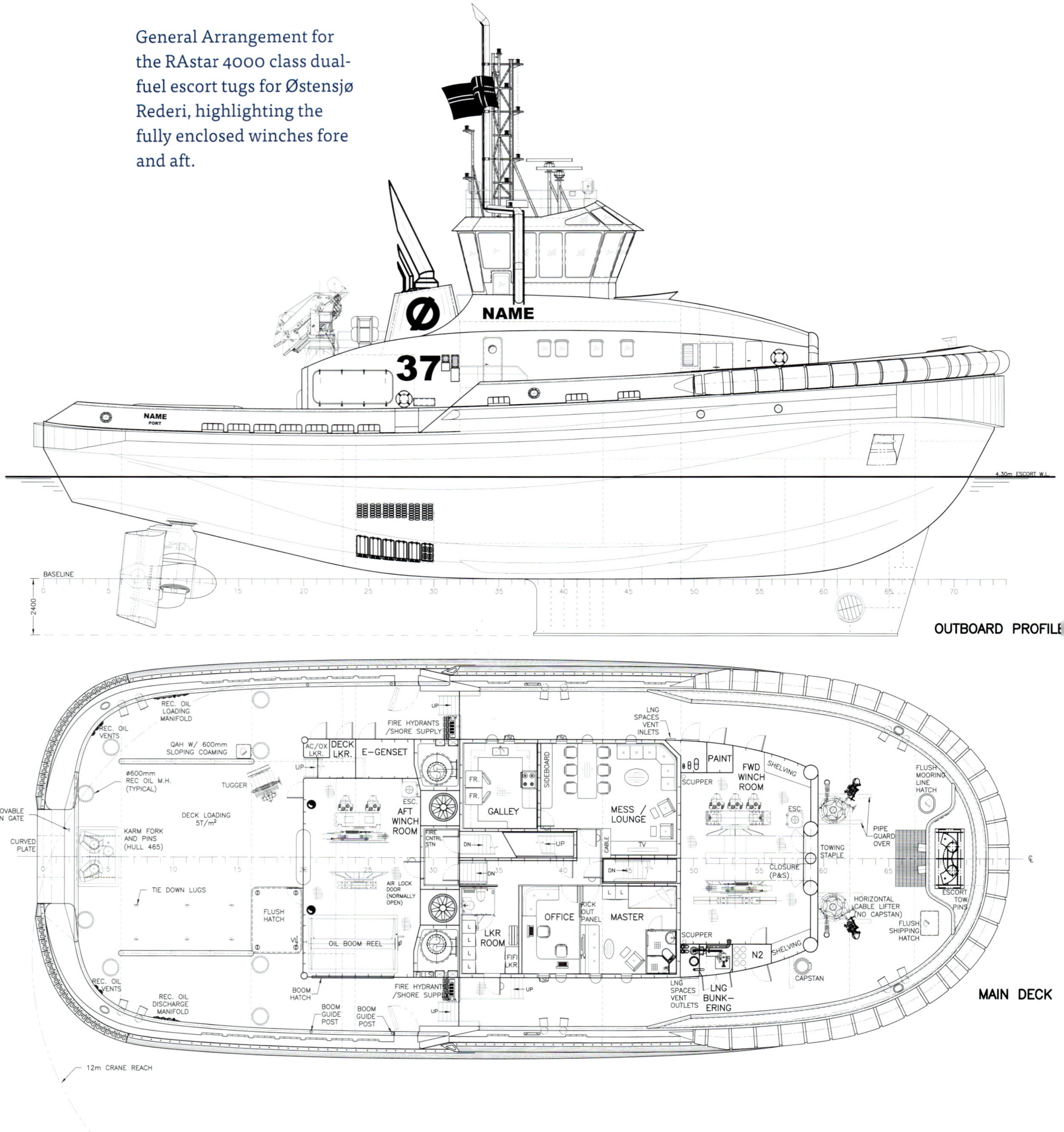

The Turkish Connection

How It All Began

If I had been asked in the early '90s where a significant portion of Robert Allan Ltd's future work would come from, I can assuredly say that Turkey would not even have been on my radar. At the time, even prospects for work in Europe seemed rather remote. But sometimes luck has a way of finding you, and the ensuing 25 years have indeed been a magic carpet ride, fuelled to a large degree by associations made in Turkey.

In 1993 the Royal Institution of Naval Architects (RINA) hosted a conference in London about the emerging technology of escort towing. I wrote a paper for that gathering entitled "Methods for Preliminary Determination of Escort Tug Requirements." In retrospect, it was not a very good paper, but these were the early days of escort towing, and everyone was still exploring different ideas. Several months later I answered a phone call from a young man attending Istanbul University who asked if he could ask me some questions about this paper and the whole subject of escort tugs. I then spent maybe half an hour on the phone talking to Ali Gurun about tugboats. I hung up and never gave it a second thought.

Top: Ali Gurun and Rob Allan at the commissioning of *Sanmar 5* in Istanbul, the first of many memorable visits; **bottom:** An 18.28-metre 1,640 kW Sanmar class twin-screw tug based on the 1,800 hp Cates tugs in Vancouver. This simple project started a lengthy and mutually beneficial relationship with this first-class family-owned Turkish tug owner and shipbuilder.

Maybe six months later, I answered another call: "Hello, Mr. Allan, this is Ali from Istanbul...do you remember me?" Ali went on to explain how the market for ship-handling tugs in Turkey was starting to change from government-owned tug fleets to private ownership. He also explained that he had neglected to tell me before that his family owned and operated some small tugs and was looking to expand its business, and did I have a simple design that they could buy from us. After we discussed his needs, I offered to sell him (for a very nominal fee and an equally nominal royalty fee) the existing design for the 60-foot 1,800 hp Cates tugs, and to do some simple sketches of what this might look like with a different deckhouse configuration. They would build these tugs themselves. This became the start of the Sanmar class tugs, and Ali's family's company, Sanmar Denicilik Ltd, built more than 20 of these simple tugs to support its venture into the local tug market. A strong and lasting business relationship and friendship was created.

As of 2022 Sanmar is our leading client. To date, it has built more than 250 tugs of all shapes and sizes to Robert Allan Ltd designs, with another 50 scheduled for delivery in the next two years, and is without question the highest quality and probably the most efficient tugboat builder in the world.

—R.G. Allan

In the mid-1990s Turkish ports were undergoing a dramatic change in their operating structure. All their ports and the associated towing businesses had been controlled by and operated by the government, but that had just changed to allow private tug and pilotage operations, and numerous companies seized the opportunity to get a piece of the action. Suddenly, the activity that had begun with Sanmar created further interest in tug designs from Robert Allan Ltd. We were approached by several other Turkish shipyards and tug owners, all looking for similar types of tugs.

The demand initially was for smaller, simple tugs, often twin-screw, as the Turkish market was not yet used to Z-drive propulsion and the Voith system was too expensive for their market at the time. We soon established good working relationships with Uzmar Uzmanlar Denizcilik Tic. ve San. Ltd ("Uzmar") and Med Marine A.Ş. Early tug designs done for this market were smaller twin-screw tugs, such as the 25-metre Doğançay class tugs for Sanmar and the 22-metre Alsancak series built by Uzmar.

Doğançay class twin-screw tug from Sanmar.

The lead quartet of the 22-metre 2,820 hp
Alsancak class tugs built by Uzmar.

It did not take long, however, for Turkish shipbuilders to realize that the future was not in these small conventional screw tugs but in larger and more powerful tugs with Z-drive propulsion. Investments in new infrastructure soon followed, and shipyards modernized at an incredible pace. Turkey rapidly took its place on the world stage as a leader in modern high-quality workboat construction.

Sanmar built the 24-metre Ulupinar class of Z-drive tugs, and Uzmar built multiple Z-drive tugs, beginning with the aptly named 32-metre *Altug* for their own account, later sold to Italian owners MOBY of Naples and renamed *Andrea Onorato*. This was followed by a series of 25-metre Ege class ASD tugs.

Sanmar in 2020 commemorated delivery of their 200th Robert Allan-designed tug.

With very competitive prices, very efficient and excellent work quality, Turkish shipyards soon found the world knocking on their doors. In 2010, on Robert Allan Ltd's 80th anniversary in business, Sanmar commemorated building 80 tugs to our designs. By 2020 that total was 200 tugs, the last being a 24-metre Boğaçay class tug. This 200th tug reflected the desire of both companies to continue to move the goalposts for advanced tug designs. The *Boğaçay XXXVIII*, the latest evolution of the highly successful RAmparts 2400-SX design, incorporated the newly developed CAT hydro-mechanical hybrid-propulsion system.

Boğaçay XXXVIII was painted with a bright green hull and colourful flowers on the deckhouse, inspired by Andy Warhol's "Flowers" print. This regalia, reflective of the trend towards reduced environmental impact from tugs, was a refreshing change in the usually moribund world of tugboat aesthetics.

As mentioned, we found ourselves with an embarrassment of riches, with several very fine Turkish shipyards calling for our design services. The Izmir-based Uzmar organization was another fine company with both tug operations and shipyard capabilities. Uzmar has, to date, built more than 100 tugs to our designs, including many of the big river tugs for South America (Chapter 14). Med Marine of Istanbul (another company that is both tug owner and builder) has now built more than 85 Robert Allan tugs. Their current portfolio offering consists of 22 different "standard" tug designs, all but two of which are Robert Allan designs. All these yards are supplying high-quality tugs to clients around the globe in a very competitive marketplace.

Ege 5, one of a successful early series of 25-metre ASD tugs built by Uzmar.

Below: The 31.8-metre *Altug*, first of many collaborations with the Uzmar organization, and later sold to Italian owners as *Andrea Onorato*.

The hybrid-powered *Boğaçay XXXVIII*, the 200th tug built by Sanmar to a Robert Allan design, showing off her Flower Power.

The Chinese Market

China is known to be a difficult market to deal with, especially on matters of intellectual property. However, our experiences there have in general been very positive, in large part due to those of our key staff who are Chinese Canadians and who are able to communicate with shipyards and owners in China in their own language. This capability has been a huge asset for our business.

Our business dealings in China began in 2002, when PSA Marine of Singapore was planning to build several of the new Z-Tech design tugs. PSA contracted with Cheoy Lee Shipyards Ltd of Hong Kong, which was just in the process of moving its new construction facilities to mainland China. We were introduced to Mr. Ken Lo of Cheoy Lee, who has proven to be a first-class client and friend. Cheoy Lee is an interesting company with a long reputation for large high-quality luxury yachts, but which also has a commercial division with a similar enviable record. Cheoy Lee built the first Z-Tech tugs, *Indee* and *Pardoo*, for Teekay Shipping's operations in northwest Australia and has since built more than 175 of our tugs for clients around the world. The larger Z-Tech story follows later in this chapter.

Indee, first of the trendsetting Z-Tech tugs built by Cheoy Lee.

Our close working relationship with a yard like Cheoy Lee is quite unique but reflects the mutual trust and respect the two organizations share. Cheoy Lee has built 13 Z-Tech tugs for the Panama Canal Authority and numerous other tug types, mostly for owners in Southeast Asia and Australia. The tugs for Panama sailed on their own bottoms from Hong Kong to Panama, all with a few extra barrels of diesel fuel on deck.

The most recent (2020) deliveries from Cheoy Lee have been for a pair of powerful RAstar 3200-CL tugs. The first boat, *Hai Kun*, achieved 88 tonnes bollard pull and a speed of 13.5 knots. To date, Cheoy Lee has delivered 21 of this class of tug, a design which was developed exclusively for the company to suit its specific production and launching facilities.

Our dealings with other shipyards in China were not quite so positive initially, but through some nurturing have grown to be an important part of our business. This is largely due to the unrelenting efforts of our senior naval architect Xuhui (Bill) Hu, who ably manages our "China Desk." Through Xuhui's efforts, Robert Allan Ltd has developed good relationships with not only several Chinese shipyards but more importantly with significant port operators who have commissioned designs directly from us. The first of these, the 32-metre 60-tonne bollard pull *Yong Gang Zhong Lian No. 6* was designed for the Port of Ningbo, part of a six-tug order.

This contract marked the first occasion when a Chinese Port Authority awarded a design contract outside of China for tugboats or similar commercial vessels, and it was a major milestone for Robert Allan Ltd in the promotion of the RAmparts series of modern ASD tug designs.

Above: *Hai Kun*, a RAstar 3200-CL class tug built for Yiu Lian Dockyards Ltd of Hong Kong; **right:** Z-Tech tugs at work in the Panama Canal; **opposite:** *Yong Gang Zhong Lian No. 6*, a Robert Allan RAmparts 3200 class ASD design, the first tug built for a Chinese Port Authority to an offshore design.

Working in collaboration with Zhenjiang Shipyard Ltd, the nominated shipbuilders for the new tugs, Robert Allan Ltd provided a complete classification society design approval package as well as a complete 3D product model for all the steel construction of these tugs.

First Venture for Chinese Owners

I remember our first project design for Chinese owners, Ningbo Port, back in 2004. Rob and I flew to Shanghai. We arrived in the evening, and because the shipyard at Zhenjiang was about 250 km away from Shanghai, and there were no more trains that night, an employee of the shipyard came and picked us up at the airport and accompanied us to the Holiday Inn near the train station. We stayed there overnight and caught the train at about 6:30 in the morning to Zhenjiang. I was really tired, and we were jetlagged. When we arrived, the Port people were waiting for us in a conference room. I think the meeting went on all day and the contract eventually got signed for a 32-metre RAmparts tug. The contract fee was very low because of Rob's vision of future business over there. Rob was right again, and the Port became a very good client, and I think they built about 10 tugboats to our design. The latest, in 2019, was a dual-fuel (diesel and LNG) tugboat.

—Xuhui (Bill) Hu

Solving Problems

Naval architecture involves really complicated engineering, yet there is always something that can be improved. And when we haven't done well, we try to fix it. I remember back in 2013, we designed a new 5,000 hp Z-drive harbour tug. It was ready for sea trials in 2015, and I went to China for the trials. The speed, behaviour, manoeuvrability, noise levels and bollard pull all met the requirements, and the owner thought it was all wonderful. Vibration was good when we were going ahead, but when we were doing bollard-pull astern tests, the Z-drive was reversed 180 degrees to push the water forward, at full engine speed, and the mast had some shaking. It was a minor problem that just happened at full speed in reverse, but still the shaking was not so good. We couldn't find an obvious fix, so we decided to take a video, and I brought all the information back to the office in Vancouver, and our people carried out an analysis. We used all our analytical tools and found an easy, simple solution. The shipyard had tried to solve the problem without success while we were doing our analysis. We got them to remove what they had done and simply weld a couple of small support rings along the mast. That fixed it like magic. This is the kind of thing we try to solve if something goes wrong.

—Xuhui (Bill) Hu

Since that 2004 project, a further 26 tugs have been built for service in China for Chinese owners, including ten tugs for the Port of Ningbo and nine for the Port of Tianjin. Since the initial order for 32-metre tugs, the size and power of the Chinese port tugs has grown steadily, and the most recent delivery is a pair of 34.7-metre 80-tonne bollard pull dual-fuel tugs for the Port of Cao Fei Dian, the *Cao Gang 30* and *Cao Gang 31*.

Jin Gang Lun 31, a 40-metre ASD tug for the Port of Tianjin.

Cao Gang 26, a RAmparts 3300 class tug for Cao Fei Dian Port.

Below: *Yong Gang Tuo 31*, a RAstar 3900 class 90-tonne bollard pull tug for the Port of Ningbo.

The European Market

Robert Allan Ltd's business grew from the local tug and barge industry of BC and from the innovative ship-handling tug work done here with the Cates organization. It seemed a bridge too far to hope that we might one day develop some business in Europe, home to the world's largest tug owners. However, the breakthroughs achieved in our collaboration with Østensjø Rederi gave our company some much-needed brand recognition, and slowly we began to see interest in our designs from some of the larger tug companies. This recognition primarily grew, however, from our almost continual presence at the bi-annual International Tug Conferences. These began in London in 1969, hosted initially by the historic company Thomas Reed Publications, of the UK, and later, after some reorganization of that company, by the ABR Company. The host of all these conferences was the late Allan Brunton-Reed, who, until his death in 2020, did a masterful job of organizing these events into what became a must for every major tug owner worldwide. These conferences featured great technical programs, excellent trade shows and an always entertaining social program. There was no better marketing and networking opportunity, and Allan was the consummate host.

These conferences were joined in 2007 by the smaller, more technically oriented Tugnology conferences, which ran in the alternate years. Allan, who had become a good friend, was very generous in accepting numerous papers from Robert Allan Ltd staff over the years, and they were all very well received as we focused primarily on issues in the industry rather than just another sales paper. Rob Allan has presented 14 papers to ITS conferences and a further six to Tugnology events. He attended more of these tug conferences than any other delegate and was on track to be acknowledged for attending his 20th conference when that event was cancelled due to the Covid-19 pandemic and the subsequent failure of the ABR company. Through these conferences, however, the Robert Allan brand and our many high-performance tug designs have become well known and well regarded. ITS was instrumental in our global success.

In My Father's Shoes

Bob Allan wrote a significant paper for the 2nd International Tug Conference in London entitled "Shallow Draft Towboats in the Canadian Northland." The conference was scheduled for the fall of 1971. I had graduated from university just a few months earlier and was in my first-ever engineering job, working for the major consulting naval architecture firm of Burness, Corlett & Partners in Basingstoke, England. Not more than two weeks before the conference I received a call from my father, saying that he was unable to attend the conference due to business conflicts and asking if I would be willing to present this paper on his behalf! That was a significant challenge: I was a neophyte in the industry and would be addressing an international audience of seasoned tug owners and operators. Dad reassured me that, as I had a few years earlier spent a summer working on the Mackenzie River (see Chapter 9), I would know more about the subject area than anyone in the room except Arthur McLaren of Allied Shipbuilders, so I should have no problems.

I reluctantly consented, obtained the necessary permission from my employer and attended this conference in the heart of London. Never have I been more nervous, but I managed to make it through the presentation, and Arthur was very kind to me in his questions and comments.

That was a baptism by fire in public speaking for me, but since then I have subsequently spoken at many of these conferences and developed some skill in making presentations of what I hope were significant and appreciated papers. Those ITS conferences became a significant event in our corporate business calendar and took us around the world, to some often quite remarkable venues, including cities in Italy, Spain, Germany, Holland and the UK. More far-flung locations included Capetown and Singapore. It also included a memorable 2010 conference in Vancouver.

—R.G. Allan

Rob and Enneke Allan at the reception for the 2nd International Tug Conference, at London's Guild Hall, 1971.

The world leaders in tug design and operation, certainly in the 1960s and '70s, had always been the Dutch, with long-established large towing and salvage companies like SMIT and Wijsmuller. UK-based companies like United Towing were also key players. There didn't seem to be a place for a small Canadian design firm in this marketplace. However, mergers and acquisitions in this arena gradually led to a significant reduction in the number of tug companies operating in the world's busiest markets. By about 2000 the largest companies operating in Europe were Svitzer (a division of the Maersk organization), Smit Lamnalco (a combination of the Dutch SMIT organization and Lamnalco, a UAE-based offshore operator), Boluda of Spain, Bourbon of France and Rimorchiatori Riuniti of Italy. An emerging dominant player, principally in Western Europe, was the KOTUG organization of the Netherlands with their unique triple-Z-drive "Rotortugs." We had slowly been meeting and establishing contacts with all these companies, both at the international conferences and also through the excellent sales of tugs to our designs from Turkish shipyards, who were certainly shaking up the tug construction market traditionally (at least through the last quarter of the 20th century) the domain of Dutch, German and Spanish shipyards.

Rob Allan presenting at Tugnology 2017.

By the end of the first decade of the 21st century, all these major towing companies were clients of Robert Allan Ltd and were growing their tug fleets to cope with the increasing demand for larger, more powerful tugs to handle the increasingly large bulk carriers and container ships now dominating the world's busiest ports. At the same time, the market for LNG was expanding, and those terminals were being located in more exposed areas to avoid the perception of risk associated with LNG operations in port. This gave rise to the new genre of "terminal support tug" that was generally much larger and more powerful than typical ship-handling tugs, in order to cope with work in rougher sea conditions, and many also had an escort role.

One of the most exciting projects of this type was the requirement, begun in 2006, for a new fleet of terminal support and escort tugs for Milford Haven, in the southwest corner of Wales, a major UK oil and gas port facility. Svitzer was the successful bidder for this contract, based on designs proposed by Robert Allan Ltd. Five tugs of the RAstar 3400 class and one RAstar 3900 class (the *Svitzer Kilroom*) were built by Freire Shipyard of Vigo, Spain. A couple of years later a RAstar 3600 class tug, the *Svitzer Pembroke*, was added to the fleet.

RAstar 3400 class terminal support tug *Svitzer Haven* in heavy weather off Milford Haven, one of a quintet of sister tugs that includes *Svitzer Caldey*, *Svitzer Lindsway*, *Svitzer Ramsey* and *Svitzer Waterston*. These very capable tugs measure 33.6 metres length by 14.5 metres beam and generate 80 tonnes of bollard pull from 4,400 kW of power.

The RAstar Tugs

My first management project was with the company's new RAstar series of tugs in about 2004. I was working under project director Rob Allan. These were to be high-performance escort/offshore terminal azimuthing stern-drive (ASD) tugs, initially designed for a client in the Middle East, but the project for them was delayed and not actually built until several years later. Instead, the first RAstar tug was built for Svitzer, the largest tug operator in the world. This was a series of RAstar 3400 tugs built to service the Port of Milford Haven, in southwest Wales. We designed three different boats for Svitzer, and six or seven were built in Spain for Milford Haven. That was a fairly defining project for us in many ways. The RAstars were among the first, biggest and most powerful ASD escort tugs that had been built to date for Europe. While Europe had Voith escort tugs before that, the RAstar utilized a hull form with sponsons as well as prominent foil-shaped escort skegs forward, developed exclusively by Robert Allan Ltd. We went on to eventually design 18 different versions of the RAstar. That was quite an important series. Since then, we've built 200 or something of those tugs. It was Rob who developed that unique hull form, but my job was to carry that one through to construction.

—Mike Fitzpatrick

The terminal support/escort tug *Svitzer Kilroom*, the "queen" of the Milford Haven fleet. At 39.1 metres long, with a beam of 14.7 metres, she generates 117 tonnes bollard pull ahead, 113 tonnes bollard pull astern, and can run at 15.7 knots with her 6,100 kW of power.

Outboard Profile of the *Svitzer Kilroom*. It is so easy to make a tug look good when one can have some length to play with!

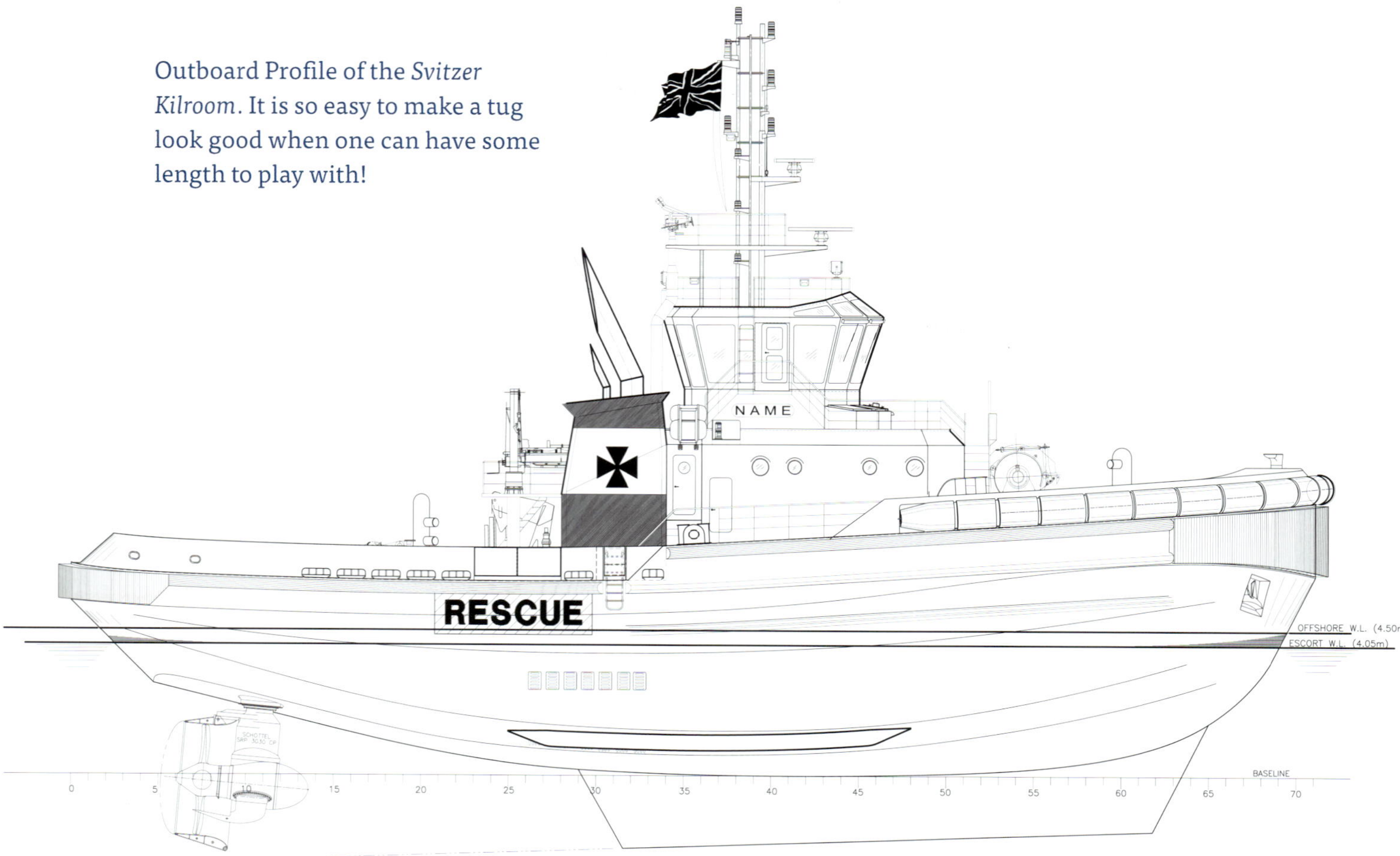

Going on a RAmpage!

Another major development for our international business was the design of the RAmpage 5000 class offshore tugs. The 50-metre 104-tonne bollard pull tug *Seabulk Angola*, built in 2003, was the first of what we designated as the RAmpage series of offshore support tugs. This series was developed to address a growing demand from operators for high-performance towing and anchor-handling tugs for critical offshore terminal and oilfield support duties, without reverting to the much more expensive full-sized Offshore Supply Vessel/Anchor-Handling Tug Supply (OSV/AHTS) vessels, which typically would be 70 metres or more in length. The RAmpage concept offered tugs of relatively compact dimensions and high power, capable of performing a wide range of duties in the offshore market much more cost-effectively than their bigger cousins.

More than two dozen of the RAmpage class tugs have been built for major operators such as Bourbon Offshore, SMIT and Seabulk. The majority have been 50-metre-long tugs, but there are derivatives ranging from 37 metres to 55 metres now in service.

Seabulk Angola, first of class for the RAmpage series of offshore tugs.

Above: Iconic photo of RAmpage 5000 class offshore support tug SMIT *Angola* with Table Mountain, Capetown, South Africa, in the background. A feature in this particular vessel is a 3-metre square moon pool on centre, with a side-launching gantry for a dive bell, etc. Adjacent is a diver decompression unit. The aft deck is dominated by the presence of a large articulated crane and a double drum towing/anchor-handling winch.

Left: *Lamnalco Griffin*, a RAmpage 5000 class offshore support tug based in the Middle East.

The Z-Tech Story: A Challenge Not to Be Denied

Over the past two decades or so there have been a few "different" tug designs that their developers claimed were going to revolutionize the ship-handling business. Some of those only had one or two copies built, and others had maybe four. None revolutionized the business as per the claims. The challenge in ship-handling is that most ports have quite different geographic and environmental characteristics, and in addition likely have different ship types and sizes and differing facility layouts. All of this really means that there is no universal tug type panacea for ship-handling, and individual operators tend to foster the types which work best for them and then don't stray too far from that success. However, there is always room for improvement, and when we as designers are given such opportunities, it can be a really fascinating challenge.

Such an opportunity arose for us in 2002 when a call was received from Peter Chew of the Port of Singapore Authority (PSA). PSA had a need for some new tugs to add to or replace older tugs in their significant fleet. They had a random mixture of azimuthing stern-drive (ASD) tugs, some Z-drive tractors and some Voith tractors. Different crews favoured different types, generally according to the type with which they were most familiar. Peter's challenge to us was simple: "We need a *single* new design which incorporates the best handling/operating features of both ASD tugs and Z-drive tractors." Voith tractors were not on the table due to cost.

Blue Sky Thinking

In response to Peter Chew's tantalizing challenge, Ken Harford and I went to Singapore in the summer of 2002. We spent several days riding on tugs, riding on the attended ships, talking to tug crews and to PSA management, gathering input about how they worked in one of the world's busiest ports, and watching their ship-handling operations. After about four days, Ken and I were having a well-earned beer after a busy day and brainstorming about how best to achieve Peter's objectives. Ken had an "ah-ha" moment and described his thinking about how to do just that, with the proverbial bar napkin sketch. I spent the evening doodling at a somewhat higher level, and in the morning we had the genesis of what was to become the Z-Tech design to discuss with PSA. This was really fun "blue sky" engineering design work!

—R.G. Allan

Ken Harford's concept was quite brilliant and simple: take a basic ASD tug, make its shape in plan more or less symmetrical fore and aft, but reverse the sheer line so it is lower "forward" than "aft" (these terms can often get quite murky in the modern omni-directional tug world), move the house much farther aft than typical, and make sure there are good sightlines in all directions. This provided a large, clear, relatively level, safe working deck around the winch when operating in the "ASD mode" (drives aft), and if towing, or when rough weather transits were required, the tug could simply run going stern-first in "tractor" mode (drives forward). Only one winch was required, saving space and cost.

The first of these tugs were the *Indee* and *Pardoo*, built by Cheoy Lee Shipyards. Although ordered by PSA Marine, these tugs went directly into service with Teekay Marine as ship-handling/escort tugs in Port Hedland, Australia. They were followed in rather quick succession by an extended series of builds for PSA Marine themselves, followed by orders from Lamnalco, Keppel Smit Towage (KST) and Adsteam Towage, among others. This first generation of Z-Techs were 27 metres long by 11.5 metres beam and had 5,000 hp producing approximately 60 tonnes bollard pull.

Design Zentrum Nordrhein Westfalen

reddot design award
winner 2005

Die Auszeichnung „red dot: best of the best" für höchste Designqualität, die in überragender Weise Innovation in Form und Funktion ausdrückt, wird verliehen an:
The award "red dot: best of the best" for the highest design quality, expressing innovation in form and function in an outstanding manner, is presented to:

Z-Tech Schlepper, 2004
PSA Marine (Pte) Ltd, Singapur
Design: Robert Allan Ltd.,
Vancouver, Kanada
Produktion: Cheoy Lee Shipyards Ltd.,
Hongkong

reddot award: product design

Der red dot design award ist einer der renommiertesten internationalen Designwettbewerbe. Seit 1955 verleiht eine Fachjury im Rahmen des Wettbewerbs Auszeichnungen für hervorragende Designleistungen, die in ihrer jeweiligen Produktkategorie neue Maßstäbe setzen. So ist der red dot heute eines der begehrtesten weltweit anerkannten Qualitätssiegel.
The red dot design award is one of the most prestigious international design competitions. Since 1955, expert juries have made awards annually in the competition for outstanding design achievements which set new standards in the relevant product categories. The red dot has thus become one of the most highly prized and globally recognised seals of quality.

Essen, 4. Juli 2005

Odo Klose
Wuppertal (D)

Giuliano Molineri
Moncalieri (I)

Kenneth Grange
Devon (GB)

Prof. Dr. Peter Zec
Geschäftsführender Vorstand President
Design Zentrum Nordrhein Westfalen, Essen (D)

The Z-Tech concept attracted a lot of attention and, jointly with PSA, Robert Allan Ltd promoted the novel design as a candidate for a variety of industrial design awards. The year after the first Z-Tech design was delivered in 2004, it was awarded the Singapore Design Award for "Design for Business Effectiveness," and in 2005 it won the prestigious German Red Dot International Design Award as the "Best of the Best" in product design, competing against the likes of Porsche, BMW and Bose. The criteria for this award include the stipulation that "only designs with excellent characteristics and maximum innovation receive the coveted 'Red Dot: Best of the Best' award."

Since the first Z-Tech series, the Z-Tech design has continued to evolve into larger and more powerful variants, and even in some cases more modestly powered variants, according to the needs of our clients. Outside of Asia, the major adopters of the Z-Tech fleet have been G&H Towing, operating tugs in Houston, Texas; the Panama Canal Authority; and the US Navy. G&H, currently operating 26 Z-Techs in the Bay-Houston area, has opted for a mixed fleet of 24-metre and 30-metre Z-Techs, with the larger tugs now developing up to 80 tonnes bollard pull.

As of February 2022, 91 Z-Tech tugs in eight design variants have been built or are under construction worldwide. These are distributed widely, but principally in Singapore, Australia, the Middle East, Panama Canal, Texas, and with the US Navy in the Seattle area and in the Pacific Islands.

Left: Z-Tech *Yandeyarra* on trials in Hong Kong before taking up duties for Teekay Marine in Australia; **above:** Z-Tech KST *Zenith* at work in Hong Kong. This view clearly illustrates the excellent "aft" visibility inherent in this design, used when running astern in tractor mode.

General Arrangement for the first Z-Tech tug, designed for PSA Marine of Singapore.

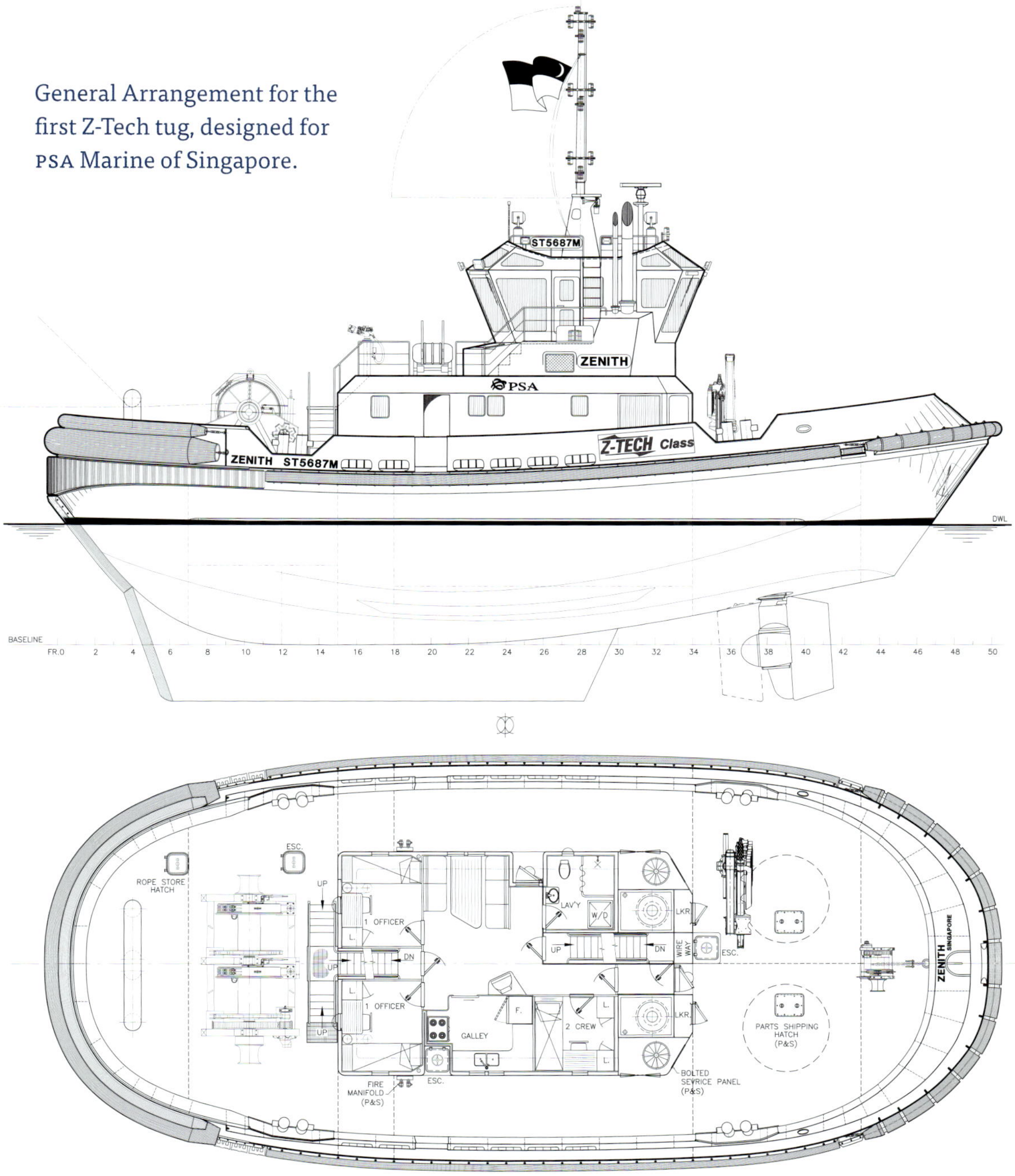

The Popular Z-Techs

The first Z-Techs were built by Cheoy Lee Shipyards in China. Rob paid a lot of attention to that project. I still remember when Rob flew to Hong Kong to visit the shipyard on mainland China for the first sea trials. When he came back, he was very happy with the result. I remember saying to Rob, "Is it perfect?" "Well," he said, "I cannot say it is perfect. We always have something to improve." This was the first project where I was project manager officially. I learned a lot through that project. It was excellent. It was a very good design and was an eye-opener to the industry. After the first tug was built, it gained immediate acceptance, and two technical papers describing the new design were presented to the International Tug and Salvage Conference in 2004 and 2008. We now have more than 90 Z-Tech tugs delivered around the world.

—Xuhui (Bill) Hu

Z-Tech 27-60 *Star Voyager* at work in the busy Port of Singapore.

Bocas del Toro, one of 21 Z-Techs in the Panama Canal system.

Z-Techs in the Panama Canal

Although their procurement processes leave a lot to be desired, the Panama Canal Authority (ACP) became a major user of Z-Techs. There are more than 20 Z-Tech tugs working in the canal system. Despite our strongest recommendations, the ACP technical team elected not to fit slipping clutches on the first series of these tugs, ostensibly to save some initial cost, and with their heavy usage soon found they were burning up clutches on the drive units. They called this a design fault. A second series however were fitted with slipping clutches and are working well. The ACP tugs are all of the 27-metre class and have either 60 or 65 tonnes bollard pull.

Tug of Choice for the US Navy

The US Navy was looking for a new fleet of tugs to handle its operations in Bremerton, Washington, when the newly introduced Z-Tech concept caught its attention. Several meetings were held with the operational group, which was convinced that the Z-Tech was the answer to its requirements. However, the procurement process for Navy assets is, to say the least, convoluted, and it took several years to get all the ducks in a row to enable this to happen. After an initial order for four Z-Techs in 2008, the US Navy is currently (2020–21) accepting its second series of Z-Techs, which were designed in 2011 but rather delayed by the bureaucratic procurement process under which the USN operates. These tugs are outfitted for docking warships and submarines and therefore are heavily fendered at and below the waterline. Because they handle rather delicate (and very expensive) ships, these tugs have only 45 tonnes bollard pull, more than adequate for the controlled moves they make.

More of a Good Thing

Even the US Navy is using the Z-Tech design, and it now operates more than 10 of these tugs at its bases in Washington State, on the US East Coast and in the Pacific. There they work with big ships such as destroyers and aircraft carriers, but when compared to commercial cargo ships, they are quite a bit lighter. The tugs don't do that much escort, but when the ships arrive and leave, they help with getting them in and out of their docks. The Navy people and ships' pilots love our Z-Tech very much. During our recent round of Navy tugs, some American design firms said their boats were just as good and that the Navy should stick to American designs. However, when the Navy asked its pilots and tugboat captains their opinions, they said Z-Tech is definitely what they want, so the Navy continued with our designs.

—Xuhui (Bill) Hu

What's That Smell?

We designed a number of Z-Tech tugs for the US Navy and we're quite proud of that. They are grey, they look good and the crews are happy with them. But we started to get these calls from the tugs based in the Navy base at Bremerton, in Washington State. They said that occasionally there was this terrible smell of sewage in the accommodation areas. Sometimes it would be there and sometimes it wouldn't. They had tried everything, and they just couldn't figure out what it was. Finally, it got to the point that Allan Turner and I figured we'd better get down there and try to solve it.

Off we went, with our coveralls, and drove down to Bremerton. We found the tugs, got on board, and the crew said, "Sorry, guys, today the smell is just not here." Nevertheless, we thought we'd better see if we could make the smell happen, so they took us for a ride and tried some manoeuvres and rolled all around, thinking maybe that would cause the smell. Meanwhile, Alan and I are crawling around with our flashlights, sniffing drains and toilets, but we got absolutely nothing. So we went back home after this really unsatisfying day.

In the end, the problem sort of solved itself. Somebody else figured it out. The galleys on these tugs were equipped with these pretty powerful ventilation fans. Normally, when you turn on a galley fan, you have an air bypass—you take air from the outside and vent it through. For some reason, the bypass hadn't been installed, and every time the crew turned on the galley fan to cook something, it created so much suction that it started sucking all the gases from the waste-water tank up through the drains in the accommodations. The story is only funny because it was such a bizarre problem, and we sort of embarrassed ourselves by crawling around sniffing at all the toilets while we were there.

—Vince den Hertog

Above: YT 802 *Valiant*, the first Z-Tech for the United States Navy, an interesting contrast to the old-style American tug in the background.

Left: YT 805 *Seminole* at work.

Z-Techs in Texas

Another major adopter of Z-Tech technology has been the G&H Towing Company of Galveston, operating tugs in most major ports on the Texas Gulf. G&H operate vessels for two groups of owners, Suderman & Young Towing Company and Bay-Houston Towing Company.

The Texas Z-Techs

In 2004 we started working with what was to become a very good client, G&H Towing, in Galveston, Texas. They do a lot of escort towing and ship-berthing in the Gulf area. They had seen one of the technical papers Rob had written about the Z-Tech design, and they approached Rob and said they liked the concept and wanted to discuss it. So Rob flew to Galveston for a technical meeting. In the past, the company had only used local designers, but the Z-Tech was so advanced that they were sold right away. I was project manager on this one, too, but Rob was standing behind my shoulders all the way through the process of design, construction and sea trials. We called the design for them the Z-Tech 7500 (75 tons bollard pull). So far, the company operates about 30 Z-Tech tugs. They also have our 30-metre Z-Tech and now our 24-metre Z-Tech tugs.

The company does a lot of escort operations. The client was planning new construction in 2017, and they loved their Z-Tech tugs. The client said that everybody was also talking about our RAstar design for escort operations. RAstar designs were characterized by having sponsons, which add buoyancy and improve stability when heeling. We recommended putting sponsons on their future Z-Tech tugs. So we did a series of simulations and showed an increase of 12 to 15 percent in escort capability. Later I went to Galveston to visit the completed new tugs, and the Z-Tech with sponsons was clearly more stable during operations. Rob proposed renaming the modified design and, because Texas is known as the Lone Star State, he suggested Z-Star. However, the client said, "No, we prefer to keep the Z-Tech name because it is so famous. We may use the new name in future."

—Xuhui (Bill) Hu

Chloe K, a 24-metre 60-tonne bollard pull Texas Z-Tech for Bay-Houston Towing.

Mercury, a 30-metre 80-tonne bollard pull Z-Tech working in the Houston Ship Canal for Suderman & Young.

Chapter 19

Rotortugs

The triple-Z-drive Rotortug was conceived by Ton Kooren of KOTUG International of Rotterdam as a unique solution to the challenges of handling ships in the very constrained waterways of the Netherlands and elsewhere in Western Europe. The first of these tugs were built in 1999 and introduced to the tugboat world at ITS 2000 in Jersey (UK). Although many were sceptical about the merits of the concept, and particularly how an operator could safely manage three sets of controls, the design has been continually refined and is a well-regarded tug type today.

The association between Robert Allan Ltd and Rotortug (KST) B.V. began in 2012 when KOTUG was reorganizing its business and wanted to engage a naval architecture firm to take on the further development of the Rotortug. The company invited Robert Allan Ltd to be the sole licensee for Rotortugs worldwide, which we did with pleasure. This has since been a very positive collaboration, particularly as we explored how the triple-drives would work in escorting applications. This has proven to be an extremely effective combination, and a series of 32-metre 85-tonne bollard pull Rotortugs, designated as the ART 85-32 design, have become the tug of choice in Port Hedland, in northwest Australia, the busiest and one of the most challenging bulk-handling ports in the world, moving more than one million tonnes of iron ore per day through its numerous tidal berths.

The *RT Evolution*, the first Rotortug designed by Robert Allan Ltd. A hybrid Rotortug built in 2015, she is shown undergoing escort towing trials in fine style.

ART 85-32 FMG *Hammerhead* working hard in indirect towing mode in choppy seas off Port Hedland, Australia.

In the Back of Beyond

Located on the northwest coast of Australia, Port Hedland is the largest bulk export port in the world. Iron ore comes in on miles-long train convoys from mines far inland. The ore is loaded onto massive bulk carriers, which deliver it directly to China. The area has a significant tidal range and a 22-mile dredged channel from the port to deep water. Due to the volume of commerce, it is critical that the number of ships get out that need to get out on the tide. That may be five or more bulkers. They need tugs the whole way, and they need tugs that can act fast if a ship loses steering or power—it has to happen. We've designed some very, very high-performance tugs for Port Hedland. Robert Allan Ltd designs are used almost exclusively to move the bulkers, with about 20 working the port at any given time; that is because they're known to be the best tugs possible for this type of work.

—Derek Noon

Bulk carrier outbound from Port Hedland with Allan-designed escort tugs.

In 2014 Seabulk Towing of Florida elected to adopt the Rotortug concept, and a trio of 67-tonne bollard pull tugs was built, entering service in 2017. These are the first Rotortugs in US registry, at 98 feet 6 inches long by 43 feet 6 inches beam.

In 2014 the Rotortug concept was selected for the Infield Support Vessels (ISV) built for Shell's "Prelude" floating LNG (FLNG) production operation in northwest Australia. Three tugs of this type were built to a Robert Allan Ltd design: the RT *Beagle Bay*, RT *Kuri Bay* and RT *Roebuck Bay*. These imposing Rotortugs are designated as the ART 100-42 class Rotortug and are the world's first purpose-built FLNG ISVs. The

Seabulk Towing's Rotortug *Trident* off the Florida coast.

Below: RT *Beagle Bay* in 2016 getting ready to go to work supporting Shell's Prelude floating LNG plant in northwest Australia.

vessels were built by ASL Shipyard in Singapore for KT Maritime Services Australia, a joint venture of KOTUG International and Teekay Shipping Australia. Measuring 41.95 metres by 16 metres, these multi-functional tugs are powered by Niigata main engines driving three Niigata ZP-21 Z-drives with fixed-pitch propellers. Each engine generates a power output of 2,220 kW, thus producing 100 tonnes of bollard pull. These are among the largest Rotortugs in service to date, though concepts have been developed for some variants up to 45 metres in length.

Tugs in Ice

Working on designs for Canada's eastern and Arctic coasts gave Robert Allan an early lead in the challenges of designing vessels for working in ice. The big tugs designed for the Irving organization in the 1960s (Chapter 18), and later the many supply vessels and tugs for the Beaufort Sea designed throughout the 1970s and early '80s, gave the firm a solid grounding in both the theory and the practicalities of ship design for heavy ice conditions. That early experience enabled us to secure a contract in 2005 for high-ice-class escort tugs to support offshore operations in the Sakhalin area of eastern Russia for the Svitzer organization, the first of what has since become the TundRA class of tugs.

Four tugs were built for the Svitzer project: *Svitzer Aniva, Svitzer Busse, Svitzer Korsakov* and the *Svitzer Sakhalin*. Two of these tugs were built in St. Petersburg, Russia, and two in Batam, Indonesia, by ASL of Singapore. Built to satisfy Russian registry requirements, the tugs had an ice class rating of Lloyd's 1A and Russian LU-5, indicating an ability to break up to a metre of ice at three knots. These tugs are 34.5 metres long with a 13-metre beam. They deliver 73 tonnes of bollard pull from a pair of 2,300 kW diesels. Their mission includes anchor-handling and buoy-handling as well as ship-docking and escort work, hence the large knuckle-boom crane on the aft deck.

One of the Russian-built tugs, the *Svitzer Korsakov*, was hijacked by Somali pirates on its long delivery voyage around Asia in February of 2008. The situation was resolved after 47 days of captivity, without the crew being physically harmed, and the tug eventually made the remainder of its voyage around to Sakhalin, a little the worse for wear.

In 2011 the design for a slightly larger ice-breaking tug was commissioned from Robert Allan Ltd by Groupe Ocean of Quebec. Their tugs had to work in the challenges of the St. Lawrence River in Eastern Canada, with strong currents and heavy winter ice. The ice-breaking hull form was extensively model-tested to ensure superior performance in ice, as well as to provide the best possible open-water and escort performance, bearing in mind the contradictory hull geometry requirements of those two functions. The resulting tugs, built in 2013 and 2016 at Groupe Ocean's own shipyard on Isles-aux-Coudres, Quebec, are the 36-metre *Ocean Tundra* and *Ocean Taiga*. These tugs produce 113 tonnes of bollard pull with 6,000 kW of power and have an ice-class rating of Lloyd's 1A Super FS. They are, at time of writing, the most powerful tugs in Canadian registry.

The launching of these heavy deep-draft tugs posed an interesting challenge for the shipyard, which it overcame with ingenuity. Due to the ice-breaking hull shape with cut-away ends, it was necessary to provide a considerable amount of extra buoyancy during the launch, as illustrated on page 468.

Svitzer Aniva working in ice at Sakhalin.

Below: *Ocean Taiga* with scenic Quebec City in the background.

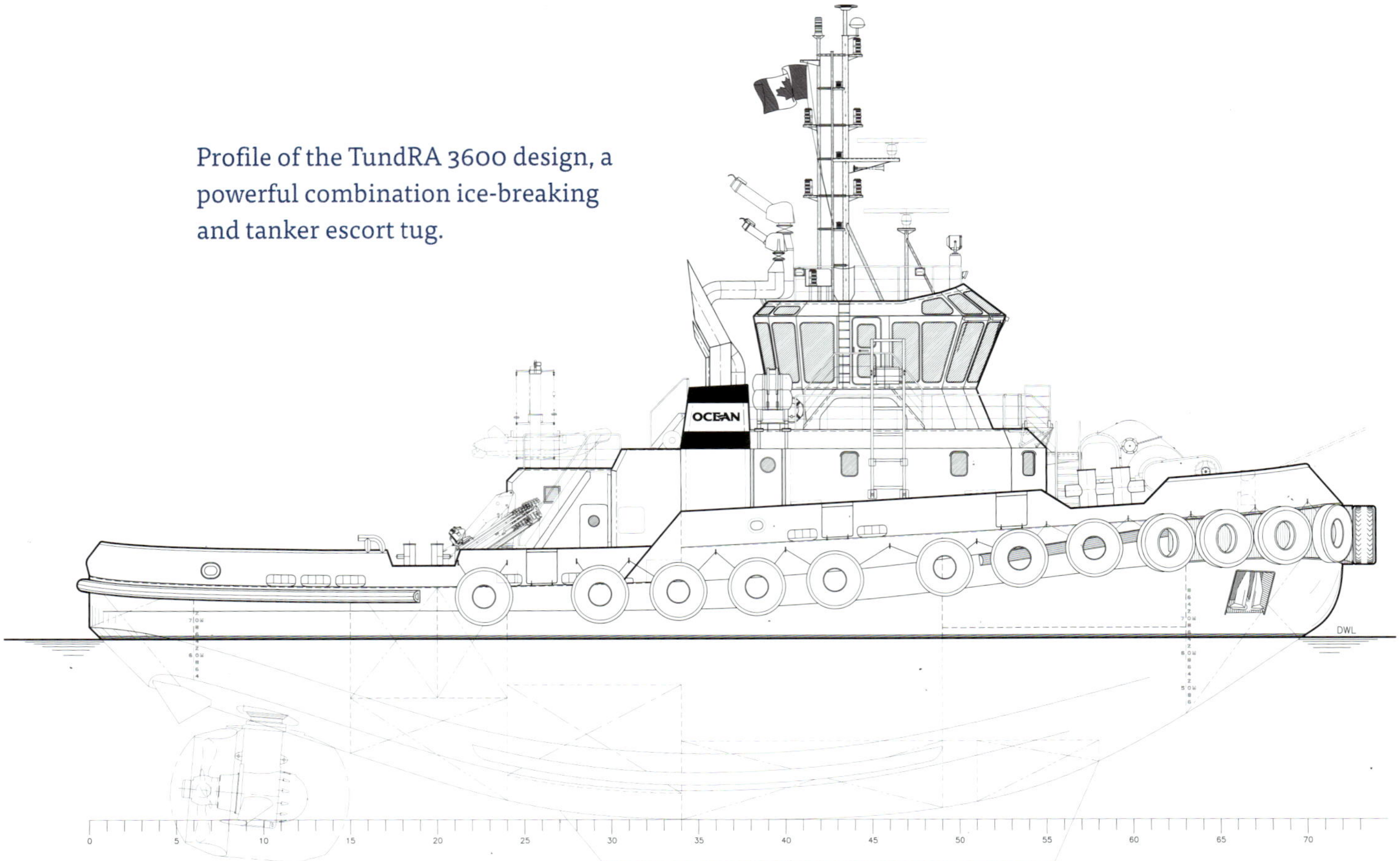

Profile of the TundRA 3600 design, a powerful combination ice-breaking and tanker escort tug.

Ice-breaking tug *Ocean Tundra* ready for launch, showing the added buoyancy bags and the protective ice-knives astern of the Z-drive units.

Ocean Tundra, first of the TundRA 3600 class designs, in the St. Lawrence River in winter.

In 2013 a commission was received for an ice-breaking tug for the US St. Lawrence Seaway Development Corporation. The model of the TundRA 3600 class tug was used, though the layout was modified to suit the operational profile and crewing requirements. The tug, christened *Seaways Guardian*, was built in Gulf Island Shipyards of Houma, Alabama, and delivered in 2020. The primary mission of this tug is the retrieval and deployment of aids to navigation in the spring and fall, for which it has a large knuckle-boom crane aft. Another of its activities is pushing a 300-ton-capacity derrick crane barge, for which it is equipped with push-knees forward.

The Great Lakes area of Canada has always been a tough market for us. Perhaps it's just the easterner's typical perception that a West Coast firm couldn't possibly provide better services than those in Central Canada, even though there is a paucity of naval architects in Ontario. In any event, in 2015 we did finally get a commission from Ports Toronto for a replacement for an aging veteran tug. The new, albeit modest, little tug was built by Hike Metal Products of Wheatley, Ontario. It was christened as the *Iron Guppy* and is 20 metres long, powered by a single 750 hp diesel engine driving a fixed-pitch open propeller, and produces a modest bollard pull of 6 tonnes. The amusing name *Iron Guppy* was selected from a long list of names developed by students from Waterfront Elementary School in Toronto.

Vilja is another 36-metre ice-breaking escort tug designed to ice class 1A Super standard, capable of breaking 1 metre of ice at up to 3 knots. Based on the success of the earlier TundRA class ice-breaking designs, Robert Allan Ltd was commissioned in 2016 to design this new powerful ice-breaking tug for the Swedish Port of Luleå. Delivered in June 2019 by Gondan Shipyard in Spain, the vessel is used for ice management, firefighting and navigation-aids service duties in Luleå, and will assist ships and conduct coastal towing in the northern sector of the Gulf of Bothnia.

This vessel is equipped with an innovative hybrid propulsion system with in-line-shaft motor generators and a pair of Wärtsilä 8L26 diesel engines, each rated 3,100 kW, driving controllable-pitch Z-drive units. The additional power boost is provided by the shaft motor/generators and from electrical battery storage. This configuration allows effective power and fuel consumption management. The tug can operate on the electrical battery power only, utilize a single main engine in the hybrid diesel-electrical mode, or on two main engines in diesel-mechanical mode with possible battery boost. The resulting operational flexibility yields significant fuel, emissions and maintenance savings. This multi-functional propulsion system produces a bollard pull of 89 metric tonnes in mechanical mode and 98 metric tonnes in boost mode. The free-running speed is 14.6 knots.

Seaway Guardian, a new multi-purpose ice-breaking tug for the US St. Lawrence Seaway, the first in 61 years.

The *Iron Guppy* at home in Toronto harbour.

Designing the TundRA 3600 Hybrid Tug

One of the most interesting ice-breaking tug projects I worked on was *Vilja*, a TundRA 3600 hybrid-powered tug for the Swedish port of Luleå. She was specially designed for ice-breaking—clearing channels—ice management, ship assist and escort duties. The Port of Luleå is one of Sweden's busiest seaports, open year-round despite the ice covering from January to May, so they must clear a channel in the harbour for vessels that come into the port. The tug is designed to break 1-metre-thick ice at 3 knots and strengthened to the highest Finnish-Swedish ice-class standard. This was to be their first compact hybrid-powered icebreaker.

We designed the TundRA 3600 with a spoon-shaped bow and shallow buttocks angles, which aid the breaking of level ice into floes with relatively low resistance; the dual Z-drives enable flushing the broken ice outboard by directing the thrust outwards to about 45 degrees.

We included a 600 kWh lithium-ion battery bank and left an allowance for an additional 600 kW hours in the future. Fortunately, Luleå has a great facility in the harbour to plug into the grid to charge the batteries. The tug can simply dock and plug in. The electric motors are mounted onto the shafts with a clutch so they can be engaged or disengaged. The hybrid system made the tugs much more fuel-efficient. The tug performs ice-breaking, escort, towing and ship-handling in diesel-mechanical mode, but she is also able to run fully in electric mode for transit and standby operations, and then charge batteries at the port or also during the mechanical operation mode.

We worked with the Richmond branch of Corvus Energy, who supplied the lithium-ion batteries. The client was thinking ahead, because they said let's put 600 kWh initially and then have allowance to double that, knowing that battery technology keeps developing and gets cheaper and cheaper, so in the future they can add more and that would cost less than doing it initially.

The challenge for the port was to train its captains on how to drive this tug, because it was the first Z-drive tug for them. I was lucky for the opportunity to go and experience the tug in operation in person. We compared their old conventional ice-breaking tug design, with classic V-shaped bow, with the new one. This is where the difference came in, because even though the classic wedge hull might be better for cutting through ridges, without possible blockage of the fixed-pitch open propeller, they now had this new boat with two controllable-pitch steerable propellers in nozzles. They had a bit of a learning curve to get used to the manoeuvrability of the Z-drives. If you did aggressive ice-milling with the nozzle propellers, you could actually suck in the pieces of ice and block the props. At one point, we had an urgent call from the Port saying the tug could not proceed at all in the ice, possibly due to blockages of the nozzles. When we went over there, one of our Finnish clients who is an ice captain expressed interest in coming along, and he showed them how to drive this Z-drive tug in the ice with more speed, to be able to generate the top of the stern wave just forward of propellers to minimize the ice suction, and we haven't had any complaints since. To avoid getting ice in the props during ice management, you had to start turning in light or more open ice, make as wide an arc as possible, and direct the thrust outwards to clear the ice. This was different compared to how they were doing it with the other tug—it simply required a different approach. Both the old and new tug are really good tools to do ice-breaking and ice management, but the new one is a multi-functional tug, combining open-water performance and ice-breaking capability as well as significant escort performance.

In diesel-mechanical mode, the new tug has something like a 90-tonne bollard pull and up to 100 tonnes with the electric motors' boost. We were especially proud of this project because we won the contract against Nordic ice-breaking experts.

—Bilyana Ivanova

In 2021 Finnish operator Alfons Håkans took delivery of two TundRA 3200 class ice-breaking and escort tugs ordered from Turkey's Sanmar Shipyards. This project gave our company the opportunity to finally collaborate with a good friend in the tug industry, Jarkko Toivola, a highly experienced tugmaster and vessel manager with whom Rob Allan has collaborated extensively on a number of international tug safety programs.

The tugs, christened *Selene* and *Helios*, purpose-designed for operations in the northern Baltic Sea, are 31.5 metres long by 12-metre beam, with a bollard pull of 67 tonnes. These tugs are classed ice class 1A FS and, in keeping with modern trends, have a Green Passport EU notation, indicating that their entire life-cycle is considered (and documented) in the design and construction processes.

Svitzer Embla is a 30-metre ice-class ASD tug designed in 2019 and built by Med Marine and the Eregli Shipyard on Turkey's Black Sea coast, delivered in 2021. *Embla*, the first woman created by the gods in Norse mythology, is the name given to the first of a series of two ice-breaking tugs for Svitzer to assist ships year-round in Scandinavian ports.

This is a rather unique design, featuring a single winch for both forward and aft towing. A special towing arrangement allows the towing/escort winch located in the forward winch room to operate through a forward escort staple, as well as over the stern, via a tunnel and through the aft towing staple in the General Arrangement drawing.

The distinctive *Vilja*. At time of her commissioning, in 2019, this was the most powerful (and perhaps the only?) ice-breaking tug afloat with hybrid propulsion.

Selene, one of a pair of TundRA 3200 ASD tugs, purpose-designed for operations in the northern Baltic Sea.

Below: *Svitzer Embla*, a TundRA 3000 class ice-breaking tug for Swedish ports in service of the Svitzer group.

General Arrangement for the ice-breaking tug *Svitzer Embla*.

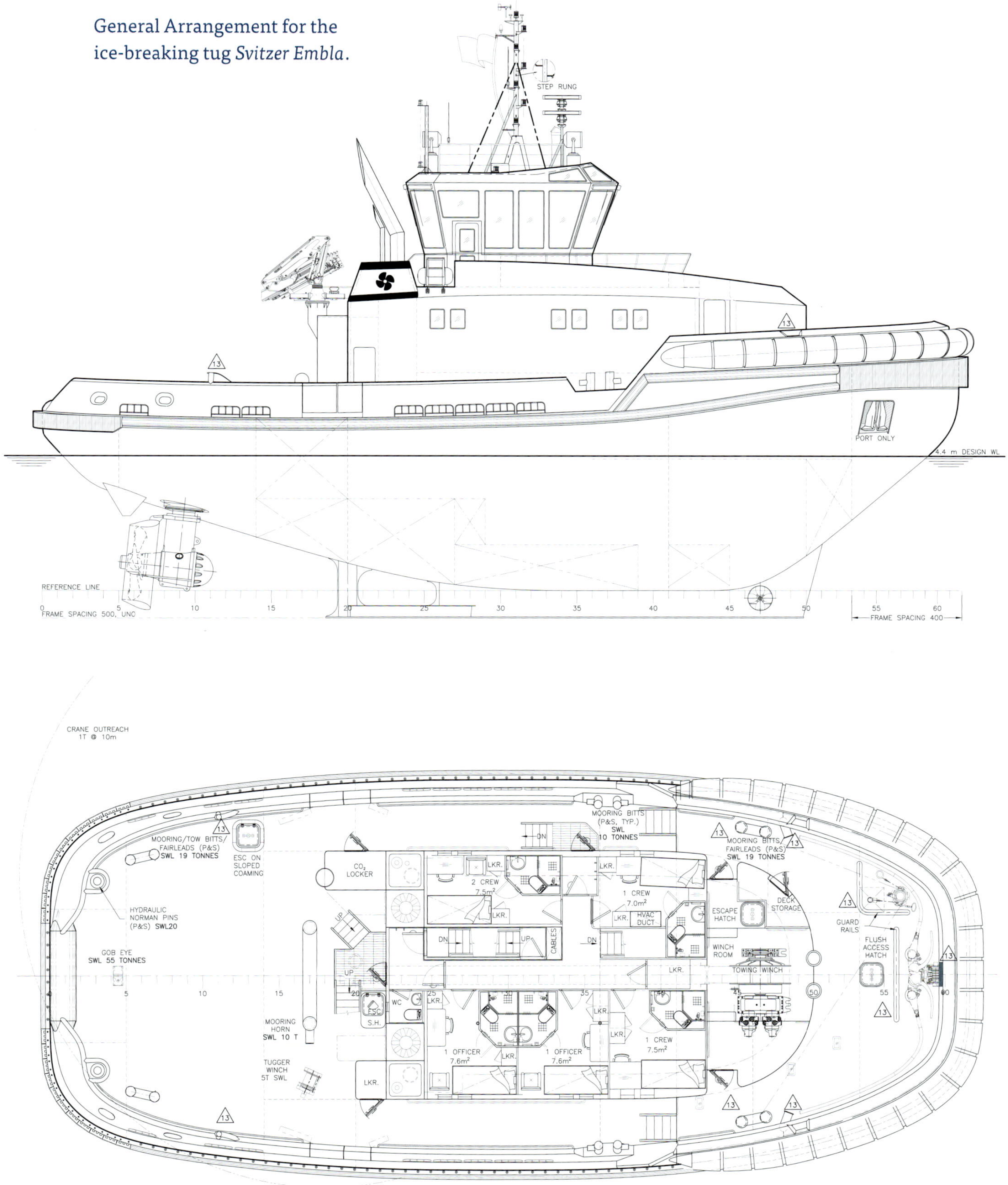

Chapter 19

Breaking into Africa

As our work spread around the globe it became clear that we didn't have much to show on the African continent. Many of our boats were working there but belonged to European or American owners. However, early in 2020 we were retained to design a new major offshore-rescue tug for the Kenyan Port Authority. The *Mwokozi II* is a 42-metre 120-tonne bollard pull salvage and harbour tug, designed to operate in East Africa and Indian Ocean regions and capable of undertaking international salvage and towage operations involving very large vessels. She will be based in the homeport of Mombasa, reflecting the establishment of an IMO Regional Maritime Search and Rescue Coordination Centre within Mombasa Port. This RAstar 4200 class ASD tug is 42 metres long, 16 metres beam, with a draft of 7.2 metres. It generates a bollard pull of 126 tonnes and has a free-running speed of 14.7 knots.

Mwokozi II, a major 42-metre rescue tug for East Africa.

World-Class Tugs for BC

In the years between 2000 and 2020, more than 1,000 tugboats were built to designs by Robert Allan Ltd, with an average of almost one tug per week being delivered. Unfortunately, it is impossible to identify all of these boats in this review, but it is now almost impossible to *not* find one of our designs in most major ports of the world.

The incredible growth of our tug design business through an array of overseas clients and shipyards has been somewhat overwhelming, particularly as we consider the modest beginnings of this company. However, it is also gratifying to see that our local BC-based clients and several of our overseas shipyard clients have since established good working relationships and some of those designs are now "coming home." As commercial shipyards in BC are almost a thing of the past at the present time, owners look overseas for industry-leading designs and high-quality construction that can still meet the often unusual Canadian regulatory conditions.

Examples of Canadian vessels built overseas to Robert Allan designs include the RAstar 2800 class *Seaspan Raven* series of four similar tugs (including *Seaspan Eagle*, *Seaspan Kestrel* and *Seaspan Osprey*) built for Seaspan Marine by Sanmar in Turkey from 2011 through 2019, and the 80-tonne bollard pull *SST Orca* and *SST Grizzly*, built for Saam-Smit Towage Canada Inc. (SST Canada) for their operations in Vancouver.

At the time of writing, the next generation of low and zero-emission tugboats for the new Kitimat LNG terminal in northern BC is under construction in Turkey by Sanmar for HaiSea Marine, a partnership between BC's Haisla Nation and Seaspan Marine. HaiSea Marine has ordered two dual-fuel (LNG-diesel) tugs and three battery-electric tugs to provide ship-handling and escort services for this major LNG export facility.

Robert Allan Ltd provided the designs for both these pace-setting vessels. With a bollard pull of about 100 tonnes, the dual-fuel tugs will be the most powerful tugs on Canada's West Coast and, with indirect escort forces of nearly 200 tonnes, will be among the highest-performance tugs of their kind in the world. The all-electric terminal tugs will develop about 70 tonnes of bollard pull and will recharge with readily available, locally produced hydroelectric power, resulting in zero emissions in battery-electric mode.

The 40-metre RAstar 4000 dual-fuel escort tugs will perform the majority of their missions using natural gas as their primary fuel. These escort tugs are also capable of pollution-response/oil-spill recovery, firefighting, person-overboard response, and emergency towing.

As a fleet, these escort and terminal tugs are expected to reduce CO_2 emissions by about 10,000 tonnes per year compared to diesel-powered alternatives, with major reductions of NOx, SOx, CO and particulate matter.

SST *Orca*, a RAstar 3200 class tug built in 2018 by Cheoy Lee Shipyard for Saam-Smit Towage (now SAAM Towage) for service in the Port of Vancouver.

Below: The 80-tonne bollard pull *Seaspan Kestrel* heading to work ship-handling in Vancouver.

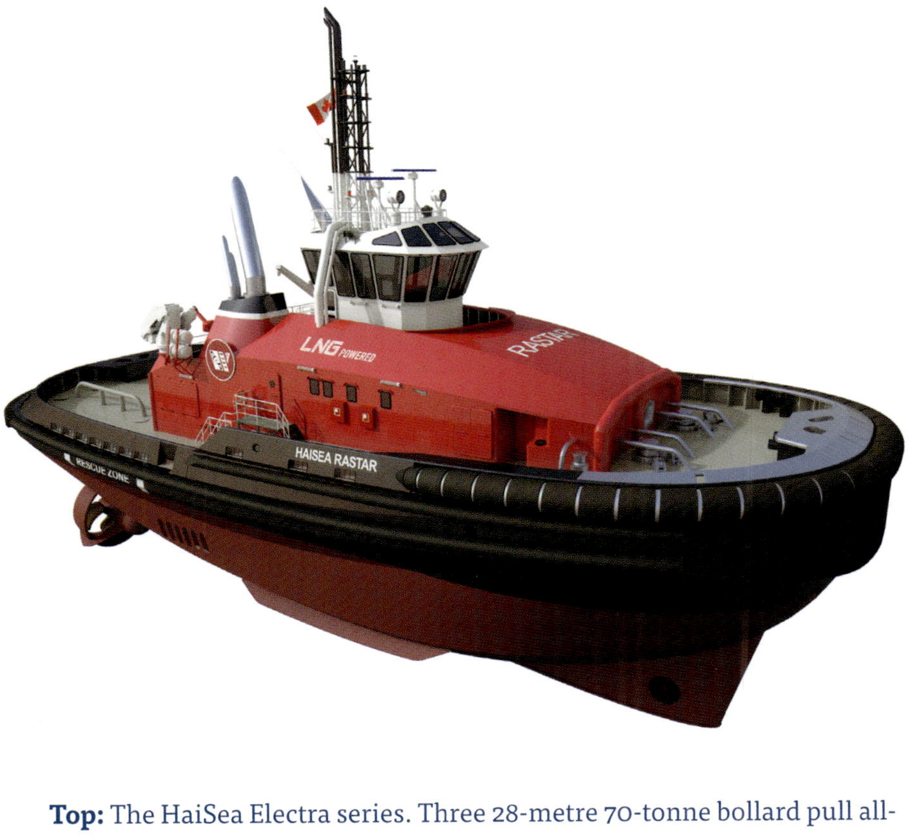

Top: The HaiSea Electra series. Three 28-metre 70-tonne bollard pull all-electric ship-handling tugs of this design are presently being built for the Kitimat LNG terminal by Sanmar in Turkey; **bottom:** The HaiSea RAstar series. A duo of these 40-metre 100-tonne bollard pull dual-fuel escort tugs, also currently being built in Turkey, will safeguard LNG tanker traffic on the northern BC coast.

PILOTS
INDEPENDENCE
BRENT
DELICAY II
PSA POLARIS
YT808
KAORI

This montage shows just the slightest impression of the range of vessels built to Robert Allan Ltd designs in recent years.

BAMSE
BREVIK
BAMSE

20

Innovations in Tug Design

We have tried hard not to be labelled as just "tug designers," but we are in fact the primary tug designer in the world at present and aim to stay there as long as our clients continue to entrust us with their future plans, and to favour the work that we do in this most rewarding area of small ship design. It is interesting (at least to me) to identify those design traits which are now often routine in our tug designs, but which were truly innovative (or at the very least creative) in their introduction, and which have shaped modern tug designs, particularly over the past 15 to 20 years.

Hull Forms

By far the most dominant form of propulsion in tugs today is the Z-drive or steerable rudder-propeller (SRP), and the most used configuration of propulsion/steering units is the azimuthing stern-drive (ASD) with twin Z-drives aft. Some of the earliest ASD tugs came out with hull shapes totally unsuited to these new drives. As illustrated, many of these, particularly in Japan, had aft buttocks with strong inflections to try to bring the drives up into the draft range of a typical twin-screw tug, but when the drives are pointed forward to go astern, a large portion of the thrust force is simply reflected off the bottom of the tug, creating a reaction force in the wrong direction. The presence of a traditional large skeg aft also redirects a lot of thrust, reducing the effectiveness of the Z-drives. Regrettably many tugs are still built today (to others' designs) with these same detrimental features.

Opposite: *Bamse*, see page 485.

The Guy with the Red Pen

Rob's ability to hand-draw is phenomenal. He's got this famous red pen. The design has to look right. You'd get started on a project, and the naval architects would do their best to draw a nice profile and what they think is right and so on, and they'd print it out. Rob wouldn't work on the computer file. He'd get a paper copy, and then he'd get his red pen out and he'd mark it up, like that line has got to go like that, with a swoop of his hand. He couldn't do that on a computer. He's capable on AutoCAD, but for that artistic part, that's done on a sheet of paper. You'll get these drawings back and, yes, the house has got to come back here, it's got to have this sort of thing on it, and then the responsible naval architect has to then put that into a CAD drawing. That creative part comes with a red pencil or pen on paper.

—Ken Harford

Far less common, but still with a loyal following among tug operators, are "tractor tugs," either with Voith Schneider cycloidal propellers (VSP) or with Z-drives located about ⅓ of the length from the bow, and Rotortugs (Chapter 19), featuring two Z-drives forward (as per a tractor) but with a third Z-drive unit aft in lieu of a fixed skeg. Due to the fitting of large drive units in the forebody of tractor tugs, this type of tug is inherently flat-bottomed, though that requirement is less pronounced with Z-drives. This presents design challenges in having a reasonable waterline entry, but that problem is also heavily influenced by bow fender geometry. Although less common, Robert Allan Ltd has a steady stream of design requests for tractor tugs (our TRAktor series) with both VSP and Z-drive propulsion.

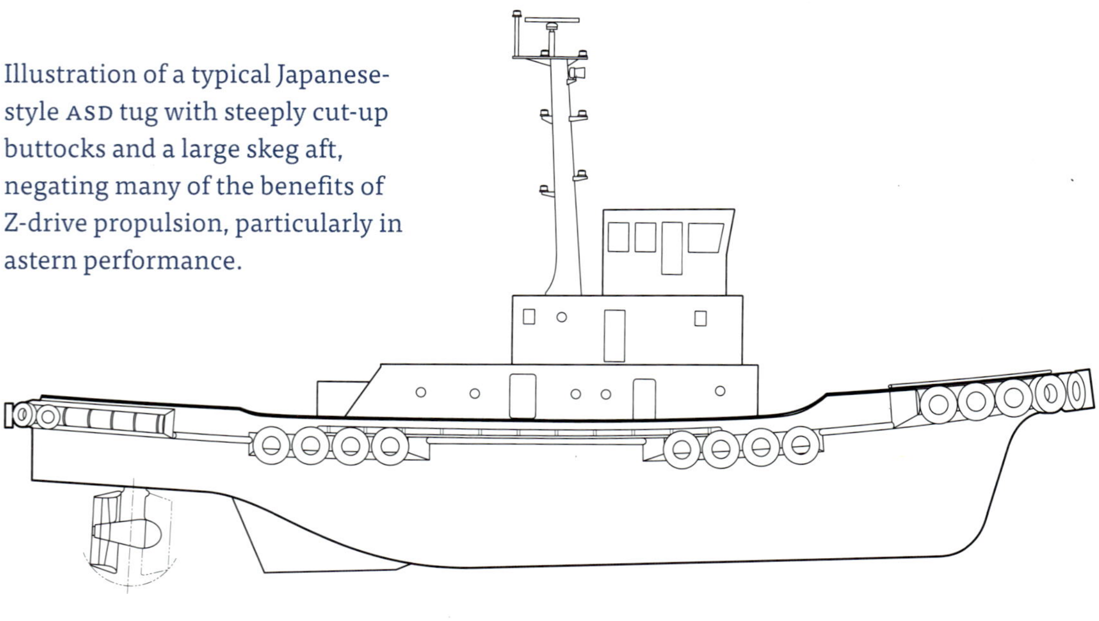

Illustration of a typical Japanese-style ASD tug with steeply cut-up buttocks and a large skeg aft, negating many of the benefits of Z-drive propulsion, particularly in astern performance.

Our ASD hull designs have always been distinctive in their shape. They have been designed from the outset to take maximum advantage of the omni-directional capability of Z-drives. That same trait can also be applied to tugs with alternate propulsion configurations, such as tractors and Rotortugs, but with obvious differences. The main objective always is to not try to force water to bend or turn corners. The hull and drive axes are aligned to maximize the delivered thrust in all directions, as shown in the early Robert Allan Ltd ASD tug design. Following that is a more recent design. The general hull shape has not changed much, but the skeg geometry reflects the general demand today for tugs with an escort rating, thus featuring a large forward-biased skeg.

Bamse, a Tier III-compliant 75-tonne bollard pull TRAktor 3000-Z (Z-drive) for Buksér og Berging AS of Norway, delivered by Sanmar in 2021.

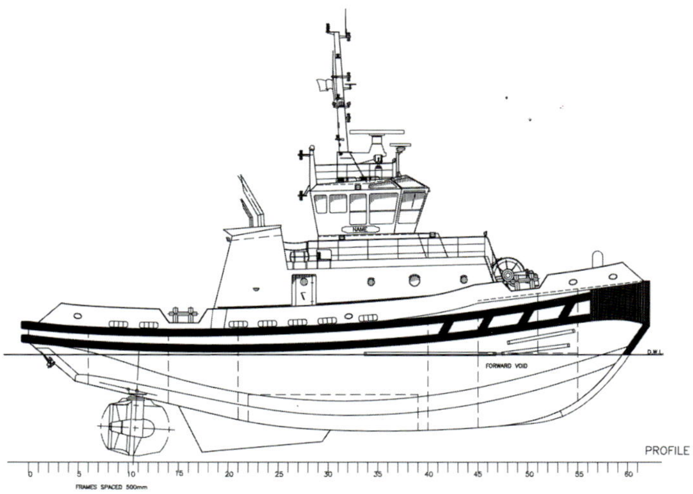

Above left: Early (1993) Robert Allan Ltd ASD tug, still with a small aft skeg, principally to facilitate docking; **above right:** A more recent ASD escort tug design with hull lines similar to earlier vessels but with a very different escort skeg configuration. Note how the flow from the drives will pass almost unimpeded under the hull of the tug, generating maximum astern thrust. Losses due to the slight tilt angle of the drives are negligible.

RAstar Hull Form for Escort Tugs

Probably the most significant development in the evolution of escort tugs has been the creation of what we now refer to as our RAstar hull form. This sponsoned shape, as illustrated here, provides many advantages: tug stability increases markedly when the tug heels over under transverse towline forces due to the increase in waterplane area; the immersed slope generates a righting force when pulled sideways, increasing both righting forces and escort forces; and the hull is narrower at the waterline when upright, reducing the power required for free-running and reducing wash. The sponson can be combined with almost any typical hull form to achieve the desired results.

Although these hull and appendage design features were developed to provide increased indirect escort performance, the characteristics of the RAstar hull form have serendipitously also shown significant sea-keeping improvements over more conventional hull forms. This form provides dramatic reductions in roll amplitude and accelerations, and the large skegs provide significant roll damping.

As part of the initial model-testing program conducted to verify the performance of the RAstar hull form, and to be able to state with confidence that the motions of this new design were a real improvement over existing tugs, a model of a similar size of well-known "standard" tug was also built, and the motions of the two hull forms were measured in identical wave conditions for a unique side-by-side, vessel-to-vessel comparison.

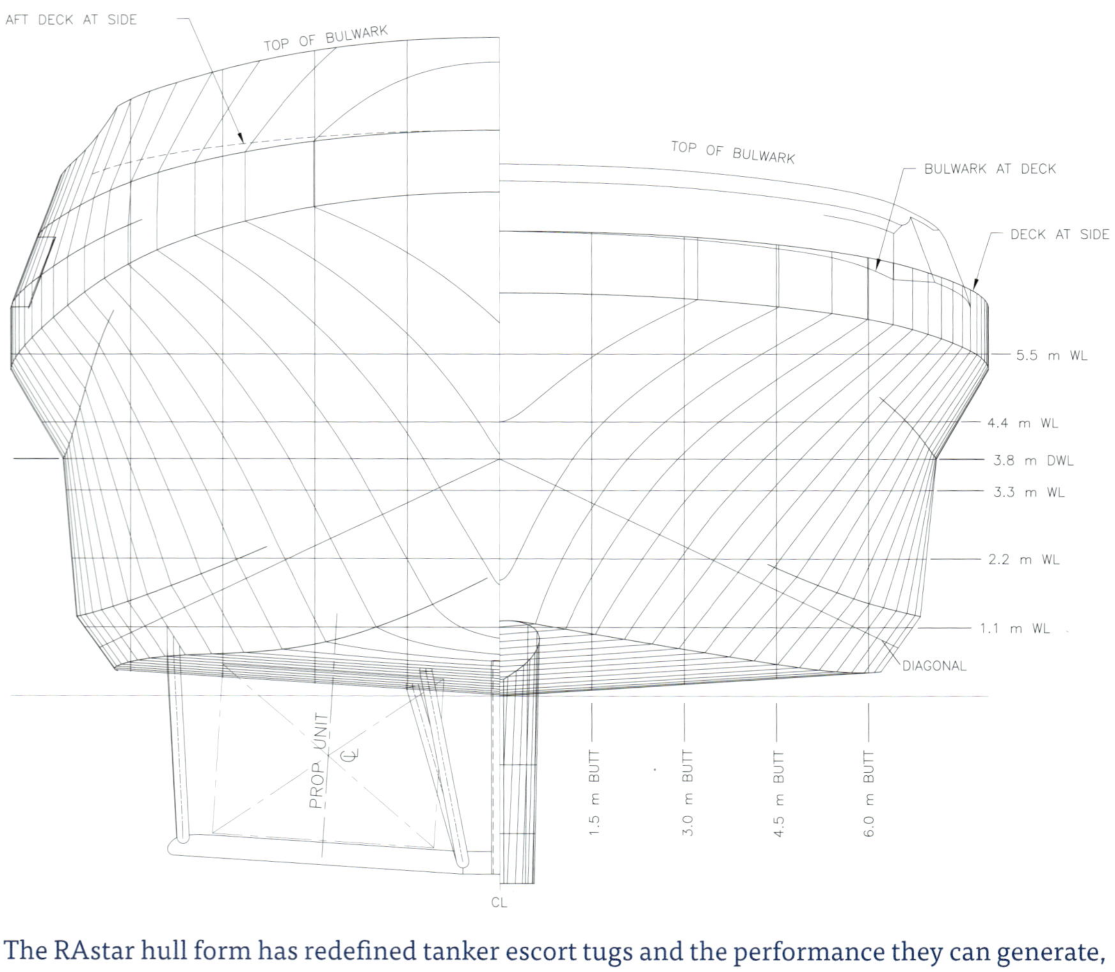

The RAstar hull form has redefined tanker escort tugs and the performance they can generate, most certainly one of Robert Allan Ltd's more significant contributions to the industry.

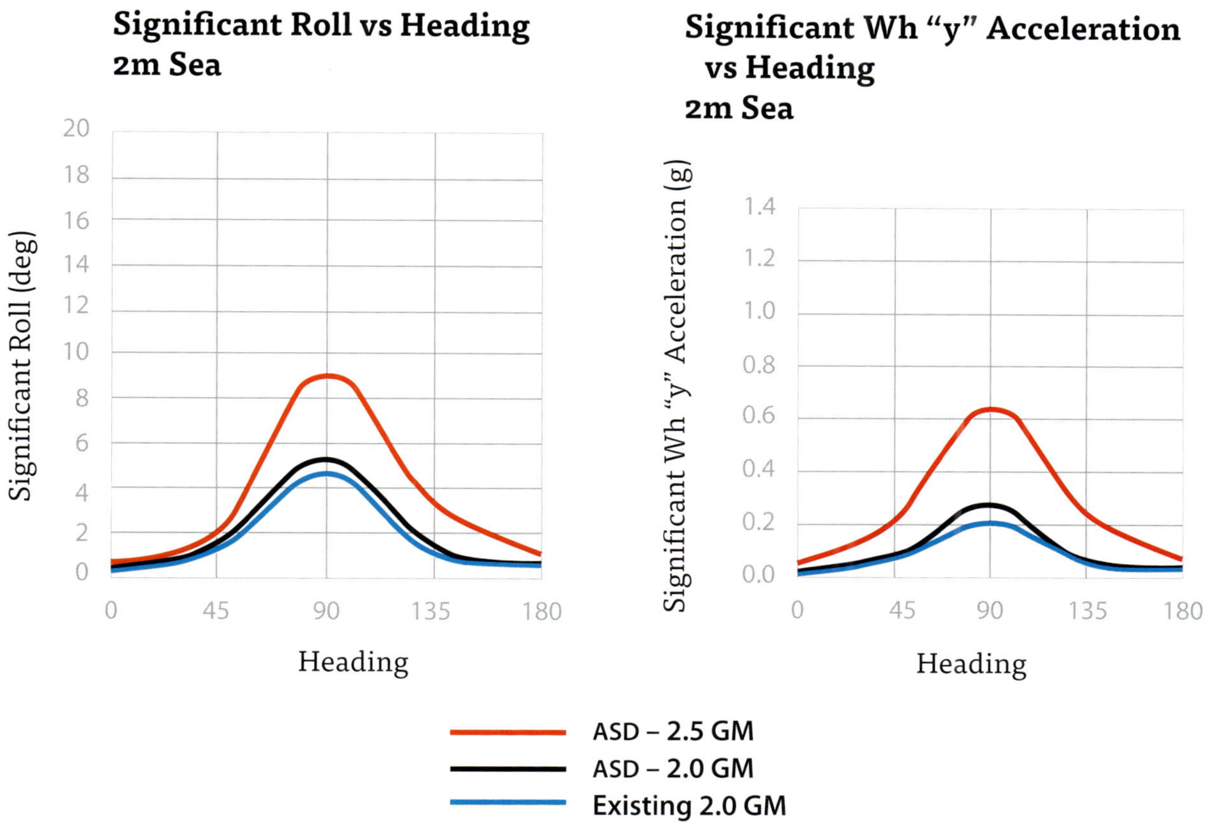

Above left: Significant roll angles of RAstar (ASD) vs. standard hull form; **above right:** Significant roll accelerations for RAstar (ASD) vs. standard hull form.

The sea-keeping predictions for the RAstar class tug clearly indicated that this type of hull form provides an exceptionally stable and comfortable platform. The substantial reduction in both motions and accelerations predicted significantly reduced crew stress and fatigue, thereby improving overall operational safety and effectiveness in comparison to the more standard tug type ("existing hull" in the graphs shown). These improved motion predictions have been thoroughly borne out in operation by crew feedback and empirical evidence.

Visibility and Safety

It should be self-evident that for tug operations to be as safe as possible, the operator, often working alone in the wheelhouse, must have the best possible view of (a) the ship he is attending, (b) his crew on the working deck and (c) his fender contact points with the ship. Most modern tugs today do a reasonable job of achieving these objectives, but frequently one will see designs that fail to respect this basic premise. Windows may be small and with wide "wooding" between glazed areas, and the pillars can be poorly placed, obstructing important sightlines.

The Occasional Good Idea

It occurred to me, when working on a sketch for a new ASD tug design in the early '90s, that the typical overhead windows then used in tugs (an example shown at bottom right) were too small to be effective in permitting the master to see clearly what was going on above him, especially when working under the flare of a ship's bow.

With one of a few occasional good ideas, the solution was surprisingly simple: slope the wheelhouse top upward from just aft of the control position and thus create a much larger viewing area for the operator. The picture shown top right illustrates the effect this has on the overall inside view from the master's position, offering a much better view of any overhead activity.

Although the old style of overhead windows still appears on some boats, the canted upper window styling and safety feature is now being emulated on many designs from other firms—the perpetual challenge of being a design leader.

—R.G. Allan

Media Luna, an AVT 3100 class VSP tug operating in Puerto de Bolivar, Colombia. The larger overhead windows and minimal "wood" between windows on this (and many similar) tugs greatly enhance operator visibility and thus operational safety.

Above: Expanded sightlines possible with canted wheelhouse top, a Robert Allan Ltd innovation.

Right: Typical old-fashioned style of small overhead wheelhouse windows offering limited visibility. Not a RAL design.

In addition to the window innovations detailed above, a common challenge to wheelhouse visibility is the size of the typical bolted framing around traditional ship's windows. Typically, the flanges around this style of window are 50–75 mm wide. Add this to the width of the structural columns between windows, and it was not uncommon for the total "wooding" between glass openings to be 200–250 mm wide. In a tug, these wide columns represent significant blind spots, and thus can impede the safety of all operations.

We began promoting the concept of using bonded (glued-in-place) windows on tug wheelhouses with the Cates-style US tug designs in the late '90s. Bonding glass to the structural columns, as is commonly done on cars and aircraft, reduces the blank area to the width of the structural columns alone, approximately 150 mm. In addition, extending the bonded glass on the outside to almost touch "pane to pane" creates a much cleaner and more modern aesthetic than a multiplicity of bolted windows, as shown in the RAmparts 2400-SX class tug *Svitzer Amstel*.

The overhead visibility issue is perhaps even more important in fireboats than in tugs. The Los Angeles fireboat *Warner L. Lawrence* illustrates how well this design feature serves that purpose.

Effective use of bonded windows in *Svitzer Amstel*, one of Sanmar's numerous Boğaçay class tugboats.

Los Angeles fireboat presents great sightlines from the control station to the firefighting streams overhead.

Noise and Vibration Control

Noise became a contentious issue on BC tugboats with the advent of the "Hot-Rods" and other higher-powered steel tugs. Obviously, steel doesn't have the vibration and noise-attenuation characteristics of a stout hull built of Douglas fir. The towing industry commissioned a study into noise issues on tugs in 1971. This work brought Ken Harford into touch with the commercial tug industry, and his influence on this critical topic is still being felt today.

Noise Control on Tugs

Ken Harford

During a significant BC towboat industry strike in the early 1970s, one of the major issues raised was noise levels on tugs. No one on either side of the table could come up with the right answer. To resolve this, the parties formed a tripartite committee to research the "Acoustic Habitability of Vessels Engaged in Towing." It consisted of the federal government (who paid for the study), the towboat companies and the union. The research project went to a company called Jackson Talbot, who hired us (Harford Kennedy Ltd) as their sub-consultant.

It was a worthwhile study. With the cooperation of the Towboat Industrial Relations Association (TIRA), detailed noise surveys were conducted on 33 tugs on the West Coast, ranging from 8.8 metres to 45 metres in length and ranging in power from 170 kW to 3,000 kW (total shaft power). The noise levels were measured in the cabins and the engine room, and detailed records were made of how much power they had, the type of construction (wood or steel), the arrangement of the cabins relative to the main machinery space, and so on. We put together a statistical model using linear regression analysis. That is a way of taking the noise data from all the tugs and saying, "Okay, what is similar about these things?" to create a mathematical relationship. We knew that the noise levels were higher on certain boats and lower on others, and so we wanted to know why. So we took all these different parameters and put them into this model and assigned factors to length, power, displacement, arrangement to some degree, and so on, to create a formula for a curve that fit the data. We had all this data on how noisy the existing tugboats were, so we used that data to create a way to predict what another tugboat's noise level might be. And it basically comes down to, if you have a smaller boat with more power it's going to be noisier, which sort of fits with one's intuitive sense. I was still using this statistical model when I wanted

Tug *Seaspan Planet* (ex. *La Belle* of Vancouver Tugboat Co.), a typical "Hot-Rod," served as the test bed for noise-control methods in tugs.

to do a quick estimate of what the noise level was going to be on new tugboats, until I retired from Robert Allan Ltd in 2016.

That was one part of the study. The other part was to use the tug *Seaspan Planet* (ex. *La Belle*) to demonstrate what could be done to reduce the noise levels on tugs. We did a lot of investigating, and the solution was pretty obvious right from the beginning. Our recommendation to the committee was to put the main engine of the *Seaspan Planet* on vibration isolators. I mean, what do you think your car would sound like if you bolted the engine directly to the frame?

I thought I was going to be tarred and feathered when I said we must put these engines on vibration isolators. To me, as a mechanical engineer, it made perfect sense. The reaction, particularly from the marine engineering community, was "Oh no, we have to have fitted bolts and we have to allow for thermal growth, and we have to do all these other sorts of things."

Basically, we had to borrow the vibration isolation technology from other applications (cars, trucks, airplanes). Fortunately, I had a very good relationship with some of my university professors, who could help figure out how to do this. Ron Hazell did the calculations to analyse the excitation characteristics of the engine and the response characteristics of the isolated engine on the vibration isolation mounts to select the right mount dynamic characteristics. Nowadays this involves doing a solid body dynamic analysis using a sophisticated computer program, but he did that analysis by hand on about five or six handwritten pages.

We selected rubber mounts with fairly limited static deflection because we didn't think we could get by with anything softer than that. It was difficult enough getting those mounts accepted. Mount selection has a lot to do with vibration frequencies generated by the engine and whether the engine excitation frequencies are lower or higher than the natural frequency of the engine on the mounts. If you're right on the natural frequency, you've got a problem. It's like Ella Fitzgerald singing a note and breaking a wine glass. If the excitation frequencies are equal to the resonant frequency of the system, the vibration is amplified. However, if the resonant (natural) frequency is well below the excitation frequency, then you get really good vibration isolation—the transmitted vibration and the resulting noise levels go down.

In order to demonstrate that the engine vibration was the main source of the noise problem in the crew cabins, we did some other experiments on the tug to show that the noise was all from structural vibration and not from the airborne noise in the engine room being transmitted through the steel bulkheads. We rented a big guitar amplifier and speaker with a pink noise generator and put them in the engine room to generate, through an airborne source without the associated structural vibration, the same sort of levels in the engine room as the engine was generating (about 108 dB). Then we did the same survey on the boat that we had done with the engine running, and it showed that if the source of the noise in the engine room was only airborne, the noise levels in the rest of the boat would be about 30 dB lower. What was happening was that the engine was vibrating the hull, and that vibration was transmitting into the rest of the structure so that all the ship's sides and panelling were just working like loudspeakers radiating the sound into the spaces. The solution was to break that path—don't let the vibration get into the hull. Engine-room soundproofing insulation and treatments like added lead barriers only affect the airborne radiation, and we proved that airborne radiation didn't really matter.

If you have two noise levels that are exactly the same, and you turn one of them off, the noise level in decibels only goes down by 3 dB, and when you turn the second noise source back on, the combined level only comes up by 3 dB. Therefore, if you have a sound path that is 30 dB below the other sound path, doing anything about the 30 dB lower noise source makes zero difference to the combined noise level.

So we did this conversion to the main engine of the *Seaspan Planet* at Allied Shipbuilders. The first issue that arose in this refit was aligning the engine after the flexible mounts were installed. The shipyard machinists were having a heck of a time, so I got called in. I scurried around town and bought dial indicators and shim stock and things like that. Two of us then went down to the shipyard. We got it aligned in a few hours, but we had an advantage because we knew the stiffness of the mounts and the weight of the engine and so on, and we knew we could figure out that if we put a 20-thou shim in there, it was only going to go up 10 thou, for example. This is not really logical to machinists used to working with solid mounts. When you put in a 20-thou shim under a solid mount, it goes up 20 thou. Since a resilient mount is relatively soft, it compresses when it takes on more weight, and the increase in height due to a 20-thou shim will be less.

We also put a new coupling between the engine and the gearbox to accommodate the expected movement of the engine on the vibration isolators. This coupling needed to also reduce potential damage from rotational (twisting) vibrations in the engine crankshaft and shafting. Caterpillar performed a torsional vibration analysis (TVA) and advised that the torsional vibration characteristics with the coupling we selected were acceptable. Unfortunately, their analysis assumed an engine in perfect running condition.

On the day of the sea trials, we got about 100 yards out of Coal Harbour and the skipper revved up the engine. The boat shook so bad he refused to take it back to the dock. We had to call Cates to take the boat back into the dock.

Of course, I'm sweating, thinking this is all my fault. Right away I thought it was caused by torsional vibration, so I got in touch with Ted Spaetgens, the torsional specialist from a local company called Lo-Rez Vibration Control Ltd, and he came down to look at it. He connected his torsiograph to the propeller shaft and measured the torsional vibration. He phoned me back the next afternoon and said there wasn't any problem with the torsional coupling. The problem was that engine was out of tune. So I phoned up Finning and said, "The engine is out of tune." And they said, "No, the engine is not out of tune, we just serviced it." However, I prevailed, and they sent a mechanic, and he and I spent a day down there with him going through the whole engine, measuring all the push-rod clearances and everything that would affect the delivery of the fuel to the engine and the firing sequence and magnitudes. We found little things here and there and eventually got the engine all back to factory spec. The engine was perfect.

We then went out for a new trial, and everything was smooth as silk and the noise level had gone down by 10 dBA. This is equivalent to reducing the sound intensity by a factor of 100. For example, the difference between 70 and 80 dB is a factor of 100 times in intensity. The decibel scale isn't linear, so the difference between 80 and 70 dB is really huge.

So the vibration isolation demonstration was a great success. The problem was that the engine would only stay in tune for six months, then the vibrations would start to develop, excited by the uneven engine-firing conditions. As soon as we started getting this recurring torsional vibration, the move was on. Seaspan insisted that the engine go back on solid mounts, despite the sound level decrease. The unions were saying just keep the engines tuned up. However, that was the end of the experiment— at least on the *Seaspan Planet*. The engine went back onto solid mounts.

I didn't learn until several years later why the tug companies were so insistent that the engine on the *Planet* had to go back on solid mounts. Capt. Arnie Davis and I had a beer at the cocktail lounge in the Empress in Victoria during a BC Towboat Conference. We got onto the subject of the *Seaspan Planet*, and I learned that in settling the original strike, the companies had agreed that whatever was successful on the *Seaspan Planet* would be implemented throughout the fleet. The *Seaspan Planet* simply could not succeed, because if it did succeed, they'd have to change the engine mounts and install flexible couplings on every boat in the fleet. So as soon as there was this glitch, the industry had a way out of their commitment. We could have changed the coupling easily and fixed this problem of the torsional vibration so the drive system would be more tolerant to the "normal" imbalances in the engine-firing energy, but that would have resulted in the need for the industry to put resilient mounts under every engine on every tug on the West Coast.

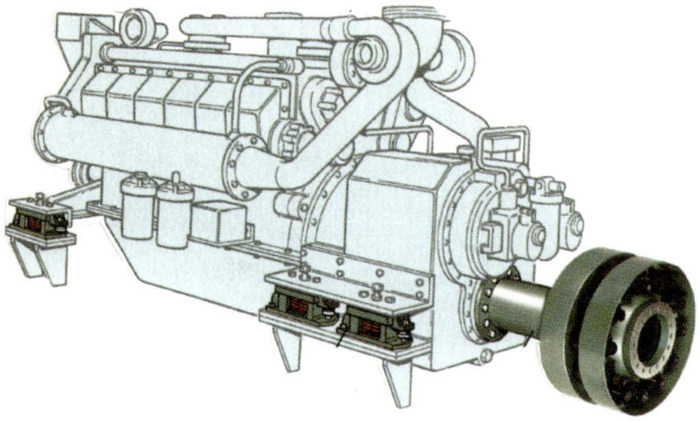

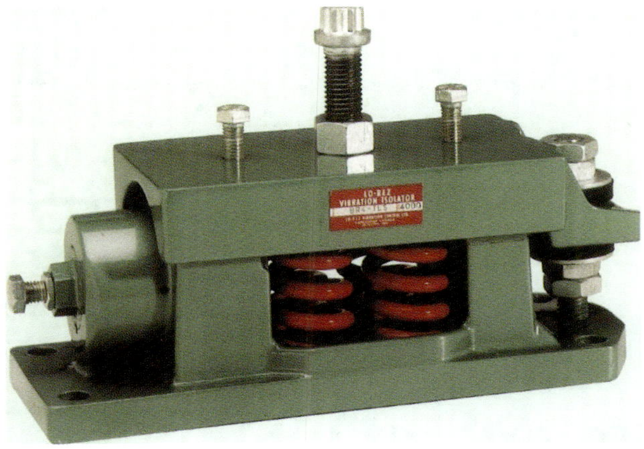

Typical vibration isolation package installed in modern workboats (left), and detail of spring isolator (courtesy Lo-Rez Vibration Control).

With the research work performed by Ken and his former associates, the writing was clearly on the wall and every new tug design developed at Robert Allan Ltd since about the mid '70s incorporated some form of vibration isolation and various other noise-control measures. Once Ken joined Robert Allan Ltd in 1988, he brought his expertise on this (and many other technical subjects) into play on virtually every project we did.

As tug power-to-size ratios increased year over year, this noise abatement treatment became increasingly important. Today it is rare to find a new Robert Allan tug with noise levels in accommodation spaces above 70 dBA, and often they are under 65 dBA. Noise levels in control spaces such as wheelhouses are generally in the range of 60 dBA.

Moving beyond the issues associated with the noise inside tugs, today the focus has shifted to the problem of underwater radiated noise (URN) from commercial vessels. In particular, the impact of URN on the southern resident killer whale population in the Salish Sea has become a major focus of research. Robert Allan Ltd is actively engaged in this issue, developing models similar to what Ken developed almost 50 years ago to analyse the external noise problem.

Smarter Shaft Lines

The use of Z-drives poses some interesting challenges to the designer. The drive-shafts no longer exit the hull underwater through a gland or stuffing box like a conventional propeller shaft. Z-drive shafts stay entirely within the hull and terminate at the top of the drive units. The typical solution to this in early designs was to use a series of "Cardan shafts," lengths of solid shafting with universal couplings (U-joints) at each end, making the transition from the engine to the drive unit in a series of linkages at various angles, as illustrated, often with multiple bearings to support the entire shafting system.

As we experimented with tipping Z-drive units forward for better flow into the nozzles, we realized that we could also, within reason, tip the engines "down at the head," and it became obvious that a much simpler straight-line shaft could be achieved between engine and drive, eliminating the complex and sensitive alignments used with U-joints, as well as reducing bearing requirements.

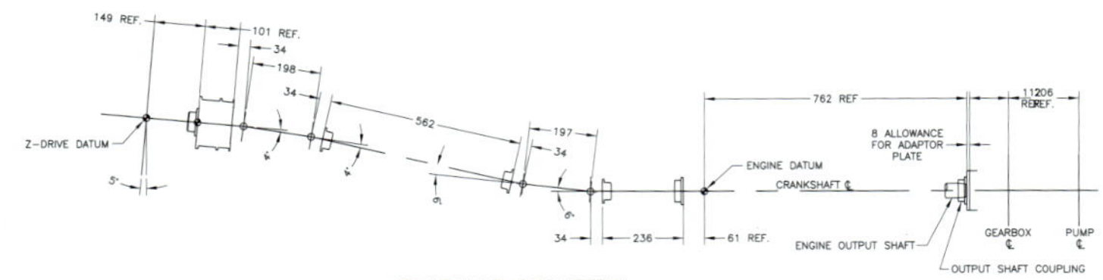

SHAFTLINE GEOMETRY

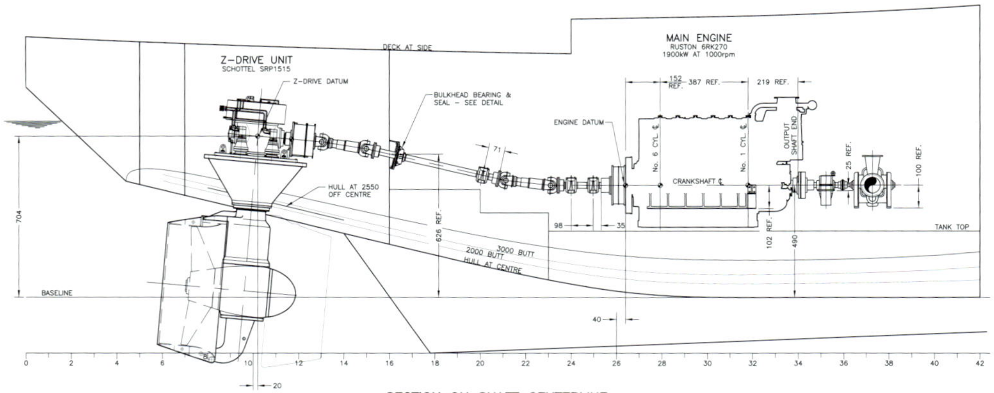

SECTION ON SHAFT CENTERLINE

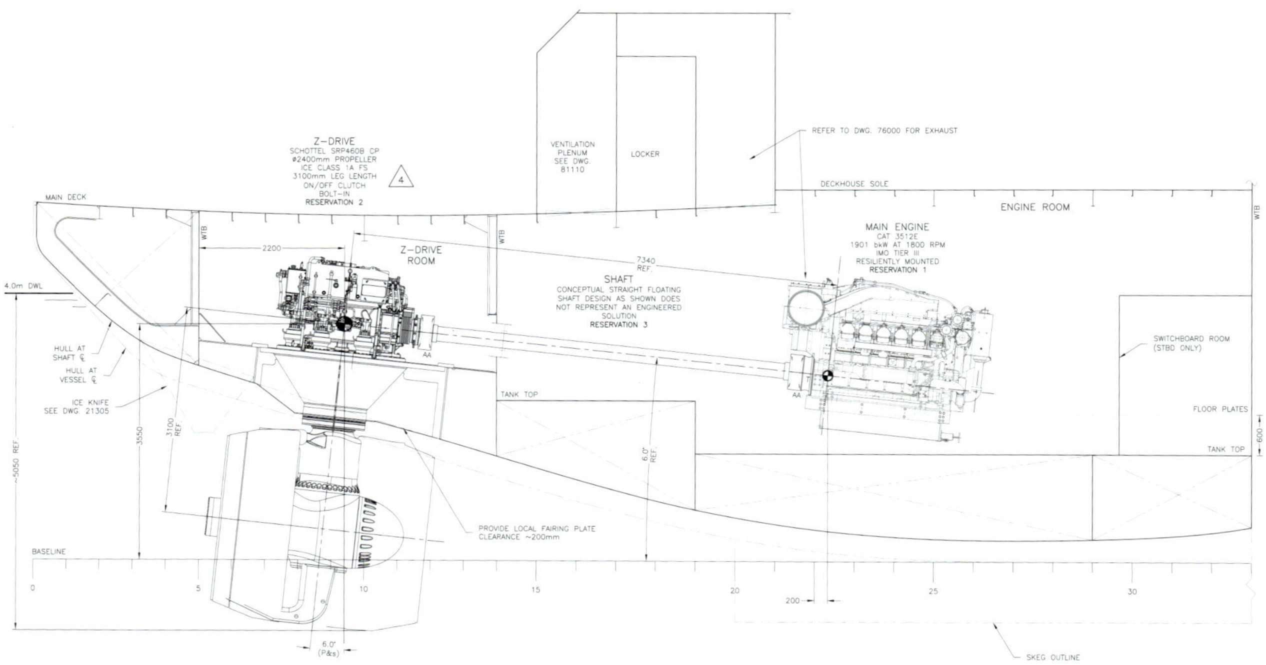

Top: Typical early Z-drive shafting arrangement with Cardan-shafting elements and multiple bearings; **bottom:** A typical straight-line carbon-fibre shaft system.

When carbon-fibre shaft technology became widely available in about the early 2000s, it was obvious that straight shaft lines could be easily adapted to this new technology. Carbon-fibre shafts are much lighter and don't require bearings, as is the case for the much heavier solid-steel shafts. Carbon-fibre shafts are expensive compared to steel, but since there are no bearings and no bearing seats, there is a substantial saving in the installation man-hours. Shipyards objected strenuously to our specifying these shafts at first but soon realized there was a direct benefit to them. The lightweight hollow shafts also don't require extensive rigging, and there are no bearings to service for the life of the vessel. Eliminating the bearing seats not only saves installation costs but frees up space to route piping in the often-congested area below the floorplates. Two people can easily pick up the shaft for a 2,000 hp engine. This also eliminates the not infrequent problems associated with shaft bearings. Today almost all our drive installations utilize carbon-fibre shafting, as illustrated.

Designing Fan Carriers

As a key part of designing mechanical systems for the tugs, one of the areas I worked on after starting with Robert Allan Ltd was engine-room ventilation. Over the years, many tugs had sunk due to down-flooding through hatches and doors left open to stop the engines from overheating. They were left open because the ventilation systems were inadequate. I had previously engineered large enclosed backup genset enclosures for commercial and quasi-military applications, some for installations in Saudi Arabia. In those designs I had to make sure they got enough air to not only cool their radiators but also cool the machinery space and, at the same time, try to keep the sound from getting out. Those principles of moving large volumes of air in the genset enclosure applications were the same principles that I applied to designing tug engine-room ventilation systems. The solution was simply to get a hell of a lot more air in and out of the engine room than had been done previously. What you want is big fans and to have the air intake as high up from the water as you can, and usually that is through the stacks. You can't have the intake air grilles at the top of the stack, as that is where the hot exhaust is, so you need the intakes lower, but not too low. Despite the importance of this work, I regularly got accused of helping design "fan carriers" instead of tugs, but these tugs could operate on hot days with the watertight closures secured and hence were inherently much safer.

Work on designing the mechanical systems of the tugs, including the ventilation systems, was one of the big areas, but I kept getting involved in the noise-level aspect of the designs as well. Noise levels were becoming more and more of a concern, but it was such a pleasure to be doing that at the paper stage, where you could actually do something sensible instead of being called in after the fact to design a "band-aid" treatment.

—Ken Harford

Finite Element Analysis

As tug powers grew, it became increasingly necessary to fully assess the strength of the fittings used in all aspects of towing and escort operations. The empirical or even rule-of-thumb estimates used in previous generations of designs simply could not be trusted to deliver correct answers about the rated strength of highly stressed fittings.

As our computer system grew, we acquired finite element analysis (FEA) software and, more importantly, the clever young engineers familiar with its operation and this type of sophisticated analysis. Very quickly FEA became an essential part of our engineering processes in all aspects of ship design, but the most critical work remains the design and analysis of highly stressed components like escort towing staples, as illustrated.

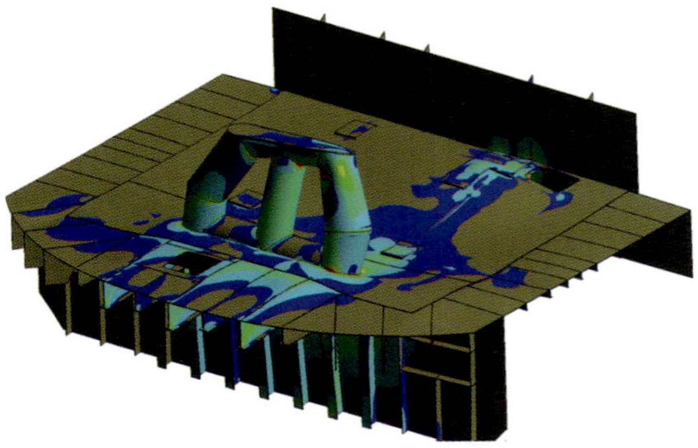

Example of finite element analysis representation of a large towing staple on an escort tug, subject to a 150-tonne sideload.

Today classification societies are demanding the submission of this type of analysis for all highly loaded structures.

Another innovative application of FEA resources is the analysis of fendering loads. With the introduction of large LNG carriers, which are somewhat lightly built, the classification societies started introducing maximum fender contact loads on the sides of these ships. Tender documents for new tugs at exposed offshore terminals suddenly began appearing, stating that tugs could apply not more than 20 tonnes per square metre (and in some cases as low as 16 T/m^2) in fender pressure to the ship hulls, even with clearly marked and reinforced "push here" or "tug" areas as shown.

Container ship with clearly marked "tug" contact area.

That is all very well in an "ideal" world where tugs might be considered as having large, flat plane bows, and which don't move around in a seaway, but the real world of tugs with large, rounded bows, and with LNG terminals located in often quite exposed areas, is quite different. In recent times terminal operators are calling for tugs with 80–100 tonnes bollard pull. Creating a sensible design with a uniform fender contact area of 4–5 square metres is next to impossible. Considerable effort has been invested at Robert Allan Ltd into identifying concepts that are workable with varying types of fender geometries and hardness, and that work continues with the development in 2021 of a new "offshore" fender design, illustrated below.

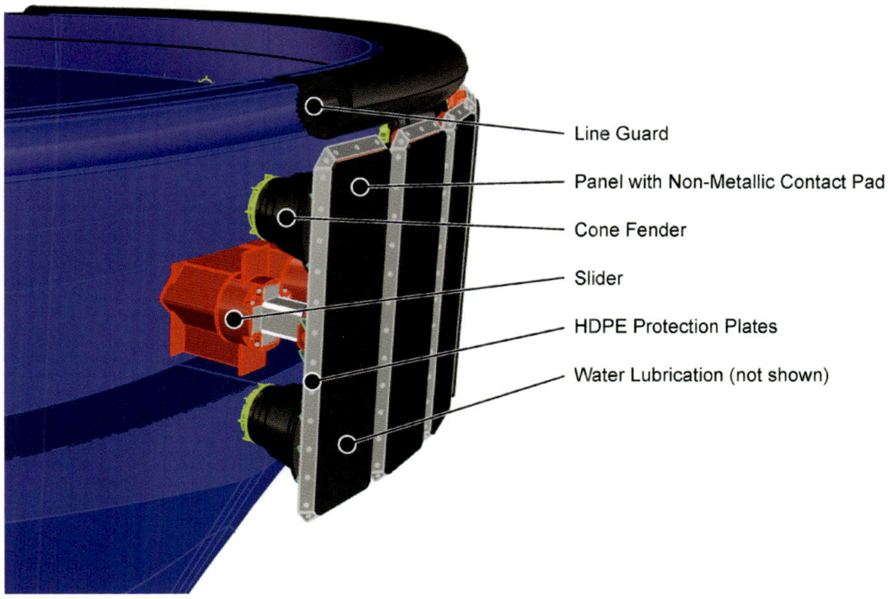

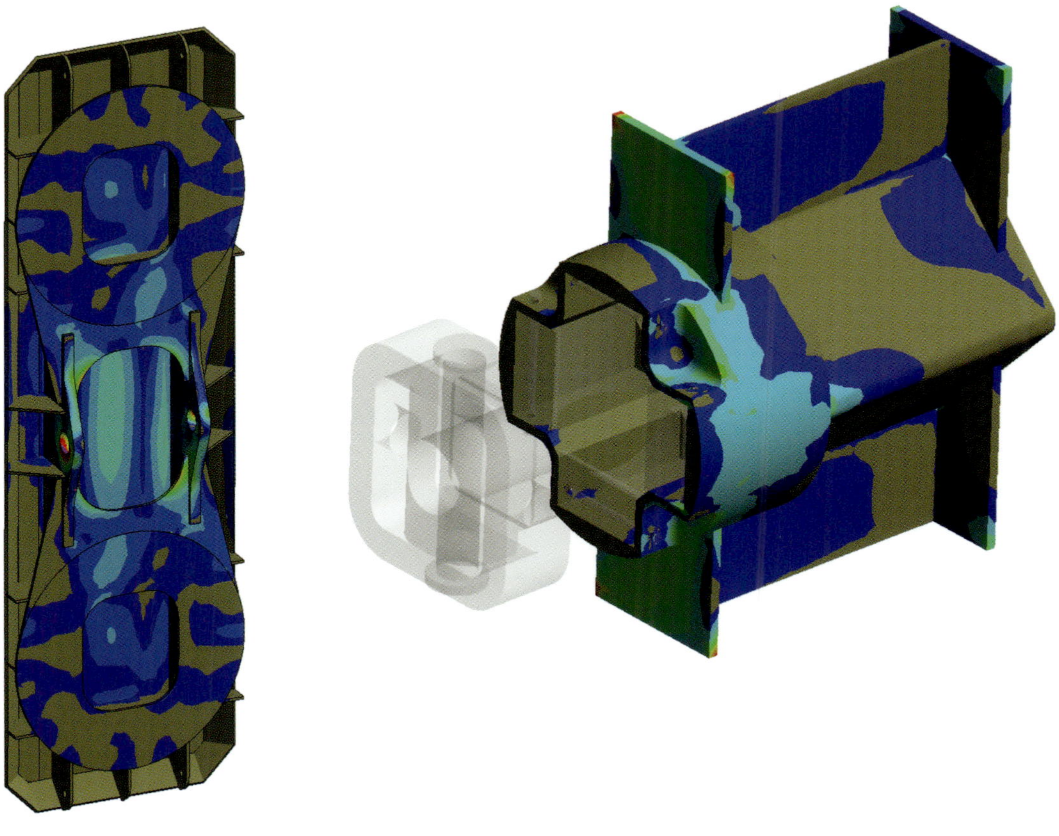

Robert Allan Ltd offshore fender design concept showing FEA analysis of major components.

Computational Fluid Dynamics

The world of workboat design in which we operate has traditionally not been a sector where a great deal of research has been involved. Before the advent of escort tugs and the need to really verify their indirect towing performance, it was in fact quite rare to model-test any tug designs for their general performance. The main operating criteria for tugs was considered to be bollard pull, and that was relatively simple to predict with a high degree of accuracy by basic propeller thrust calculations.

With escort towing, all of that changed and it became essential to do an increasing amount of scale-model testing to verify that the required indirect forces could be generated. Model-testing, as illustrated, is a great tool and, in addition to verifying performance, can also give a good impression of overall vessel behaviour, even if some of those characteristics are not being measured in the test program. Since the mid-90s, model-testing has become a critical tool in our R&D efforts. We have worked extensively with facilities such as MARIN, in the Netherlands, and the Vienna Model Basin in particular, the latter being a very nimble and cost-effective organization. Our work there includes a lot of escort testing, and as illustrated, even some wind-tunnel testing on an offshore support vessel where the exhaust-distribution characteristics were critical to the project. However, model-testing is time-consuming and quite expensive, and doesn't lend itself well to much design iteration in the test process.

In the early 21st century, the use of computational fluid dynamics (CFD) was beginning to work its way into the design of more general ship types, rather than just the more exotic. The cost of commercially available CFD programs, though still very expensive, was coming down into "consideration" territory. In 2009 the management at Robert Allan Ltd was convinced to acquire Star-CCM CFD software, and to use this as an important tool in our arsenal of escort tug performance prediction. Having several examples of quite

Top far left: Topsides model of an ART 4000 Rotortug, ready for wind-tunnel testing; **top left:** Resistance tests for a large semi-displacement fireboat hull; **bottom far left:** Self-propulsion tests on a quadruple-screw river pushboat; **bottom left:** Indirect towing tests of the Carrousel RAVE tug.

detailed model test results, such as those for *Ajax*, enabled us to benchmark the CFD predictions against model test results for the best possible accuracy.

CFD requires a very large computer capacity for number-crunching, and we soon realized that in order to do this work cost-effectively, we had to invest a great deal more into our computer systems, with much faster processing capability. By 2020 we had invested nearly a million dollars in one of the most capable CFD systems in our field of business. We had also recruited a team of very talented engineers, who today are working almost exclusively on CFD for a multiplicity of projects.

These figures illustrate some of the diverse projects on which CFD has been used to either predict performance or to troubleshoot performance issues.

One of the most important aspects of the use of CFD in escort performance analysis has been the acceptance by classification societies of this methodology as proof of vessel performance rather than relying, as previously, on full-scale testing to determine the rated escort performance of a tug. Robert Allan Ltd was the first organization worldwide to have its escort-towing analysis methodology approved by classification societies for this purpose. Using CFD and the supporting stability analysis, it is possible to accurately represent what is truly achievable in terms of indirect steering and braking forces for any particular design. This contrasts with conducting a full-scale trial on a new boat with a master who is unfamiliar with the tug, an attended large ship which costs a great deal to charter and may not represent the actual type of ship usually attended, and all the vagaries of wind and weather and wave conditions at the time of trial. The CFD predictive method represents greater accuracy of results and a significant cost-savings to the tug owner.

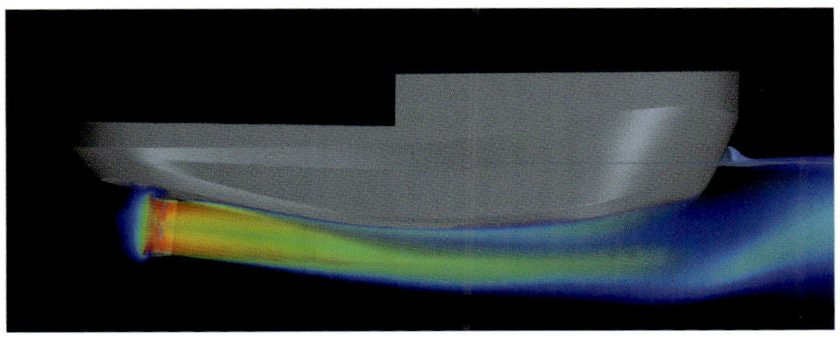

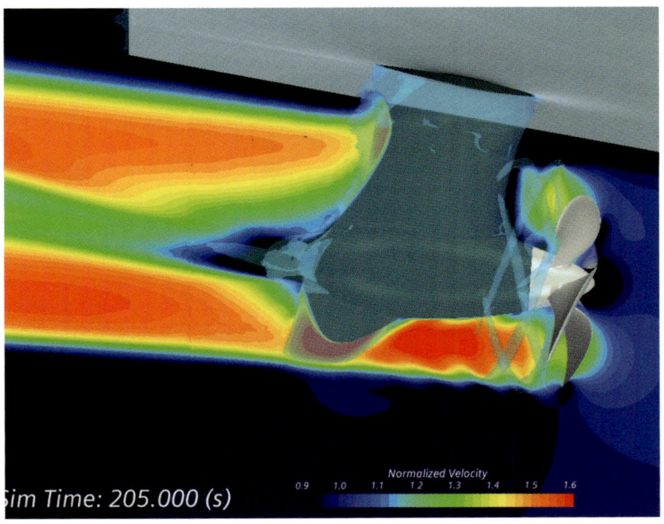

Top left: Modelling of a Rotortug performing indirect escort manoeuvres. The figure indicates both the flow pressures from the drives and the hull (red being highest pressure), as well as the wave profile developed around the hull in this model; **top right:** Modelling of the wake from a Canadian Coast Guard search-and-rescue lifeboat; **bottom left:** Modelling flow around a high-speed propeller; **bottom right:** Modelling thrust of a Z-drive in reverse, showing how little the thrust flow is affected by this combination of hull shape and drive orientation.

CFD Versus Tank Testing

Computational fluid dynamics (CFD) has largely replaced tank-testing as a means of verifying vessel performance. It is a really neat tool, great for high-speed vessels. We have six PhD types in the office, hydrodynamics experts who do nothing but CFD. It started out being a bit of a novelty—and a very expensive one at that—but now this team is really busy and booked months in advance. We use CFD for all our new designs. Model-testers wouldn't admit it, but it is actually better than model-testing. It is expensive but much cheaper than model-testing and much quicker. We can get the answer on performance before we get even halfway through the design. With model-testing, you usually get the answer a few months after the design, even when the boat is already under construction, and if things don't work properly, well...! The other thing is that the answers are very precise. If you change the rudder or change some little thing, CFD gives you a precise new number. The actual value may not be precise compared to true life, but compared to the old number, it is very precise. Change the chine to look like this and, yes, we know if it made it slightly better. We're sure of that. It is very useful when working on high-speed boats, when you're not sure if it will work and nobody will pay for the model test but you're committed to meeting a certain trial speed. CFD lets us try ideas that maybe we have been thinking about for years and would have liked to have tried but had no data to tell us if it was better or worse, you just thought it might be better. Now we can try stuff like that and run it. That said, every now and then we do a physical model test to make sure that the data agrees with the CFD.

—Hans F. Muhlert

Influencing the Rules

Over the length of a long career, it has become obvious that one of the more significant contributions one can make to the marine industry, in addition of course to developing ground-breaking designs, is to use one's influence to try to make the regulatory framework in which we all must work more logical and effective: "effective" in terms of its use by designers, builders and owners in making vessels safer and more cost-effective. Unfortunately, many regulations are created for historical reasons which today simply present unjustifiable barriers to sensible and efficient ship design. Some of these have been mentioned previously: the 24-metre registered length boundary for application of Loadline rules; the 15 GRT and 150 GRT rule limits for vessel inspections in Canada; and the 500 GRT limit for vessels, above which they must comply with SOLAS regulations. In a world in which we must strive to design the most efficient vessels possible, any regulation that constrains dimensions can be deemed to be counter-productive, particularly when one of those dimensions is length. Dimensions (usually depth) which are constrained by attempts to squeeze under a GRT limit ultimately seem to limit freeboard, which has a direct impact on range of stability and thus overall ship safety. While attempts to change these international regulations might seem quixotic, there is a need for change and, through published papers and articles, we continue to make the case for more sensible regulations.

Harmonizing Tug Regulations

In 2006 I presented a paper to the International Tug and Salvage Conference (ITS 2006) entitled "A Proposal for Harmonized Regulations for the Design and Construction of Tugboats." The premise of this quite extensive paper was simple:

> There is a diverse array of rules and regulations that governs the design and construction of tugs around the world today. These include the often-conflicting requirements of the various IACS members and a multitude of individual National regulations and standards. Under SOLAS and Loadline rules, many regulations that were conceived for full size ships present major challenges to efficient tug design and operations, with results that frequently do NOT necessarily improve the safety of the boat.
>
> Tugboats are mobile assets that work in waters around the world, generally performing the same sorts of tasks, regardless of geography. It stands to reason therefore that the same regulations and safety standards should apply to them, regardless of where they operate, or in which country they were built. When a tug is bought and sold on the international market, owners often have to spend large amounts of money to prove compliance with a different set of rules to those under which it was constructed. The current system defies all logic.

This paper examined the discrepancies between rules for basic tug design and construction of some of the major International Association of Classification Societies (IACS) members and proposed a more uniform rating standard based on voyage and the nature of operations, and not just on size.

It was very gratifying to see that within a year or so, the arguments raised in this paper were receiving real attention from at least some of the classification societies, with a lot of support from tug owners. The creation of the international SAFETUG Joint Industry Project in 2005 was enabling more concrete discussions about the regulatory environment, and we actively participated with IACS members in these discussions. Largely through the singular efforts of Gijsbert de Jong of Bureau Veritas, all the stability regulations for both conventional and escort tugboats were presented to ITS 2010 in draft form, and then further enhanced and summarized in a thorough paper by de Jong at ITS 2014. The process of formal adoption of these new rules by IACS is ongoing, but many classification societies have nonetheless adopted the same new stability standards. Other regulatory harmonization trends are also ongoing, and Robert Allan Ltd is still actively involved in encouraging and supporting these changes.

I consider this initiative one of my more significant career accomplishments: a meaningful regulatory change impacting vessel design for much enhanced safety under new operating conditions.

—R.G. Allan

Escort Tug Stability Indicators

In reviewing the challenging issues of stability of escort tugs, Robert Allan Ltd independently developed an escort safety placard for posting in the wheelhouse to convey limits for heeling angle and speed to the tug master as unambiguously as possible. Escort tugs operate in inherently dynamic situations, and the challenge is to convey heeling angle limits in a way that takes normal day-to-day dynamics into account while not unreasonably curtailing performance when static heeling limits are momentarily exceeded due to small oscillations. This objective triggered a closer look at the basis for current escort tug stability criteria as stipulated by classification societies, with an eye to (a) understanding how existing stability criteria accommodate the real hydrodynamic situation during the extreme dynamic scenarios potentially encountered by escort tugs (i.e. loss of thrust, failure of towline, etc.) and identifying the scenarios that pose the greatest risk, and (b) proposing an intermediate heeling angle range or "amber zone" to accommodate dynamic heeling oscillations without prejudicing tug safety.

The escort placard developed by Robert Allan Ltd achieves this by conveying recommended limits for escort operations to the tug master in terms of green, amber and red heeling angle zones, plus other information, such as rated escort speed. This placard is custom-prepared on the basis of the specific stability characteristics of each individual escort tug. This simple innovation has been widely praised by operators.

In order to translate this information into a real-time readout for the master, we collaborated with manufacturers of inclinometers to develop a suitable instrument, as illustrated.

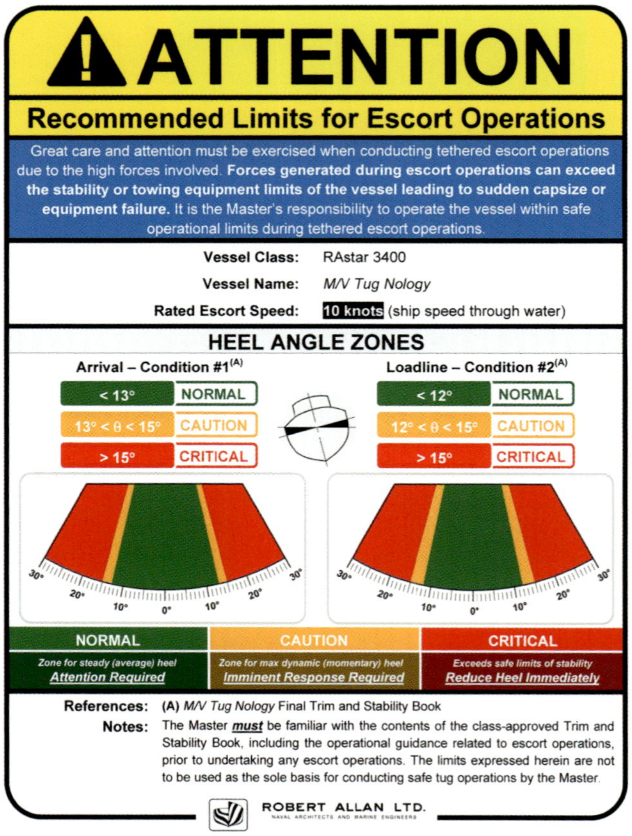

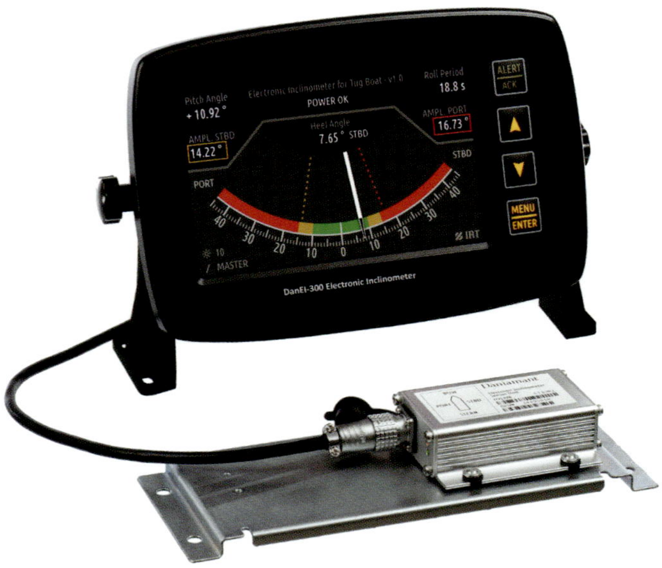

Left: Example of the escort tug safety placard concept developed by Robert Allan Ltd; **above:** Inclinometer customized for escort tugs, showing "amber zones" to alert operators to higher heel angles of potential risk.

Tug and Barge Relationships

In 1990 we obtained a contract from Transport Canada to conduct a detailed study into the present state of the marine towing industry in Canada. The resulting study, entitled "The Canadian Towing Industry: Tugs, Barges and Their Relationships," was published in December 1991 and represented the most comprehensive study of the industry ever performed. It was conducted at the time when ASD tugs were just starting to enter the Canadian industry and so today is rather dated, but many of the vessels in the 1990 database are still in service and in fact are still exposed to the same risks identified in this study. One of the key findings in the study was the identification of "safe" industry practices with respect to the size and power of tugs used to tow varying sizes of barges. This power/size relationship has been an issue in the industry for decades and can be shown to occasionally have deadly consequences. The sinking early in 2021 of the small tug *Ingenika* in northern BC, with the loss of two lives, is a case in point.

Transport Canada has never acted on the recommendations in the 1990 study.

Tug and barge transportation systems are a vital link in BC's coastal economy, and Robert Allan Ltd designs have been instrumental in introducing many innovations for this trade.

ST1591I
Z-TECH
PSA TAURUS
YS03
ST1591I

21

Game-Changers and Big Sellers

Documenting this history of both family and company has been challenging at times, but exceptionally rewarding and incredibly interesting, as it revealed numerous vessels about which I had no prior knowledge. However, one great frustration is the fact that it is impossible to show the reader every important boat built to our designs, now spanning almost a century of output. In an effort not to omit any really important projects in this record, the following is a series of short summaries illustrating what are believed to be the really ground-breaking or game-changing designs produced by Robert Allan Ltd, particularly in the past 25 years. (The most notable earlier designs have been well described in the relevant previous chapters.)

Opposite: Latest evolution of the Z-tech, PSA *Taurus*, see page 520.

Designs That Made a Difference

Charles H. Cates II

The 2,400 hp *Cates II* was the first Z-drive tug designed and built in Canada, and it became the template for an entire generation of simple, highly efficient and cost-effective tugs which are still being built today. This design really became the launch-pad for exporting our design services to the rest of the world.

Charles H. Cates II.

ASD Tugs for Atlantic Towing

The 30 tugs built to this 30.8-metre 4,000 hp design for Atlantic Towing (and others) marked a significant milestone in our progression into larger ASD tug designs and played a key role in our expansion into the development of comprehensive production designs for smaller shipyards. The most important aspect of this very successful design, however, was the connection it made for us in Scandinavia with the Østensjø Rederi organization, leading to some of the world's most truly innovative tug designs and opening the European market for us.

The *Atlantic Spruce*, one of the first of a series of 30 tugs built by East Isle Shipyard in Prince Edward Island to a comprehensive production design.

Ajax

Described previously, this powerful VSP tug for Østensjø Rederi represented a key step for Robert Allan Ltd into large high-performance escort tugs. This first use of the sponsoned RAstar hull form represented a pioneering, truly innovative design in the development of high-performance tanker escort tugs.

Svitzer Kilroom

This big, handsome and powerful tug was one of the earliest true offshore terminal/ escort tugs with Z-drive propulsion, setting performance standards for a whole generation of ASD escort tugs. It was also an early project for Robert Allan Ltd with Svitzer, the world's largest towing company, now a very good and regular client.

The RAmparts Series of Tugs

The use of RAmparts as a marketing brand under the Robert Allan banner began after a flurry of requests between 1998 and 2000 for similar tug designs in the 30-to-32-metre range with about 60 tonnes bollard pull. Our objective has never been to produce "standard" designs, but obviously when one has handfuls of comparable designs and continuing demand for more, it makes sense to at least have a baseline design as a common starting point, which can then be customized to suit individual client requirements. The RAmparts label started a trend in which we named almost every distinct class of tugs with a name beginning with (or at least incorporating) the initials "RA," a branding strategy which has been most successful.

Tirreno, a RAmparts 3200 built by Sanmar.

Al Fenci, a RAmparts 3200 built by Cheoy Lee.

Z-Tech

The Z-Tech story is one of true design innovation and has created a completely unique configuration for modern ship-handling tugs, in both award-winning style and in function. The widespread adoption of this distinctive tug configuration, with a low bow and higher stern, and with bi-directional operational capability, is a testament to its worldwide success.

ART Series of Rotortugs

Designed in 2012, RT *Evolution* was the first Rotortug built to a design by Robert Allan Ltd and is in itself a unique hybrid-powered tug. KOTUG's contract with Robert Allan Ltd to be their sole design agent globally was a significant event for our company and cemented a long-standing relationship among some of the more creative people in the tugboat world. The triple Z-drive Rotortug concept has proven to be exceptionally capable in both ship-handling and escort operations.

The 80-tonne bollard pull Rotortug RT *Evolution*.

The RAVE Tug Concept

The RAVE (Robert Allan Ltd Voith Escort) tug is a concept introduced in 2013 based on the brainwave of our VP of Naval Architecture, Dr. Oscar Lisagor, for a highly manoeuvrable, high-performance escort and ship-handling tug, jointly developed by Robert Allan Ltd and Voith Turbo Marine. The most unique characteristic of the RAVE concept is the longitudinal alignment of the two VSP drives, in contrast to the more conventional transverse tractor drive configuration.

The performance and configuration optimization of the RAVE tug has been thoroughly documented through an extensive series of model-tests and computational fluid dynamics analyses. The RAVE concept can be applied to any size of VSP-powered tug, depending upon the application and power requirements. Intuitively, one might think that this axial alignment of drives would result in a lot of interference on the downstream drive, but the research has demonstrated that with more than four diameters' separation there is very little interference between the drive units, and thus the overall performance is considerably better through 360 degrees than the conventional side-by-side tractor arrangement.

In 2014 the RAVE design group teamed with Multratug of the Netherlands to introduce a further development of this design, combining the patented rotating "Carrousel" towing system with the RAVE design. Orders for the first Carrousel RAVE tugs were announced in November 2015.

The RAVE concept applied to a 37-metre escort tug.

As noted above, the first vessel built to the RAVE concept also embodied the unique circular Carrousel towing system. The first of class, *Multratug 32*, was delivered in 2017, followed by *Multratug 33* in 2018. Both have been performing exceptionally well. The Carrousel towing system significantly reduces heeling forces, and the RAVE drive arrangement enables precise position and orientation control. This configuration is most useful in constricted waterways such as canals and locks, where it is able to generate high steering and braking forces at relatively low speeds.

Multratug 32, first of the Carrousel RAVE tugs built.

Developing the Carrousel Tug

In 2013 we teamed with the Dutch company Novatug B.V., a subsidiary of Multraship Towage and Salvage. They had developed a concept for a movable winch system which they called the "Carrousel winch." Instead of the standard arrangement of a winch being hard-mounted on the deck with a fairlead/staple, the idea was to design a winch that would be mounted on a circular track around the deckhouse. This would allow the freewheeling winch to rotate 360 degrees so there would always be direct pull on the winch drum regardless of direction. Novatug had the patent for this concept and had tried it on an older existing tug, but they needed someone to integrate it into a modern high-performance tug.

After many discussions, Rob Allan and Oscar Lisagor decided to take on the project. Oscar was the project director, and I was the team leader of naval architecture and the project manager. Bart Stockdill was the team leader for marine engineering. It was called the Carrousel RAVE Tug (CRT). We designed the 32-metre hull to have a raised circular ring that spanned almost the entire beam. The winch would rotate around the ring so that the towline would always align with the direction of pull. The structural support of the carrousel ring consisted of a matching supporting ring underneath, with radial webs tying into the main hull structure. Norbert Schumacher completed a lot of FEA work (finite element analysis) checking structural loads.

Because of the Carrousel winch, all other equipment on deck had to be either inside the ring or well below its height, otherwise it would obstruct the towline. This meant no crew could be allowed on deck any time the winch was ready to move. We made the deckhouse circular and placed it inside the ring to maximize the accommodation space.

Propulsion was another important design feature. In 2010 Oscar and Rob, in collaboration with Voith, had developed the RAVE concept. Unlike a traditional Voith water tractor, the RAVE configuration features two Voith Schneider propellers (VSPs) mounted fore and aft along the centreline of the tug; the standard Voith configuration has the two units mounted side by side near the bow. So this CRT was a marriage of these two different unique concepts—Novatug with their winch patent and us with the RAVE tug—but a lot of work needed to be done first to convince ourselves that the combination would work.

A significant amount of model-testing was undertaken starting in 2013, both at the Vienna Model Basin and at the Port Revel Ship Handling Training Centre in France. Port Revel is a ship's master and pilot training facility. They have scale-model manned cargo ships and bulk carriers and tugs, and complete the tests on a man-made lake. Everything is at a 1:25 scale. The Carrousel RAVE tug is now one of the models being used there to familiarize the pilots with its methods of operation.

Novatug built two of the 32-metre CRTs operating in the Netherlands and in Belgium.

—Andra Papuc

The All-Electric Tugboat

The concept of an all-electric tugboat has been around for several years, though initially hampered by the lack of suitable battery capability, not to mention high cost, but as of 2022 it is a reality with increasing acceptance. While still much more expensive than a diesel-driven equivalent, the social contract for cleaner vessels of all types is being realized, and it is obvious in an era of rapidly rising fuel prices that many more such vessels will follow.

The first built to a Robert Allan Ltd design will be the three ElectRA series tugs currently under construction in Turkey for HaiSea Marine. These boats, scheduled for delivery starting in late 2022, are perhaps the most environmentally friendly tugs ever to be built. With battery-electric propulsion delivering up to 6,102 kWh, these tugs will be the first in the world with ABS ENVIRO+ and Under Water Noise (UWN) notations.

Similar designs for all-electric tugs both smaller and larger are currently under development. The all-electric tug makes good sense where clean shore-power is readily available and where voyage transits and the uses of higher powers are not overly long.

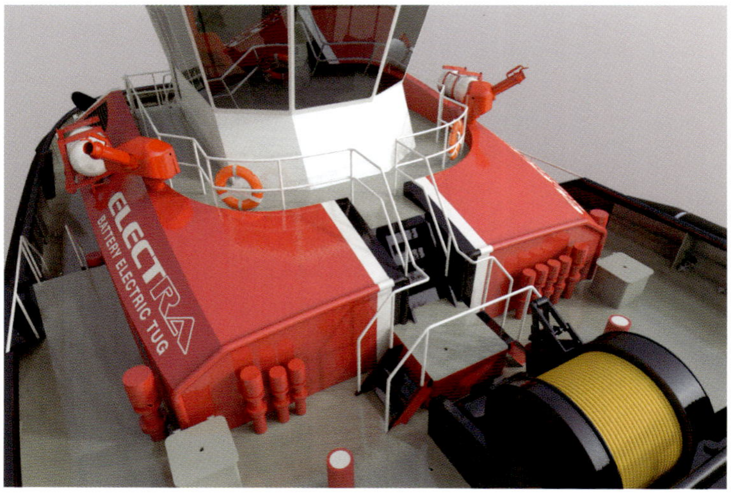

Left: ElectRA 2800 class all-electric tug being built for HaiSea Marine in 2021; **below:** ElectRA 2500 design proposal.

The Big Sellers

The market for new tugboats has been virtually insatiable for nearly two decades as ships have increased to unanticipated sizes and new markets, such as offshore LNG terminals, have emerged, all demanding tugs of greatly increased power over previous generations. In this milieu, a few of our designs have emerged as market leaders and have been built in quite significant numbers.

RAmparts 3200 Class

Our most popular design is the RAmparts 3200, an ASD tug typically with about 70 tonnes bollard pull. This size of tug measures 32 metres length overall by 12.4 metres beam, and generally is configured to measure just under 500 Gross Tons. Installed power in these tugs ranges from 3,600 to 4,800 kW, delivering a bollard pull from 60 to 80 tonnes, depending on the application. More than 180 variants of this design have been built to date.

An early RAmparts 3200 built by Sanmar.

Ocean Legacy, a RAmparts 3200-CL design, a variant developed exclusively for Cheoy Lee.

General Arrangement of an early RAmparts 3200 standard design version developed for Sanmar.

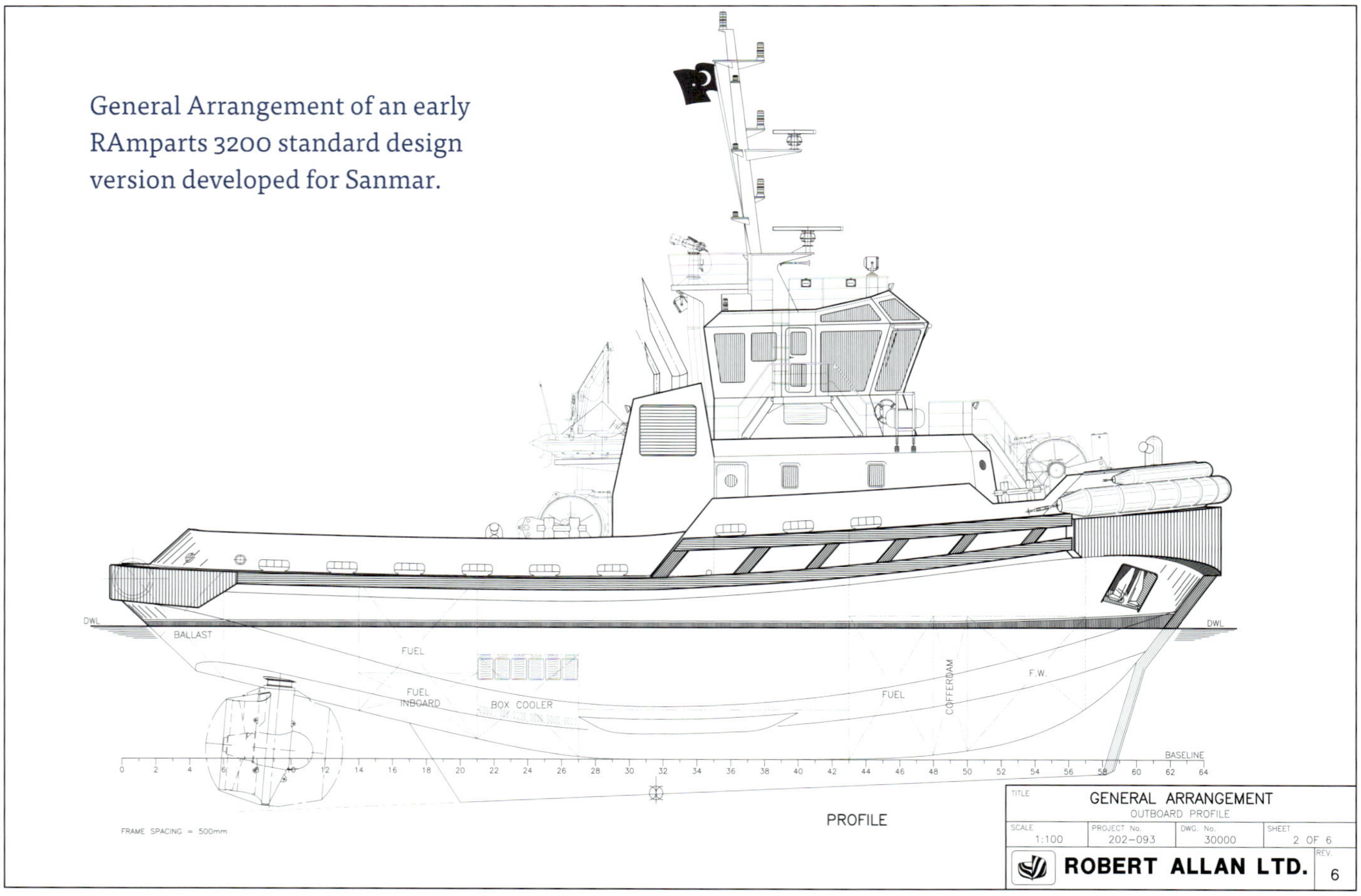

RAmparts 2400 Class

Close behind in popularity is the RAmparts 2400 class design (sometimes also referred to as a RAmparts 2500), and particularly the variant built by Sanmar and marketed as their Boğaçay class tug, of which their 50th edition was under construction in late 2021. These boats vary slightly in dimensions, depending on power, but typically are 24.4 metres long by 11.25 metres beam, designed specifically to measure under 24 metres in registered length to avoid Loadline and SOLAS regulations. Installed power typically ranges from 3,600 to 4,700 kW, delivering 60 to 80 tonnes bollard pull.

Tai Pari, a RAmparts 2400-W class tug in the Port of Tauranga, New Zealand.

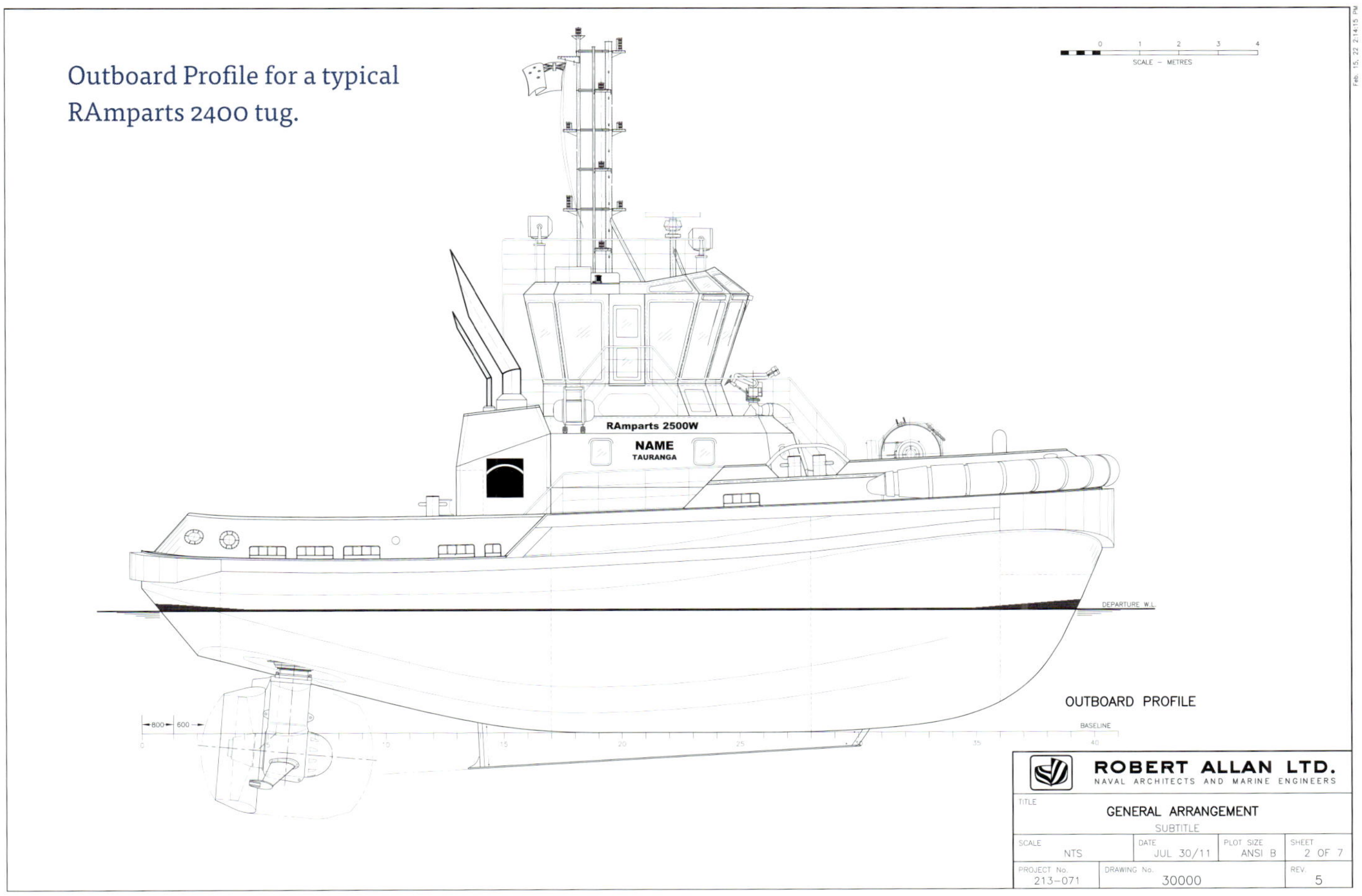

Outboard Profile for a typical RAmparts 2400 tug.

Z-Tech Class

And finally, with 91 versions currently afloat, the Z-Tech has been widely accepted around the globe as a highly versatile and efficient ship-handling tug, produced in several size and power variants. This is also a design concept being shamelessly copied.

PSA *Taurus*, the latest version of Z-Tech tugs built for the Port of Singapore by Cheoy Lee.

1,000 Tugs and Counting

From the very first Robert Allan tug design, *Weaver Lake*, and on through the past two incredibly busy decades, the history of Robert Allan Ltd is quite inextricably linked to the evolution of modern tugboat design. In 2017 it became obvious that our total tug output (in terms of vessels actually built, not just designs) must be getting close to 1,000. The development of ever-larger container ships, major LNG terminals, expanding bulk carrier ports such as Port Hedland, and, of course, the rapid evolution of tanker escort technology, demanded a whole new generation of specialized high-performance tugboats, and our office was responsible for a significant portion of the global total. A "head count" indicated that since 2000 more than 720 tugs had been built to Robert Allan Ltd designs worldwide, an average of more than 42 per year (and in some years an average of one per week!). When added to the historical database, the total number of tugs delivered by 2017 stood at 1,005! Since that time another 259 tugs have been built, so at time of writing, in late 2021, the total number of tugs delivered to a Robert Allan design stands at 1,264 and counting. The following figure illustrates this growth quite dramatically, as well as reflecting the past glory of the 1960s.

Tugs Built to Robert Allan Ltd Designs by Decade

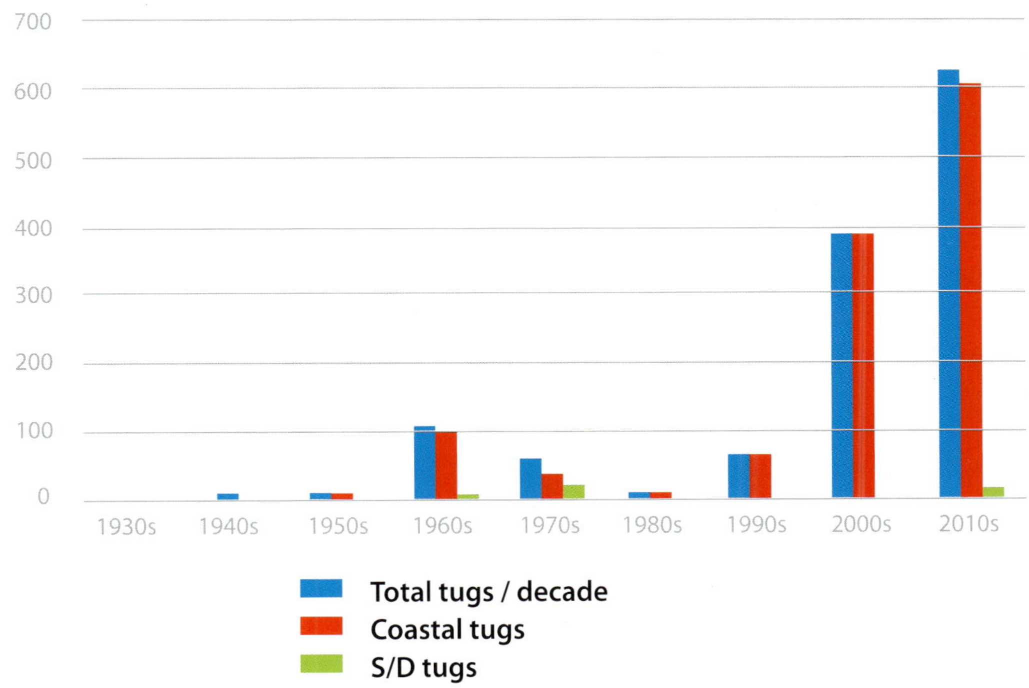

As previously noted, the honour of the 1,000th Robert Allan tug went to *Dux*, the first of three ultra-high-performance dual-fuel escort tugs built by Gondan Shipyard in Spain for Østensjø Rederi of Norway. *Dux* and her sisters were the first dual-fuel escort tugs in the world.

For the past few years, we have been using AIS data to track where boats to our designs are working around the world, and it is both amazing and gratifying to see how widespread these vessels are.

The yellow dots show the working locations of more than 1,000 vessels built to Robert Allan Ltd designs as of January 2022. Many are not visible on this map due to being in close proximity to each other, in which case the icon is a yellow dot with a small halo.

A Word of Thanks

We are indebted and more than grateful to those many clients who have shown their continued faith in our design work and who have supported Robert Allan Ltd for many years. A special tribute is owed to Terry Waghorn and to the late Claire Johnston of C.H. Cates & Sons, whose vision of the perfect tug (for Vancouver) and whose initial trust in our design capabilities helped to spawn the whole modern Z-drive tug movement. To commemorate the 100th anniversary of the Cates organization in 2000, Claire asked me to sketch what the "tug of the future" might look like. We chatted about some ideas, and I created a sketch based on that concept. At their centennial anniversary celebration, Claire gifted me with a large framed print of that sketch (which had been nicely enhanced by a talented airbrush artist). The inscription on this picture reads "Let Us Not Restrict Our Imagination." I hope we have honoured that objective!

—R.G. Allan

Cates z-2000, a vision of the tug of the future as conceived in the year 2000.

BRAtt

22

The Odd-Job Jar

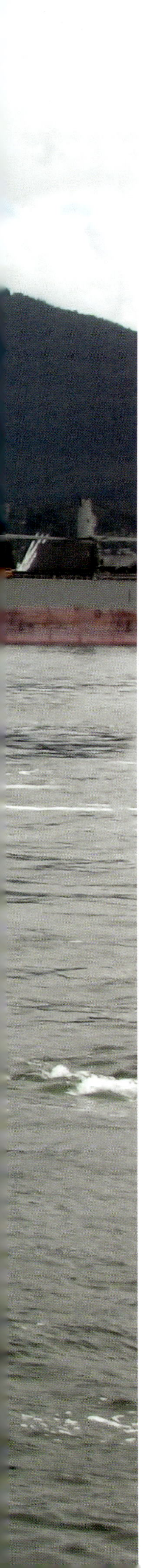

With the dominance of our design work in the world of tugboats since the mid-90s, it has been difficult not to be identified as simply a tug designer. However, the actual breadth of our portfolio rather belies that fact. It is also often the most unusual projects which challenge our collective imagination much more than typical routine work. The following are some of the more unusual projects which Robert Allan Ltd has authored over the years.

Training Tugs

As tugs grew in size and power to match increasing ship sizes, the number of qualified operators in the industry was on a downward trend. It was widely recognized that better training techniques for tug masters was required. Using full-size tugs with full crews is a very expensive way to train people and puts a multi-million-dollar asset in the hands of an inexperienced person, representing a lot of potential risk. The use of a small-sized training tug with realistic performance and handling characteristics was considered an ideal way to train people and at much lower cost and risk.

The BRAtt was built by Adrenalin Marine of Delta, BC, as a demonstration of what such a vessel would look like. Despite its small size, the main components are still quite expensive, and towing companies haven't rushed to embrace this concept. However, the idea did attract interest from companies looking for a "micro-tug" to do specialized work in some areas, and some customized little tugs have subsequently been built on this general model.

Opposite: The BRAtt training tug.

One company that did adopt the concept of a training tug was KOTUG, who commissioned the design of the RT *Borkum* as a combination training Rotortug and demonstration vessel. With its yacht-like finish, this lovely little craft has been used for general KOTUG promotion and for training Rotortug masters, as well as for demonstrations of how unmanned tugs might be used in port.

A Small Idea

We designed and built an 8-metre 500 hp aluminium tug with two Cummins diesels, called the BRAtt (the Burchett–Robert Allan Training Tug), specifically as a Z-drive training tug. It was built in time for the 2010 International Tug and Salvage Conference held in Vancouver. Ron Burchett, a local scale-model tug builder, was involved, as he was also a local Z-drive tug training skipper. Driving a Z-drive tug is completely different than a conventional tug. I was onboard a Z-drive tug one day when the trials skipper ended up wrapping the towline around the whole wheelhouse, taking out railings and such. He thought driving a Z-drive tug was easy, but it only takes a split second of inattention unless you have many operating hours under your belt. It's got to be intuitive, like backing up a trailer. You just have to do it enough so that it comes naturally. The concept was that this little tug that only cost $1 million would be better to use for training than a company's $10-million tug—and doing real damage to things. I think the industry figured since they already had the big tug, and they hadn't had too many bad accidents, they'd just carry on the way they had been going. We ended up selling the BRAtt to Harley Marine in Seattle, both of us thinking that because it was under tonnage, it was small enough to get around the Jones Act. However, some of their competitors saw this little tug arrive on site and raised all hell. It ended up that they couldn't even touch a barge with the BRAtt, not even for training purposes. They still used it for training and displays at launching ceremonies and stuff like that. But Harley Marine recently changed hands. They're now trying to sell it to the US military training academy, which is really where it belongs. If they can do that deal, I'll be happy.

—Ken Harford

Above: *Suncor V*, a "micro-tug" workboat.

Opposite: The BRAtt taking on the big guys!

RT Borkum, a sweet little 10-metre-long Rotortug for training and demonstration purposes.

Fuel Stations

Vancouver Harbour at one time had four or five marine fuel stations at the entrance to Coal Harbour. These served the pleasure craft community as well as commercial workboats. Several of these were designed by Bob Allan in the early 1950s, including barges for Standard Oil and Home Oil.

Over time, all but one of these disappeared from the harbour, and in 2007 the latest edition of a floating gas station, the *Chevron Legacy*, was designed, 50 years after her corporate predecessor. Carrying approximately 1,200 tonnes of diesel fuel, 130 tonnes of gasoline and 70 tonnes of lube oil, this unique craft was built by Alaska Ship and Drydock in Juneau, Alaska (sadly reflecting the fact that few builders in BC were able or willing at the time to bid on even a relatively simple project such as this).

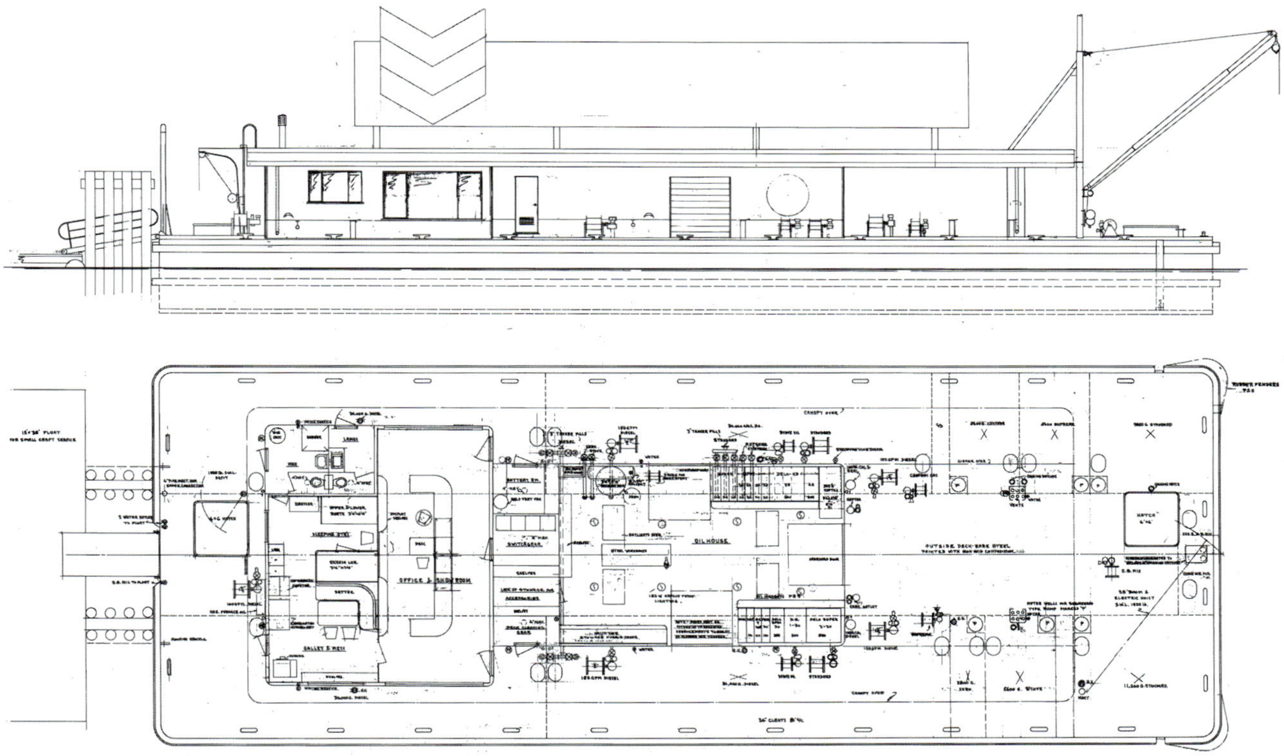

Above: Standard Oil *Barge #22*, designed and built in 1958.

Left: *Chevron Legacy*, the latest iteration and the only surviving Coal Harbour fuel station, a vital feature of the Vancouver waterfront.

McBarge

Few floating objects in the vicinity of Vancouver are more familiar than *McDonald's Friendship Restaurant* (popularly known as *McBarge*!), built for Expo '86. We don't claim authorship of the design, which was developed by architects Waisman, Dewar Grout & Carter, but Robert Allan Ltd was heavily involved in the project, with responsibility to make sure that it floated and would be level, attributes with which none of the other architects and engineers involved were familiar, other than the builders of the reinforced-concrete pontoon, Cefer Systems of Richmond. Cefer was a firm with whom we had worked on numerous other projects, such as chip-unloading pontoons at pulp mills. We also designed all the access ramps for *McBarge*, as we were the only part of the team familiar with designing for tidal variation.

This was a unique project and received the highest attention possible from McDonald's head office in Canada. It was most interesting to see how a large multinational organization like this works from the inside, and how loyal they were to even small contractors who did good work for them, thus allowing their suppliers to benefit from very long-term business.

McDonald's had entered into contracts with the City of Vancouver leading up to Expo on the understanding that they would have access to a "permanent" waterfront mooring site once the world's fair was over. The city apparently reneged on that promise, and the unique facility that is *McBarge* has been left to languish into disrepair ever since. Various potential buyers have looked at it for alternative uses, and we have assessed the risks of towing it to other locations, all to no avail. What could have been a unique attribute to the Vancouver waterfront has become a piece of waste.

Robert Allan Ltd was an integral part of the design team for *McBarge*.

Submarines for West Edmonton Mall

In 1984 Robert Allan Ltd received one of its more whimsical commissions. International Submarine Engineering (ISE) was a highly regarded local company with whom we had worked on specialized projects on various occasions over the years. ISE had won the contract to supply "tourist submarines" to operate in a "lake" inside the then very popular West Edmonton Mall. The subs were to be about 40 feet long and carry 24 people. They would run on a rail at the bottom of the lake, so no real control system was required, but they would be fully submerged (to a depth of about 8 feet) and, accordingly, ISE and the mall owners wanted these subs to be fully certified as submarines by the American Bureau of Shipping (ABS).

Robert Allan Ltd was engaged to provide the complete design for ISE, and we collaborated closely with them throughout the process. Ultimately, four subs were built and went into service entertaining tourists in northern Alberta. At the time, these were the only legitimate submarines in Canada, and numerous cartoons appeared drawing the attention of the Canadian Navy to this fact. By 2005 the novelty had worn off, and the mall owners mothballed the submarines in favour of some new attraction. The subs were deconstructed in 2012.

These fully certified tourist submarines for the West Edmonton Mall ran on tracks beneath an artificial lake.

Floating Drydocks

Over the years, Robert Allan Ltd has designed a number of drydocks for various clients. The earliest of these was a small dock for John Manly Ltd, and the latest is a much larger dock for Hidrovias do Brasil, to support their new fleet of river tugs (Chapter 14). This dock, shown opposite, has a lifting capacity of 1,600 tonnes on a platform measuring 54 by 28.8 metres.

Trans-Shipment Barge

Quite frequently the company has been engaged to conduct detailed transportation systems analyses, for example comparing tug-barge transportation systems to conventional shipping, or simply identifying the optimum tug-barge system to move specific cargoes on new routes. Less occasionally, these studies lead to full design projects for components of the nominated transportation system. A most interesting example of such a project was in 2016, for an Indonesian company. This involved identifying an optimal system of tugs and barges necessary to move wood chips from a shallow river estuary into an outer deep-water harbour, where the chips would be transferred to a waiting bulk carrier. The vessel needed to trans-ship the chips from the barges to the ship, and this obviously became a key component of the system, needing the capacity and the crane reach to move all the cargo within a limited time-frame. Due to the shallow waters at the chip storage site, the individual barge capacity was limited, so multiple trips were necessary for each barge to comprise a shipload. Having devised and optimized the tug-barge system, we were then engaged to define and design the trans-shipment barge. That involved evaluating a number of cargo transfer options, from conveyors to cranes. The Belgian-manufactured e-Crane was identified as the most cost-effective contender for this operation, having the operational speed and bucket capacity necessary to meet the required throughput. The vessel, shown in profile in the figure below, was built at ASL Shipyard in Singapore/Batam and put into service late in 2018.

Chip trans-shipment barge for Indonesia.

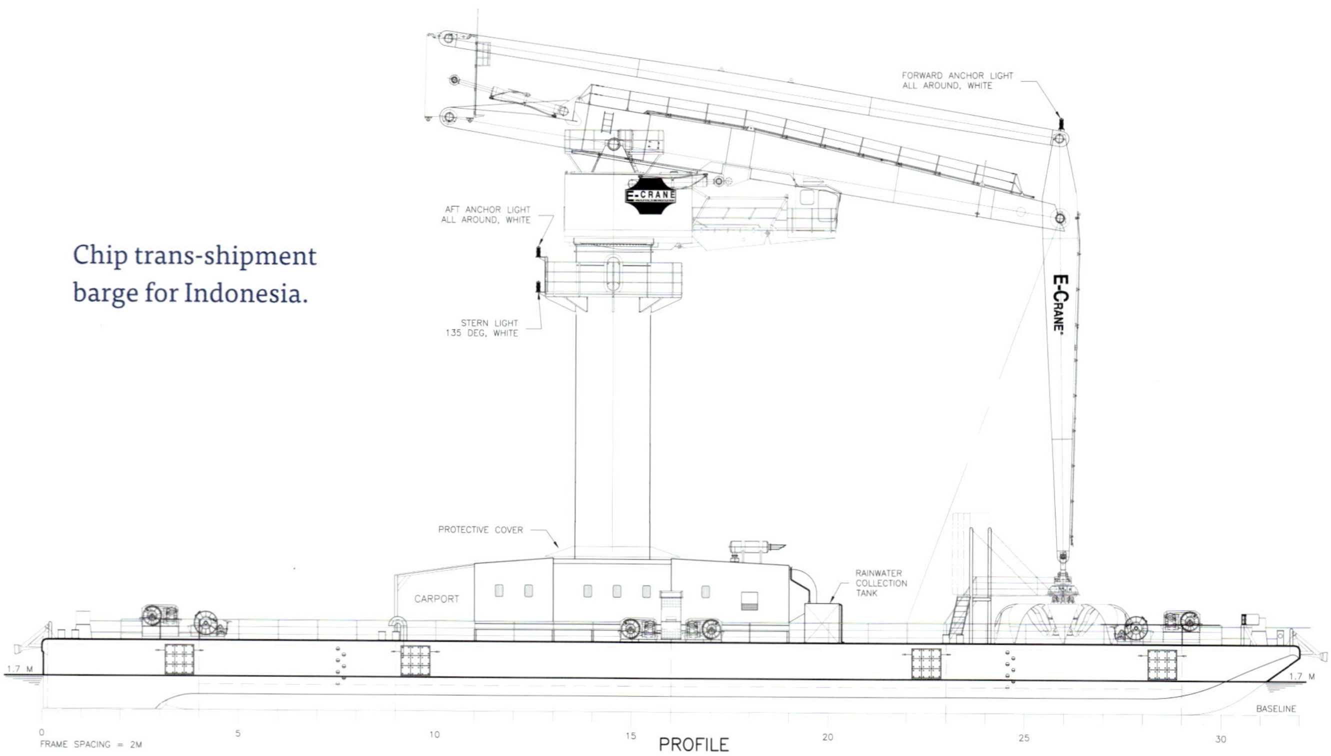

Floating dock for Hidrovias do Brasil in the Río de la Plata.

Canadian Navy: Torpedo and Ship-Ranging Vessels

Naval vessels are generally not in the lexicon of Robert Allan Ltd, though we have been involved in the design of several of their auxiliary fleet, such as harbour tugs (Chapter 17) and fireboats (Chapter 10). Another example was the Sechelt class Torpedo and Ship-Ranging Vessels (TSRV) built by West Coast Manly Shipyards in 1991. Four vessels of this type were built: the *Sechelt, Sooke, Sikanni* and *Stikine*. Their function was to recover expended torpedoes at the Nanoose test range on the east coast of Vancouver Island. (Two were later converted to dive tenders.) The concept design for these boats was developed by others for the Navy. Robert Allan Ltd was retained by the shipyard to develop the full design and prepare a full suite of production working drawings. They are 108 feet long by 38 feet 6 inches beam.

The four Torpedo and Ship-Ranging Vessels (TSRV) designed for the Canadian Navy.

Major Vessel Conversions

Not every project begins with a blank slate. Sometimes we are challenged to "make a silk purse from a sow's ear." On numerous occasions, we have been commissioned to design modifications for adapting a vessel to a completely new operational profile. One of the most significant of these was the 2008 conversion of a relatively standard offshore supply vessel into a major seismic research vessel for a Singapore-based client. The changes involved designing almost a complete new topsides superstructure for the ship to accommodate the extensive seismic winch array and associated equipment required, plus a heli-deck and numerous other alterations. The addition of this upperworks necessitated adding side sponsons to the hull to satisfy operational stability requirements. The resulting 64.5-metre-long ship, named *Pacific Titan*, was converted at Todd Pacific Shipyard in Seattle and entered service in 2000, working principally in Southeast Asia and Australasian waters. The new arrangement for this very extensive conversion is shown below.

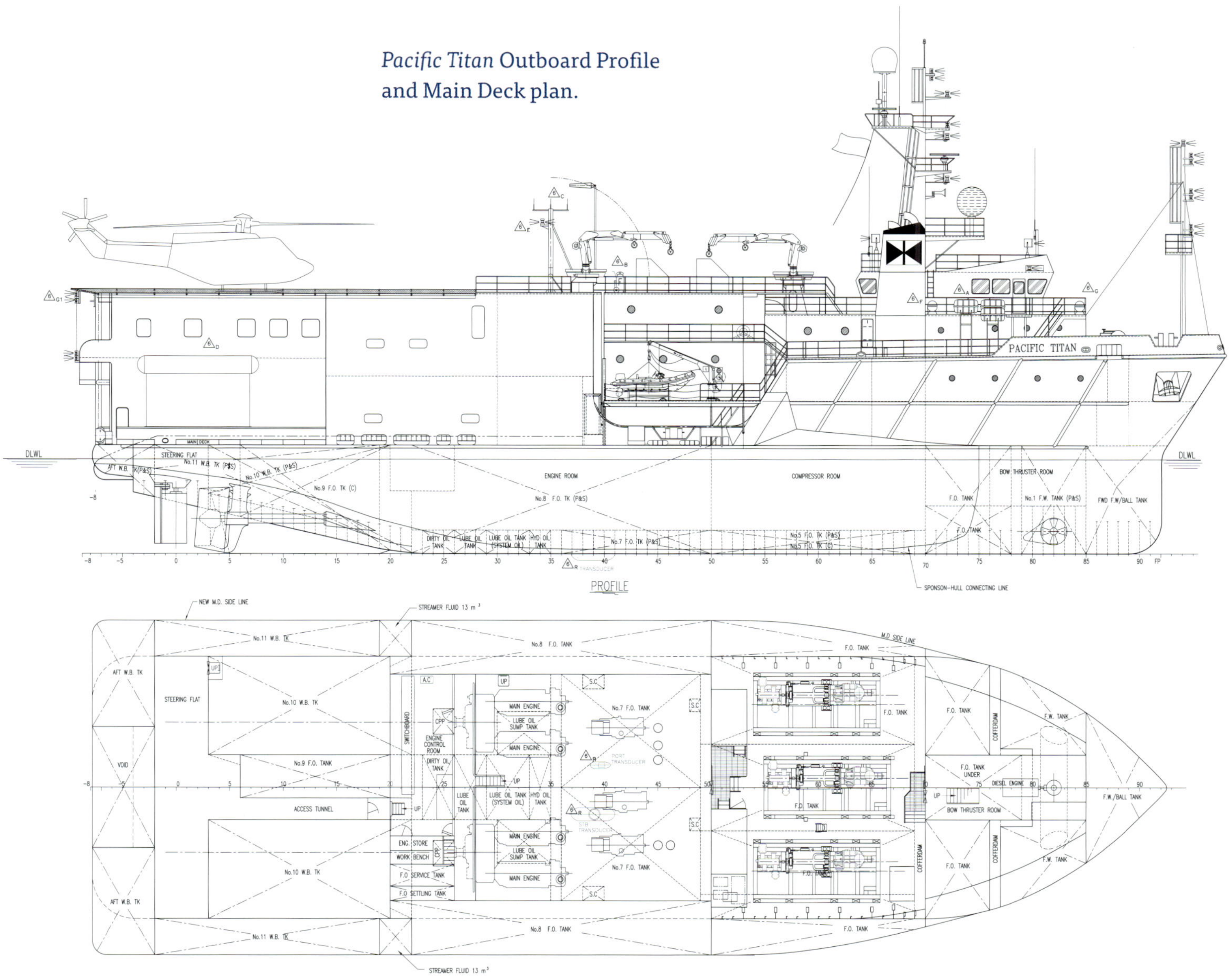

Pacific Titan Outboard Profile and Main Deck plan.

Pacific Titan reborn as a seismic research ship.

Other repurposing examples include converting Ultra Large Crude Carriers (ULCCs) to Floating Storage and Offloading (or Off-take) (FSO) vessels for Teekay Shipping, one of the world's largest tanker operators. One of the most recent of such conversions was the *Suskan Salamander*, a 229-metre FSO.

In 2013 Teekay Corporation and Salamander Energy of Thailand reached agreement on the supply of an FSO unit at the Bualuang field in the Gulf of Thailand. The 1993-built shuttle tanker *Navion Clipper*, which had been operated by Teekay for 20 years (typically the customer-imposed operating life of such a vessel), was nominated for the conversion. To ensure that the FSO remains safely on location in all expected conditions, it is moored by an external heavy-duty turret, allowing the vessel to weather-vane through 360 degrees. The scope of work from Robert Allan Ltd included the design of the crude-oil import and export systems, a new heli-deck capable of landing a Sikorsky S61N helicopter, accommodation and systems modifications to suit an increased crew complement, water-handling and delivery systems, materials handling systems, offshore crane support structures, upgrades to the fire and foam system and safety systems, and ongoing construction support services.

Suskan Salamander on her heavy-duty turret mooring.

Bidding Contracts in the International Market

Another interesting project involved helping Teekay Shipping with tenders to Petrobras of Brazil for converting older single-hulled ULCCs into Floating Storage and Offloading ships (FSOs) for the offshore oilfields in Brazil. Single-hull tankers were being phased out of worldwide oil transportation in favour of double-hull tankers, to reduce the risk of oil spills in the event of a collision, so conversion to an FSO was an excellent business opportunity. These ships would be moored on the oilfields and use submerged pumps and piping systems to pump the oil from the ground structure underneath the ocean into these big floating tanks. It was an interesting tender process. Petrobras looked at all the bids and said, "We think everyone complies." They then asked if any of the bidders wanted to object. All the bidders had one or two hours to look at each of the other bids. Some of the bidders claimed that some of their competitors didn't comply and protested. They kicked out a few bids based on that first little review and the protests that resulted, but there were no protests of our bid, so we survived that first round. The surviving bidders were then given a full day for a further review of the competitors' bids, and time to submit a critique of the other bids in writing, stating why Bid A or Bid B didn't comply. They were getting very qualified people to do all of this. Basically, they were getting the bidders to do all the tender review work for them. We survived the whole process. Our joint bid involved Teekay giving Petrobras a price for the costs associated with keeping one of those ULCC ships on station for 20 years. That included all the capital costs and operating costs, and they had to crew the ship for 20 years. All of that had to be boiled down to a per diem bid. We had to get the ship conversion costs and so on together, and the other guys at Teekay were looking at the manning costs. When the bids were opened, Teekay's was the second-lowest bid and lost the contract. We were a bit concerned that we had overestimated the conversion costs, but in the end it was revealed that the difference was that Teekay was looking for a higher return on investment than the lowest bidder.

—Ken Harford

Going Autonomous

Autonomous control drone/Seismic Drifter for seismic research.

We started to explore potential opportunities for autonomous vessels in the early years of the 21st century, principally in harbour towage and firefighting applications, and a real unmanned vessel opportunity arose in 2009. The requirement was for an "Autonomous Recording Vessel" (a.k.a. Seismic Drifter), which would be used as a remotely operated control drone for positioning both ends of an Ocean Floating Cable (OFC) for deployment of seismic acoustic receptors. The concept was designed for International Submarine Engineering (ISE), who were in turn contracted for this development by the French firm Kietta (now FreeCable). Although our involvement stopped with the completion of a highly detailed concept design, including evaluation of several different hull forms and model-testing of those, four of these unique craft were built by Dutch shipyard De Kaap BV (now Thecla Bodewes Shipyards). These are an important example of the trend toward autonomous vessels, and certainly similar applications will follow.

23

People Who Made a Difference

Robert Allan Sr. largely worked alone almost his entire career. That suited his temperament well. My father, Bob Allan, recognized that to grow the business he needed support, and he slowly built the staff up to about a dozen people. After I assumed control of the business in 1981, personnel levels grew to keep pace with our expanding business, and today there are about 90 permanent staff.

Opposite: R.F. (Bob) Allan, the person who made the biggest difference, building a company based on integrity of design and business ethics alike.

As tempting as it is to try, it would be too challenging to identify every person who has worked for Robert Allan Ltd over the many years of its operation; many do deserve recognition, but that might become another book! However, there are a few key people, especially from the most formative years of the company, who must be acknowledged for their truly significant contributions to our long success. Following is a brief tribute to each of them.

Jack Macfie

Jack Macfie was born in Prince Rupert and apprenticed there as a shipwright in his teens. He worked in local BC shipyards, joined the Canadian Navy and eventually served as the senior construction overseer on HMCS *Bonaventure*, the last aircraft carrier in service with Canada's armed forces, built by Harland & Wolff in Belfast during the 1950s. He later worked at Burrard Dry Dock in a superintendent capacity.

Jack joined Bob Allan in the mid-1960s in the capacity of a shipyard overseer. There was a lot of new tug construction underway in local yards, and Robert Allan Ltd was given the responsibility of overseeing that work. Jack was a big man and had a rather gruff demeanour, which hid the kindest disposition one can imagine. He was the perfect person to stand up to shipyards that were trying to cut corners, as he knew shipbuilding inside and out.

When Robert Allan Ltd was managing the design and construction of the four icebreaker/supply vessels for Dome Petroleum in 1974/75 (Chapter 12), Jack led a team of four other shipyard overseers in ensuring that these complex vessels were well built and, just as importantly, that they were built on schedule. There was a tight weather/ice window during which those ships could get into the Beaufort Sea to support the planned drilling schedule, and it was punishing work both in the office and in the three shipyards involved. During one working lunch at a local restaurant, Jack (who was then probably in his late 60s) suddenly complained of chest pains. My father quickly drove him to the nearest medical office and was told to get him to hospital ASAP—he was in the midst of a fairly severe heart attack. Fortunately, Jack survived that incident, but it was a very close call and signalled the end of his long and busy working life.

Jack was a superb mentor to those of us who were working on the drawing boards in those days. He would guide us gently away from designing things that couldn't be built and show us how it should be.

Carol Nilson

Carol joined Robert Allan Ltd in mid-1972, about six or eight months before I joined my father in the business. She was an extremely proficient secretary at a time when every letter, report and specification that we produced went through her fingertips. Her speed and accuracy were legendary, and she wasn't afraid to point out to us the odd grammatical error either.

In the late '70s, my mother was stricken with cancer, and that removed her from her role as company bookkeeper. Carol took it upon herself to get some training in that skill and took over those duties as well. The early 1980s were very difficult years. I was dealing with the death of two parents and trying to learn how to manage the company at a time when work was scarce. Carol was an absolute rock during those years, and an incredible support to me personally as she calmly (at least outwardly) steered the company away from the risks of financial ruin.

As we slowly emerged from those difficult years, Carol took on increasing administrative tasks, as well as bringing our clerical and management systems into the emerging computer age. She was later made our office manager, the position from which she retired after 38 years of incredibly dedicated service. I am eternally thankful for Carol's unwavering support and commitment to this company.

Peter Woodward

Peter was one of a generation of young British naval architects who came to Canada in the early '60s and worked their way through various shipyards, including Burrard Dry Dock in North Vancouver. Many of those people, including Peter, worked periodically for Robert Allan Ltd on a "moonlighting" basis, putting in some extra hours in the evenings to help get the workload out and supplement their incomes.

Peter eventually went to work for Transport Canada as a steamship inspector. In 1974 he was hired by my father to work as one of the shipyard overseers on the

Peter Woodward (right) with a young Hans Muhlert at the launch of *Eleu* for Hawaiian Tug & Barge at J.M. Martinac Shipbuilding Corp. in Tacoma, Washington, in 1989.

Dome Petroleum project, under Jack Macfie. Peter is a really personable individual, and we became good friends. After a few short years at Robert Allan Ltd, he was lured away to become the president and executive director of the nascent Council of Marine Carriers (CMC), an organization representing the interests of the majority of tug and barge operators on the BC coast to various levels of government. Given his technical and government background, this was an ideal position for Peter's talents.

In about 1983 I was looking for someone to look after our shipyard overseeing work, and fortunately Peter was also looking for a change. We agreed to form a joint company for this purpose, called Allan-Woodward Ltd, of which Peter was president. We worked these two businesses side-by-side until late in 1987, when the local work for these shipyard services was becoming scarce, and Peter was invited back to head up CMC. However, his presence in our office during those very difficult years was a huge support to me personally. His connections within the towing industry gave us some much-needed credibility at a time when I personally was not well known in the industry.

Peter served CMC extremely well until his retirement in 2005. Unfortunately, he suffered a severe stroke shortly after, which has left him somewhat debilitated. However, we remain in touch, and he is always glad to hear how well the company is doing. That success could not have been possible without his friendship and support at a very critical time.

Claire Johnston

Claire was another who was hired to support the construction of the Dome Petroleum offshore supply vessels in 1974/75. He was one of the many prairie boys who joined the Navy to see the world and ended up as a well-qualified marine engineer. He ended his naval career as a senior officer aboard HMCS *Provider*. I believe he joined Robert Allan Ltd directly out of the Navy. As well as being a highly competent engineer, Claire was a jovial and good-natured individual who was full of stories, many that couldn't be repeated in polite company.

After the Dome project, he was given the responsibility of overseeing the construction of a fleet of small ferries for the Government of Manitoba (Chapter 7), a pusher tug for Guatemala (Chapter 14) and several 60-foot twin-screw harbour tugs for Cates Towing (Chapter 17). The latter projects, done close to home in North Vancouver rather than in some remote location, really captivated Claire's imagination, and he was offered the position of vice-president at C.H. Cates & Sons by Terry Waghorn. It was sad to see him leave Robert Allan Ltd, but we also didn't have a great deal of local work at the time. He remained a good friend. On Terry's retirement, Claire became president of the Cates organization.

Not long after he joined Cates, Claire ran the program of adopting Z-drive tug technology into the Cates fleet, and we became the consultants/designers for that work. It was a distinct pleasure to work with Claire in developing these concepts. The contracts with Cates were rather informal: Claire would write a purchase order to us stating "For the design of one Z-drive Tug," to the agreed price, and that was it! What a pleasure it was to have the degree of trust that we all shared in each other's capabilities and business ethics, and to be able to do that. We worked with Claire on the full series of the three 2,400 hp Z-drive tugs and on the *Cates X*, a slightly smaller 1,800 hp vessel, as well as a few larger, but eventually unrealized, tug concepts.

Sadly, Claire suffered a fatal heart attack in 2001 and his loss was widely felt throughout the BC maritime community.

Hans Muhlert

Hans joined Robert Allan Ltd directly out of his graduate studies at the University of Michigan in 1970. He has remained in the employ of this company ever since, and still today, well past the age at which many have retired, he puts in full weeks of work.

Hans was running our small office in Singapore in 1981 when my father died. We agreed that it would be difficult to manage both that office and the main Vancouver office at that time, noting that there was not a great deal of work arising during the global recession. Hans returned to Vancouver and assumed the responsibilities of managing the design office, while I concentrated on finding work and managing the business. As the business grew, Hans' responsibilities did as well, and when Ken Harford joined the business in 1988, we made a pretty good, though very overworked, management trio. My job was getting the work in the door, and Hans and Ken were responsible for getting it out the door.

Hans Muhlert, Ken Harford and Rob Allan in 2005.

Hans has been a quiet strength at Robert Allan Ltd for virtually all of his 50-plus years with this company. He is an exceptionally talented and meticulous engineer and designer, and for many of his latest years has been using those skills to mentor a new generation of naval architects in our business.

Ken Harford

Ken joined the company in 1988 after a successful career as a consulting acoustical engineer and then managing a business designing and building marine control systems. His initial role at Robert Allan Ltd was to oversee new vessel construction, which he did ably, but his engineering talent was also invaluable in the design office. Ken's impact on our understanding of noise and vibration control has been instrumental in our

becoming a leader in the design of quieter vessels of all types but especially tugboats. As engineering manager and later as vice-president of engineering, he has also ensured that all of our machinery systems are properly designed.

For all his many contributions in engineering, however, Ken's single most significant contribution to our corporate success has been his management of the transition of the business from private ownership to an employee-owned business. Ken steered the new owners through the acquisition process and, on realization of that, became president of Robert Allan Ltd. He then carefully managed the company through its first few years of operation under employee ownership and created a template for sound operations and future growth. That was not an easy task, as the first year under the new ownership structure coincided with the global recession of 2008/09, and although we had a reasonable backlog of work, there were many challenges to overcome. Ken's business acumen was fortunately a very close second to his numerous engineering talents, and the company emerged from that recession relatively unscathed.

After ensuring himself that the ship was ready to sail, Ken retired from the company in 2016, after 28 years of invaluable service and commitment to Robert Allan Ltd. He remains a good friend and a highly valued member of the board of directors. I am forever grateful to Ken for the role he played in making this transition happen in the entirely successful manner in which it did.

Mike Fitzpatrick

Mike joined Robert Allan Ltd in 2003 after studying and working in Australia. He is a very talented naval architect and contributed significantly to many projects in his early years at the company. However, it was also clear that Mike had many talents in managing people and establishing good relationships with clients.

As we worked through the process of the company transition to employee ownership, Mike clearly rose to become an obvious choice to succeed Ken Harford as the president, which he did in 2016. In the years since, the company has grown from strength to strength under Mike's direction, and he has put his own distinct stamp on the company's future direction and reputation. Seeing the company continue to grow and prosper in Mike's capable hands has made it immeasurably easier for me to gradually step away from daily involvement with the business and simply witness, with considerable pride, the esteem in which this once "family basement business" is now held around the world.

S.S. MASTER
EOS DIGITAL
STONE GRILL

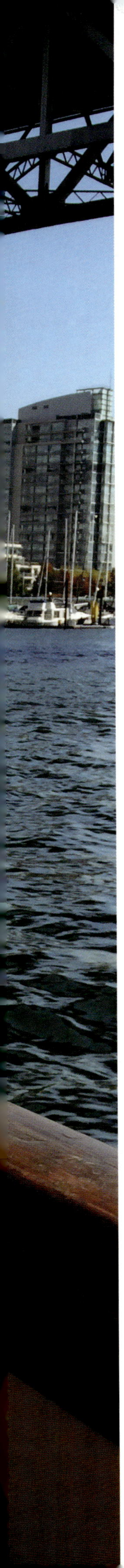

24

The Ownership Transition

In 2006 I was almost 60 and had been working long and hard throughout my entire career. Frankly, I was feeling the effects of 40 years of relentless 55–60-hour work weeks. Enneke and I had been blessed with three very talented and capable sons, whom we carefully (a) never named Robert and (b) never pressured to consider naval architecture as a career (but if they had expressed interest they certainly would have been encouraged). That was not to be the case; they each elected to follow their own diverse interests and embarked on much different and very successful career paths in film, art history and geology. As retirement age (if not the actual event) was looming, I had to consider both my personal future and the future of the family business.

Opposite: In retirement Rob Allan is able to devote time to pursuits such as the preservation of notable heritage vessels such as the steam tug *Master*.

I was being approached at the time by a couple of large offshore consulting companies who were in the process of buying up several small naval architecture offices in BC and elsewhere. As many of the principals of those companies were at or near retirement age, this was a nice opportunity for several of them to realize some gains and ease out of their careers. Nigel Gee was a former colleague of mine in the UK and a prominent multi-hull designer who had built a very successful practice in Southampton, England. He had recently sold his practice to one of the companies approaching me, so I took advantage of a conference in the area to visit and discuss how that had gone for him. His recommendations were generally positive, and he gave me some good pointers on how to make a mutually beneficial deal. I subsequently went quite far down that road in meeting with the prospective purchasers and understanding what foreign ownership would mean for our business.

Before I made any commitment, I also met with some industry colleagues in Seattle who were a generation ahead of me in the ownership transition process. The Glosten Associates is a highly regarded naval architecture company founded by Larry Glosten, who was a contemporary of my father. He had sold his successful consulting business some years earlier to a small team of his employees who were my age or slightly older. The principals of The Glosten Associates were very open and helpful in showing me how they had set up their transition plan, enabling an ongoing process of employee ownership. This seemed to be another good option.

I held a meeting with a group of more than a dozen senior staff and outlined the basics of the two options I was considering. As this was a plan that would significantly impact all of them, both professionally and personally, these people had to have a say in their own futures! After outlining the options, I then met personally with each of them, went over the options and asked which they felt would be of most interest. A very few were somewhat ambivalent, but almost all felt that they would prefer to try the employee ownership option. The die was cast.

In discussing this future option, I had assumed (without consulting them) that my two most senior colleagues, Hans Muhlert and Ken Harford, who were both somewhat

Rob Allan receiving RINA Small Craft Group Medal in London in 2005 from former colleague Nigel Gee, "for significant lifetime achievements in Naval Architecture."

Rob Allan in 2008, contemplating life outside the office after the sale of Robert Allan Ltd to key employees.

older than me (Ken by at least three months!) would not want to be involved in what was destined to be probably a decade-long buy-out program. I had underestimated Ken's potential interest, but fortunately he was approached by the others to join the buy-out group. Ken became the manager of the entire transition process and did a superb job of guiding all of us through what was a rather complex process with many meetings with lawyers and accountants. This part of the process is best told in his own words:

> In my side business in doing land development and stuff like that while working for Rob, I had fair experience with company organizations and putting deals together, and I knew some accountants and lawyers who were fairly creative. This was to be a transition from a family-owned company—a company run by one individual—to an organization that would be a going concern in itself and not dependent on one individual. Rob wanted to stick around and still be involved on the board, which we definitely wanted as well, there was no question about that.
>
> In 2007, when we started, I agreed to lead that process, so that really became my job. The actual transition was in 2008. I was intimately involved in all the contracts and draft agreements. That process took about 18 months. Rob was very generous. He had some prime objectives. He wanted to get enough money that he, his family and kids could be comfortable, but he wanted the company to survive and thrive to carry on the traditions his family had built.
>
> Rob was familiar with a naval architecture firm in Seattle that had been sold to its employees, and that model seemed to fit here. They had done it, and it seemed to be successful, and they're still carrying on as an employee-owned company. They have different tax laws in the US, so the details are different, and we spent a lot of time structuring it from a taxation point of view, both for Rob and for the shareholders. His main accountant wasn't that keen on the whole valuation we agreed on—he thought Rob was giving away the farm. The lawyers we had were quite open and constructive in the process, but it was the accountant's son-in-law Ron Batty, who worked for Rob's accountant, who read the tea leaves and saw what Rob wanted to do, and he really helped us structuring it.
>
> In negotiating things with Rob, one of the key elements that I think Rob got, and that his accountant didn't get, was that none of us potential shareholders were independently wealthy. None of us had another revenue stream which could pay Rob close to what he expected. The business was going to have to buy Rob out of the company, and that's basically how it worked. We each put up a very nominal amount of money, just to show we had some skin in the game. The way it works is, you are invited to be a shareholder and the company essentially pays for your shares. So, really, the company bought Rob out. You have to be an employee to be a shareholder, but not every employee is a shareholder. Without getting into the share structure and so on, basically we would keep carrying on working for our salaries; he'd carry on taking all the profits until he got a certain amount of money. You can't put too many constraints on something like this if your objective is to make it work instead of being hard-nosed. I think Rob got more out of it than he ever dreamed, financially. He got what he was asking for the company and has continued to actively participate in the designs without

> worrying about running the business. Where his contemporaries weren't able to do so, Rob has carried on doing the things that he likes to do—the creative stuff, and now the history stuff, and he's still executive chairman of the board, which is kind of a figurehead title. He attends all these conferences, and now he can present papers criticizing the Coast Guard and classification agencies without worrying about the consequences. So there is a lot of freedom in his new role, and I sure hope that he is happy.
>
> When it was all over, and I became president, it was like we let the genie out of the bottle. All of these guys who were working as employees all of a sudden became owners, and the mindset was a complete change. Rather than going to look for somebody else to do something, they became more creative in terms of marketing, ideas and responsibilities. We used to be limited by how much travel Rob could manage, now many of us travel and cover a lot more ground. Bill Hui now pretty well looks after China, and other people have new relationships with key clients. As one guy, there's only so many relationships that you can build and maintain, so it started going off from one person to 10, and then it's gone from there to, I think, 26 shareholders, and that's what's created the growth in the business that one guy could never handle.

This change of ownership was not an easy process for me on a personal level, as it was hard to give up the family business, even though I had complete faith in the take-over team. However, I soon became used to the idea, and it was actually a very smooth process. Ultimately this has been the very best possible solution to ensure the long-term viability and success of this historic company. I believe that everyone involved has benefitted significantly, and none more than me and my family, seeing that there is life after work! It is particularly gratifying to see how the younger generation of shareholders are playing an active role in the future of Robert Allan Ltd. I am confident that this will continue to be an extremely successful and highly regarded ship-design business.

Views of the New Owners

During the course of compiling this book, my collaborator, Peter Robson, interviewed a number of the Robert Allan Ltd naval architects and engineers who became shareholders following the sale of the company. Here are their stories in their own words.

Mike Fitzpatrick, *President and CEO*

When Rob decided to sell the company to a group of senior staff in 2008, I was included in the first ownership group. The way I describe it is that we needed several people to replace Rob. It is amazing how Rob did all those roles by himself before the transition. I don't know how he did it. Of course, as the company grew, it was even more important to split the roles. I don't think it could all be done by one person anymore. Now that I'm the CEO and president, and we have 90 people, I certainly can't do it all.

Ken Harford became president in 2008 and initially took over the administration and business development role, and he did it all very well, but I'm not sure how much

Ken really liked the business development side of things. What Ken liked best was the actual process of designing vessels. To this end, he carved off the business development side of things to me.

The eight others were all really good engineers, and their skills would be wasted in management. My skills were a little more in negotiation and business, so during the ownership transition, it was generally Ken and myself in the negotiations with Rob.

When the project started moving to completion, I hired Rollie Webb to come in and manage the transition. We were busy supervising construction of all these vessels for South America, and we had projects with shipyards scattered all over the world. Rollie came in for what was expected to be a two-year job. The idea was that he would share an office with me for a couple of months while I passed the ownership transition project off to him. He came to work in 2008 as a contractor and advisor, but he is still here today, currently as senior vice-president.

Ken was in his 60s when he took over as president, and it was pretty well the plan from the beginning that I would follow as president when he chose to retire. He wanted to retire earlier, but I enjoyed the role I had in just doing the business development and I didn't really want to take on the other bits of being the president. However, it was a long planned for and quite a smooth transition. Ken hung around for quite a while afterwards to help me. He was a big help. I took over officially after the second ownership group was brought in, in 2012.

Mike Fitzpatrick was born in Toronto and studied naval architecture in Australia. He worked with InCat Designs in Sydney before returning to BC to work with Robert Allan Ltd in 2003.

Derek Noon, *Project Performance Manager*

The ownership change was an interesting time. I was one of the first group of ten. What an offer that was from Rob! It was interesting because there's a delineation in our business. There's the technical side and then there's the administration/business side, and pretty much everybody in our industry, myself included, loves the technical.

Technical issues have clearly defined problems to solve, with answers that tend to be absolutes, often to several decimal places. The business side doesn't give you that. You can look at the financials, historical comparisons, but it doesn't tell you what the future is going to be, so it was a little nerve-wracking for a lot of us who had little business experience. At the time, I would say only a few of the new owners had any business background. Running a business was new to many of us; it was concerning because we hadn't done it before and there was no certainty.

The management team looks after the day-to-day operations, and the shareholders are kept informed on the state of the business. The shareholders are regularly provided with the financial statements, etc. The board of directors provides overall governance and looks after the interests of the company and the shareholders.

The transition of ownership was significant in the sense that the company was dealing with three generations of the Allan family up to that point. How does it end in a way that works? I give full credit to Rob on this one, because he was being romanced by all sorts of international conglomerates. That happened to several other contemporaries, who built businesses up and sold out. Finding a way to exit for owners is not easy. I was talking to someone around that period, and he said: "I've got this business that's a success and none of my employees want it, my family doesn't want it. What do I do? I'm not going to just walk out and leave it." He couldn't get out of it, couldn't disengage in any way.

Rob made it happen. At the time of the ownership change, myself and others of my vintage had relatively young families. My kids were teenagers. We had no money. I'd just come through a divorce, so I definitely had little money to invest. So the only money we were going to get, the only revenue stream, was going to be from the company, and the purchase was structured around that.

Three of the original shareholders have since sold their shares entirely, and there are four of us at present who are selling our shares over a 10-year period. These shares are being sold to new shareholders, mostly the next generation who will carry the company forward.

Australia-raised Derek Noon began work at a shipyard in Western Australia right out of high school and earned a diploma in mechanical engineering through night school. He moved to Canada and worked as a naval architect and design engineer for BC's S class ferries. He joined Robert Allan Ltd in 1993, almost 30 years ago.

Vince den Hertog, *Vice-President of Engineering, Director of Research & Development*

I was a bit perplexed as to why Rob wanted to sell the company to the employees. I didn't know what to make of it, to be honest, and felt Rob should be free to sell it if he wanted to. Wasn't the better option from his own personal making-money point of view to sell the company outright? If he did, someone else would own it and we would likely all be fine. I didn't quite get it and, as a result, I declined the offer to be part of that first owner's group. Fortunately, I got another chance later on and became an owner as part of the second group, after everything settled down.

Back when I started at Robert Allan, there were only about two dozen employees. Before and after the ownership transition, business was good and the company started to expand—to where there are now close to 90 employees. Previously, everything ran through Rob Allan, Ken Harford and Hans Muhlert. One of the major changes during that time was that we began splitting projects under separate project directors and separate project teams that ran autonomously. With the new group of owners, it was very much the way they thought the company should, and needed to, run. That became a bit more enshrined in the day-to-day work we were doing. With Rob no longer the owner, and with business increasing steadily, it was easier and more logical to run things that way.

Fortunately, the company, from an ownership point of view, continues to be in the hands of technical people—naval architect and engineer types—so that is a very positive thing. It's not bean-counters running the company. We're all doing it because we love designing boats. That's always the marvel for me. I walk through the office, and it hits you when you see all these screens of hull forms and structure and all that and know that people are able to go to work and just sit down and design boats and really like their work.

Vince den Hertog was born in Edmonton, Alberta, and worked for many years as a marine engineer with International Submarine Engineering, where he designed robotics for Autonomous Underwater Vehicles and other sub-surface vessels. He joined Robert Allan Ltd in 2001.

Xuhui (Bill) Hu, *Senior Naval Architect/Vice-President, Asia*

Before the company transitioned to the employees, Rob could have sold his company to a larger corporation, and he would have made a lot more money. But he didn't take that way. He still wanted his family name—the Robert Allan name—to carry on, and he wanted to reward his hard-working employees. He helped us fund the transition process, and that made the transition very successful. I was proudly among the first group of new owners, and I do appreciate it and thank Rob for the opportunity.

Dr. Xuhui (Bill) Hu was born in Wuhan, China, and lived with his family at a Communist forced-labour camp for several years during his childhood. Later he was able to study naval architecture at the Huazhong University of Science and Technology. He emigrated from China and came to work for Robert Allan Ltd in 1997.

Grant Brandlmayr, *Manager, Design Services*

The company is made up of very talented people, and the abilities of individuals make the collective whole. Rob Allan, Ken Harford and Hans Muhlert set that standard for the core group that became the first shareholders group.

I was lucky to be part of the first shareholders group. I appreciated the fact that when Rob was looking at finding an appropriate method to allow the company to continue, in such a way that he could wind down from a financial point of view and ensure a sense of security for the employees, he didn't simply go to the marketplace and sell to some big entity that would take it over and likely people would lose their jobs, or whatever role they had in building the company would disappear. I respect Rob very much for that, and I appreciate the fact he included me as one of those individuals in the first group. I hope that I have been able to contribute and help make the company stronger as we progress.

From what I see, the transition was successful, in that it follows Rob's culture of treating our employees with respect, and as a result there is very little turnover and retention is very good. We continue to strive to advance the technologies in what we do, to ensure safe vessels for people to go and spend their lives working on.

Vancouver-born Grant Brandlmayr attended Douglas College and Simon Fraser University in Burnaby, BC, and prior to joining Robert Allan worked for Brandlmayr Marine, Spencer Boats and Shore Boats. He joined Robert Allan Ltd as senior naval architect in 1995.

Andra Papuc, *Project Manager, Naval Architect*

It was monumental when Rob sold the company. I became an owner in the second intake, in 2012. Once the company brought on extra shareholders, a concerted effort was made to advertise to the industry that the new owners were here for the long-term.

As an owner, the company made sure I travelled more, attended more trade shows, presented papers, and generally had more interactions within the industry. There was a change, but in terms of employment, it was made very clear that my job as an employee was completely separate from ownership. Being an owner means you have shareholdings, but it is not related to your job description. In terms of the day to day, nothing changed.

When the company was smaller, it was possible for one person to make all the decisions, but that would be challenging, if not impossible, with 80+ employees. I think Rob really felt that—as the one person in charge, and as the payroll got bigger, he felt this weight on his shoulders.

Rob picked the first owners group, and they in turn picked the second group; but Rob was, and is, very much still around. He was the heart of the company, and though there are now several owners, we try to maintain that familial environment.

Andra Papuc was born in Romania and her family immigrated to Canada in 1995. Andra graduated from the Mechanical Engineering and Naval Architecture program at the University of British Columbia and joined Robert Allan Ltd in 2008.

Bilyana Ivanova, *Senior Naval Architect*

I was invited to be an owner during the second intake, in 2012. It was a smooth process and a very exciting opportunity that Rob provided to us as employees. Being an owner has really kept us even more involved. While we all have a passion for working in this field, being an owner gives you a bit of a different perspective, like more motivation for the success of the company. It was a great opportunity, and we're really thankful to Rob for providing it to us. We're happy to keep the Robert Allan family legacy going and to work for a company with such a long history.

Before, Rob Allan, Hans Muhlert and Ken Harford were finding and signing contracts and made all the decisions. Now, there are more people like them. We have a larger group doing business development, and that's definitely helped the growth of the company. I think we've grown more than 50 percent since then. Rob has been a really great mentor to everybody, and I can't thank him enough for the mentorship, for the opportunity we have, and we're really proud to carry the legacy.

Bilyana Ivanova was born in Bulgaria and graduated as a naval architect from Technical University of Varna. She worked with then socialist-government-owned Ship Research and Design Institute before immigrating to Vancouver and joining Robert Allan Ltd in 2003.

Jim Hyslop, *Director of Project Development*

The ownership transition was fairly seamless. I don't think anyone outside of the company would have really noticed a whole lot. I can't say things really changed a huge amount for me, apart from the fact that I was now an owner, with the benefits and the worries. But the company has done quite well over the past few years, so we shareholders and employees have also done quite well.

Starting in about 2006 Rob began to gradually transition out, while others were gradually transitioning in. It was clear that now that the company had grown so much, we needed to divvy up some of Rob's workload. Rob had been doing the job of about five people. He ran the company, he looked after the accounting, he did all the marketing, he did all the business development and he did a lot of the design work. I enjoyed the marketing, so I said, I'd look after that. Now most of my time is spent on business development, but I still do some design work as needed. The management group and the accountants and other employees that work behind the scenes are the ones who actually run the business side of the company, and that gives me a little freedom to do what I consider the fun stuff.

Vancouver-born Jim Hyslop studied naval architecture at the Fisheries and Marine Institute of Memorial University of Newfoundland and worked in the design office of Vancouver Shipyards before joining Robert Allan Ltd in 1994.

Lines

Once the deal was done in finalizing the sale of Robert Allan Ltd, it was time to celebrate. All the new owners and their partners were invited for a day of celebration at our favourite place, our summer cabin on Bowyer Island in Howe Sound. It was a glorious summer day, and we had a really lovely full day relaxing together, as we don't often. The owners presented me with some lovely gifts to mark the occasion, including these special words penned by Vince den Hertog:

A picture may say a thousand words, but the humble line has the power to bring ideas to reality.

Lines are about ideas. And they are about ideals.

A few well-placed lines reveal an idea without the clutter of detail.

Beauty can be in the lines themselves, or what's between the lines.

As you have taught us Rob, a small move or inflection can have big consequences.

Lines need to be fair.

There are blue lines and there are red lines.

Blue lines are about new ideas; inspiration on the right course.

Red lines are where things aren't right and a course correction is needed.

Lines are powerful. Lines can float a boat, and indeed a company. If they are right, they assure graceful movement and safety in rough seas. Pleasing to the eye, they instil pride for all on board.

Thank you for the blue lines and for the red lines, Rob, and for all they have given us.

And thank you for placing us on a line that leads from your grandfather, to your father, to you, to us, and to future generations.

It is truly an honour.

—R.G. Allan

The New Owners

The following are those people who have become shareholders since the initial sale of the company by Rob Allan in 2008.

Ryan Ayres

Todd Barber

Phil Barron

Lawren Best

Grant Brandlmayr

Dave Christopher

Vince den Hertog

Mike Fitzpatrick

Evan Gatehouse

Ken Harford

Darren Hass

Xuhui (Bill) Hu

Jim Hyslop

Bilyana Ivanova

Erik Johnston

Dmitry Kapiturov

Carolyn Kirkham

Oscar Lisagor

Jamie McCarthy

Oben Naki

Derek Noon

Andra Papuc

Mike Phillips

Henry Reeve

Robin Stapleton

Bart Stockdill

Allan Turner

Philippe Whyte

Mavis Ye

Hongling Zhang

ROBERT ALLAN
Naval Architects & Marine Engineers

25

The View from Inside: An Outstanding Workplace

At Robert Allan Ltd, it isn't all about designing the world's most innovative and scientifically advanced tugs and other workboats, it's also about the people, the workplace environment and what many employees and owners call a culture of respect and sharing, something they credit largely to Rob Allan's leadership. Here's what they have to say.

Opposite: The present offices of Robert Allan Ltd in Vancouver.

Working at Robert Allan Ltd, I always felt valued and appreciated. I think there must be something inherent in anything to do with boats that attracts the kind of people that it's a pleasure to work with, whether it's fellow employees, clients, subcontractors, suppliers or service providers.
— *Carol Nilson, retired office manager*

We've got brilliant designers in our company, just absolutely brilliant people, both the young and old. The young kids who come out of university these days—Memorial University, the University of Victoria, the University of

British Columbia, or whatever—as far as their knowledge is concerned, they are heads and shoulders above anybody I ever knew in my day. Their experience is lacking, of course, but they eventually get the experience to go with the qualifications and education. That's what Robert Allan Ltd is. It's full of very smart and clever people.
— Rollie Webb, senior vice-president

There have been a lot of exciting things going on in the past decades with Robert Allan Ltd. The company has a super group of people who collaborate very well together. Staff turn-over is rare. Part of the reason for the low turnover is the people we hire. We've done a lot of hiring out of the co-op programs out of the University of British Columbia and Memorial University, in Newfoundland and Labrador, which have naval architecture programs and co-ops. We've hired several people who came to us first as co-op students, liked the work and stayed on. Our office manager, Carol Nilson, who retired in 2010, had worked for Robert Allan Ltd for 38 years, and one very long-term employee, (our senior project director), Hans Muhlert, is in his 50th year of continuous employment—longer even than Rob.
— Ken Harford, former president

The work has been very interesting. The company has treated me very well. I've got no complaints. We all get along, and there's a steady train of really interesting projects coming in. For the past 10 years or so, we've been getting projects from all around the world and, until Covid, we were able to go to those places. It's a very good group and the people are good to work with. I learned a lot and have been given a free hand in a lot of stuff—and when I stumbled and made mistakes, they didn't jump on me too hard. I'm a bit lazier these days, and most of the work gets done by other people, but we've got some really bright young people who are a lot of fun to work with. They may not have much experience, but they are very smart, very fast thinking and hard-working. I don't supervise so much as I mentor the younger guys and just watch that things don't go off the rails.
— Hans Muhlert, senior project director

When I came to Robert Allan Ltd in 1994, I was young and looked at Rob and Ken and Hans—they were the three main guys—as almost father figures. They taught me a lot about how to conduct myself professionally in this engineering world. They were very willing to pass on their knowledge to the younger generation, and that is still happening with other mentors. Today, I'm very, very grateful to have been lucky enough to work for those three. I consider them equally as important in my development as a professional naval architect and in instilling in me a philosophy that the design comes first: "Let's do it right or don't do it at all." I try to emulate them by being professional and careful.

At Robert Allan Ltd, it has always been very much a culture of doing the right thing. It was never about how to make more money. It wasn't even talked about. It is about being professional, and that came from Rob, Ken and Hans, and it has really rubbed off on the company as a whole—and it carries on today. I think it was that professionalism that gave our company such a good reputation. Our engineering was exceptional. There are very few mistakes. Everything is done well, done properly and carefully mapped out. *— Jim Hyslop, director of project development*

Because of Rob, we keep challenging ourselves to make sure we are technologically above our competitors—to be better than others. We keep having to improve ourselves and also keep our eyes on new technology. I agree with Rob that nothing is perfect and there is always something to improve. That is one reason for our success and being recognized as the most technologically advanced tug designer in the world today. *— Xuhui (Bill) Hu, senior naval architect/vice-president, Asia desk*

Rob led the way by developing a culture of individuals respecting each other for what they are contributing to the company, and also of relying on each other where there is a knowledge base and sharing that knowledge and experience. That adds to the strength of the company as a whole. He also grew the company by looking for people who have experience that not everybody else has within the office. But one of the keys is to always respect each other.

In a company such as ours, where you are selling your engineering/design services, people are your strength. People with the right skills—complementary skills—and the ability for groups of people to function and work together and support each other, that's really a key element of success. This is one of the things I respected about Rob, Ken and Hans before and after I started full time at Robert Allan. All three of these people understood the value of the people and the need to respect each other.

In about 2005 or 2006 I started working more closely with Rob to interview and hire students on a regular basis. At one point Rob said to me that we've had occasional students in here, but we're going to embark on a program of regular interviews with students, to consistently bring students here and to show them what a naval architecture office is like. That has grown. I get a huge amount of personal satisfaction working with these young people, and many of them have become long-term employees, some are now shareholders.
— Grant Brandlmayr, manager, design services

I've been working at Robert Allan Ltd for more than 20 years, and among the reasons I've stayed so long is the combination of the work itself—which for sure is interesting—

but it is largely the work culture. Everyone here is constructive and respectful. It is a great bunch of people to work with. I give Rob a lot of credit because he really set the stage for that culture. Rob's approach to his work and his attitude towards the clients set the standard for everyone else. It also comes from the way he respects people's work/life balance and appreciates their input. Rob is a listener. He's not just in transmit mode like a lot of bosses—do this, do this, do this. He actually does listen and weighs what he hears. He started that whole culture, and it has carried through to today, even though he is no longer directly involved. I don't think that has really changed since he transferred ownership. The players have changed, but that same attitude is still there.

You walk into the office, and everybody has their head down, working quietly and doing their thing. It is easy to misconstrue that as being like a sweatshop. However, the reality is, people are simply left to get on with their work. These are people who really enjoy what they're doing and who are given the tools and a fair amount of latitude to get on with it. A lot of responsibility is given to individuals. If you have to do this drawing, then you own that drawing.

Robert Allan Ltd is not a top-down type of company. There's a lot of consensus within the company. We don't have the kind of organization where there are a bunch of rungs on a ladder that you move up. People here are happy enough just to broaden their own experience and do different projects. A lot of what people end up doing, they end up doing because they're interested in it. Jobs are usually created around the people, rather than the other way around. Every time someone leaves, there's this big shuffle. We never seem to be able to replace one person with someone who is an exact fit. It is more like a gap has been created and so-and-so will do more of this and little less of that, and this other person will do this instead of that. It kind of works like that, and I think people like that. We try to accommodate that from a management perspective: we talk to the staff, find out what their interests are, and we try to cater to that to some extent.

It's a place that is remarkably free of conflict or backstabbing or people that are overly ambitious.

It is a pretty egalitarian sort of place, where people feel fairly free to express their opinions and the conversations tend to be very technical. We don't have non-technical people as bosses, who are just trying to meet the budget and get things out on time. We are never put into situations where we have to get it out by Friday; none of that "screw it and ship it out regardless." That was never there. Employees are never put into the position where they feel their work has been wasted, or that poor quality is going out the door. I've always felt part of an organization that is trying to put out really high-quality work. That was, and is, the number-one thing.

Like the way it was with Rob, we all feel that our president, Mike Fitzpatrick, always has our back. If something doesn't work out with a client, he will be on your side—he won't throw you under the bus. Mike really cares about the employees, and if somebody has some troubles inside or outside of work, he'll do whatever he can to help fix things. Employees who feel that caring respond well, and I think that is part of the culture and one reason there is such little turnover.

— Vince den Hertog, vice-president of engineering

I would like to emphasize how important Rob has been in my life. Rob is quite an interesting person, and especially related to the selection of people. He has this kind of vision to get the right people for the right job and get the right working environment for these people. The culture of the business is coming from Rob's head. This is what he kind of settled or arranged. And the culture that we have right now pretty much sits on this base of what Rob developed. And I would say it is amazing, because what we have in our office is cooperation. We try to help each other with knowledge, with ability, with personal connections, and so it's really different than places where the employees are mostly in a competition and survival mode. We get everything done and cooperate and help each other in positive ways.

When you give your employees freedom—an element of ownership of their work—that helps with this environment. This is what Rob was doing and what I try to do with my design team—giving them the feeling of developing something themselves. Rob gave me freedom. When I came to him with an idea, he would most often say: "It's fine, just work it." Was his opinion always the same as mine? It is difficult to say, but he gave me freedom to create and get ownership. This culture is what Rob developed, and it continues now. I feel I am lucky to be there in such an environment. This is another element of Rob's legacy.

— Dr. Oscar Lisagor, vice-president of naval architecture

The best naval architects have some shipyard experience. Working in a design office in a shipyard is certainly valuable experience for a naval architect. You get to see what you're drawing actually get built. If you design something that's impossible to weld, the welder will come and drag you by the collar and say, "You weld this." Having naval architects with shipyard experience is a big strength for us. We had a client the other day, an owner of a small tugboat company in Bulgaria, contact us and say he was interested in a tugboat design. Well, I got one of our Bulgarian naval architects to respond to them in Bulgarian. We do a lot of work in Asia, so the ability to communicate in Mandarin is quite valuable. Because our clients are all over the world, it makes sense to have that diversity of people here. We are certainly aware of that. If there are countries that we do a lot of work in, we like to have a few people from that country also working in the office. It gives us insight and better communication when needed. I can sort of understand what is going through the mind of a Brazilian, but not as much as having a Brazilian on staff. That allows me to interpret what's going on in their minds so much better. Our diversity in the workplace not only reflects the diversity in Vancouver, but also the diversity of where we work around the world.

— Mike Fitzpatrick, president and CEO

It is amazing how diverse we are at Robert Allan. We have people from more than 15 different countries and cultures, and that's helped us a lot with international business

development. We have dedicated business development managers that have come to us with different backgrounds. It is good to know the culture and know the languages. One is originally from China and is handling all our Chinese clients. Another came to us from a Turkish shipyard, so he's our main person for our Turkish clients. A colleague from Brazil is dealing with all Brazilian clients. And now we have a Finnish person looking after the Scandinavian and European clients. That diversity may have something to do with naval architecture being such a narrow field in comparison to the civil architecture; there are many Canadian-educated architects, but not so many naval architects.
— *Bilyana Ivanova, senior naval architect*

Because our client base is so diverse and spread all over the world, it is really useful to have someone with local knowledge, so we have a pretty significant business development team. Instead of commercial agents, our client relationships are managed by working engineers who take part in designing as well. We have people doing business development from many countries, and they understand the needs of clients in those areas.

Another reason we've been so successful is because we not only have our group of business development guys, but we work with the shipyards' departments. There are four or five shipyards out there that do almost nothing but build our designs. They have their own marketing departments out there to sell our tugs, and basically when they're successful, we're successful. It is very much a two-way street with the best of our shipyard clients, where our role is often just to support their business development efforts.

For me, developing relationships with shipyards around the world is key. In most cases, I don't even need a handshake. We have a very close symbiotic relationship with our shipyard clients. We maintain our independence, because our clients are competitors with each other, but we have a lot of trust in one another. I think in the time that I have been responsible for the relationships with our best clients, we've really deepened those relationships. That is based on mutual trust. That's the big thing here. We don't always do everything perfectly, but when there is a problem, the client knows that we're going to be there to fix it. We're going to be there with them and find a way no matter how long it takes or how much it costs.
— *Mike Fitzpatrick*

Everybody's very open: if you've got a question, there will be full disclosure and a willingness to help. It's a very lovely place to work, it really is. I've been here now for about 30 years, though it doesn't feel like it. There is minimal staff turnover.
— *Derek Noon, project performance manager*

One great thing about Robert Allan is that he and Hans always kept challenging me and others to expand our professional knowledge, because if you always do the same thing, that's boring. To me, Rob was so wise and generous. In addition to his very sharp eyes and incredible knowledge as a naval architect, he had a strong feeling of where the business should go.
— *Xuhui (Bill) Hu*

I came in as a contractor just after the company was sold to a key group of senior employees in 2008. The strength of the company today is that because we have an awful lot of good, capable people, we can handle so many things at one time. We are twice the size we were when I joined. We've got all of 90 people here, and it's grown every year as we hire more and more people. The problem is always finding the right talented person. It's easy to get bodies, and maybe they're going to be alright, but it's another whole ball game to select capable people and attract them to the company and attract them to Vancouver. A kid may go to Memorial University and then, after graduating, has to move across the country to Vancouver for a job and, wow, that really sets them back on their ears, especially with the cost of housing and everything else, but they always find a way.

One way we're making it work is getting people to come and get used to the company, and then move back to where they came from and connect virtually. We've got quite a few employees that are resident elsewhere on the globe: the US, Newfoundland and Labrador, Australia, Finland, and so on, and more locally, on Vancouver Island and the interior of BC. They're all over the place, and it isn't causing problems. We still get the same or more volume of work done in a unit of time, and we're getting more and more projects completed. In recent years, at least 90 vessels of our design have been completed every year. That's almost two a week built somewhere around the globe. A lot of those are tugs, but a lot are something else, and we do all kinds of weird and wonderful things.

We have a dozen employees from Asia, plus people from Australia, Turkey, South America, Holland and other places, and many of those people came from where the markets are today. While there is different learning from different countries, the rules of physics are pretty well the same anywhere; it's just the practices that are different. Perception is an issue when it comes to what is important—degrees of quality, functionality, appearance. We have to mix and match the personalities to get the right skill sets and to make things not only function but look good.
— *Rollie Webb*

At Robert Allan Ltd, gender has not been an issue for me. I found the company to be very familial and protective of me. I remember the first time I travelled to Turkey for a

project meeting; it was my first long work trip, and I was quite anxious. I went with Mike Fitzpatrick and Hans Muhlert. I heard afterwards that Rob had spoken to them about taking care of me on the trip; probably because I was 24 and the only woman on site when we arrived at the shipyard. The trip went very well and was a very positive experience for me, from which I learned a lot. My manager, Grant Brandlmayr, is very conscious of not putting me in a situation where I would be uncomfortable. In my early travel days, I feel like he always ensured the right kind of buffers were in place, and I have been lucky to have such positive travel experiences.
— *Andra Papuc*

I'm not sure exactly of what our male/female breakdown is, but the reality is that naval architecture is still a male-dominated profession. I would say since I've been in charge of hiring new naval architects, about 40 percent of new hires have been female. It isn't a conscious decision, but just that there are more women going into the field today.

We're a Vancouver company and I think the diversity in our office reflects the diversity of the city. Just over 50 percent of our workforce was not born in Canada, and I think that's pretty well the same as Vancouver. We appreciate the diverse knowledge that has come to us from all over the world.

The reality is that there are not many shipyards in Canada and especially at the time, say 10 years ago, when we were hiring a lot of people, so getting people who had shipyard experience from all over the world was quite valuable to us. We didn't go out and hire these people in other places. They came to Vancouver to work here, and we were more than happy to hire them.
— *Mike Fitzpatrick*

Over the years the business has really changed. When I first started, we did a lot of odd jobs here and there, maybe three new vessel designs in a year. Now it's completely the other way. We do so much new design work—20 or so new vessel designs a year. If I think of the '90s, our business for new designs were more for an owner in a smaller business. It wasn't big corporations or big tug companies. And we almost exclusively worked for owners. These days, pretty well two-thirds of our contracts are for shipyards, and many of those shipyards are building on speculation.

If you walked into our office before Covid, when it was full of people, one of the first things you'd notice is how quiet it is. Everyone is so engaged in what they're doing. I think we're unique in that way. I think we have that going for us. It's just very, very interesting work. We have a unique work environment. That is the key, and I think we do interesting and challenging things. That combination really works, and there is very little staff turnover.

Probably the most unique thing about Robert Allan is that people genuinely enjoy

what they're doing. When you talk to people in different businesses, they say they are just passing time at work and getting paid for it. We're really lucky in the sense of what we do is really, really enjoyable. It's a lot of fun: we get to design boats all day. It is challenging, it's enjoyable, it's rewarding. We have strived for many years to keep the culture. It's a very good work environment.
— *Derek Noon*

We all really like working here. Even when I first started, fresh out of university, I was the only person I knew (outside of this company) who genuinely liked their job. My friends that I graduated with had the mentality of "I'm just trying to do this until I can get something better—trying to make money." It was never about that with me, because the work is very interesting and it is a good work environment.

I think the work environment has only improved over the years. Now we have a social committee, and they plan extracurricular fun social events—even during Covid. For example, they recently hosted a virtual chocolate-tasting night with the chocolate kit delivered to everyone's homes ahead of time. Overall, it's a very nice atmosphere. I am very grateful that my colleagues are not only strong technically but are also genuinely nice people.
— *Andra Papuc*

ROBERT ALLAN L

26

Looking to the Future

Having looked back at nearly a century of work done by this company, it is hard not to be impressed, even if the viewer has innate biases. However, it is also intriguing to look ahead and try to anticipate what the future holds for the design of working vessels in a rapidly changing world.

As this conclusion is being written, the world is still gripped in a pandemic that has dramatically and probably irrevocably changed the way in which businesses will work for years to come. Fortunately, the management of Robert Allan Ltd had in 2018/19 looked closely at our contingency plans for how the business would survive a major disruption, thinking mainly of the potential of fire risks, and had invested significantly in major computer upgrades and off-site data-storage options. When the Covid-19 pandemic hit in early 2020, the office moved seamlessly into a work-at-home mode and barely skipped a beat. Work continued, albeit with lots of "Teams" meetings, and all deadlines were met. As the pandemic lingers, it is obvious that the workplace will never be quite the same, as people have adapted to the work-at-home lifestyle and proven that it is workable. What is not so obvious, though, is how to preserve the incredible team spirit that is the hallmark of this most able group of people. The management is working hard to ensure that it is not lost. It is clear that the office environment that emerges from this disruption will look quite different, but the collaboration and communication between the staff are still strong and must remain so.

Opposite: CEO Mike Fitzpatrick—shown receiving the export business award in 2016 from Canadian Manufacturers & Exporters of British Columbia, in partnership with the BC Government Ministry of International Trade—has led the company to new successes.

At this point in history, it is also abundantly clear that the world is finally engaged with the concept of decarbonization, and many sectors of the maritime world are working hard to define what the course to a greener future looks like. While trans-oceanic shipping is by far the largest consumer of fuel oils in the marine world, every sector, regardless of how small, needs to look closely at what the alternatives are for their own future, including the commercial workboat sector in every part of the world.

The year 2050 is only a generation away, and significant efforts towards the "clean and green" future are already underway in many areas.

In the tugboat world, the first tentative steps to reducing emissions had already been taken more than a decade ago, and Robert Allan Ltd has been heavily involved with almost every major tug project on that front since. For example:

2010	*Carolyn Dorothy*	world's first hybrid-powered tugboat
2014	*R.T. Evolution*	world's first hybrid Rotortug
2017	*Dux, Pax* and *Audax*	world's first dual-fuel (diesel/LNG) escort tugs
2020	*HaiSea ElectRA*	world's second and most powerful all-electric tugboat
2021	*Svitzer*	Robert Allan Ltd commissioned to design the world's first fuel-cell tug

At the present time, Robert Allan Ltd is working on contracts for vessels embodying almost every potential alternative fuel/energy type currently available, including methanol, ammonia, hydrogen, battery-electric and hybrid-electric. These are not speculative designs but actual paying contracts from our far-sighted clients who can see what they must do to secure their own business futures, even if some of these options are not economically competitive at present. We are gratified that they have confidence in the abilities of Robert Allan Ltd to assist them in getting to the goal of a low or zero-carbon future. The entire marine industry is on the threshold of an energy system renewal, not unlike what was seen in the days of switching from sail to steam or from steam to diesel. The outlook is both challenging and exciting, with many opportunities to use intelligent design to make really significant contributions to a healthier planet.

The autonomous tug is almost certainly a thing of the future.

Autonomous Vessels

As we examined options for the "tug of the future" over the past few years, it became obvious that the use of unmanned autonomous vessels of many different types will be a factor in the world's maritime future. It is increasingly difficult to attract personnel to a life at sea on large ships, and operational concepts for autonomous shipping are already being tested. It is natural that this same technology will ultimately be applied to tugs, though their operations are very different than the simple "Point A to Point B" runs of a deep-sea ship. We have defined and evaluated several autonomous tug concepts. These are technically viable, but clearly present human-factor challenges to regulators, owners and crews alike. Our first published concept for an autonomous tug, illustrated at left, was for a more or less conventional "master" tug working with an unmanned "slave" tug, the RAmora 2400, controlled by an operator on the lead vessel. The unmanned tug could be operated in the more high-risk aspects of ship-handling.

While contemplating similar unmanned applications it became obvious that using autonomous vessels in any high-risk application would be a wise move, avoiding putting human lives at risk. That led us to identify similar autonomous vessels for firefighting and SAR duties. There is significant interest in these concepts, and the first full-scale versions are likely just a few years away.

—R.G. Allan

Rob Allan was recognized by the Engineers and Geoscientists of BC in 2017 for "significant technical achievements."

As it rapidly approaches the landmark of a century of business, Robert Allan Ltd is well poised to continue doing excellent and innovative design work in support of our clients around the world. The business is well served by a unique combination of highly experienced senior staff and an extremely bright and energetic younger cadre of naval architects and engineers who see their own futures and personal success within this organization. The reputation of the company for producing the highest standards of design work is secure in the hands of the new owners and managers of the business.

It has been a distinct privilege to have been a part of this story, and I have never regretted a moment of the hard work invested into helping to make this family company succeed. I have garnered many more personal credits for the work of the Robert Allan team than I deserve, but I share all those awards with everyone who thinks of this company as theirs and who works every day to make the company a continuing success.

It would be wonderful to know what my father and grandfather would think of what has been accomplished in their names.

EPILOGUE

This story is ongoing, as Robert Allan Ltd goes from strength to strength under the new ownership and management team. The number of vessels built on the world stage to our designs currently numbers close to 100 every year, which at almost two deliveries per week is a significant accomplishment.

The complete list of vessels built to Robert Allan designs is impossible to include in a publication such as this; however, work on that complete listing is underway and will hopefully soon be available on the corporate website (www.ral.ca).

On the occasion of the 75th anniversary of the company in 2005, Enneke and I thought it would be a nice touch to present every employee of the company with a suitably nautical memento. We approached the late Don Hutchison, a noted ceramic artist and sculptor whose studio was next to the one that Enneke shared with several weavers on Granville Island, with the idea of creating about 70 "boat-like objects" for this project. Don jumped at the chance but convinced us not to have a mass production of essentially identical objects but to allow him to create a series of individually unique "vessels." We were happy to give his talents free rein, and he and an associate produced an amazing array of these commemorative craft, some very obviously boats but many also based on sea animals or mythical sea-creatures. They were glorious.

As a "thank you" to us for this very significant commission, Don created this wonderful work, the "Three Generations Vessel" and the accompanying text. The closing thoughts of this narrative are perhaps best expressed in Don's kind words.

—Rob

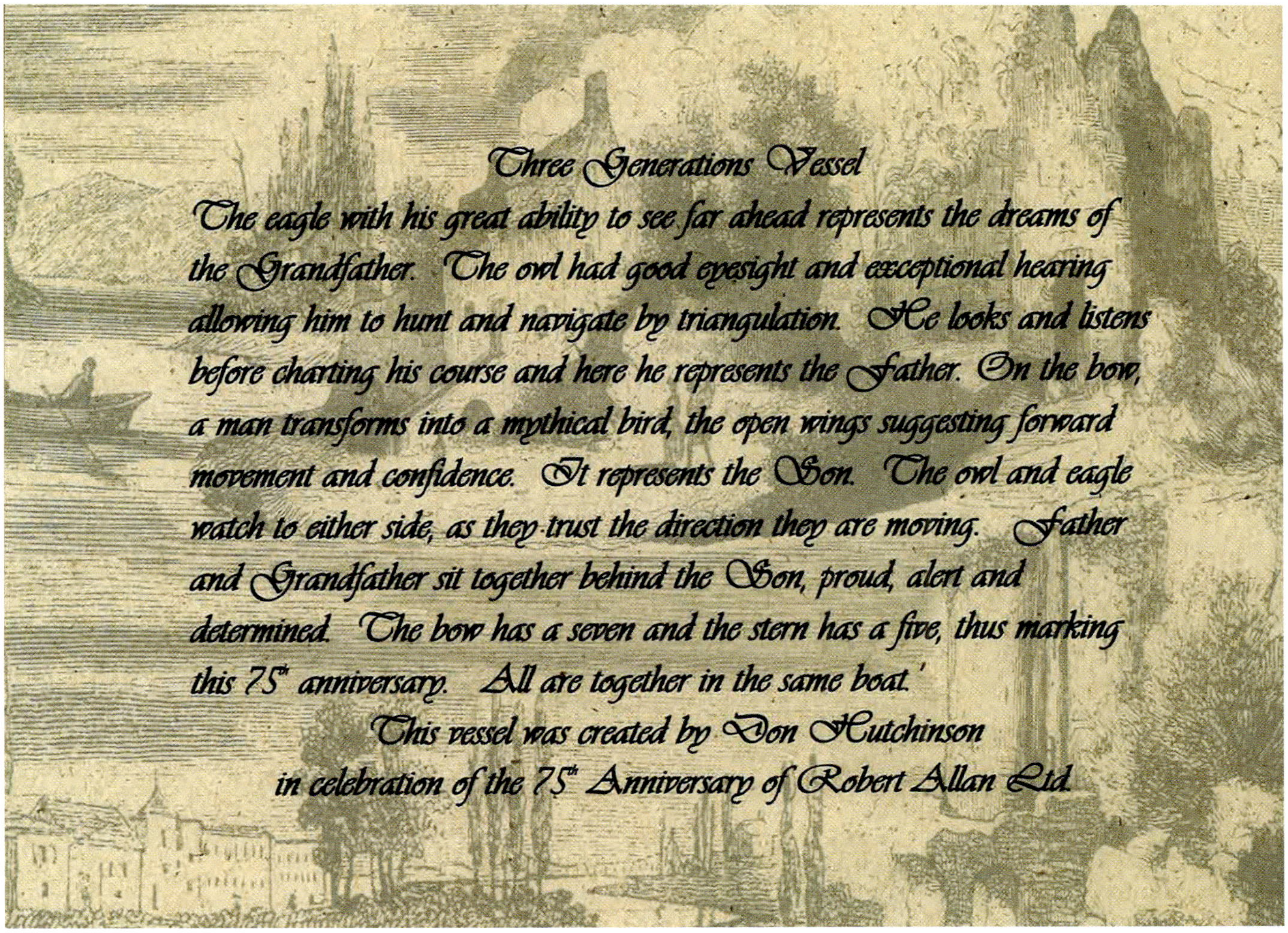

Three Generations Vessel

'The eagle with his great ability to see far ahead represents the dreams of the Grandfather. The owl had good eyesight and exceptional hearing allowing him to hunt and navigate by triangulation. He looks and listens before charting his course and here he represents the Father. On the bow, a man transforms into a mythical bird, the open wings suggesting forward movement and confidence. It represents the Son. The owl and eagle watch to either side, as they trust the direction they are moving. Father and Grandfather sit together behind the Son, proud, alert and determined. The bow has a seven and the stern has a five, thus marking this 75th anniversary. All are together in the same boat.'

This vessel was created by Don Hutchinson
in celebration of the 75th Anniversary of Robert Allan Ltd.

Image Credits

All photos, tables, graphs, renderings and drawings not credited are courtesy Robert Allan Ltd files.

Dustjacket front: Robert Allan Ltd files
Dustjacket back: Ernst Schneider

v, vi (left and right), Allan family archives; viii–ix (background), City of Richmond Archives, #1985 4 70; x, courtesy Peter G. Noble; xii, R.G. Allan; 1, 2, 3, 4 (top and bottom), Allan family archives; 5, Foncie Pulice/Allan family archives; 6, 8, 9, 11 (top and bottom), Allan family archives; 12 (top), Vancouver Maritime Museum, LM2018.999.038; 12 (bottom), 14, Allan family archives; 15, NWT Archives/Henry Busse fonds/ N-1979-052: 0642; 16, BC-1532-457-2/ Fisherman Publishing Society Fonds/UBC Library Rare Books and Special Collections; 17, Allan family archives; 19, Murray McLellan, Vancouver Maritime Museum, VMM55.Neg.832.9; 22, Murray McLellan, Vancouver Maritime Museum, VMM55.01.01E.05; 29, Peter Vassilopoulos; 31, Vancouver Public Library, 15923, Frank Leonard; 35, courtesy Directorate of History and Heritage, DND, Government of Canada; 38, Ernst Schneider; 40, City of Vancouver Archives CVA 260-909, photographer James Crookall; 42, Allan family archives; 44, Peter Vassilopoulos; 48 (top and bottom), Allan family archives; 49, Vancouver Maritime Museum, *Harbour & Shipping* magazine, March 1941; 50, Finning Tractor/Commercial Illustrators; 52, Allan family archives; 53, City of Richmond Archives, #1985 4 70; 55 (top left), BC-1532-1199-1, Fisherman Publishing Society fonds, University of British Columbia Library Rare Books and Special Collections; 55 (top right), City of Richmond Archives, 2001 34 9; 55 (bottom), City of Richmond Archives, #1985 4 55; 56, BC-1532-540-1, Fisherman Publishing Society fonds, University of British Columbia Library Rare Books and Special Collections; 58 (top), courtesy Nauticapedia.ca/ John MacFarlane; 59, Finning Tractor/Commercial Illustrators; 61 (left), City of Richmond Archives, Graham Elliston #1999 6 1603; 61 (right), R.G. Allan; 62, Finning Tractor/Commercial Illustrators; 63, Doug Harrison painting; 66, courtesy George Bird; 67 (top left), BC-1532-554-1, Fisherman Publishing Society fonds, University of British Columbia Library Rare Books and Special Collections; 67 (top right), courtesy Don Macmillan; 67 (bottom left), BC-1532-548-11, Fisherman Publishing Society fonds, University of British Columbia Library Rare Books and Special Collections; 67 (bottom right), BC-1532-603-1, Fisherman Publishing Society fonds, University of British Columbia Library Rare Books and Special Collections; 68, Calvin J. Charlton; 70, Bernard Remmem; 71 (left), Gerry Boudreau; 71 (top right), Mike Rydqvist-McCammon; 71 (bottom right), Paul Paterson; 72 (top), BC-1532-486-18, Fisherman Publishing Society fonds, University of British Columbia Library Rare Books and Special Collections; 72 (bottom), Murray McLellan, Vancouver Maritime Museum, VMM55.01.02A.56; 75, BC-1532-1142-3, Fisherman Publishing Society fonds, University of British Columbia Library Rare Books and Special Collections; 84 (top), Nauticapedia.ca; 84 (bottom), courtesy Barrie Fairall; 87 (top left, top right), Rolf Leben; 87 (bottom), 88, courtesy Fisheries and Ocean Canada; 89 (top right), Murray McLellan, Vancouver Maritime Museum, VMM55.01.01N.02; 91 (left), Murray McLellan, Vancouver Maritime Museum, VMM55.01.01T.03; 91 (right), Murray McLellan, Vancouver Maritime Museum, VMM55.01.01K.06; 92, 96, R.G. Allan; 97, courtesy Charles David Yachts; 101, Murray McLellan, Vancouver Maritime Museum, VMM55.01.02D.08; 102, R.G. Allan; 105, Gabriola Historical & Museum Society archives; 106 (bottom), MONOVA: Archives of North Vancouver, #15806; 109, Murray McLellan, Vancouver Maritime Museum, VMM55.01.02J.01; 114, Murray McLellan, Vancouver Maritime Museum, VMM55.01.01B.33; 115, courtesy Creston Museum; 116 (top and bottom), 118, courtesy La Crete Mennonite Heritage Village; 122 (top), Rick Horne; 122 (bottom), courtesy Pacific Boat Brokers; 123, NWT Archives, Henry family, N-2008-016: 0026; 124, R.G. Allan; 125, Murray McLennan, Vancouver Maritime Museum, VMM55.01.01L.13; 128 (top), Murray McLennan, Vancouver Maritime Museum, VMM55.01.02D.20; 128 (bottom), Don Brown; 131 (top and bottom), courtesy Alberta Transportation, Government of Alberta; 136, Alex McCluskey; 137, courtesy Harbour Cruises; 138, Dave Leyden; 139, Murray McLellan, Vancouver Maritime Museum, VMM55.01.01B.26; 140, courtesy Canadian Coast Guard; 143, Murray

McLellan, Vancouver Maritime Museum, VMM55. Neg.19.8; 146, Nauticapedia.ca/John MacFarlane; 150, Robert Etchell; 151, R.F. Allan; 155, R.G. Allan; 156, courtesy Mr. Salim Duzgit; 157, courtesy International Submarine Engineering; 158, courtesy Canadian Coast Guard; 161, D. Chamard, courtesy Fisheries and Oceans Canada; 163, courtesy Canadian Coast Guard; 166, NWT Archives, Jean Boulva, N-2021-002: 0174; 167, NWT Archives, Jean Boulva, N-2021-002:0270; 168 (top), Murray McLellan, Vancouver Maritime Museum, VMM55.Neg.55.4; 168 (bottom), Russ Lloyd; 171 (bottom), courtesy Alaska Department of Fish and Game, Kyle Hebert; 173 (left), courtesy Fisheries and Oceans, P. Dionne; 173 (right), courtesy Fisheries and Oceans Canada; 175, courtesy Seaspan Shipyards; 177, courtesy the Commonwealth Scientific and Industrial Research Organisation, Australia; 184 (right), Murray McLellan, Vancouver Maritime Museum, VMM55.01. 01U3.34; 186, courtesy Henry Reeve, Robert Allan Ltd's Long Beach Fireboat Team project leader; 188 (right), Martin Nate Rawner; 196 (left), courtesy Hike Metal Products & Shipbuilding; 199, Matt Tomlin; 200, Blvdone/stock.adobe.com; 201, Brian Gauvin; 211, courtesy U.S Coast Guard, District 8; 212, courtesy Khulna Shipyard; 214, Murray McLellan, Vancouver Maritime Museum, VMM55.01.01.U3.03; 216, Murray McLellan, Vancouver Maritime Museum, VMM50.Neg.10164.16; 217, courtesy Canadian Coast Guard; 220, Murray McLellan, Vancouver Maritime Museum, VMM55.01. 01.U3.03; 223 (bottom); 226, Murray McLellan, Vancouver Maritime Museum, VMM55.01.02B.12; 229 (bottom), Murray McLellan, Vancouver Maritime Museum, VMM55.Neg.540.10; 236, www.fleetmon.com/Svetlanka; 244 (top right), Alex McCluskey; 245, courtesy Remontowa Shipbuilding; 248, Murray McLellan, Vancouver Maritime Museum, VMM55.Neg.173.10; 255 (top), courtesy Seaspan Marine; 255 (bottom), Murray McLellan, Vancouver Maritime Museum, VMM55.Neg.6; 257, courtesy Seaspan Marine; 259, Murray McLellan, Vancouver Maritime Museum, VMM55.01.02G.31; 260, R.G. Allan; 263, Murray McLellan, Vancouver Maritime Museum, VMM55.01.02S.02; 265, Murray McLellan, Vancouver Maritime Museum, VMM55.01.02E.27; 266, Murray McLellan, Vancouver Maritime Museum, VMM55.01.02L.36; 271, Murray McLellan, Vancouver Maritime Museum, VMM55.Neg.778.7; 272 (top), Murray McLellan, Vancouver Maritime Museum, VMM55. Neg.935.4; 274, Murray McLellan, Vancouver Maritime Museum, VMM55.Neg.173.10; 277 (top), R.G. Allan; 277 (bottom), Don Brown; 281, NWT Archives/Henry Busse/ N-1979-052: 0655; 286, Michael Cooper; 287 (top and bottom), R. F. Allan; 290 (top), Murray McLellan, Vancouver Maritime Museum, VMM55.Neg.389.5; 306 (top), courtesy Seaspan Marine; 306 (bottom), K.F. UK/Alamy Stock; 309, Hans Esveldt; 312, Murray McLellan, Vancouver Maritime Museum, VMM55.01.02H.01; 315, courtesy Seaspan Marine; 317, Vancouver Maritime Museum, LM2017-999-006; 318; City of Vancouver Archives, CVA 260-888, photographer James Crookall; 319 (bottom right), Murray McLellan, Vancouver Maritime Museum, VMM55.01.02S.102; 320, courtesy SS Master Society; 321, 322, 323, Don Brown; 324, Murray McLellan, Vancouver Maritime Museum, VMM55.01.02J.05; 325, Murray McLellan, Vancouver Maritime Museum, VMM55.01.02M.03; 326, 327, Don Brown; 329 (top), Murray McLellan, Vancouver Maritime Museum VMM55.04.03.16; 329 (bottom), Murray McLellan, Vancouver Maritime Museum, VMM55.01.02O2; 331 (top), Paul Dixon; 331 (bottom), Dianne Reid; 332, Paul Dixon; 333 (top), Murray McLellan, Vancouver Maritime Museum, VMM55.01.02E.30; 333 (bottom), Murray McLellan, Vancouver Maritime Museum, VMM55.01.02E.33; 334, Murray McLellan, Vancouver Maritime Museum, VMM55.01.02L.14; 335 (top left), Murray McLellan, Vancouver Maritime Museum, VMM55.01.02B.20; 335 (top right), Murray McLellan, Vancouver Maritime Museum, VMM55.01.02S.58; 335 (bottom), Murray McLellan, Vancouver Maritime Museum, VMM55. Neg.745.4; 336 (top), Mike Zelt; 337, Murray McLellan, Vancouver Maritime Museum, VMM55.01.02I.32; 338 (bottom), Murray McLellan, Vancouver Maritime Museum, VMM55.01.02I.23; 341 (top), Nauticapedia Collection, John Henderson; 341 (bottom), Don Brown; 342, Murray McLellan, Vancouver Maritime Museum, VMM55.Neg.106.9; 343 (bottom), Nauticapedia Collection, James Picard collection; 346 (top), Dianne Reid; 346 (bottom), Murray McLellan, Vancouver Maritime Museum, VMM55.01.N47; 348, Murray McLellan, Vancouver Maritime Museum, VMM55.01.02H.36; 349 (top right), Nauticapedia Collection, John Macfarlane; 349 (top right), Murray McLellan, Vancouver Maritime Museum, VMM55.01.02N.07;

349 (bottom), Murray McLellan, Vancouver Maritime Museum, VMM55.Neg.97.11; 351, Murray McLellan, Vancouver Maritime Museum, VMM55.01.02Y.01; 353 (top), Murray McLellan, Vancouver Maritime Museum, VMM55.01. 02Y.01; 354, Murray McLellan, Vancouver Maritime Museum, VMM55.Neg.457.3; 355, Don Brown; 356 (top), Murray McLellan, Vancouver Maritime Museum, VMM55.Neg.451.8; 356 (bottom), Murray McLellan, Vancouver Maritime Museum, VMM55. Neg.495.11; 371, Rick Horne; 372, Murray McLellan, Vancouver Maritime Museum, VMM55.01.02C.95; 373, Murray McLellan, Vancouver Maritime Museum, VMM55.01.01N2.11; 374–375, Murray McLellan, Vancouver Maritime Museum, VMM55.01.01C05; 377, Murray McLellan, Vancouver Maritime Museum, VMM55.01.01N2.10; 378, courtesy Schottel GmbH; 378 (top), Murray McLellan, Vancouver Maritime Museum, VMM55. Neg.959.3; 379 (bottom), Murray McLellan, Vancouver Maritime Museum, VMM50.Neg.950.4; 384 (top), Murray McLellan, Vancouver Maritime Museum, VMM55.01.01C.51; 385, Murray McLellan, Vancouver Maritime Museum, VMM55. 01.01C.35; 388, Kurt Redd, DMI; 390, Vancouver Maritime Museum, LM2016.004.068; 393, ART Collection/ Alamy Stock; 398, courtesy Foss Maritime; 399, Dave Miner; 400, courtesy E.N. Bisso & Son, Inc.; 402, Murray McLellan, Vancouver Maritime Museum, VMM55.01.02I.18; 408, Jean Hemond; 413 (bottom), Andre Castonguay; 414, Paul Clarke; 418 (bottom), courtesy Seaspan Marine; 423, Helge Hansen courtesy Østensjø Rederi; 427 (top and bottom), courtesy Østensjø Rederi; 429 (top), Øyvind Saetre, courtesy Østensjø Rederi; 429 (bottom), Jack Gaston, courtesy Østensjø Rederi; 430, Clive Thomas; 436, 440, courtesy Sanmar Shipyards; 441, Al Lindner; 449, Alex Brown; 450, Bart Stockdill; 451, Gwenolé de Kermenguy; 452 (top), Glenn Kasner; 459 (top), Brian Gauvin; 462, 464, M. van Hoorn; 474 (top), courtesy Alfons Håkans; 478 (bottom), Rick Voice; 489 (top), courtesy Sanmar; 489 (bottom), Wikimedia Commons, Ein Dahmer; 492, Don Brown; 495 (left and right), courtesy Lo-Rez Vibration Control; 498 (bottom), Lonnie Wishart, www.lonniewishart.com; 505, courtesy Seaspan Marine; 508, Murray McLellan, Vancouver Maritime Museum, VMM55.01.01C.36; 514, courtesy Multratug; 529, courtesy Parkland Fuel Corp; 531, Danita Dellmont/ Alamy Stock; 534, Murray McLellan, Vancouver Maritime Museum, VMM55.01.01U3.25; 536 (top), Photo 224768433, John Krutop, Dreamstime.com; 536 (bottom), courtesy Teekay Corporation; 537, Used by permission of KIETTA; 540, Ernst Schneider; 542 (bottom), courtesy *Western Mariner* magazine; 556, R.G. Allan; 560–561 (background spread), Murray McLellan, Vancouver Maritime Museum, VMM55.01.02W.19; 564–565 (background spread), Murray McLellan, Vancouver Maritime Museum, VMM55.01.02A.56

Index

Note: Vessel names are listed alphabetically after topics in subentries. Page numbers in *italics* refer to photographs or illustrations.